Agri-Food Quality
An Interdisciplinary Approach

Agri-Food Quality

An Interdisciplinary Approach

Edited by
G. R. Fenwick
Institute of Food Research, Norwich

C. Hedley
John Innes Centre, Norwich

R. L. Richards
John Innes Centre, Norwich

S. Khokhar
Haryana Agricultural University, India

THE ROYAL
SOCIETY OF
CHEMISTRY
Information
Services

The proceedings of the International Conference 'Agri-Food Quality' organised by the RSC Food Chemistry Group and the RSC Agriculture Group, held 25–29 June 1995 at Norwich

Special Publication No. 179

ISBN 0-85404-711-5

A catalogue record for this book is available from the British Library.

Published by The Royal Society of Chemistry,
Thomas Graham House, Science Park, Milton Road,
Cambridge CB4 4WF, UK

Printed by Bookcraft (Bath) Ltd.

Preface

Agri-Food Quality '95 is the latest in a series of international meetings involving the Food Chemistry Group of the Royal Society of Chemistry. Like the earlier meetings, Nutrient Bioavailability (1988), Dietary Fibre (1990) and Food and Cancer Prevention (1992), Agri-Food Quality was held in Norwich at the University of East Anglia.

However, unlike the previous meetings, Agri-Food Quality '95 was jointly hosted by the Royal Society of Chemistry's Food Chemistry and Agriculture Groups, and organised by members of these bodies as well as staff at the Institute of Food Research and the John Innes Centre, situated adjacent to the University.

This local combination of educational and research organisations embracing the agri-food sciences was but one reason for the initiation and development of this meeting. A second was the enthusiastic support expressed within the Food Chemistry and Agriculture Groups of the Society. Yet a third reason was the emphasis of the UK Technology Foresight programme, the UK's Biotechnology and Biological Sciences Research Council and, not least, the European Commission, on 'quality', and the realisation that this is an issue at all points along the food chain, and that improvements in the latter could be most (cost-) effectively addressed by considering the totality of that integrated chain.

Having successfully created fora for chemists, nutritionists and clinicians to discuss problems and opportunities in the first three meetings, it was the organisers' hope that these discussions could be extended to include plant breeders and molecular biologists. As a starting point the concept of 'quality' was described and discussed by experts from across the agri-food sciences, as well as representatives from funding bodies, industry and national government.

The hardest part of any inter-disciplinary collaboration is frequently identifying partners from other disciplines, developing a common interest and harnessing unfocussed enthusiasm. It is to be hoped that the meeting contributed to such processes, and that the publication of papers from the conference will stimulate additional, and wider, contacts and collaboration.

The Food Chemistry Group have been encouraged that two of its earlier meetings have developed into series. The second Bioavailability conference was held in Germany with the next planned for The Netherlands, whilst it has been agreed that the Food and Cancer Prevention series will be held alternately in Norwich and Wageningen, with the second being held in May 1996. Such developments are to be encouraged, and contribute significantly to the development of contacts – and friendships – within European science. We very much hope that Agri-Food Quality would develop in a similar manner.

Such contacts must not, however, be exclusive and inward-looking; the organisers recognise the changes which have occurred within Europe since this series of meetings began. One of the pleasures of Agri-Food Quality lay in the participation of colleagues from Central and Eastern Europe and from the republics of the former Soviet Union. A number of participants also attended from further afield and it is important to remember that quality is an emerging issue in developing countries, both in terms of export potential and produce for local markets.

A great many colleagues assisted in the organisation and running of this meeting. We would like to take this opportunity to acknowledge their help, enthusiasm and dedication. Finally, we would like to thank the contributors to this volume for respecting deadlines and for assisting the editors throughout in a positive and constructive manner.

Preface

Contents

Section 6 **Post-Harvest Effects and Quality Determinants**

Section 7 **Anti-oxidants in Food Plants**

Section 8 **Enhancing Nutrients and Nutritional Quality**

Section 1: Quality Enhancement by Plant Molecular Biology and Conventional Breeding

PLANT MOLECULAR BIOLOGY: NOVEL FOODS AND WHOLESOMENESS

I. Knudsen

Institute of Toxicology
National Food Agency of Denmark
19 Mørkhøj Bygade, DK 2860 Søborg

1 INTRODUCTION

There is an exponential increase in the number of food plants presently under modification by modern biotechnology and ready to enter the field trials and subsequently the test diets within the next few years.

Such food plants raise several questions for the plant breeders, the consumers and the government officials: Are they novel? Are they safe? Are they nutritionally adequate? Legislative approaches are presently being considered and developed to make sure any uncertainties are addressed in national and international settings.

At the same time scientific endeavours are taking place within WHO[1-3], OECD[4,5] and in national settings[6-9] to create the scientific basis and strategy for the safe introduction of novel foods in the broad sense including genetically modified food plants.

The food from food plants are complex mixtures characterised by their content of both nutrients, antinutrients, natural inherent toxicants, and complex constituents such as fibers, which protect and support good health. Due to the occurrence and interference between health protection, toxic properties, even health promotional capabilities of the food items, the term "safety" is too narrow. Therefore, the term *"wholesomeness"* is used throughout this paper to highlighting such complex properties of our most common dietary ingredients.

2 THE STEPWISE APPROACH

In assessing the wholesomeness of food plants modified through modern biotechnology the attention needs to be focused on primary effects of new functional genes inserted with recognized cellulary functions or extracellulary gene products. At the same time there might also be secondary and/or pleiotropic effects leading to substantial increases in the content of natural inherent toxicants beyond limits of natural variations, and substantial decreases in the content of nutrients important for the overall dietary intake beyond limits of natural variations.

Several approaches for evaluating the safety have been suggested by international (WHO[1-3], OECD[4-5]), regional organizations (Nordic Council of Ministers[10]) and national authorities (UK[6], USA[7], Canada[8], Japan[9]) as well as by non-governmental organizations

(IFBC[11], ILSI[12-14]). No agreed methodology in the evaluation of wholesomeness exists yet. The Nordic approach - to which the present author is most close - is presented in *figure 1*. The clearing of the food item submitted for evaluation can take place from step 1, step 2 and step 4 depending of the nature of the food item, the data submitted and the results obtained. The information relating to substantial equivalence is used at step 1 and at step 2 in the procedure suggested by the Nordic Group on Food Toxicology and Risk Assessment[10].

Fig. 1. Evaluation strategy for novel foods: Necessary data (modified from the NNT, 1991[10]).

First step

Data on

1) Origin
2) Composition
3) Exposure

Second step

Data on

1) New traits/effects
2) Secondary compositional changes

Third step

Data from non-human studies

1) Nutritional/metabolic studies in rats
2) Testing extracts for genotoxicity
3) Comb. 90 days/reproduction study in rats
4) Comb. chronic tox/carcinogenicity study in rats

Fourth step

Data from human studies

1) Single meal
2) Studies of 4 weeks
3) Special groups

Marketing

Data from marketing surveys

3 SUBSTANTIAL EQUIVALENCE

The report of the Joint FAO/WHO consultation, "Strategies for assessing the safety of foods produced by biotechnology"[1], underlines that the comparison of the final product with one having an acceptable standard of safety provides an important element of the safety assessment. The OECD[4] has elaborated this concept and advocated that "the concept of substantial equivalence is the most practical approach to address the safety evaluation of foods or food components derived by modern biotechnology". To establish substantial equivalence between a novel food products and its traditional counterpart data on origin (genetic background/method of production), composition (degree of alteration/complexity) and exposure (never eaten before/higher intake than before) is needed.

It is important to keep in mind that the establishment of substantial equivalence between a novel food product and its traditional counterpart implies:

- that the safety or unsafety of the novel food to human health is equal to its unmodified counterpart,
- that when the conventional food product has a history of safe use, the analogous new product can be considered safe to the same extent ,
- that the establishment of substantial equivalence in itself is not equal to establishing the wholesomeness of the food as such.

It is also important to remember that even if substantial equivalence cannot be established this does not in itself indicate potential health problems connected to the new product.

In October/November 1994 WHO arranged a workshop on "Application of the Principles of Substantial Equivalence to the Safety Evaluation of Foods or Food Components from Plants Derived by Modern Biotechnology" in Copenhagen[3]. Substantial equivalence is here seen as a dynamic, analytical exercise in the assessment of the relative safety of a food or food component to an existing food or food component. The comparison might be a simple task or be very lengthy depending on the extend of experience and the nature of the food or food component under consideration.

Three scenarios are identified:

- substantial equivalence between the new food and a traditional counterpart can be established,
- substantial equivalence between the new food and a traditional counterpart can be established <u>except</u> for the inserted trait,
- substantial equivalence between the new food and a traditional counterpart cannot be established.

Clearance through the first scenario leads back to the first step in *figure 1,* while clearance through the second or the third scenarios equals clearing at the second step or the fourth step in *figure 1,* respectively. The workshop considered that the determination of substantial equivalence can be carried out at the level of the food plant or of the specific plant products. The determination needs to take into account the molecular characterization of the new plant line, its agronomic traits, critical nutrients and critical toxicants. The identification of appropriate reference characteristics for comparison needs to take into consideration variability caused by genetic factors, environmental factors, post harvest handling and processing, intended use and overall role in the diet at the regional level. The comparison of the new line to the traditional food crop plant species should take into account the natural variation and range for parent variety and species based upon proper statistical analysis. Data for this exercise is to be identified both in the regional and international setting. The reference characteristics used for substantial equivalence comparison need to be flexible and changing over time in accordance with changing needs from producers and consumers and with gained experience. Thus a genetically modified food plant including one or more new traits, demonstrated to be wholesome, will become the new reference point for future assessment of substantial equivalence of newly developed varieties.

The WHO workshop made clear that molecular characterization, including knowledge concerning the source and function of the introduced DNA as well as DNA sequence data, does not in itself demonstrate substantial equivalence, but is of value in pointing towards relevant parameters which need to be examined. Consideration concerning the level and function of the products of the introduced gene in the plant may be useful in judging substantial equivalence. For example, if an amino acid pathway has been altered, then the assessment of the amino acid profile is necessary.

Agronomic traits can form a good starting point for evaluating substantial equivalence of the genetically modified plants to their conventional counterparts. These traits are measured by plant breeders for a given crop, often in accordance with national variety registration requirements and commercial needs. For example, in the case of potatoes the measurements do include yield, tuber size and distribution, dry matter content, disease resistance and uptake of environmental contaminants.

The chemical characterization does include the determination of critical nutrients and critical toxicants. Critical nutrients are those components in a particular food product which have a substantial nutritional impact in the overall diet. They may be major constituents (fats, proteins, carbohydrates) or minor compounds (minerals, vitamins). Critical toxicants are those toxicologically significant compounds known to be inherently present in the species, i.e. such compounds which in toxic potencies and levels may be significant to health (e.g. solanine in potatoes if the level is increased). According to the workshop in Copenhagen experiences with genetic modifications do support the idea of focusing on the important components of the food plant. In addition to analysis of critical nutrients and toxicants the extent of analysis for unintended effects will in part be determined by the nature of the intended alteration and by the data from molecular and agronomic characterization. The workshop recognized that differences among consumption patterns and practices in various cultures and societies may change the identity of the key nutrients and toxicants to be examined in different food plants. Therefore, the critical nutrients and toxicants to be addressed should be determined based upon consumption data from the target region.

The WHO workshop also discussed the need for a stepwise approach in analyzing the individual key components, first by comparing the new line to its parents, secondly to other edible varieties of the species, all the time ensuring growth under comparable environmental and agronomic conditions. Also evaluation of substantial equivalence should typically be performed on the unprocessed food product. Progeny derived from food varieties shown to be substantially equivalent would be expected themselves to be substantially equivalent. Analytical methods for key substances and methods to access agronomic traits should be selected according to standardized methods validated in terms of accuracy and reproducibility. Relevant data for use as reference standards need to be retrieved from internationally available databases containing only validated data on the nutritional and especially toxicological components of commonly consumed plant varieties. Some information can already be obtained from the international centres of the

Consultative Group on International Agricultural Research (CGIAR).

The report[3] from the WHO workshop in Copenhagen is expected in June 1995.

Establishment of substantial equivalence of food plants modified by molecular techniques to their traditional counterparts does include considerations concerning the primary changes due to the products of inserted genes (proteins new to the organism, proteins already found, but more, or no protein [anti-sense techniques]), secondary changes in specific, phenotypic traits (changes in synthetic pathways, changes in biochemically related compounds) and unspecific, pleiotropic changes at the phenotypic level (the function/malfunction of "mutated" gene(s), exhaustion of metabolic pathways and/or alteration in down stream gene functions).

Again referring to the evaluation strategy outlined in *figure 1* the first step in the procedure depicts the situation, when substantial equivalence for the whole plant or edible parts thereof can be established based upon molecular characterization, agronomic traits and chemical characterization (critical nutrients and critical toxicants), reflecting both the primary, secondary and unspecific pleiotropic changes.

In performing this exercise it is important to remember that the word "substantial" in the term substantial equivalence allows for insignificant differences in the molecular make-up, agronomic traits and key substances, differences which by experts in the field are considered to be of minor biological importance for the overall wholesomeness of the food product. Substantial equivalence may thus be established not only when the genetic modification results in the production of the protein that already can be found in the parent organism in the same amount, but also when it is found in higher or lower quantities. Also if no protein is formed due to the application on the anti-sense technique, substantial equivalence might be established at the first step.

At the second step in figure 1 the term substantial equivalence applies to the situation where substantial equivalence can be established between the new food plant product and the traditional counterpart except for the inserted trait. Here the safety assessment or wholesomeness is focused on the product of the new trait. This may have an established use as safe "additive" to the food item in question or comparable foods, based upon safety assessment of data from traditional toxicological tests. When the primary product of the newly inserted gene has no established history of safe use in food this does indicate the need for an evaluation based upon toxicological data relating to the primary product itself.

This type of reasoning implies that when a food plant with a new, phenotypically expressed marker gene is presented for the first time, the food plant will need an evaluation by a step two procedure based upon new toxicological and nutrition data on the marker gene. If the new marker gene is accepted as a safe marker gene, it might next time be accepted in a different food plant by a more easy step two procedure taking advantage of the already presented toxicological and nutritional data on the primary product of the marker gene.

4 NON SUBSTANTIAL EQUIVALENCE

For those plant foods, which cannot be cleared by the step one or the step two procedure, the NNT has recommended a testing programme which includes a combination of nutritional, metabolic and toxicological studies *in vivo* and *in vitro*. These studies are to be performed in a logical sequence taking into account the results from preceding studies in the selection and design of the following studies. Importantly the chemical composition of the novel food products should be taken into consideration for the design of a meaningful nutritional screening study. The results of such a screening study are needed before initiating the toxicological studies in order to formulate a semisynthetic or synthetic or human type animal diet from individual ingredients, including the test item itself, which is expected to fulfil the dietary requirements for the rodents in the toxicological study. This strategy will ensure that adverse differences in response between the control group(s) and the test groups in the toxicological studies can be explained by inherent, non-nutritional adverse properties of the novel food or food component. Products from the parent plant may be included in the diet as a "positive control".

Two dietary levels of the test item should be included in the animal study:

1) the "natural" or expected level in the human diet, and
2) an exposure level as high as possible without distorting the animal diet from a nutritional point of view.

Selected extracts of the test material should be tested in *in vitro* mutagenicity tests in order to identify any unexpected genotoxic capability. *In vitro* digestibility studies of the new gene product(s) indicating normal metabolic behaviour may contribute significantly to the assurance of safety.

Unexplainable adverse effects in the combined 90-days/2-generation study or in the genotoxicity studies may lead to the rejection of the product or more likely to the recommendation to reconsider or reexamine the chemical characteristics of the non-substantial equivalent plant food product before initiating the long-term animal study.

NNT recommends that the wholesomeness envisaged from the 90 days/2-generation study is confirmed by testing the same dosages in a long-term rat study in order to ensure absence of adverse chronic effects.

Finally limited short-term human studies are needed to clarify the potential of the new food item to introduce acute/subacute adverse effects in humans such as allergy, intestinal disturbances as flatulence or to identify - if possible - special sensitive groups of people.

5 THE CONCEPT OF FOOD SAFETY

In the report OECD-report from 1993[4] the concept of food safety is the following: "*The safety of food for human consumption is based on the concept that there should be a reasonable certainty that no harm will result from intended uses under the anticipated condition of consumption. Historically foods prepared and used in traditional ways have been considered to be safe on the basis of long-term experience, even though they may have contained natural toxicants or anti-nutritional substances. In principle food has been presumed to be safe unless a significant hazard was identified*".

The novel food era based upon plant molecular biology opens up for more nutritious, more tasty, more good looking foods with longer shelf lives and lower prices for a growing world population, but also warrants attention towards the unintentional introduction of adverse toxic and/or antinutritional effects.

The OECD concept of food safety illustrates that the assessment of wholesomeness of novel foods will need to take into account both testing and assessment procedures, which are different from those traditionally used for defined chemicals. This new challenge includes both a technological problem, a conceptual problem and an ethical problem.

The technological problem: This term refers to the limited possibilities for the use of present nutritional and toxicological methodology for the testing of bulky food items. The traditional approach for the establishment of the *Acceptable Daily Intake, ADI,* used for food additives applies a 100 safety factor to the *no-effect-level* in the long-term animal studies. This is difficult or impossible to apply in the safety evaluation of major food items. Food items making up more than 1 per cent of the human diet will have to be fed to the animals as the only food source to give the safety factor of 100. If such testing is to be enforced, the experimental animals will have to eat a completely distorted and unbalanced diet creating misleading adverse effects not related to any inherent toxic properties of the tested food item. Therefore, the overall strategy of the testing sequence, its benefits and limitations and the size of the safety factors selected for establishing the and safe usages does need to be discussed in the scientific community to reach for - if not scientific consensus - at least a clear scientific message to the public.

The conceptual problem: A growing number of scientific data demonstrates that many food components have dual biological effects, both beneficial and adverse. That means that too little is bad for the health and so it too much. Even socalled natural toxicants may have beneficial influences, e.g. by increasing the activity of detoxifying liver enzymes. From a health point of view the many compounds of a complex food both strengthen and counteract each other in a complex manner. Thus the biological or "health" effects of a complex food are different from the "arithmetic sum" of the nutritional and toxic effects of the individual, biologically active compounds of such a food. So far no evaluation strategy to summarize the pluses and minuses of a complex food leading to a "health sum" derived from the biological properties of its individual components has become widely accepted. For instance certain inherent natural toxicants need to stay in the plant foods for the reasons of plant protection, e.g. solanine in potatoes. Even if it could be taken away by modern gene technology, it needs to stay or be replaced by a product with similar antimicrobial activity. Therefore, it will be important to make it clear to and accepted by the public that no zero risk can be achieved for food, not even for novel food.

The ethical problem: If human experience or animal experiments demonstrate that a plant food has a toxic potential, maybe as an initiator or a promoter for carcinogenicity and that this negative effect cannot be abolished by physical, chemical or biotechnological means, the question arises whether or not such a plant food should be promoted for use. The question is even more difficult if the toxic constituent also is present in the naturally existing food plant, and modern techniques make it possible to get rid of or just diminish its presence. Then the question arises whether lower levels of this compound below the level of natural occurrence should be pursued. This problem may become even more difficult from an assessment point of view if such a compound in low quantities has a nutritional value rendering it an optional nutrient. A scientific option may be to pursue research in the beneficial, nutritional properties of the compound and at the same time to clarify its toxic properties thereby thriving at establishing a balanced view of the positive and negative characteristics of certain inherent toxic compounds in the complex diet.

Another approach to addressing the ethical problem is to introduce a more open labelling policy in a world wide sense allowing for the announcement of both potential health benefits and potential adverse health effects substantiated through thorough scientifically based testing. The public awareness of the pro's and con's for the food and the different food items will then determine the future role of the individual novel foods in the human diet.

6 CONCLUDING REMARKS

The scientific understanding of nutritional needs for food to maintain and promote a good health has increased tremendously through the last two hundred years. During the last fifty years the knowledge concerning the negative consequences of a distorted or unbalanced diet not only in regard to general health but also in regard to specific diseases such as cardiovascular diseases and cancer has increased.

In the same period the potential of man to introduce intentional and unintentional changes in individual food items has increased exponentially.

The interaction between these technological developments and the increased scientific knowledge about the health impact of individual foods and the diet as a whole creates a number of questions concerning the overall wholesomeness of the individual food item in relation to the complex and variable diet. At present the scientific community is not equipped to deal with this complicated situation either from a scientific or from a resource point of view.

A world wide scientific debate reaching out for identifying and defining scientific and regulatory tools for assessment of tomorrow's food supply is urgently needed. The time is short, if we are to avoid non-customrelated barriers in the national and international food trade.

7 REFERENCES

1. FAO/WHO, Report of a Joint Consultation, Geneva (1991): Strategies for assessing the safety of foods produced by biotechnology. WHO, World Health Organization, Geneva (59 pages).
2. WHO, Report of a Workshop (1993): Health aspects of marker genes in genetically modified plants. World Health Organization. Food Safety Unit (32 pages).
3. WHO, Report of a Workshop (in press): The application of the principles of substantial equivalence to the safety evaluation of foods or food components from plants derived by modern biotechnology. World Health Organization. Food Safety Unit, Geneva.
4. GNE, Group of National Experts on Safety in Biotechnology (1993): Safety evaluation of foods derived by modern biotechnology: Concepts and principles. The OECD Environment Directorate and Directorate for Science, Technology and Industry, Paris (79 pages).
5. OECD, Report of a Workshop (in prep.): Food safety evaluation. Environmental Health and Safety Division, OECD, Paris.
6. ACNFP, Advisory Committee on Novel Foods and Processes (1991): Guidelines on the Assessment of Novel Foods and Processes. The Department of Health and Social Security, London (30 pages).
7. FDA, Food and Drug Administration (1992). Statement of policy: Foods derived from new plant varieties, notice. Federal register **57,** 22984-23005.
8. Food Directorate (1993). Draft Guidelines for the Safety Assessment of Novel

Foods. The Health Protection Branch, Health Canada (39 pages).

9. Anonymous: Guidelines for Foods and Food Additives (1993): Produced by recombinant DNA-Techniques. The Ministry of Health and Welfare, Japan (129 pages).
10. NNT, Nordic Working Group on Food Toxicology and Risk Assessment (1991): Food and New Biotechnology - Novelty, safety and control aspects of foods made by new biotechnology. Nordic Council of Ministers. Nord 1991:18 (247 pages).
11. IFBC, International Food Biotechnology Council (1990): Biotechnologies and Food: Assuring the Safety of Foods produced by Genetic Modification. *Regul. Toxicol. Pharmacol.* **12**(3), (part 2 of 2 parts).
12. ILSI Europe, International Life Science Institute, the European Branch (1989): The Assessment of Novel Food. A discussion paper. The Europe Technical Committee on novel foods, Bruxelles.
13. ILSI Europe, International Life Science Institute, the European Branch (1990): Reevaluation of current methodology of toxicity testing including gross nutrients (Eds: R. Kroes and R.M. Hicks). *Food and Chemical Toxicology,* **28,** 733-790.
14. ILSI Europe, International Life Science Institute, the European Branch (1993): Nutritional Appraisal of Novel Foods. (Eds. A. Bruce et al.). *International Journal of Food Sciences and Nutrition,* **44,** suppl. S1-S100.

CARBON FLUX IN TRANSGENIC SUGAR BEET

W.Monger[1], P.J.Harrington[1], M.A.Acaster[1], N.G.Halford[2], C.C.Ainsworth[3], T.H.Thomas[1].

[1] IACR-Broom's Barn, High, Bury St Edmunds, Suffolk IP28 6NP
[2] IACR-Long Ashton Research Station, Long Ashton, Bristol BS18 9AF
[3] Wye College, University of London, Wye, Ashford, Kent TN25 5AH

1. INTRODUCTION

Our current knowledge of carbon flux in plants has been founded on earlier biochemical and enzymological studies. With the advent of transformation technology it has been possible, at least in theory and increasingly so in practice, to make specific changes to key enzymes of biological pathways (see for example Ref. 1). A more detailed understanding of carbon metabolism in major agronomic crops such as sugar beet could have a significant economic impact if the crop could be manipulated to produce added value products e.g. trehalose[2].

Currently sucrose is the sole carbohydrate stored in the tap root of *Beta vulgaris*. It can accumulate to very high levels, up to 0.8M, accounting for at least 65% of the dry weight of the harvested beet. Sucrose is produced in the leaves and transported via the phloem to the tap root, where it is stored without further modification[3]. In this respect sugar beet is unusual amongst higher plants which more generally utilize starch as the principle storage carbohydrate. The storage organ of beet contains no starch despite starch being used as a transient storage compound in the leaves.

1.1 Modification of Enzymic Pathways

1.1.1 SNF1. We have sought to understand the carbon fluxes in beet initially using a model system in hairy roots to evaluate the effects of any biochemical modification before progressing to whole plant studies. The first gene we have attempted to modulate is a homologue of the sucrose non-fermenting (SNF1) gene that was first isolated from yeast[4]. This gene which encodes for a protein kinase activity, is an essential part of a signal transduction pathway controlling repression and de-repression of glucose responsive genes.

Homologues of SNF1 have been isolated from rye, barley, arabidopsis, potato and now sugar beet. These related protein kinase genes show approximately 50% sequence homology with yeast SNF1. These plant genes may serve a similar function in plants to that of SNF1 in yeast. Expression of the rye SNF1 homologue in a yeast snf1-minus mutant restored SNF1 function, and the ability to utilize carbon sources other than glucose[5].

A sugar beet SNF1 homologue has been identified using PCR primers designed against the rye SNF1 homologue. This PCR reaction produced a 150 bp fragment which was cloned into pUC18 and sequenced.

Table 1. *Comparison of SNF1 peptide sequences*

	210					260
Arabidopsis	D	GHFLKTSCGS	PNYAAPEVIS	GKLYAGPEVD	VWSCGVILYA	LLCRRLPFDD
Potato	D	GHFLKTSCGS	PNYXPEVVS	GKLYAGPEVD	VWSCGVILYA	LLCGTLPFDD
Rye	D	GHFLKTSCGS	LNYAAPEVIS	GKLYAGPEID	VWSCGVILYA	LLCGAVPFDD
Sugar beet	D	GHFLKTTCGT	XNYAAPEVIS	GKLYAGPEVS	.WSCGVILYA	LLC . . .
Yeast	D	GNFLKTSCGS	PNYAAPEVIS	GKLYAGPEVD	VWSCGVILYV	MLCRRLPFDD

Anti-sense expression approaches are being taken to investigate the role of plant *snf1* homologues in carbohydrate metabolism in both potato and sugar beet. A 503 bp fragment isolated from potato has been used to prepare antisense expression cassettes using either a constitutive (35S) or a tissue specific promoter (patatin). The constructs have been introduced into potato and sugar beet hairy roots using *Agrobacterium*-mediated infection and a Bin 19-based binary vector. To identify transgenic plants more rapidly we are currently cloning the sugar-beet constructs into a plant transformation vector containing Gus-int and kanamycin resistance genes; again the 35S and patatin promoters are being used.

It has proved more difficult to establish sugar beet hairy root cultures containing the 35S driven construct than with the patatin driven version; similar difficulties have been found in the potato transformation studies[6]. Some of the hairy root cultures exhibited an altered morphology in that they were thicker and more nodular than wild-type hairy roots. This altered morphology was more apparent in the cultures that have been produced with the 35S driven construct. These cultures are being screened for protein kinase activity using a synthetic peptide derived from the rat acetyl-CoA carboxylase phosphorylation site and given the acronym SAMS.

Once transgenic cultures have been identified they will be analysed for alterations in the sucrose metabolising enzymes, invertase, sucrose synthase, sucrose phosphate synthase and hexokinase to determine the effects of SNF1 down-regulation in sugar beet. We shall also determine the effects of any such changes on the levels of glucose, fructose and sucrose.

1.1.2 Starch. Another area of our work concerns starch biosynthesis or more accurately the lack of starch accumulation in the tap root of sugar beet. Despite very high levels of sucrose no starch accumulates in the tap root, whereas transient starch is made in the leaves. The biochemical reason for the lack of starch accumulation could include the possibility that no starch is made, or that it is rapidly degraded?

We have used heterologeous probes from various sources, including pea, potato and wheat, and have used Southern blotting to ensure these probes recognise the sugar beet genes. Using Northern blots to identify which genes are transcribed has shown that the first enzyme in the dedicated pathway leading to starch biosynthesis, ADP-glucose pyrophosphorylase (AGPase), is not transcribed in the tap root but is in the leaves. Furthermore, using enzymic assays AGPase was detected in the sugar beet leaves and in potato tubers, but was not detected in the sugar-beet tap root itself. In mixing the sugar beet tap root extract with either potato or sugar leaf extracts no evidence for a root AGPase inhibitor was found.

Table 2. *ADP-Glucose pyrophosphorylase activity*

Source		ADP-glucose pyrophosphorylase activity n moles min^{-1} g^{-1} FW
Sugar beet	Leaves	7-11
	Beet	Not detected
Potato	Tubers	10-25

It seems likely that the reason why sugar-beet tap roots do not accumulate starch is because the first enzyme necessary for starch synthesis is not transcribed. To test this assumption we have prepared hairy root cultures which have been transformed with a bacterial form of AGPase (Glg-C). This gene is driven from a patatin promoter and has the Rubisco small subunit chloroplast targeting sequence upstream of the start codon. These cultures are being analysed to determine the effects of re-expressing AGPase or Glg-C in sugar beet.

2. ACKNOWLEDGEMENTS

We acknowledge the gifts of a plant transformation vector from British Sugar Technical Centre, Norwich, U.K. and the gift of the Glg-C construct from Advanced Technologies Cambridge (UK) Ltd., Cambridge, U.K.

3. REFERENCES

1. U. Sonnewald, M. Brauer, A. von Schaewen, M. Stitt and L. Willmitzer. *The Plant Journal*, 1991, **1**, 95-106.

2. Mogen Press Release No. 1994-03.

3. R. Wyse, *Plant Physiol.* 1979, **64** 837-841.

4. J.L. Celenja and M. Carlson, *Science*, 1986, **233** 1175-1180.

5. A. Alderson, P.A. Sabelli, J.R. Dickinson, D. Cole, M. Richardson, M. Kreis, P.R. Shewry and N.G. Halford. *Proc. Natl. Acad. Sci.*, 1991, **88**, 8602-8605.

6. N.G. Halford, J.H.A. Barker, P. Purcell, A.L. Man, W.A. Monger, M.A. Acaster, T.H. Thomas and P.R. Shrewry, *Biologia Plantanum*, 1994, **36** 568.

7. M.C. Elliot and G.D. Weston. In: *The Sugar Beet Crop*. D.A. Cooke & R.K. Scott (Eds). Chapman & Hall 1993.

IMPROVEMENT OF THE QUALITY OF BARLEY FOR BEER PRODUCTION

R.J. Henry

Queensland Agricultural Biotechnology Centre
Gehrmann Laboratories
University of Queensland 4072
Australia

1 INTRODUCTION

Barley is the traditional raw material for beer production. Advances in barley transformation[1], barley gene mapping[2] and brewing technologies[3-5] offer new opportunities to improve the quality of barley as a raw material for brewing.

Despite the wide potential to use barley for food, feed and industrial purposes[6], beer remains the end-use of barley that is most demanding of quality. Possible improvements in barley for brewing will be reviewed in this paper.

2 PRODUCTION OF BEER FROM BARLEY

Barley is the major raw material used in the traditional production of beer (Fig. 1).

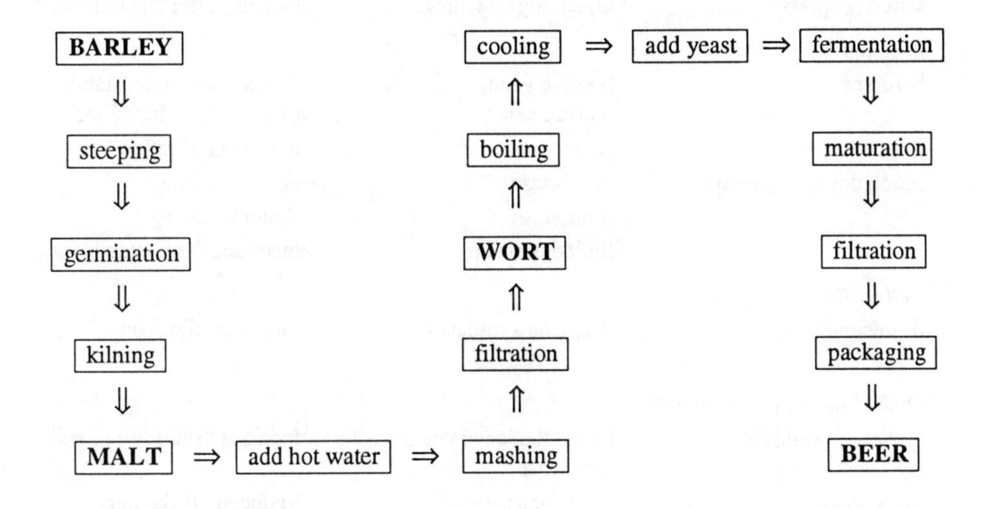

Figure 1. *Traditional beer production*

The barley grains are germinated under controlled conditions (malted) to generate hydrolytic enzymes. The malt, is ground and extracted with hot water under conditions favouring the hydrolysis of starch to produce a solution with a high sugar content (wort). The wort is cooled and yeast is added to convert sugar to alcohol by fermentation. The starch, proteins, cell walls and other components of the barley grain all influence the malting and brewing process and may contribute to beer quality.

3 IMPROVEMENT OF CARBOHYDRATE CONTENT

The carbohydrates of barley include sugars, starch and cell wall components[7]. The sugars and starch contribute fermentable substrates for brewing.

Starch is the major component of the barley grain and the greatest source of carbohydrate as substrate for fermentation in brewing. A high starch content in the barley will contribute to a high malt extract in brewing. Starch properties desirable in brewing include a low gelatinisation temperature and a starch granule size distribution and amylose/amylopectin ratio that will allow rapid conversion to give a wort with the fermentability required for the specific beer product to be produced. Genetic engineering of barley could be used to improve the carbohydrates of the grain as substrates for malting and brewing[8](Table 1).

Table 1. *Possible improvements in the malting and brewing characteristics of barley using genetic engineering.*

Characteristic	Genes	Influence on quality
Carbohydrates		
starch content	ADP-glucose pyrophosphorylase	Higher malt extract
starch properties	branching enzymes	Increased fermentability
Proteins		
hordeins	hordein genes (antisense)	Greater beer foam stability Longer shelf life of beer
β-glucanases	β-glucanases	Improved filtration
starch degrading enzymes	α-amylase	Faster mashing
	β-amylase	Faster mashing
	limit dextrinase	Increased fermentability
Cell walls		
β-glucans	β-glucan synthases (antisense)	Improved filtration
Other (minor) components		
proanthocyanidins	biosynthetic enzymes (antisense)	Increased shelf life of beer
lipoxygenases	lipoxygenase	Reduced off flavour

4 IMPROVEMENT OF PROTEIN CONTENT

The proteins from barley may influence beer quality and the brewing process in several ways. Proteins make a positive contribution to beer quality by stabilising beer foam but may also cause undesirable chill hazes. Specific proteins may be associated with malting quality[9]. The starch, protein and cell wall degrading enzymes of barley play an important role in beer production. The stability of the enzymes at the temperatures used in brewing may be limiting in many cases. Considerable research has been devoted to the development of improved temperature stability in the starch and β-glucan degrading enzymes in barley [10, 11]. The two main approaches have addressed the manipulation of the structure of barley proteins to improve their stability at high temperature and the identification of thermostable enzymes from other sources that might be introduced into barley. These approaches are both likely to succeed technically but may be perceived differently by beer consumers.

5 IMPROVEMENT OF CELL WALL CONTENT

The polysaccharides of the cell wall of barley, especially the β-glucans may influence the brewing process and the quality of the beer. High malt β-glucan can contribute to filtration and stability problems. Reduced β-glucan in malt may be achieved as easily by high β-glucanase activity as by low barley β-glucan content although both characters must be considered desirable in barleys to be used for beer production.

6 IMPROVEMENT OF MINOR GRAIN COMPONENTS

Elimination of proanthocyanidins may be worthwhile. This could allow the production of beers with a greater shelf life. However in some experimental brews a failure of protein precipitation in the kettle in the absence of proanthocyanidins has resulted in inferior quality beers. The sugars of barley represent only a minor part of the potentially fermentable carbohydrate. However they are highly fermentable and changes in the level of sugars in the grain would not be of great advantage if the starch is able to be readily converted to sugars. The husk of the grain provides a natural filter bed in traditional brewing processes. Elimination of the husk would allow barley to contain a greater proportion of fermentable material but requires adjustments of some brewing processes. Oxidation of lipids may be important in the development of off flavours during the ageing or staling of beer. This has resulted in attempts to modify lipoxygenases by genetic engineering[12](Table 1).

7 IMPROVEMENT OF THE MALTING AND BREWING PROCESS

Alteration of the biochemical constituents of barley may allow improvements in the efficiency of the brewing process and improve the consistency of the quality of beer.

Increased starch content and altered starch properties could improve the yield of fermentable substrate available for brewing and allow better control of the fermentability of worts for specific beer styles. Changes in brewing facilities may permit the widespread use of naked or hullless barley with the associated potential for increased extract yield.

Improved thermostability in starch and β-glucan degrading enzymes might permit kilning of malt and mashing at higher temperatures with potential savings in processing time. Enhanced expression of β-glucanase may reduce filtration problems. Expression of appropriate enzymes in the mature grain rather than during germination could allow the malting process to be avoided altogether. Alteration of the storage proteins may achieve the desirable combination of good foam stability together with enhanced shelf life in cold storage.

Major redesign of the traditional beer brewing process (Figure 1) could be achieved in conjunction with the introduction of genetically engineered barley specifically designed for efficient processing to produce beer.

References

1. Y.C. Wan and P.G. Lemaux *Plant Physiology*, 1994, **104**, 37.
2. P.M. Hayes, B.H. Liu, S.J. Knapp, F. Chen, B. Jones, T. Blake, J. Franckowiak, D. Rasmusson, M. Sorrells, S.E. Ullrich, D. Wesenberg and A. Kleinhofs *Theor. Appl. Genet.*, 1993, **87,** 392.
3. J. Kronlof and M. Linko *J. Inst. Brew.*, 1992, **98**, 479.
4. R. G. Peel and R. Glastonbury 'Proceedings of the Twenty-Second Convention, The Institute of Brewing, Australia and New Zealand Section', 1994, p 99.
5. T. D'Amore *J. Inst. Brew.* , 1992, **98**, 375.
6. Alternative End uses of Barley, D.H.B. Sparrow, R.C.M. Lance and R. J. Henry, eds, 1988, Royal Australian Chemical Institute, Melbourne.
7. R.J. Henry *J Inst Brew* ,1988, **94**, 71.
8. J. Preiss, D. Stark, G.F. Barry, H.P. Guan, Y. Libal-Weksler, M.N. Sivak, T.W. Okita and G.M. Kishor, 'Improvement of Cereal Quality by Genetic Engineering' R.J. Henry and J.A. Ronalds,eds, Plenum, New York, 1994, p. 115.
9. P. R. Shewry, A.S. Tatham, N.G. Halford, J. Davies, N. Harris and M. Kreis, 'Improvement of Cereal Quality by Genetic Engineering' R.J. Henry and J.A. Ronalds, eds, Plenum, New York , 1994,p. 79.
10. G. B. Fincher, 'Improvement of Cereal Quality by Genetic Engineering' R.J. Henry and J.A. Ronalds, eds, Plenum, New York, 1994, p.135
11. J.E. Vickers, S. Weining, S.E. Hamilton, A. Shaw, A. Wamsley, J. deJersey, P.A. Inkerman and R.J. Henry *Chemistry in Australia*, 1994, **61**, 504.
12. S. Hoekstra, M. van Zijderveld, S. van Bergen, F. van der Mark and F. Heidekamp, 'Improvement of Cereal Quality by Genetic Engineering' R.J. Henry and J.A. Ronalds, eds, Plenum, New York, 1994,p 139.

IMPROVEMENT OF NUTRITIONAL QUALITY OF WHEAT THROUGH SOMACLONES AND MUTANTS

R.K.Rana, Subhadra and Shashi Madan

Wheat Research
Department of Plant Breeding
CCS HAU, Hisar-125004 (India)

1. INTRODUCTION

Most of the world population depends upon cereals as their staple food. Cereals generally suffer due to low protein content and other quality traits. Bread wheat serves as a major food crop for the human population. The nutritional improvement of this cereal is quite important to provide balanced diet to avoid malnutrition. Improvement of bread wheat through plant and seed manipulation has resulted into genotypes which provide better nutritional qualities as food.

The primary objective of this research was to evaluate the usefulness of variability induced through tissue culture & mutagens for genetic improvement of rust resistance and nutritional quality of wheat.

Somaclonal variation was coined by Larkin and Skowcroft (1981) for the genetic variation induced by passage through tissue culture. It is a novel source of genetic variation and has been documented in many major crop plants such as potato, sugarcane, rice, oat and soybean. Genetic variation has been created through tissue culture in wheat (Larkin et al. 1984; Maddock and Semple 1986; Carver and Johnson, 1989).

2. MATERIAL AND METHODS

Cultivar WH 147 was chosen based on its high productivity and widely adaptive value in Northern plains of India. This variety was released by wheat scientists of Haryana Agricultural University, Hisar two decades ago and it has brought green revolution in the region. But this variety needed further improvement with respect to rust resistance and grain quality and nutritional improvement. Therefore, research work was undertaken to improve this variety through non-conventional breeding methods like tissue culture and mutation.

2.1 Somaclone variants

The tissue culture experiment was initiated with the help of Biotechnology Laboratory, Deptt. of Genetics. From the parent variety WH147 the plants were examined for their genetic purity and marked for their dates of anthesis. The immature caryopes were taken out of the spike 18-20 days after anthesis and sterilized in mercuric chloride followed by washings with sterilized distilled water. For callus initiation, the immature embryos were

were dissected out and placed on basal MS medium supplemented with 2,4-Dichloro-phenoxy acetic acid. Callus initiation was observed in all the embyros. The reduction in level of 2,4- Dichloro-phenoxy acetic acid induced the development of somatic embryoids in high frequency after 8-10 days of incubation of cell suspension. A typical well developed normal embryoids showed bipolar structure tapering into root and shoot. These somatic embryoids formed plantlets after 3-5 days of further incubation either in fresh liquid or on agar based medium. A large number of plantlets were obtained and some of them were transferred to the pots which survived to maturity (Subhadra et al. 1995).

The mature spikes were labelled on each plant to maintain pedigree and threshed separately to provide a total of 60 lines . These were propagated in the field next season (1992-93) at HAU Hisar Research farm and observations were made for various traits. The lines which showed field resistance to leaf rust were marked and harvested separately. The resistant lines were further evaluated in the wheat experiment field during season of 1993-94 for various traits and then for quality analysis in the wheat quality laboratory.

2.2 Mutants

The mutants were developed by treating the samples of wheat variety WH147 with 25 Kr doses of Cobalt 60 mutagen. The advanced material was studied for various traits in the field and the lines with better agronomic traits were evaluated for quality analysis in wheat quality laboratory during 1993-94.

3. RESULTS AND DISCUSSION

The data on some quality traits of some somaclonal variants and mutant lines have been given in Table 1 and 2, respectively. The data of table 1 showed that some somaclonal variants exhibited improvement in grain size and all excelled in grain score over the check variety. All the somaclonal variants were found superior in protein percent (11.92 to 13.16) as compared to check variety (10.12%). Some lines had more sedimentation value than the check variety. The range was from 27 ml to 34 ml against 31 ml of parent variety.VarSV-1had low 1000 grain wt. (30.3 g) and high protein percent (13.16) and moderate sedimentation value (31 ml)while another variant SV-3 had high grain wt. (42.7g) and moderate protein percent (12.87) but had low sedimentation value (27 ml). The other variants had shown moderate values of grain wt., protein content and sedimentation value and thus have medium-strong bread score.Shashi et al. (1981) had also observed+ve correlation between protein content and sedimentation value. It was further noted that these somaclonal variants were resistant to the leaf rust disease which causes great yield losses. These data reveal that through tissue culture technique genetic variants can be got for various traits and some desirable variants can be selected which show improvement in some quality traits like grain size, grain score, portein percent, sedimentation value and other important characters like resistance to diseases which lead to the superior agronomic performance (Rana et al. 1995).

The data in Table 2 showed that some mutant lines exhibited improvement in grain size and grain score while other lines showed improvement in protein percent and sedimentation value over the check variety. The range in protein percent was from 9.22 to 12.94 against 10.12% of the check variety. The sedimentation value ranged from 29 ml to 37 ml as compared to 31 ml of the check variety. From data it was also observed that a mutant line WH147-4 had lowest 1000-grain wt (30.0g) but highest value of protein percent (12.94) and highest value of sedimentation (37 ml) while in contrast other mutant line WH 147-5 had highest 1000-grain wt. (39.2), highest grain score (7.0), but had low protein percent (9.96) and low sedimentation value (29 ml). Other mutant lines had shown the moderate values. Improvement in rust resistance was not achieved in the mutant lines as compared to the parent variety.

The present study has demonstrated that it is possible to generate fertile plants through embryogenic cell suspensions. This technique offers a potential means to effect genetic amelioration in wheat. Further research is needed on tissue culture-induced variability to evaluate the worth of this novel technique to improve the crop varieties for various desirable traits.

Table 1: Comparison of some quality traits of somaclonal variants of Bread wheat variety WH 147

Line	1000 Gr. Wt (g)	Gr.Score (Max 10)	Protein (%)	Sedimentation value (ml)	Leaf rust reaction
SV-1	30.3	5.0	13.16	31	TMR
SV-2	38.8	5.5	11.92	27	TS
SV-3	42.7	6.5	12.87	27	TS
SV-4	33.4	4.0	12.24	30	TMS
SV-5	38.9	5.5	12.72	34	TMR
SV-6	31.3	5.0	12.87	29	TS
WH 147 (Check)	34.4	4.0	10.12	31	100S

Table 2 : Comparison of some quality traits of Mutants of Bread wheat variety WH147.

Line	1000 Gr. Wt (g)	Gr.Score (Max.10)	Protein (%)	Sedimentation value (ml)	Leaf rust reaction
WH 147-1	32.3	4.0	9.96	30	80S
WH 147-2	30.8	4.0	10.28	32	80S
WH 147-3	33.4	3.5	10.60	32	70S
WH 147-4	30.0	4.0	12.94	37	80S
WH 147-5	39.2	7.0	9.96	29	70S
WH 147-6	30.5	5.0	11.54	32	90S
WH 147-7	30.8	4.8	9.22	30	1005
WH 147 (Check)	34.4	4.0	10.12	31	1005

References

B.F.Carver and B.B. Johnson. TAG 1989, **78**, 405-410

P.J.Larkin and W.R. Skowcroft , TAG, 1981, **60**, 197-214.

P.J.Larkin, S.A. Ryan, R.S.Bretell and W,R.Scowcraft, TAG, 1984, **67** , 433-455.

R.K.Rana, S.S.Karwasra, R.B. Srivastava and M.S. Beniwal, Indian Wheat Newsletter, 1995, **1**, 6.

S.E.Maddock and J.T. Semple, J. Exp. Bot., 1986, **37**, 1065-1078.

Shashi Taneja, K.S. Dhindsa and S. Popli, Indian J. Biochem. & Biophys. 1981, **18**, 84.

Subhadra, J.B. Chowdhury and Randhir Singh, Ind. J. Exp. Biol. 1995 **33**, 142-149.

T. Murashige and F. Skoog. Physical plant, 1962, **15**, 473.

SAFETY ASSESSMENT OF THE *Bacillus thuringiensis* INSECTICIDAL CRYSTAL PROTEIN CRYIA(b) EXPRESSED IN TRANSGENIC TOMATOES

H.P.J.M. Noteborn, M.E. Bienenmann-Ploum, G.M. Alink[2], L. Zolla[3], A. Reynaerts[4], M. Pensa[5] and H.A. Kuiper

Department of Risk Assessment and Toxicology, RIKILT-DLO, 6700 AE Wageningen, [2]Department of Toxicology, Agricultural University Wageningen, 6700 EA Wageningen, [3]Department of Environmental Sciences, University La Tuscia, 01100 Viterbo, [4]Plant Genetic Systems, B 9000 Gent, [5]SME Ricerche SCPA, La Fagianeria, 81015 Piana di Monte Verna (Caserta).

1 INTRODUCTION

Recombinant DNA technology bridges barriers between widely different species. In applying modern biotechnology in plant breeding, breeders will be able to introduce traits in food crops that were never part of it before. The experience with transgenic food crops for human and animal consumption is still in its infancy. Public concern over genetically modified food reflect a need for a systematic assessment of potential risks to human and animal health.

1.1 Insecticidal Activity in Transgenic Tomatoes

Transgenic insect-resistant tomato plants (variety RLE13-0009) were obtained from the parental line TL001, a round type tomato, by transformation with a disarmed *Agrobacterium tumefaciens* strain using cotyledons as the explants. Thereto, the vector pPSO216 was used that comprised two chimeric genes between the T-DNA border repeats. The chimeric *neo* gene consisted of the promoter of the T-DNA TR1' gene and the coding region of the *neo* gene encoding neomycin phosphotransferase II (NPTII) from transposon Tn5. The chimeric *Bacillus thuringiensis* (*Bt*) gene consisted of the wound stimulated promoter of the T-DNA TR2' gene and the coding region of the C-terminal truncated *Bt*2 gene, called *IAb*6 derived from the coding region bt884 encoding the insecticidal CRY-type ICP, CRYIA(b) toxin (MW 66-68 kDa) (1).

1.1.1 Chimeric gene expression. The nature of the genetic modification was analyzed with respect to size and number of gene inserts as well as to size and levels of transcripts and proteins in different organs of the tomato. In the harvested transgenic variety RLE13-0009 as analyzed by ELISA the CRYIA(b) toxin was expressed in fresh fruit (i.e. non-induced) at levels of about 7.5 ng/mg of protein, and in induced tomato fruit at levels of about 25.4 ng/mg of protein with a nominal content of 0.8% protein of wet weight.

1.1.2 Recombinant CRYIA(b) toxin. As sufficient amounts of recombinant DNA proteins can not be easily extracted from transgenic tomato tissue, a number of toxicologic studies were performed with CRYIA(b) toxin isolated and purified from the overproducing recombinant *Escherichia coli* strain K514 (pGI502) as previously described (2).

2 SAFETY ASSESSMENT STRATEGY FOR TRANSGENIC Bt-TOMATOES

The strategy for food safety evaluation of transgenic Bt-tomatoes was designed in a way that answers could be given to questions like: (i) does CRYIA(b) toxin exert a similar toxicity in mammals as observed in larvae of susceptible lepidopteran insects; (ii) do the deliberately introduced proteins cause systemic adverse effects in mammals, and if so, can a No Effect Level be established; (iii) are the novel dietary proteins immunogenic and/or potentially allergenic, and (iv) does the transgenic phenotype lead to any significant changes in the (bio)chemical composition of the food crop that would compromise the safety of the product as human and animal food.

2.1 Insecticidal Activity of CRYIA(b) in Mammals

The possibility of receptor binding in tissues of the mammalian gut was studied extensively, since CRY-type ICPs (i.e. CRYIA(b)) bind to specific receptors in the midgut of susceptible insects that seems essential prior to the onset of toxicity.

2.1.1 Immunochemical localization. After feeding rats single doses of CRYIA(b) toxin, corresponding to a human daily intake of 2000 kg of Bt-tomatoes, no binding of CRYIA(b) toxin or pathological effects could be detected in any tissue of the segmented gut of treated animals. No specific binding of CRYIA(b) toxin was observed in intestinal tissues of mice, rats, rhesus monkeys and humans after incubation of biopsies with CRYIA(b) *in vitro*. Whereas, midgut tissue sections of larvae of the lepidopteran *Manduca sexta* incubated with CRYIA(b) toxin showed an uniform binding in the brush border membrane over the entire length of the epithelium (i.e. peritrophic membrane and apical microvilli). After ingestion by susceptible larvae, CRYIA(b) toxin provoked a rapid swelling of columnar cells and vacuolization of cytoplasm and hypertrophy of the epithelial lining (3).

2.1.2 Proteolytical processing. The degradation of CRYIA(b) toxin studied in simulated human gastro-intestinal fluids indicated clearly a two-step process. At pH 2 in the presence of pepsin (1 : 100, w/w) the protein was rapidly cleaved to yield 15 kDa polypeptides. After continued incubation at pH 8 with chymotrypsin and trypsin (1 : 25, w/w) an extensive fragmentation to smaller fragments (<< 10 kDa) was seen as analyzed by gel permeation HPLC and western blot. On the contrary, marker NPTII appeared completely digested at pH 2 in the presence of pepsin. The digestibility of CryIA(b) toxin studied in rats with an iliac fistula followed by western blot analysis of collected chymus 5 to 7 hrs after feeding, revealed a rapid and extensive breakdown of CRYIA(b) toxin.

2.2 Systemic toxicity of CRYIA(b) in mammals

Studies of the histopathology and mode of action of CRY-type ICPs on mammals and other non-target species are very limited and have all used crude preparations of spore-crystal mixtures. Therefore, rodents were exposed to CRYIA(b) toxin to test for systemic and immunotoxic effects upon passage through the intestinal wall.

2.2.1 Sub-acute exposure. Daily doses of purified CRYIA(b) toxin were fed to mice and rabbits via drinking water *ad lib.* for 28 and 31 days, respectively. Controls received tap water only. At dose levels corresponding to a human daily intake of 60 to 500 kg of transgenic Bt-tomatoes, no differences were observed in all clinical and hematological parameters tested. In serum of treated rabbits no antibodies against

CRYIA(b) toxin could be detected, nor was the total immunoglobulin (IgG) content elevated with respect to controls. No pathological alterations were found in organs and intestinal tissues of treated animals. Microscopic examination revealed no indications for immunotoxic effects of CRYIA(b) toxin as judged by examination of spleen, lymph nodes and Peyer's patches.

2.2.2 Hemolytic activity. Human red blood cells incubated with CRYIA(b) toxin demonstrated no hemolysis as analyzed by monitoring the osmotic fragility (4). RBCs entrapped with CRYIA(b) toxin showed also negligible hemolysis.

2.3 Food Safety Assessment of Transgenic Crop

The field trial manifested no significant differences in vegetative growth and harvest characteristics between transgenic Bt-tomatoes (RLE13-0009) and the control. The question whether the transgenic phenotype may result in any secondary or unintentional changes in the metabolism of the food crop which could be of toxicologic significancy, was addressed by two approaches.

2.3.1 Compositional analysis. Chemical analyses were made of a number of components related to the nutritional composition of representative samples of freeze-dried transgenic Bt-tomatoes (RLE13-0009) harvested after field testing. The selected nutrients were compared to the non-transformed control line TL001 and found to be unchanged and within published ranges. Moreover, levels of the glycoalkaloid α-tomatine were similar in control and transgenic Bt-tomatoes (range: 1.4-1.7 mg/kg). Other known solanaceous alkaloids were not observed or below detection limit.

2.3.2 90-Day feeding trial. Three groups of 12 male and 12 female weanling Wistar rats were fed, respectively, a control semi synthetic animal diet (Muracon SSP TOX), the same diet supplemented with 10% (w/w) of freeze-dried transgenic Bt-tomatoes containing 40.6 ng CRYIA(b) toxin/mg of protein, or with 10% (w/w) of freeze-dried control tomatoes. Insufficient data on amino acid composition of tomato proteins and the observation of increased potassium urinary concentrations in a 13 week dosed range-finding study (1) limited further dose escalation. The macro- and micronutrient composition was equalized in all test diets: 7.5% E(energy) fat, 20% E protein, 41.5% E carbohydrate and 10.4% fibre (w/w). The amounts of supplementary minerals and vitamins were deduced from the actual levels in tomatoes. The average daily intake of tomatoes over the 91 day period corresponded to a daily human intake of 13 kg of transgenic Bt-tomatoes. There was no feed refusal or unusual behaviour in any animal. No significant differences in clinical and biochemical parameters under study were noticed between the different diet groups, and no hematologic and macroscopic abnormalities were found. The relative weights of liver, kidneys, spleen, and thymus did not show differences in treated animals compared to controls. Serum immunoglobulin levels (IgA, IgG, IgM) in transgenic tomato-fed rats did not differ from those in controls. Microscopic examination did not reveal signs of pathologic effects related to transgenic Bt-tomatoes feeding. However, the analysis revealed a calcareous debris situated at the corticomedullary junction of female kidneys with a slightly higher incidence of focal intratubular lithiasis found in tomato-fed animals. These findings are believed to be of toxicological insignificance. Mineral deposits are often found within the renal tubules of young female laboratory rats, and their presence correlates with the dietary calcium/phosphorus ratio, while carbonic anhydrase inhibitors may enhance this process (5).

3 CONCLUSIONS

In summary, it was determined that oral exposure to transgenic Bt-tomatoes poses no additional risk to human and animal health. However, a number of aspects concerning the safety assessment of transgenic Bt-tomatoes will require further study. First, studies performed up till now were carried out with CRYIA(b) toxin purified from *E. coli*. Differences in post translation modifications in *E. coli* and tomatoes may influence the toxic potential of introduced proteins. Second, studies on the allergenic potency of the CRYIA(b) protein should be considered. In particular, because of the increase in general exposure of the population to CRY-type ICPs via the diet. However, predictive and validated *in vivo* models are largely lacking. Third, it is generally recognized that animal feeding trails with complex products suffer from distinct disadvantages like relatively low exposure levels due to adverse effects from multi-component exposure, occurrence of natural toxicants, or nutrient imbalances. Thus, limited ranges of safety factors can be applied due to these dietary restrictions. Therefore, a semi-synthetic animal diet with interchangable nutrients, adapted for the nutritional requirements of rats and deduced from the actual levels in tomatoes, was used. Although this approach appeared feasible, there were still additional deficiencies discovered. On the other hand, the lack of both appropriate analytical methods and knowledge of the composition of traditional foods in order to make comparisons effectively limits chemical analyses of single nutrients and natural toxicants.

New ways of food analysis should be explored, focusing on the characterization of whole crops with respect to profiles of secondary metabolites (i.e 'metabolic fingerprinting') rather than on determination of single compounds. To this end, analytical techniques for multicomponent analysis could be applied like Capillary Zone Electrophoresis (CZE), LC-NMR and LC(-GC)-MS(-MS). In this respect, a differential analysis at the DNA or mRNA level would be, at least in theory, of great value. However, it should be emphasized that the identification of possible differences in recognition patterns depends to a considerable extent on the background of the natural variation in composition of food crops. Nevertheless, analytical approaches may offer possibilities to refrain from animal feeding trials.

This work was carried out in the framework of the EU Food-Linked Agro-Industrial Research (FLAIR) program (contract no. AGRF-CT90-0039).

References

1. H.P.J.M Noteborn, M.E. Bienenmann-Ploum, M.J. Groot, G.M. Alink, A. Reynaerts, G. Malgarini and H.A. Kuiper, *Med. Fac. Landbouww. Univ. Gent (MFLRA3)* 1994, **59(4a)**, 1765.
2. C. Hofmann, H. Vanderbruggen, H. Höfte, J. Van Rie, S. Jansens and H. Van Mellaert, *Proc. Natl. Acad. Sci. USA*, 1988, **85**, 7844.
3. A. Bravo, S. Jansens and M. Perferoen, *Invertebr. Pathol.*, 1991, **60**, 237.
4. L. Zolla, G. Lupidi, M. Marcheggiani, G. Falcioni and M. Brunori, *Biochim. Biophys. Acta*, 1990, **1024**, 1.
5. P. Greaves and J.M. Faccini. 'Rat Histopathology", Elsevier Science B.V., Amsterdam, 1992, p. 171.

Genetically engineered food: The effects of product exposure on consumer acceptability

Dr L.J. Frewer, Dr R. Shepherd, and Ms C. Howard.

Department of Consumer Sciences;
Institute of Food Research,
Earley Gate,
Whiteknights Road,
Reading, RG6 2EF

1 ABSTRACT

Previous research into consumer perceptions of genetic engineering and subsequent consumer acceptance of resulting products has of necessity focused on attitudes where respondents have not been exposed to the products of the technology. Public attitudes towards genetic engineering remain uncrystallized, and are likely to be influenced by whatever information becomes available. This study examines the influence of realistic product exposure on (i) purchase intention of genetically engineered food products and (ii) attitudes towards the technology overall. The results indicate that respondents perceive genetically engineered products to be less natural than their conventional counterparts. Purchase intention was lower for genetically engineered products, although there was some indication that purchase intention increased in instances where there was some tangible benefit to the consumer. Realistic product exposure did not influence perceptions regarding genetic engineering in general.

2 CONSUMER ACCEPTANCE OF GENETIC ENGINEERING

Public acceptance of genetic engineering is an issue of great concern to both scientists and industrialists involved in the development and the application of the technology. Public attitudes may determine the future development of the technology, as well as the success or failure of products reaching the market place. Previous research has shown that there are currently low levels of public understanding regarding the technology [1], although this does not prevent the formation of risk perceptions regarding its application [2]. Thus the introduction of the novel products of the technology into the supermarkets may influence attitudes towards the technology, particularly since this is likely to represent a major source of information about the possibilities of genetic engineering for many consumers.

3 RISK PERCEPTIONS AND GENETIC ENGINEERING

The traditional view found within the literature is that the lay public is ill-informed regarding their perceptions of the risks associated with food-related hazards, and indeed risks in general [3]. Problematically, this perspective does not take account of the viewpoint that risk is "socially constructed" [4]. Risk perceptions are likely to be multidimensional in nature, and it is the psychological constructs which determine risk perceptions which also determine public reactions and behaviours.

Principal components analysis has been used to determine how people characterize different types of potential hazard relative to each other. It has been found, for example, that different hazards are characterized in terms of three factors, "dread", "familiarity" and the "number of people exposed" [5]. The psychometric approach has also been used to assess risk perceptions associated with different food-related hazards.

In this study, respondents were asked to rate different food-related characteristics on issues such as the seriousness of the hazard, control, voluntariness and whether the hazard is likely to be fatal [6].

The mean responses were analyzed by principal components analysis. The first component accounted for 45% of the variance and was labelled as "severity", while the second component accounted for 33% of the variance and was labelled "known risk". Figure 1 illustrates how genetic manipulation is characterised differently from other food-related hazards.

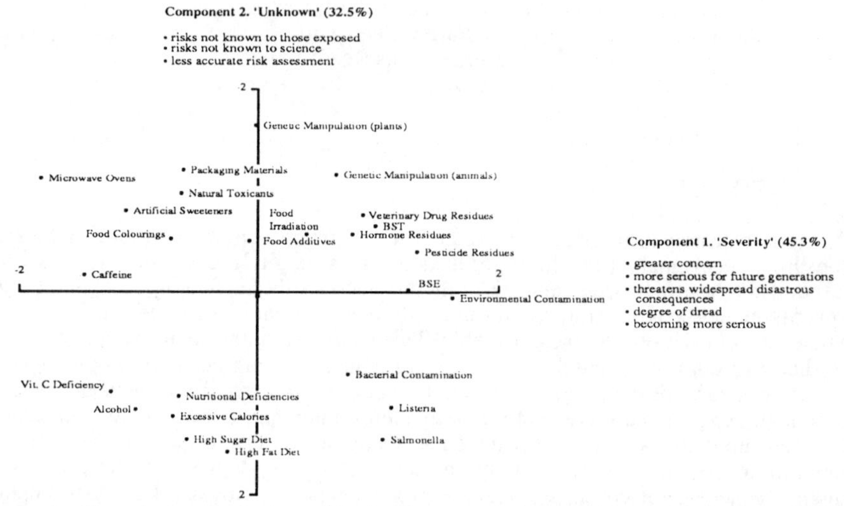

Figure 1. *Genetic manipulation of both plants and animals is seen as being a highly unknown hazard of moderate severity (see [6] in reference list)*

Ethical concerns also play an important role in determining public reactions to the application of genetic engineering in food production, and must be considered when assessing consumer reactions to particular products and applications [7]. Research has shown that ethical concerns are greater for food-related applications of genetic engineering than medical applications, and that greatest ethical concern is associated with transfers involving animals or human genetic material [8].

4 THE EFFECTS OF REALISTIC PRODUCT EXPOSURE ON ATTITUDES

Knowledge appears to increase acceptance of genetic engineering, at least in the United States [9]. However, previous survey work has either tended to consider people's attitudes to the technology in the abstract [10], or presented potential future applications in a hypothetical context [11 12 13].

The introduction of the novel products of the technology into the supermarkets may influence attitudes towards the technology, particularly since for many consumers labelled foodstuff may be one of the most important sources of information about genetic engineering and its application. Against this, research has indicated that both risks and benefits of genetic engineering are seen by the public to apply mostly to future generations, as opposed to the average person or the self [14]. Thus public attitudes towards the technology appear to assume that genetic engineering is not a viable technology at the present time. The availability of the products of the technology may be highly influential in crystallizing public attitudes towards the technology, as well as providing information about potential benefits.

Public acceptance of food products may also depend on the type of organism being manipulated, with transfers involving plants being more acceptable than those involving microorganisms, which are, in turn, more acceptable than those involving animals [15 16]. However, other product characteristics (differences in quality, price, and so forth), may influence consumer acceptance of particular products of genetic engineering, because the tangible benefits to the consumer may offset perceived risks.

The effects of realistic product exposure (where products have been produced using genetic engineering) on consumer attitudes have been examined experimentally. Likelihood of purchase and perceived naturalness of particular products were examined, as well as the effects of product exposure on general attitudes towards the technology. The research took into account the possible effects of differences in product type in terms of potential tangible benefits of the technology. A second group of subjects were included in the experimental design, who were not exposed to realistic products.

The first group of respondents were asked to make assessments of likelihood of purchase and perceived naturalness for a range of genetically engineered products. Comparisons were made with ratings obtained following exposure to matched conventional products. All respondents were asked to answer standardized questions which had been used to assess attitudes towards genetic engineering in previous research. Whilst these questions have been shown to differentiate genetic engineering as applied to food production from other food-related hazards in a hypothetical context [17], the apparent availability of realistic products may influence attitudes towards the technology as the implications of the technology are made more tangible. Questions relating to perceived risks and benefits of genetic engineering applied to food production, as well as perceived control over, and knowledge about the technology were asked. Questions were directed towards the self, other people and society. Genetically engineered products were perceived as significantly less natural than conventional counterparts, and respondents were also less likely to purchase the genetically engineered products compared to conventional counterparts. However, likelihood of purchase of one type of product, genetically engineered tomatoes, increased if the modification was associated with health and environmental benefits, but not cost or increased shelf life. Realistic product exposure had no effects on attitudes towards genetic engineering in general.

Consumer acceptance of genetic engineering may be greater for products of similar quality to conventional products but with reduced cost, but not for genetically engineered products of improved quality but greater cost [18]. Benefits to health and to the environment may represent more acceptable modifications than reduced cost or increased shelf life. Likelihood of purchase of genetically engineered products may be linked to the perceived "naturalness" of the products. If food produced by genetic engineering is seen to be "unnatural" by the public, consumer acceptance may be low.

5 PERCEIVED CONTROL AND GENETIC ENGINEERING

The public feel that genetic engineering as applied to food production is seen to be at the level of society rather than under individual control. Other hazards (for example, the high fat diet) are characterised as being under personal control [19]. Increasing perceptions of control (for example, labelling the products of the technology) may increase consumer acceptance of the technology, since consumers perceive that they have choice over whether or not to consume its products.

6 CONCLUSIONS

A general decline in perceptions of quality of life in the UK related to the application of the technology has been observed [20]. However, the availability of the products of genetic engineering on the supermarket shelves may provide evidence of the tangible benefits of the technology to the consumer and it is likely that they will have a great impact on consumer perceptions. Other strategies (for example, effective information strategies) can be designed to increase public awareness and understanding of the technology, although due account must be taken of other factors which may determine public reactions, (for example, trust in the information provided [21]).

Continuing research has indicated that, when considering consumer acceptance, specific applications are idiosyncratically linked to perceptions of perceived need and benefit, as well as risk [22]. It is likely that consumer acceptance of the products of genetic engineering will be highly dependent on the nature of the application itself, rather than on attitudes to the technology overall.

6 REFERENCES

1. E. Marlier, Eurobarometer 35.1: Opinions of Europeans on biotechnology. In 'Biotechnology in Public: A Review of Recent Research', ed. J.Durant, Science Museum, London, 1992, Chapter 4, p.52.
2. L.J. Frewer, R. Shepherd and P.Sparks, *British Food Journal*, 1994, **96**, 26.
3. B. Fischhoff, B., S.R. Watson and C. Hope, Policy Sciences, 1984, **17,** 123.3.
4. Dake, K. and Wildavsky, A, Individual differences in risk perception and risk-taking preferences. In 'The Analysis, Communication and Perception of Risk, ed. B.J. Garrick and W.C. Gekler, Plenum, New York, Chapter 2, p. 15.
5. P.Slovic, B. Fischhoff and S. Lichtenstein, Facts and Fears: understanding perceived risk. In Societal Risk Assessment: How Safe is Safe Enough? eds R.C. Schwing and W.A. Albers, Plenum, New York. p. 181.
6. P. Sparks and R. Shepherd, *Risk Analysis*, 1994, **14**, 799.
7. P. Sparks, R. Shepherd and L.J. Frewer, *Journal of Basic and Applied Social Psychology*, 1995, **16**, 267.
8. L.J. Frewer, and R. Shepherd, *Agriculture and Human Values*, in press.
7. T. Hoban, E. Woodrum, and R. Czaja, *Rural Sociology*, 1992, **57**, 476.
8. L.J. Frewer, and R. Shepherd, *Agriculture and Human Values*, in press.
9. Zechendorf, B, *Bio/technology*, **12**, 870.
10. E. Marlier, Eurobarometer 35.1: Opinions of Europeans on biotechnology. In 'Biotechnology in Public: A Review of Recent Research', ed. J.Durant, Science Museum, London, 1992, Chapter 4, p.52.
11. A.M. Hamstra, Biotechnology in foodstuffs: Towards a model of consumer acceptance Research Report 105, 1991, SWOKA, Institute for Consumer Research, The Hague.
12. A.M. Hamstra, Consumer acceptance of food biotechnology: The relationship between product evaluation and acceptance. Research Report 137, 1993, Institute for Consumer Research, The Hague.
13. T. Hoban, E.Woodrum and R. Czaja, *Rural Sociology,* 1992, **57**, 476.
14. L.J. Frewer and R.Shepherd, *Public Understanding of Science*, 1994, **3,** 385.
15. T. Hoban, E.Woodrum and R. Czaja, *Rural Sociology,* 1992, **57**, 476.
16. P. Sparks, R.Shepherd and L.J.Frewer,*Agriculture and Human Values*, 1994,**11** 19.
17. L.J. Frewer, and R. Shepherd, *Agriculture and Human Values*, in press.
18. T.J. Hoban and P.A. Kendall, 'Consumer Attitudes about the use of Biotechnology in Agriculture and Food Production', 1992, North Carolina State University, Raleigh, N.C.
19. L.J. Frewer, R. Shepherd and P. Sparks, *Journal of Food Safety*, 1994, **14**, 19.
20. Shepherd, R., Sparks, P. and Howard, C. Changes over time in public attitudes to the application of gene technology to food production, in preparation.
21. L.J. Frewer, R. Shepherd and C. Howard, What factors determine trust in information about technological hazards, Proceedings, Society for Risk Analysis Annual meeting, Baltimore, 4th-7th December.
22. L.J. Frewer, C. Howard and R.Shepherd, Public Perceptions of Genetic Engineering - The Importance of Tangible Benefits, Proceedings, 4th NIAB International Forum, Cambridge, 2-3rd February 1995.

IMPROVEMENT OF QUALITY AND NUTRITIONAL VALUE OF SPICE RED PEPPER (*Capsicum annum L.*) BY CROSS BREEDING OF HUNGARIAN AND SPANISH CULTIVARS

H.G. Daood, P.A. Biacs,
CFRI - H-1022 Budapest, Hungary

T.T. Huszka
Paprika Processing RT Szeged, Hungary

and **C.F. Alcaraz**
CEBAS - Murcia, Spain

1. INTRODUCTION

Seasoning paprika is one of the economically most important vegetables in Hungary and Spain. Factors most likely to determine quality of paprika may include flavour, color content and antioxidation potency. It has been demonstrated that color stability in paprika products is extremely dependent upon carotenoid composition and antioxidant content (Biacs et.al. 1992). Within the framework of scientific co-operation project Hungarian and Spanish researchers started a cross-breeding experiment aiming at production of new hybrids with outstading quality attributes exploiting the advantages of Spanish round lilac-colored and Hungarian red longum cultivars. This article deals with the results of this experiment.

2. MATERIALS AND METHODS

A 5-year cross breeding experiment was carried out in the experimental station of the Szegedi Paprika RT. - Szeged, Hungary. The Spanish Negral (unrestriced in growth and lilac, round fruit yielding cultivar) was crossed with Szegedi genotype (semi-determinate in growth and red longum fruit yielding cultivar). The seeds of the first generation (F-1) were planted in the next year to produce 12 plant combinations (F-2) varying in plant growth and shape and color of the fruits (Figure 1).

The fully ripe fruits of the parent, F-1 and F-2 were harvested and analysed for their carotenoid, tocopherol and ascorbic acid content using modern high-performance liquid chromatographic (HPLC) methods (Biacs and Daood, 1994; Daood et.al. 1994). To prepare paprika powders the fruits were cut into slices and dried at ambient conditions. The dried paprikas were ground by a coffee mill to powders of 0.5 μm particles which were packaged in nylon sacs and stored in a locker at room temperature for six months.

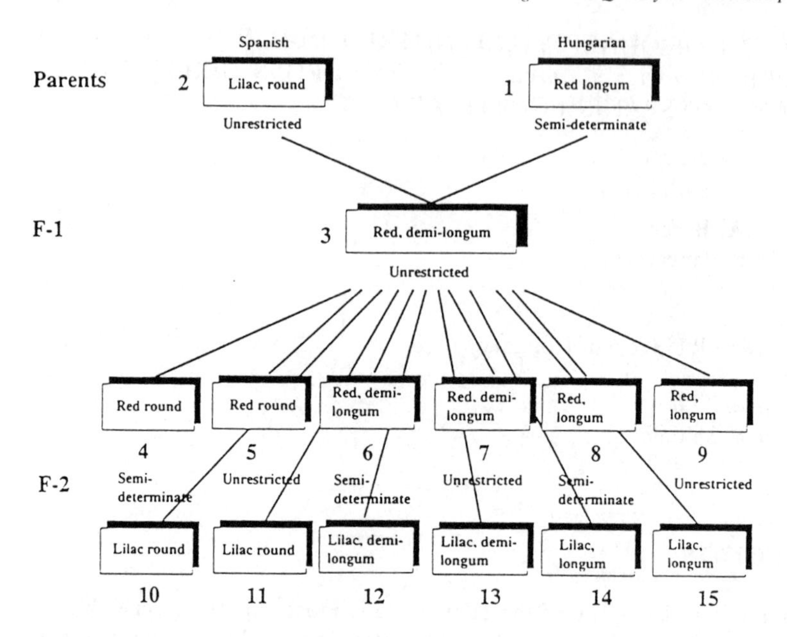

Figure 1 *Characters of Spanish and Hungarian cultivars and their generations*

3. RESULTS AND DISCUSSION

3.1 HPLC of quality components

Diversity of carotenoid extract from paprika is due to the presence of four carotenoid classes: xanthophylls, hydrocarbons, monoesters and diesters which are formed by acylation reactions with fatty acids.

Figure 2 shows the HPLC profile of paprika pigment separated on C-18 column with a quaternary mobile phase. The analytical procedure provided excellent separation of 40-45 carotenoid compounds in less than 50 minutes. Unsaponified paprika pigment can be characterized by the high proportion of red and yellow carotenoids such as capsanthin, capsorubin, cryptocapsin and violaxanthin, in the from of fatty acid mono- and diesters in addition to unesterifed carotenoids, mainly β-carotene.

According to the HPLC profile of extract Hungarian red longum paprika could be distinguished from that of Spanish lilac round by (1) high capsanthin concentration (2) absence of chlorophyll type pigments at fully ripe fruits. The lilac color of some cultivars is attributable to the relatively high capsorubin content and to the presence of chlorophylls beside carotenoids (Table 1).

As for endogenous antioxidant content, pericarp of paprika distribute α-, β-, γ-tocopherol and ubiquinone-10 as oil-soluble and ascorbic acid as water soluble antioxidants concentration of such effective metabolites ranged from 750 and 1600 μg/g dry matter (for tocopherols) and 4.5 to 20.5 mg/g dry materr (for ascorbic acid) in the tested samples.

3.2 Effect of cross breeding on quality component

Data on Table 2 implied that the carotenoid and antioxidant content of the F-1 was lower than that of both Hungarian and Spanish parent, but most of the fruits of the F-2 compared strongly with their origins in the color and antioxidant content. It is of great interest that such cross breeding was able to create new hybrids of better color intensity and antioxidant potency. Ascorbic acid and tocopherol content of lilac, demi-longum fruits from semi-determinate plants was 3 and 1.5 times higher than that found in the parent respectively. The best red colored fruits were of longum shape and obtained from semi-determinate plants. Although carotenoid content of such a hybrid was not so outstanding, it distributed much higher antioxidant that the origins.

It is very interesting and surprising that the carotenoid composition of lilac colored hybrids was somewhat similar to that of Hungarian origin, but with chlorophyll retention character gained from Spanish origin.

Table 1 *Carotenoid composition and content of two fresh spice red pepper varieties as estimated by HPLC*

Carotenoids	μg/g f.wt.	
	Spanish Negral	Hungarian Szegedi
Capsorubin	2.23	1.73
Violaxanthin	13.12	10.20
Capsanthin	8.30	25.3
Antheroxanthin	3.42	6.9
Mutatoxanthin	5.25	5.49
Lutein	7.3	12.87
Neolutein	5.57	1.94
Cryptocapsin	5.43	6.87
β-cryptoxanthin	15.08	4.72
Capsorubin ME	21.9	23.6
Violaxanthin ME	26.4	16.5
Capsanthin ME-1	14.97	24.1
Capsanthin ME-2	26.95	44.38
Cryptocapsin ME-1	18.13	9.52
Antheroxanthin ME	22.9	33.93
Cryptocapsin ME-2	4.25	2.70
β-cryptoxanthin	9.77	5.72
β-carotene (all trans)	71.49	63.6
β-carotene (cis)	16.6	28.7
Capsorubin DE-1	3.34	4.43
Violaxanthin DE-1	3.01	1.72
Capsorubin DE-2	10.77	4.81
Capsanthin DE-1	7.54	14.31
Violaxanthin DE-2	11.65	2.64
Capsanthin DE-2	23.67	28.45
Capsanthin DE-3	24.51	19.81
Violaxanthin DE-3	12.28	6.96
Capsanthin DE-4	0.75	4.53
Lutein DE	2.18	1.45
Zeaxanthin DE-1	7.61	5.56
Zeaxanthin DE-2	4.46	7.50
Chlorophyll b	1.18	-
Chlorophyll a	1.80	-
Pheophytin a	7.15	-
Total chlorphyll	13.56	-

3.3 Storage stability of pigments

Cross breeding manifested itself in the stability of carotenoids during storage of ground paprikas. Due to its higher capsanthin, ascorbic acid and tocopherol content powder of Hungarian Szegedi genotype showed better color stability than Spanish Negral when stored at ambient conditions for 6 months. The highest color retention was 69.5% for powder from red longum fruits from semi-determinate plants. These fruit, however did not contain the highest level of antioxidants emphasising on the fact that stability is extremely correlated to the high concentration of capsanthin particularly in the from of diesters in addition to the high antioxidation potency (Figure 2 and 3). This was conformed once again when the lilac, demi-longum hybrid, that had the highest antioxidant content, showed moderate color stabitity during storage.

Table 2 *Effect of cross breeding of Hungarian red long with Spanish lilac round on the quality attributes of paprika*

Cultivars and crosses		Quality parameters			
		Carotenoid * mg/g DM	Ascorbic acid mg/g DM	Tocopherol µg/g DM	Colour ** retention %
Parents					
	1	5.96	5.9	1010	48.9
	2	8.54	4.8	869	35.8
F-1	3	6.10	8.1	760	44.3
F-2 red					
	4	5.46	16.1	875	47.9
	5	5.54	15.4	865	50.3
	6	5.82	17.2	809	43.3
	7	5.42	13.8	764	25.1
	8	5.58	18.0	925	69.5
	9	7.26	13.2	686	22.7
F-2 lilac					
	10	7.82	16.2	1468	47.3
	11	8.22	15.4	1340	26.2
	12	8.62	20.3	1539	44.7
	13	9.12	16.8	1336	30.1
	14	9.96	12.5	884	17.8
	15	8.36	18.7	1350	36.8

* Carotenoid content of dried fruit to grinding

** The values represent % of pigment remaining after a 6 month storege. Numbers in the first column are described in Figure1.

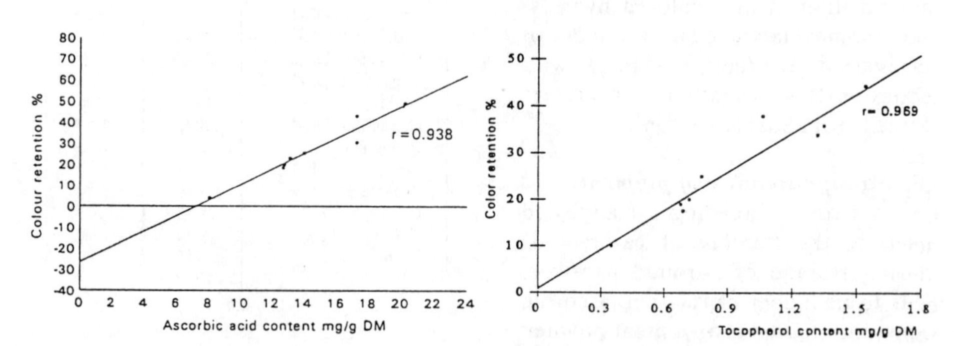

Figure 2 *Correlation between ascorbic acid and colour retention by paprika powders*

Figure 3 *Correlation between tocopherol and colour retention by paprika powders*

References

1. P.A. Biacs, B. Czinkotai and A. Hoschke, *J. Agric. Food Chem.*, 1992, 40, 463
2. P.A. Biacs and H.G. Daood, *J. Plant Physiol.*, 1994, 143, 520
3. H.G. Daood, P.A. Biacs, M. Dakar and F.Hajdu, *J. Chromatogr. Sci.*, 1994, 32, 484

THE POTENTIAL FOR IMPROVING ORGANOLEPTIC QUALITY IN APPLES THROUGH MARKER ASSISTED BREEDING RELATED TO CONSUMER PREFERENCE STUDIES

F.H. Alston & K.M. Evans
Horticulture Research International
East Malling
West Malling, Kent, ME19 6BJ

G.J. King
Horticulture Research International
Wellesbourne
Warwick, CV35 9EF

H.J.H. MacFie
Institute of Food Research
Earley Gate, Whiteknights Road
Reading, RG6 2EF

P.K. Betys
Sensory Dimensions
P O Box 68, Whiteknights
Reading, RG6 2BX

1 INTRODUCTION

Until recently, apple breeding programmes have not been aimed specifically towards the improvement or modification of flavour, but rather towards overall improvements related to the development of more efficient production systems (1). Many of the important components involved in such an approach can be assessed objectively and some, such as disease and pest resistance and tree habit, are determined by single genes with effects that can be readily recognised shortly after germination. Flavour, the result of a complex group of chemical interrelationships, is difficult to assess objectively among the large populations necessary for successful breeding programmes. It is considered at a late stage after the more easily recognised characters have been accounted for. Currently, it is usual to aim towards maintaining flavour standards based on empirical references to the flavour of certain traditional apple varieties, such as 'Cox's Orange Pippin'.

Despite the potential offered by some new varieties of high fruit production with reduced chemical inputs, an assured market is needed for their fruits. It is thus important to identify the characteristics which determine consumer preferences, which are undoubtedly related to sensory perception, and ensure that they are also present in new varieties. The critical sensory components and their genetic determinants must be identified and selection techniques devised to ease their methodical combination. It is expected that, after the key components have been related to molecular markers (2), pre-selection in the years preceding fruiting, will be possible.

2 APPLE FLAVOUR

A combination of acids, sugars and volatiles form the basis of apple flavour, but the textural condition of the flesh has a significant effect on overall flavour appreciation (3). Flesh colour, skin colour and skin finish also appear to have a significant effect. The acidic component, which is mainly malic acid, is controlled by a single dominant gene (4), with most varieties being heterozygous.

Empirical testing for 'good aroma' during tasting of 'Cox' derivatives suggested that the distinctive 'Cox' flavour component is determined by a single dominant gene

(5). Despite surveys in progenies derived from 'Cox' (6), it has not yet been possible to relate 'Cox' flavour to a single volatile component, that can be usefully selected for in progenies.

At present, little is known of the inheritance of textural components, principally firmness and crispness, and selection depends on a crude assessment of 'texture' by penetrometer after storage. The important components of texture need to be identified and quantified so that clear selection techniques can be developed and applied in breeding programmes.

3 BREEDING METHODS

New apple varieties are usually derived from segregating F_1 generations. Like most tree crops, they need to pass through a long, non-productive or juvenile period varying from 4 to 8 years prior to cropping.

Although for most features the selection of fruit quality has to await the onset of fruiting, attempts have been made to locate correlations with the vegetative characters of 1 and 2 year-old seedlings. Although the correlation was not strong, good prospects were found for selecting for fruit acidity within large populations on the basis of seedling leaf pH (7). This approach appears sufficient to discard most seedlings with the potential for bland fruit, but not sufficient for precise discrimination on the basis of acidity.

More precise pre-selection procedures are being developed for apple based on molecular markers, which have the potential to provide a higher level of precision than pre-selection criteria based on seedling/mature plant correlations (8). Initially, success was achieved through the location of isoenzymic markers for pollen incompatibility (9), mildew resistance (10), scab resistance (11) and fruit colour (12). In addition, DNA (RAPD) markers for scab resistance have been located (13). This work continues and includes investigations on the relationship of molecular markers to fruit characters, including flavour. This is being facilitated through the construction of a reference genetic linkage map for apple, in the European Apple Genome Mapping Project using isoenzyme genes, and DNA (RAPD and RFLP) markers (14). It is hoped that it will eventually be possible to relate specific components of organoleptic quality to molecular markers expressed in young seedlings and thereby facilitate early selection for these prime selection criteria.

4 CONSUMER PREFERENCE PROFILES

While several surveys of consumer preference to apple varieties have been made, there has been little attempt to relate particular preferences to specific characteristics of the fruit. Moreover, there is little information on which characteristics of apples are preferred by consumers. The first significant steps to such a goal were made through conducting hedonic evaluation by 60 consumers of the varieties most widely available on the retail market in the UK. This exercise was carried out separately for peeled and unpeeled apples. Internal Preference Mapping (15) was used to resolve the consumer scores into a set of preference dimensions that represent the differences among the varieties and a set of vectors, one for each consumer, that show the individual directions of increasing preference.

Two tests were carried out, one in June with fruit from southern hemisphere sources and another in November, with fruit from northern hemisphere sources, including UK-produced fruit. Concurrently, a comprehensive quantitative descriptive sensory profile of each variety was developed using a trained sensory panel. Principal Components Analysis was used to develop product and attribute maps to summarise sensory differences between varieties. Sensory attributes, scored by the trained panel, showing significant correlations with the primary consumer preference dimensions were identified to gain insight as to which attributes were driving consumer preference.

In the June tests (16), the majority of consumer vectors lay in the direction that is positively associated with firmness, crispness, juiciness and sweetness. A second important consumer segment was identified as those preferring a hard, crisp, juicy but more acidic apple with 'Cox' like flavour - a combination of attributes not satisfied by any of the varieties under test, thus exposing a new market opportunity.

The results of the November survey, while showing strong agreement with some features of the June survey, identified strong preferences for crispness and juiciness combined with a moderate acidity and a moderate sugar content ('Cox', 'Fiesta', 'Empire', 'Braeburn', 'Jonagold') and the hard, 'fresh' tasting, highly acidic fruit of Granny Smith. Varieties with a lower acidity and a lower sugar content ('Red Delicious', 'Golden Delicious', 'Gala') were less popular. Closer examination of the results showed that a 'Cox' flavour characteristic was detected in all the leading varieties of the November survey, except 'Empire'. In addition, the perception of 'Cox' flavour appeared to be enhanced by the presence of cream or yellow flesh. 'Cox' flavour perception was also stronger in unpeeled fruit, suggesting an association between flavour and skin appearance. The results from the November survey suggested a new marketing opportunity for an apple with sensory properties intermediate between 'Granny Smith' and 'Braeburn', in consumer preference terms very similar to that identified in June. It is natural to approach such predictions with caution when dealing with a crop like apple, where it takes 15-20 years to develop and market a new product, but evidence from other products shows that the general direction of consumer preference trends are consistent for relatively long periods.

5 THE APPLICATION OF PROFILES IN BREEDING

Having constructed an apple consumer sensory profile map, it is then possible to overlay it with the results of sensory panel assessments on fruit of individual selections or varieties. By examining a series of selections from one progeny in this way it will be possible to identify genes determining key features of consumer preference and to relate them to molecular markers in the apple genetic linkage map. Some features have already been related to characters which appear to be under simple genetic control, such as acidity, aroma, flesh colour and skin russet (5). However, it is expected that the inheritance of many will prove to be complex. Nevertheless, analytical techniques are available to relate such groups of genes to genetic maps.

The recognition of key components of consumer preference also necessitates instrumental studies of physical and chemical features in order to develop clear specifications that can be applied in fruit selection and, at a later stage, easily related to molecular markers applied for selection in the two years following germination.

ACKNOWLEDGEMENTS

The authors are grateful for financial support for this work from the Ministry of Agriculture, Fisheries and Food and from the CEC through Contract AIR3-CT920473.

REFERENCES

1. F.H. Alston, in 'Improving Vegetatively Propagated Crops', eds. A.J. Abbott and R.K. Atkin, Academic Press, London, 1987, p. 113.
2. F.H. Alston and I. Batlle, *Phytoparasitica*, 1992, **20**: Suppl., 89.
3. F.H. Alston, in 'Bioformation of Flavours', eds. R.L.S. Patterson, B.V. Charlwood, G. MacLeod and A.A. Williams, Royal Society of Chemistry, Cambridge, 1992, p. 33.
4. N. Nybom, *Hereditas*, Lund, 1959, **45,** 332.
5. F.H. Alston and R. Watkins, 'Proceedings Eucarpia Fruit Section Symposium V', Top Fruit Breeding, Canterbury, 1973, p. 14.
6. D.F. Meigh, 'Report of the Ditton Laboratory for 1967-68', p. 53.
7. T. Visser and J.J. Verhaegh, *Euphytica*, 1978, **27**, 753.
8. G.J. King, F.H. Alston, I. Batlle, E. Chevreau, C. Gessler, J. Janse, P. Lindhout, A.G. Manganaris, S. Sansavini, H. Schmidt and K.R. Tobutt, *Euphytica*, 1991, **56**, 89.
9. A.G. Manganaris and F.H. Alston, *Theor. Appl. Genet.*, 1987, **74**, 154.
10. A.G. Manganaris and F.H. Alston, *Theor. Appl. Genet.*, 1992, **83,** 354.
11. A.G. Manganaris, F.H. Alston, N.F. Weeden, H.S. Aldwinkle, H.L. Gustafson and S.K. Brown, *J. Amer. Soc. Hort. Sci.*, 1994, **119,** 1286.
12. N.F. Weeden, M. Hemmat, D.M. Lawson, M. Lodhi, R.L. Bell, A.G. Manganaris, B.I. Reish, S.K. Brown and G.-N. Ye, *Euphytica*, 1994, **77,** 71.
13. B.L. Koller, L. Gianfranceschi, N. Seglias, J. McDermott and C. Gessler, *Plant Mol. Biol.*, 1994, **26,** 597.
14. King, G.J., *Euphytica*, 1994, **77,** 65.
15. J.D. Carroll, in 'Multidimensional Scaling: Theory and Applications in the Behavioural Sciences', Vol.1, eds. R.N. Shepherd, A.K. Romney and S.B. Nerlove, Seminar Press, New York, 1980, p.105.
16. B. Daillant-Spinnler, H.J.H. MacFie, P.K. Betys and D. Hedderley, *Food Quality and Preference (in press)*.

GENOTYPE AND ENVIRONMENTAL INFLUENCE ON ROOT QUALITY AND SHELF LIFE OF PHILIPPINE SWEETPOTATO

A.L. Acedo Jr., E.S. Data and M.A. Quevedo

Postharvest Technology Laboratory
Department of Horticulture
ViSCA, Baybay, Leyte 6521-A
Philippines

1 INTRODUCTION

Sweetpotato, *Ipomoea batatas* (L) Poir, is a crop that produces more food than any of the other major root crops, legumes and cereals.[1] Its fleshy roots provide a good supply of energy and contain potentially high levels of pro-vitamin A and ascorbic acid.[1] Roughly 80% of the world's sweetpotato production are in Asia.[2] The Philippines is the 9th leading world producer. However, Philippine sweetpotato production is highly seasonal and yields remain low (5 tons/ha)[3] in spite of the development of new varieties than can produce 15-20 tons of roots per hectare.[4] Snags to farmers' adoption of the high yielding varieties (HYVs) include poor eating quality and short shelf life of the fresh roots.

Only recently, sweetpotato breeding works focused not only on the improvement of yield but of quality and shelf life as well. As part of the varietal development effort, this study investigated the genotypic variations in some chemical, sensory quality, and postharvest characteristics of the fresh roots produced from the dry and wet season plantings.

2 MATERIALS AND METHODS

Genotypes OPS 37, OPS 56, OPS 99, V15-70 and V37-151 were used and hereafter referred to as G1, G2, G3, G4, and G5, respectively. These genotypes are potential HYVs which comprised the elite selections from progenies of the polycross breeding nurseries and entered in the seedboard trial - the final phase in varietal screening. In addition to the seedboard genotypes, four existing HYVs (VSP1, VSP4, VSP5 and VSP6) were evaluated.

The different genotypes were grown during the dry and wet season, each lasting for about 4 months. During the dry season plantings, the average 4-month rainfall and daily temperature and relative humidity were 78.6 ± 5.3 cm, 28.0 ± 3.8 °C, and 73.6 ± 8.6 %, respectively, while during the wet season plantings, 145.2 ± 14.3 cm, 26.7 ± 3.2 °C, and 77.6 ± 11.0 %, respectively. Each genotype was planted in three 2 m x 6 m plots, each plot representing a replicate.

From each replicate plot, 3-6 uniform-sized roots were harvested for dry matter, starch, sugar and protein analysis. Dry matter was determined using a forced draft oven at 70°C. Starch and sugar were analysed by the modified anthrone method[5] while protein, by the microkjeldahl method.[6] For sensory quality, the overall flavour of the steamed roots was scored by trained panelists using a 9-point hedonic scale.

For shelf life evaluation, 10-30 uniform-sized and defect-free roots from each replicate plot were stored for 3 months at ambient (24.2-30.6°C, 66.0-82.2% relative humidity). Weight loss, degree of shriveling and sprouting, and percentage decayed roots were measured monthly. Root shriveling was subjectively rated using a scale of 1 (none) to 5 (severe). A scale of 1-5 was also used to rate the degree of sprouting in which 1 = no sprout, 2 = 1-3 sprouts, 3 = 4-6 sprouts, 4 = 7-10 sprouts and 5 = more than 10 sprouts per root. Shelf life was quantified as less than 1 (<1), 2 (<2) and 3 (<3) months if after the first, second and third month of storage, respectively, at least 50% of the stored roots became unmarketable. A shelf life of more than 3 months (3+) was given to genotypes with more than 50% of their roots remaining marketable after the third month of storage.

3 RESULTS AND DISCUSSION

3.1 Chemical Attributes

Dry matter, starch, sugar and protein contents of the fresh roots varied with genotype (Table 1). Dry matter content (DMC) was highest in G4 and lowest in G2, VSP1 and VSP6. The other genotypes had moderately high DMC (30-35%). The starch and sugar contents did not correlate well with DMC. G5 and G3 were the high-starch but low-sugar genotypes while VSP1, the low-starch but high-sugar genotype. The other genotypes had moderately high levels of starch (60-65%) but differed in sugar content,

Table 1 *Chemical and Sensory Quality Attributes of Sweetpotato Roots at Harvest*

Genotype	*% Dry Matter*	*% Starch*	*% Sugar*	*% Protein*	*Overall Flavour*
Dry Season					
G1	32.3cde	62.2de	6.4c	3.31abc	7.0a
G2	25.6g	61.5def	12.1a	3.94a	6.9a
G3	31.9def	62.9de	4.8de	3.31abc	6.4ab
G4	44.1a	66.2bc	6.6c	2.95bcd	6.9a
G5	32.7cde	72.2a	3.7e	3.62a	6.8a
VSP1	27.4fg	48.2g	12.0a	3.46ab	5.4c
VSP4	29.7ef	62.4de	5.8cd	3.15abc	6.8a
VSP5	32.0de	63.9cd	3.5e	3.68a	6.9a
VSP6	27.5fg	62.1de	8.3b	2.20e	6.3b
Wet Season					
G1	29.5ef	62.9de	3.7e	3.30abc	6.2b
G2	25.0g	61.0def	8.0b	3.15abc	6.6a
G3	33.1cd	72.2a	2.7f	3.15abc	7.0a
G4	39.6ab	58.2f	6.4c	2.32e	6.9a
G5	35.5bc	67.6b	3.6e	3.38abc	6.6a
VSP1	26.8fg	59.6ef	6.7c	2.84cd	6.0b
VSP4	31.6de	63.2cd	7.6bc	3.15abc	6.8a
VSP5	27.0fg	64.1cd	3.1ef	2.20e	6.1b
VSP6	29.2ef	61.4def	5.4cd	2.68de	6.1b

Mean separation by Duncan's Multiple Range Test (DMRT), 5%.

with G2 having the highest sugar content and VSP5, the lowest.

The environmental condition during growth of the crop had contrasting influence on starch content. While it did not affect the starch content of 5 genotypes, wet season growing increased the starch content of G3 and VSP 1 but decreased that of G4 and G5. On the other hand, the sugar content consistently decreased in wet season-produced roots of G1, G2, G3, VSP1 and VSP6. These changes in starch and sugar content did not effect corresponding changes in DMC. The DMC did not vary with planting season except for VSP5 whose wet season-produced roots had lower DMC than the dry season-produced ones implying increased root water content.

Protein content was low and ranged from 2.20-3.94% (Table 1). G4 and VSP6 had much lower protein content than the other genotypes. Wet season planting reduced the protein level of G4 roots as well as that of VSP1 and VSP5 roots. The protein content of the other genotypes did not vary with growing season.

3.2 Sensory Quality

The overall flavour of the steamed roots of most genotypes from the dry season plantings was rated comparably desirable (6.4-7.0) (Table 1). Only the VSP1 and VSP6 roots had markedly lower ratings due to their wet texture. When grown during the wet season, flavour ratings decreased for G1 and VSP5 roots which became wet-textured and less sweet. This seemed to reflect the decrease in DMC and sugar content of these two genotypes. In contrast, the wet season-produced roots of G3 and VSP1 had better flavour than the dry season-produced ones. This was due to improved mealiness, a possible result of the increase in starch content of the roots.

3.3 Shelf Life and Postharvest Characteristics

The fresh roots of some genotypes (G3, VSP1 and VSP5) had short shelf life (<1 month) while others (G5) had long shelf life (<3 months) regardless of growing season (Table 2). The other genotypes (G2, VSP4, VSP6) showed season-dependent root shelf

Table 2 *Shelf Life of Sweetpotato Roots During Ambient Storage*

Genotype	*Shelf Life, mo.*		*Decay[1] %*		*Degree of Shriveling[1]*		*Weight Loss, %[1]*		*Degree of Sprouting[1]*	
	DS	*WS*	*DS*	*WS*	*DS*	*WS*	*DS*	*WS*	*DS*	*WS*
G1	3+	<1	20	80	2.4	[a/]	27.9		3.1	
G2	<3	<2	67	60	4.0	3.7	33.3	41.1	2.2	1.0
G3	<1	<1	67	67						
G4	<2	3+	60	20	2.2	3.4	31.3	27.7	1.0	1.0
G5	<3	<3	73	73	1.8	3.0	30.7	38.7	2.8	1.0
VSP1	<1	<1	100	100						
VSP4	<1	<3	100	72		1.5		27.3		1.5
VSP5	<1	<1	100	60						
VSP6	<1	<3	67	43		4.3		36.7		2.1

[1]*Data taken on the first, second and third month of storage for genotypes with shelf life of <1, <2, and <3 to 3+ month, respectively. (DS-dry season; WS-wet season)*
[a/]*Data not taken for genotypes with <1 month shelf life.*

Table 3 *Chemical and Sensory Quality Attributes of Sweetpotato Roots after 3 Months of Ambient Storage*

Genotype	*% Dry Matter*	*% Starch*	*% Sugar*	*% Protein*	*Overall Flavour*
Dry Season					
G1	32.4b	51.4	9.4b	3.78b	7.2a
G2	32.4b	57.5	12.4a	3.15bc	6.8a
Wet Season					
G4	43.7a	55.0	9.1b	4.72a	6.9a
VSP6	30.0b	54.4	9.8b	2.84c	6.3b

Mean separation by DMRT, 5%.

life. Only two genotypes had a root shelf life of more than 3 months (3+) but this was dependent again on the growing season. These were the G1 roots produced from the dry season plantings and the G4 roots from the wet season plantings. Root quality deterioration was due mainly to decay and partly to shriveling, weight loss and sprouting (Table 2). Shriveling appeared to be favored in wet season-produced roots while sprouting, in dry season-produced ones.

After 3 months of storage, sound roots of some genotypes were subjected to chemical and sensory quality analysis. Table 3 shows that the root DMC and overall flavour was unaffected by storage as these compared well with that of unstored roots (Table 1). However, starch content decreased while sugar content increased indicating starch breakdown to sugars during storage. No adverse changes in protein content was obtained. It even increased in G4 roots after 3 months of storage.

Most of the seedboard genotypes outperformed the existing HYVs in terms of root quality and shelf life. Furthermore, the results demonstrate the high possibility of developing varieties of good root quality and long root shelf life for a particular growing season (e.g. G1) or for all seasons (e.g. G5).

References

1. J.A. Wolfe, 'Improvement of Sweet Potato in Asia', International Potato Center (CIP), Peru, 1989, p. 167.
2. D.E. Horton, 'Maintenance and Utilization of Sweet Potato Genetic Resources', CIP, Peru, 1987, p. 17.
3. Center for Research and Communication (CRC), 'Philippine Agribusiness Factbook Directory', CRC, Philippines, 1993.
4. F.G. Villamayor, *Root Crops Digest*, 1988, **3**, 1.
5. G.B. Cagampang and F.M. Rodriguez, 'Methods of Analysis for Screening Crops of Appropriate Qualities', Institute of Plant Breeding, College, Laguna, Philippines, 1981.
6. Association of Official Analytical Chemists (AOAC), 'Methods of Analysis of the AOAC', Washington D.C., 1980.

Section 2: Effects of Pre-harvest Treatments on Quality

FOOD QUALITY - THE CHALLENGE TO AGRICULTURE

D W F Shannon

Chief Scientist
Ministry of Agriculture, Fisheries and Food
Nobel House
17 Smith Square
London SW1P 3JR

1 AGRICULTURE AS AN ECONOMIC ACTIVITY

Let me start with a truism; farmers farm to make a living. They provide a supply to meet a demand. In the past their first challenge has usually been one of supply. In the post war period, expansion of the volume of agricultural output was given priority in agricultural policy, but no longer. In the past 30 years, the volume of UK agricultural output has more than doubled and our self sufficiency in food has increased, though it has eased back recently due in part to set aside and other quantitative restrictions. We now produce 56% of all our food and feed requirements and some 73% of indigenous products.

As a result, a more complex supply and demand pattern is emerging. In this more competitive business, the farmer needs to be more aware of customer demands. Some of these are no longer simply for food; the farmer of today is often faced with social rather than commercial demands, such as for environmental goods. Further, in the face of greater competition in food markets, farmers are looking for new outlets in non-food markets, ranging from fuel and fibres to pharmaceuticals and novel industrial feed stocks. However, these challenges are not the prime issue for this paper, which addresses one of the key issues for farmers engaged in food production today, namely the difficult concept of "quality".

The issue of quality is defined by demand and is thus a key to competitiveness, but demands are not simple or uniform. They encompass the views of primary and secondary purchasers (processors, wholesalers, retailers and importers in other countries), as well as consumers, and also the legislature, lobbyists and public interest groups. Quality may be thought of as the sum total of physical characteristics embodied in the product and may also include ethical issues in relation to animals. At best, these characteristics will be only imperfectly known to the potential customer/consumer; improving the flow of information about these characteristics will play a major part in commercial success. Investment, regulation and marketing all have their part to play in meeting quality demands but I am going to concentrate on the role of research.

I want to tease out four categories of quality issue facing the producer and his customers and look at each in the context of the demands being made. These are:-

- Fitness for purpose
- Diet and Health
- Safety and Hygiene
- Ethics and perceptions

2 FITNESS FOR PURPOSE

I place fitness for purpose first because it forces our thinking to address the whole food chain. The ultimate purpose may be that of the consumer, but to achieve it, every player along the chain of purchase and sale will have their own requirements.

The expectations of consumers in respect of the food they eat are changing and broadening; they now look for food which is more attractive, more varied, better flavoured, nutritionally improved, safer and more "natural" (i.e. minimally processed). All of these expectations are placing increasing demands on food processors. The processors will need to be able to employ techniques which retain the attributes of nutrition and freshness, ensuring a high level of product consistency and meeting stringent requirements of product safety and shelf life. The processing industries have always had their specifications on quality; high Hagbergs for bread-making wheat, low nitrogen for malting barley, low impurities in sugar beet, have long been targets both for producers and processors. These are now being extended to new targets.

To meet their requirements, farmers need to produce not just the right quantity but also meet demands relating to timing (despite the seasons) and quality; above all for consistency in supply. This concerns not just high-volume processed foods but also produce consumed directly. Requirements of multiple retailers of horticultural produce for planned, consistent availability of the high quality produce may imply the need for season-extension (which may be achieved by UK growers extending production in the UK or even setting up farms elsewhere (as,for example, with lettuce in Spain). Shelf-life may become less important with more rapid transport from field to supermarket; freshness is likely to become the predominant trait, complemented by colour, freedom from blemish, texture and flavour.

The role of the biological sciences in addressing these demands extends from breeding, to their growing, harvesting and storage of crops. Breeders are increasingly incorporating the techniques of the molecular geneticist into their tool-kits. Introduction of desirable genes from other species is now possible for many crop plants although the efficiency of the transformation techniques needs to be improved and is the subject of research funded by MAFF and others. Marker-assisted selection, the capability to follow desirable qualities through a series of inter-cultivar or inter-specific crosses by genetic fingerprints, avoids the need for progeny to be grown to full phenotypic expression and significantly improves the speed and precision of the early stages of breeding programmes. Bulking and production of the finished variety remains a season-dependent and therefore somewhat protracted process but the final product can be specifically tailored to meet a particular quality demand which might not have been possible without the initial molecular methodologies: the Flavr-Savr tomato is an obvious example. Others, for the future, might include crops made resistant to pests and diseases which cause blemishes, or barley bred to produce the precise malt needed by the brewer of a particular English ale.

I have talked mainly of crops here but one only has to look at the way carcass conformation has changed to meet new demands in recent decades to realise the impact of new developments in scientific animal breeding. Similarly, healthy animals and birds help to ensure efficient and consistent growth rates to match processors' demands. The scope for new biotechnologies to deliver improved immunological control of endemic diseases is becoming apparent and in the future there is potential to develop animals resistant to particular diseases. Progress is likely to be manifest first in pigs and poultry because these animals are kept most intensively. The result will be more control over delivery of a high quality product.

It is not only biology that is essential here. Having selected his variety for its preferred qualities, the farmer needs to grow the crop and the equipment used can have a crucial effect on product quality from the start. For field grown crops this means care is needed in seed-bed preparation and crop establishment which will affect the quality and uniformity of the product. Similarly, accuracy of placement of chemicals is vital, to optimise the supply of nutrients and to protect the crop against pests and disease. In controlled environments in greenhouses, the quality of the crop depends on the accuracy with which temperature, humidity and carbon dioxide levels can be maintained, but this must be done at a viable cost.

It is equally important to ensure that machinery causes minimum damage during harvest and transport and, where necessary, reduces product temperature quickly and maintains it in a cool chain to the point of sale. Many food products must be stored for long periods of time. Accurate environmental control in stores is, therefore, vital and as stores become bigger it becomes increasingly important to understand the air flow through the store and the heat and mass transfer processes between the air and the stored product to produce optimum designs. Computational fluid dynamics is proving to be a valuable tool in this respect.

Grading and packing equipment can add value by removing damaged or blemished items and presenting uniform products in attractive packaging. The human operator used for picking, grading and packing has many advantages in terms of flexibility and the ability to make judgements but these are increasingly being outweighed by increasing cost, lack of availability and variable performance due to fatigue or distraction. Industry is now developing robotics and image analysis techniques to produce a more consistent performance.

Engineering science can also help improve the quality of animal products through the maintenance of livestock health, particularly of those animals (pigs and poultry) kept in controlled environments. Adequate control of temperature, ventilation and dust levels help to ensure healthy animals. For example, the use of image analysis for monitoring growth rates and conformation of pigs has potential as an aid to stockmanship in providing intelligent 24 hour a day surveillance of stock. It enables growth rate and conformation to be monitored continuously, and could permit identification of abnormal behaviour allowing appropriate action to be taken earlier than would normally be possible from intermittent and intrusive human observation.

Before leaving fitness for purpose I must mention our Agro-food Quality LINK Programme. This is a programme which is jointly funded by MAFF and DTI with industrial partners, started in 1991 and currently is expected to be worth £16m by 1997.

3 DIET AND HEALTH

I have been talking of the general concept of "fitness for purpose" and stressing the wide range of "purposes" consumers, and the industry which supplies them may have. I now want to get on to a series of specific issues which are influencing consumer choice and need to be taken into account by producers. Firstly, among the key food quality criteria now used by the modern consumer are those relating to diet and health.

Diet is now seen not only by the medical and scientific fraternity but also by the consumer as a major determinant of health. Consumers are seeking foods which are not only satisfying and pleasurable but which are "healthy" as well. Recent government initiatives reflect the importance of the role of diet in relation to health.

The Department of Health's Committee on Aspects of Food Policy (COMA) has produced a recent report which clearly indicates the extent of desirable change in the nutrition of the UK population to decrease the prevalence of cardiovascular disease. COMA has set new dietary reference values (DRVs) for the UK population. The Government White Paper, "*The Health of the Nation*", has set targets to reduce the percentage of energy derived from total fat from about 40% in 1990 to no more than 35% by 2005, and saturated fat from 17% in 1990 to no more than 11% by 2005. These targets are going to be hard to achieve; the eventual target for total fat is lower than that currently pertaining to any developed country other than Japan.

The 1994 COMA Report on Cardiovascular disease gave significant emphasis to the roles of specific fatty acids, pointing to a need for higher levels of n-3 fatty acids and limits on n-6 fatty acids. The Report recommended that the long chain n-3 fatty acids, usually associated with fish oils, should increase from 0.1 to 0.2 g/d and that the n-6 fatty acids should contribute not more than 10% of the energy intake.

The significant problem with these recommendations is that they are based on epidemiological evidence with the inherent difficulty of establishing cause and effect. Over the past 5 years MAFF has been building up a significant human nutrition programme with the aim of establishing the causal mechanisms underlying the supposed relationships.

MAFF's nutrition research programmes aim to identify which constituents of the diet are protective against these diseases. There is evidence that certain groups in the population are more susceptible to these diseases than others. It is only by identifying the factors in the diet and how they influence the disease process that recommendations can be made on what constitutes a healthy food. This could lead to the design of tailored foods and ingredients for specific populations or health benefits. This will be an expanding area for the food industry in the future.

The potential implications for the livestock industry of the various dietary recommendations are quite profound. If we assume (a) that the fat content and fatty acid composition of livestock products were to remain at historic levels; (b) that the major thrust of dietary advice to meet the targets was focused on animal products; and (c) that the targets were actually met, then the profile of animal and vegetable products would change significantly. Implied changes in food supplies to meet these targets are reductions in the consumption of ruminant products - milk butter, cheese, beef, mutton and lamb of between 10 and 45% with consequent increases in cereals, potatoes, other fresh vegetables. Poultry meat would increase by 25% and pork would decrease by 40%.

These projections assume no change in the composition of the animals from which the products are derived, or indeed in the preparation of specific products. Reducing the fat content of some farmed animal species has been a commercial breeding objective for some time and in relation to some fat deposits this has proved a straightforward task. The most striking example is the pig, but selective breeding and breed substitution are also having an impact on the ruminant species.

Surprisingly when we considered the COMA recommendations on n-3 and n-6 fatty acids we found that there was relatively little information on the detailed fatty acid composition of the major meat species. We commissioned Dr Jeff Wood to undertake some detailed analysis with some surprising results. The analyses were performed on fifty samples each of beef sirloin steaks, lamb chops and pork steaks/chops. Some of the results were as one might have predicted. The major polyunsaturated fatty acid (PUFA) linoleic acid (n-6) was highest in pork and the polyunsaturated:saturated (P:S) ratio was therefore higher in pork. Stearic acid (C18:0) was high in lamb as might have been expected. However the surprising finding was that significant amounts of the α linolenic acid (C18:3

n-3) in the diets of grass fed ruminants escapes hydrogenation in the rumen and is present as a desirable attribute of the fat. The other surprising finding was the significant levels of C20:5 n-3 and C22:5 n-3 fatty acids in the pork fat. This means that beef and lamb fat have a desirable ratio of n-6 to n-3 fatty acids and pork fat, whilst having higher than expected levels of long chain n-3 fatty acids, also has high levels of linoleic acid and consequently an adverse ratio.

Estimates based on these findings (and assuming that they are representative of the meat of each species) show that the contribution to the intake of long chain n-3 fatty acids from red meat is at least as great as from fish. Furthermore opportunities exist to improve the ratio of long chain fatty acids in animal products through dietary manipulation and MAFF has just started research on the possibilities in sheep, beef and pigs.

I have tended to focus on animal products but the current epidemiological evidence suggests that as well as fish and marine products, fruit, vegetables and cereal products are protective against cardiovascular disease and cancer. Opportunities exist to breed crops and fruit for improved vitamin and anti-oxidant content and indeed to step up the supply of fish at reasonable prices through aquaculture.

Another interesting study is one undertaken by Professor Tilo Ulbricht in which he has attempted to carry out an audit of fat entering the human food chain. For example the switch from full fat to low fat milk has had a significant effect on milk fat consumed in the liquid form, but where has the butterfat gone. The study required to take into account exports and imports of fats, and inevitably required some heroic assumptions about how much fat is actually consumed and how much is left on the plate. He has carried out the audit using data for 1982 and 1992. A number of interesting points emerge. Firstly, whilst the overall fat intake has declined a little over this period, the intake of fat from particular products has changed very significantly. Consumption of butterfat as butter and as milk has declined very significantly, but there has been a tendency for slight increases in consumption as cream and in other products. Consumption of fat in red meat and eggs has declined substantially. Consumption of fat as nuts and chocolate and as shortenings has increased. That from poultry meat has gone up partly because of the increase in poultry meat consumption. Perhaps the point that is most outstanding is that the reduction in fat intake from animal products has been matched by an increase in the consumption of fats and oil in other products.

The reduction in the fat content of animal products has been quite correctly directed at the attractiveness of those products to the consumer. It is slightly paradoxical that at the same time the consumer is choosing other products so that the overall fat intake is very similar. Two further points. This study again pointed to the importance of red meats in making a major contribution to desirable or protective fatty acids. The study stops at 1992 and the recent National Food Survey data suggest that there have been important changes since 1992 which are obviously not reflected here.

The consumer preference for "healthy" foods would, in a rational world, help to maintain general well-being and help protect against some of the major diseases of our time, such as cardiovascular disease and cancer. In practice, the consumer is often faced with a bewildering array of information from the genuinely medical to the fashionably faddish; "slimming" products sometimes seem to take priority over balanced diets. Many treat vitamins as magic pills and rank them alongside doubtful herbal remedies. There is clearly an important role here for health education, both of adults and in schools, and for reputable food producers to inform their customers and for research into how best to target it.

Diet and health is clearly a debate which has yet a long way to run. However, I want to move on now to an issue related to diet and health, namely food safety and hygiene.

4 SAFETY AND HYGIENE

One of the principal aims of MAFF is to protect the public by promoting food safety, the subject of much recent public interest. MAFF's food safety research is focused in particular on contamination by bacteria or chemicals; a more recent interest is in ensuring the safety of novel approaches to food production, including genetic modification.

Food poisoning is still a problem in the UK; last year, over 40,000 cases of food poisoning were reported in England and Wales. New organisms, such as *E. coli 0157* in beef products cause concern because of their virulence. However, the main problems relate to *Campylobacter* and *Salmonella* infections in poultry products. It is MAFF's policy to implement measures which will reduce the number of cases of food poisoning substantially.

Wherever possible, research focuses on eliminating food poisoning bacteria from foods, particularly those unprocessed or lightly processed. Where this is not practicable, the focus is on controlling food poisoning bacteria to ensure that numbers are not a hazard to health. It is important to learn more about the ways in which the physical and chemical environments of food poisoning bacteria can influence their presence, growth and survival in foods and the food processing environment and to improve our methods for detecting these organisms, to permit development of new methods of elimination and control. MAFF actively encourages the involvement of relevant industry expertise to ensure appropriate targeting, cost-effectiveness and technology transfer.

There is public pressure to reduce food poisoning due to microbiological contamination of meat. At the same time, there is public concern for the welfare of animals on farms and in abattoirs, which affects control practice. The availability of new scientific techniques in molecular biology are already producing improved vaccines e.g. for control of salmonellosis, and there is scope for further significant advances in immunological control. Improved knowledge of how infections spread and, in the long term, a better understanding of genetics in farm animals may lead to animals genetically resistant to these infections. Technological advance in abattoir practice and processing could significantly reduce bacterial contamination. The introduction of Hazard Analysis Critical Control Points management is becoming widespread and there is scope for improving processing hygiene with improved machinery, automation and robotics. The use of decontamination methods for meat and meat products are becoming accepted particularly in North America. These scientific advances are supporting policies to reduce microbiological contamination of meat and improve product quality including shelf life to the benefit of public health and product sales.

Beyond the problems of microbiological safety, there are also those of chemical safety, related both to the use of chemicals on the farm and to pollution from external sources. "Chemicals" can be present in food as the result of contamination, but some may be added, and, of course all food is essentially a complex mixture of "chemicals"; a fact which is often forgotten in the public debate on food safety. Some of MAFF's recent research is focusing on the beneficial effects of the chemicals naturally present in food, which seem to have been overlooked in the debate on chemicals in food.

Farmers need to be able to protect their crops or animals against pests and diseases but, as science continues to learn more about the factors affecting epidemic development and to use this knowledge in the design of decision support systems, it becomes ever more possible to predict outbreaks. This makes it possible to restrict the use of pesticides or medicines to those circumstances where they are really necessary. Such an approach is the best way to ensure freedom from residues in food and in the wider environment. However, much remains to be done in this area and, as long as prophylactic use is a common practice, regulatory authorities must be in a position to police the levels of residues in our food. Research is essential to develop low cost and effective analytical techniques for cost effective surveillance, investigate the nature and distribution of residues in edible tissues, the effect of food processing on the fate of residues and to assess exposure to such residues on dietary intakes..

On the strategic level, it must be right to replace chemical management with alternative means of pest and disease control. In the medium term with the advent of molecular biology, and better understanding of immune systems, more effective strategies for animal disease control will become available using new vaccines. In the long term control of some plant and animal disease conditions will be controlled by genetical resistance. Biological control of pests and weeds remains an important goal.

There is also a continuing need to protect agricultural land and the food chain from pollution. With the ban on the dumping of sewage sludge at sea by 1998, and increasing pressures on the disposal of wastes to landfill we are likely to see an increased use of these materials in agricultural production. The Government encourages the beneficial recycling of sewage sludge and other organic wastes to agricultural land provided such recycling ensures that soil microbial processes and human, animal and plant health are fully protected. There is a clear need for continued research and monitoring on safe practice in waste handling to ensure standards of food safety to meet consumer needs and retailers demands.

The probable future importance of genetically modified organisms (GMOs) and their products in the food chain also raises safety issues. Genetic modification is a relatively new technology which may become a key element of future developments in food and agriculture. Facilitating the introduction of this technology is seen as consistent with the Ministry's aim of promoting a fair and competitive economy in which efficient industries can flourish. However, although it may offer many advantages, it can also present both real and perceived risks for the consumer.

The safety evaluation of "novel foods" raises issues not encountered in the safety evaluation of, for example, food additives or contaminants as such foods are frequently compositionally complex and may be consumed at high intake levels. Strategies for the safety assessment of all novel foods, including GMOs have been developed in the UK by the Advisory Committee on Novel Foods and Processes. Strategies relevant to the safety assessment of GMOs and other novel foods produced by biotechnology have been developed by the FAO/WHO and the OECD. There is a continuing need for research to support the development of these strategies and also to provide the background information against which the safety aspects of specific novel foods, particularly those obtained from GMOs, can be assessed. MAFF currently supports a strategic research programme aimed at meeting this need.

5 ETHICS AND PERCEPTIONS

In the preceding parts of this talk, I have talked not only of the "objective" issues of taste, safety and so forth, but also referred to criteria of food selection that relate to consumer perception. Consumers have a wider range of expectations about food than hitherto, encompassing in particular views about the ethics of production and the protection of the environment which are mediated in part through markets. Consumer perceptions colour those of the immediate purchasers and processors of farm produce.

Supermarkets are responding to consumer concerns by setting their own standards for animal welfare for their premium products, such as "free-range" eggs. Indeed, research has shown that there is often a relationship between welfare and subsequent meat quality. Pigs and poultry meat, for example, is particularly susceptible to the results of handling and temperature stresses on the animals. The strength of public opinion has led to the UK banning veal crates and the impending ban on the use of stalls & tethers for housing of sows. There is also pressure to ban battery cages for laying hens and there is growing concern about the use of farrowing crates. We need to be clear how strong this preference is, as there is also an issue of cost. Consumers probably prefer a range of quality at a range of prices, and do not have the information available to them to distinguish the welfare costs of different products.

Clearly, a problem for policy makers and scientists is the definition of animal welfare, in particular reconciling the objective assessments arising from scientific research and the subjective assessments made by the public. To try and bridge this gap, MAFF is embarking on research to provide a new scientific perspective on the subjective assessment of animal suffering.

In recent years we have seen the development of "ethical shopping" and this concept has spread from the manufacture of jeans to banking and even the disposal of oil platforms. The Freedom Foods campaign run by the RSPCA and various assurance schemes being proposed or implemented by the supermarkets give the consumer added reassurance that the products have been produced under humane conditions. It is interesting to speculate whether the trend to ethical shopping is going to be a major force in animal welfare. It has both pluses and minuses. Firstly if applied by supermarkets the farmer can be rewarded directly (if only by guaranteed sales) for the welfare improvements being made. Secondly supermarkets can act fairly by only buying from those producers who meet their standards and independently by taking national or even local views into account. Thirdly supermarkets have to reconcile the wishes of their customers with the welfare prescriptions they impose and if they misread the wishes of their customers they will lose out to the competition. In this way they will find out how much consumers are willing to pay for enhanced welfare standards.

Against the use of ethical shopping it may be questioned whether consumers want to be faced with ethical choices when they are shopping. There is also the danger that perceived welfare is not identical with real welfare (Hughes 1995). The welfare of hens in a poorly run free range system may well be inferior to that of a well controlled cage system.

By contrast differing views on perceptions on welfare across the EU make changing the regulatory framework extremely difficult and raising standards above some minimum level almost impossible. Policing in the absence of rewards is likely to prove problematical and controlling cross border trade with non-EU members is also difficult.

The range of welfare issues about which concerns are being expressed continues to widen. Long standing concerns about mutilation, stocking density, and welfare at slaughter have been augmented by welfare concerns associated with the use of products such as BST, newer reproductive technologies and even recently the extended use of conventional selection techniques. Further increases in the growth rates in broilers and distended udders in dairy cows are seen as undesirable.

Modern biotechnology has also faced significant resistance despite the fact that it may well enable selection to improve animal welfare such as broilers with improved gait and fewer leg abnormalities, animals better adapted and less stressed by domestication and by the long term selection for resistance to particular infectious diseases.

However, utilisation of the techniques of biotechnology may be limited to some extent by public perceptions that these techniques are "unnatural" and potentially hazardous. These concerns may or may not have substance but they are real and must be addressed. MAFF has sponsored research into public perceptions of the hazards associated with genetic manipulation in food production. This suggests that consumers do not see a real need (in terms of **their** priorities) for the technology, but become less worried about these techniques as they learn more about them. It is clear that in the future all sections of the food chain need to take account of such consumer perceptions in the planning and development of new products and processes. Since the research also showed that Government is heavily distrusted as a source of information in such matters it is similarly clear that other sectors need to assume their share of the responsibility for educating the public and providing appropriate information.

Customers are also increasingly concerned that food production is conducted in an environmentally responsible way. There is increasing emphasis on the sustainability of agricultural production both economically and environmentally. The industry's interdependence with the environment is probably unique. Farmers rely on the fertility of their soils and an adequate water supply for their crops and livestock. Landowners have a long-term interest in the productive capacity of their land. Nevertheless this tradition of stewardship has not always been sufficient to ensure protection of the environment. Government is playing its part in encouraging environmentally friendly farming through the introduction of environmental schemes, encouraging reform of the CAP, and by provision of advice, research, monitoring and, where appropriate, regulation. The "green consumer" is likely to play an important role in influencing the way in which agriculture operates, but it is important that he or she is well informed about the issues - which are often complex. It is clear that, on the one hand, retailers can obtain a premium for "natural" products, despite the fact that naturalness is at best a hazy concept of farming more related to "All creatures great and small" than to modern reality, while at the same time, the organic movement is still struggling to obtain the premiums it needs to support lower input farming.

The farming industry is rightly concerned about the use of husbandry systems and technologies with which consumers might feel ill at ease. The lobby in the UK against the use of BST has been strong in the farming community itself. Farmers do not wish products such as milk, which have a wholesome image, being linked in the consumer's mind with hormones and animal stress.

It is of course difficult to know at what point and to what extent consumers will accept particular technologies; increasing care is needed so that production systems stay within the bounds that the majority of consumers will accept. Perhaps we owe it to animals that we exploit for food to ensure that they have a reasonably comfortable life and a humane death. Getting the balance right will be a continuing challenge to agriculture and ultimately the food industry.

6 CONCLUSION

In the time available, I have been able to do little more than brush the surface of the issue of quality. I hope I have established that this is an issue facing the entire production chain, not just the final retailer of food. It is especially important for farmers to take on board the issues of fitness for purpose, which relate particularly to the requirements of the immediate purchasers of their produce and of the consumer issues of diet and health, safety and , increasingly, ethics. R&D programmes need to be re-oriented, to accommodate not just the practicalities of production of quantity but to address market requirements in all their variety so as to help production relate to actual demands whether rational or apparently prejudiced. R&D will be a key source of information to help consumers make rational purchasing choices.

MAFF continues to be pro-active in seeking the support of industry for pre-competitive research promoting wealth creation both directly with industry and their organisations and indirectly via a very wide range of research contractors. New LINK initiatives are being sought to target sustainable animal farming and horticulture. In addition to promoting participation by primary producers and their levy organisations, these should prove attractive to the pharmaceutical industry in relation to control of endemic diseases and to the supermarkets and large retailers in relation to quality, hygiene and animal welfare.

Because it is essential to understand the basis of quality attributes, to be able to measure them quantitatively and to know exactly what one needs to alter in order to achieve a particular desired change in a product or process, MAFF, together with industry, is also sponsoring major LINK programmes on "Agro-Food Quality" and "Advanced and Hygienic Food Manufacturing".

The increasing variety of food products is leading to the creation of many niche markets ideally suited to exploitation by SME companies. These are numerous in the U.K. food industry at all stages of the food chain but the population is very variable in terms of its technological awareness and capabilities. Technology transfer, to bring the bulk of SME manufacturers up to acceptable levels of best practice is likely to be critical for the future success of this sector of industry. For this reason MAFF is now supporting an increasing programme of technology transfer support measures (such as Regional Technology Transfer Centres, information clubs, and the MAFF Advanced Fellowship) aimed at food industry SMEs.

AN OVERVIEW OF INTEGRATED PRODUCTION OF POME FRUITS AND ITS IMPACT ON FRUIT QUALITY

J.V. Cross and A.M. Berrie

Horticulture Research International
East Malling
West Malling
Kent ME19 6BJ

OVERVIEW OF INTEGRATED PRODUCTION OF POME FRUITS

1.1 The Concept of Integrated Fruit Production

The concept of Integrated Production or Integrated Crop Management is well established. In simple terms, it involves combining together or 'integrating' all the most effective, most environmentally sensitive growing practices for a particular crop, within the constraints of profitability. These best practices are specified by a set of rules or standards with which the grower undertakes to comply.

The development of the concept of Integrated Production was spearheaded on pome fruit crops (apples and pears) in Europe, where the practice has been widely adopted and is at its most mature. The International Organisation for Biological and Integrated Control of Noxious Animals and Plants (IOBC) and the International Society for Horticultural Science (ISHS) have jointly set guidelines and standards for Integrated Pome Fruit Production (IFP) (Cross and Dickler, 1994), the so called 'IFP Euro-Guidelines'. Each county or region has its own IFP Guidelines which are local interpretations of the IFP Euro-Guidelines and are administered by a national or regional IFP organisation. At least one IFP organisation is established in most western European countries, and in some (e.g. Italy, Germany) there are several. Recently it has spread further afield to America and South Africa. IFP is operated as a Quality Assurance scheme and the IFP organisation or an independent body monitors the growers performance in upholding the standards. A points system for non-mandatory aspects encourages a higher performance than the minimum. Growers who meet the required standard are permitted to claim their fruit has been produced according to IFP standards and may display an appropriate produce label or logo.

1.2 Reasons for IFP

There are several reasons for the widespread adoption of IFP in Europe. Firstly, overproduction of pome fruit has lead to a fiercely competitive market. Certain fruit producing regions adopted the practice to gain a competitive edge and as a public relations exercise. Others were forced to follow suit. Secondly, because of the widespread public concerns about environmental and food safety issues, and because of due diligence requirements of food safety legislation, supermarkets were looking for greater assurance that pome fruits (and other fresh produce) had been grown legally and in an environmentally sensitive way. IFP passed on the costs of such assurance to the producer. Thirdly, in several countries substantial grants (500-600 ECU/ha) were available for IFP from the EC under regulation EC2078/92.

1.3 Technical Difficulties

There have been a number of technical difficulties with IFP. The first has been deciding how high to set the standards in the IFP Euro-Guidelines. It would have been all too easy to set idealistic, challenging, standards but these would have led inevitably to substantially increased production costs and would have at best been adopted only by a minority so having little general impact. It has not proved possible to obtain a substantially increased price for IFP produced fruit, and so greater production costs in a competitive market would render the production process unprofitable. The alternative would have been to set low easily achieved standards, but this would have undermined the value of the scheme. In practice, a middle course has been chosen. IFP involves adoption of a number of practices distinctive from good agricultural practice.

The second difficulty has been with problems of a local nature, usually associated with the climate or cultivars grown in a particular region. Every region seems to have its own special problems. In the Netherlands, sandy soil types in some regions are prone to root lesion nematodes which can only be controlled by soil sterilisation, a practice not permitted in IFP; in England, the cultivar Bramleys seedling is prone to the physiological disorder superficial scald which is most easily controlled by post harvest treatment with an anti-oxidant, a practice also not permitted in IFP. Superficial scald can also be controlled by high quality controlled atmosphere storage, but this considerably increases storage costs. Because of local problems, it has been difficult to obtain a consensus across Europe about the minimum requirements of IFP, but such agreement has been reached in the IFP Euro-Guidelines.

1.4 Achievements

Despite these difficulties, IFP has scored some important successes. It has proved a valuable tool for introducing better horticultural practices. Improvements in crop protection practices have been its greatest success. It has been shown that IFP leads to a *circa* 30% reduction in pesticide use compared to conventional production and to use of a safer range of pesticides. IFP has also been shown to have environmental benefits, though these are often difficult to quantify. It has also had a significant impact on horticultural practices and on fruit quality. In the rest of this paper we review the practices in IFP which have an impact on fruit quality and discuss how effectively they have been implemented.

FRUIT QUALITY

Fruit growing is so competitive that it is only profitable to produce high yields of first quality fruit. Quality classes are defined by EC grading standards. Production and marketing costs for second quality fruit exceed market value. The impact of IFP on three main aspects of fruit quality 1) external quality (size, shape and appearance) 2) internal quality (taste, texture, firmness, shelf-life) 3) extrinsic quality (method of production, freedom from pesticide residues and contaminants etc) is discussed below.

2.1 External Fruit Quality

2.1.1 Colour, Shape, Size and Skin Finish

For any particular cultivar, choice of clone and its health status, orchard site, tree training and pruning, and soil management practices are the factors which have the greatest influence on external fruit quality. If IFP is operated strictly, the highest standards of horticultural practice are required. The best clone with the best fruit quality and health status may only be selected. Sites with a favourable aspect and good soil must be used. Trees must be trained and pruned to allow good light penetration to encourage good fruit colour. Irrigation and mulching which enhance size, shape and skin finish, are encouraged.

These requirements are important in conventional production and are not exclusive to IFP. However, IFP should require a generally higher standard. The main question is whether or not these aspects have genuinely been improved in practice by operation of IFP schemes. The answer to this question varies from one region to another. In most regions it has proved difficult to exclude orchards on the basis of their site or to exclude orchards already planted. However, several regions have placed emphasis on a high standard of tree and soil management in existing orchards and this has certainly benefitted fruit quality though such benefits are difficult to quantify. Higher standards for planting new orchards are demanded in most regions.

2.1.2 Pest and Disease Blemishes

In IFP much emphasis is placed on high standards of pest and disease control with minimal use of pesticides. The rule is to use the safest products only when necessary. To achieve this aim, sophisticated methods of pest, disease and weed assessment have been devised (e.g. Cross and Berrie, 1994) and lists of acceptable pesticides prescribed.

Pest and disease management programmes used in IFP are highly developed. It has been demonstrated frequently that a higher standard of pest and disease control can be achieved if these systems are operated correctly. However, it has been found that such programmes are often more difficult to manage and can be more costly than routine spray applications. This has been one of the more successful aspects of IFP.

2.2 Internal Fruit Quality

Taste quality is highly subjective and varies greatly between cultivars which each have their own taste characteristics. Certain general aspects of taste quality (e.g. sugar content and acidity) can be measured, but it is difficult to set universally acceptable standards because of differences in personal preference. However, there is a need for growers to produce a consistent product. To this end, some IFP organisations (notably Covapi, France) have defined taste quality standards for certain varieties by setting acceptable ranges of certain easily measured general taste characteristics including sugar content and acidity. Firmness and texture are measured easily. For each cultivar there are defined standards for marketability of fruit. They are controlled by quality control staff of the co-operative packhouse and the supermarket as well as by Government inspections. Fruit may be of poor quality because of a wide range of physiological disorders. Two of the most important are bitter pit (caused by nutrient deficiency) and senescent breakdown (caused by overmaturity), though there are many others.

Many of these internal fruit quality factors are affected by horticultural practices, the most important of which are fruit nutrition, picking date and storage regime.

Participants in IFP are required to achieve the highest standards of fruit quality. To ensure that standards are consistently attained, regular monitoring of soil fertility and plant mineral composition are required. Fruit must be harvested at the optimum time for the purpose for which it is intended and once in store must be monitored carefully and regularly to ensure that is marketed at the correct time.

All these requirements are mandatory for participants in IFP. In non-IFP production, many growers also meet these requirements, but there is a significant minority who do not, and who occasionally sell sub-standard fruit which undermines the reputation of the cultivar and the market for apples in general.

2.3 Extrinsic Fruit Quality

Until recently, appearance and taste were the sole quality factors of importance to consumers. However, increased awareness of food safety and environmental issues has led the consumer to expect fruit to be free from pesticide residues and other contaminants and to have been produced in an environmentally sensitive way. Consumer concerns about some of these aspects may be irrational, but nevertheless must be addressed. This is the main *raison d'etre* for IFP.

IFP includes a number of measures designed to reduce significantly the frequency of pesticide residues compared to that expected under good agricultural practice. These include choice of short persistence pesticides, minimised dose rates and maximised harvest intervals. The approach has to be balanced so that the required quality can still be achieved. IFP also requires measures to enhance the orchard environment. There is only limited scope for improvements in the orchard itself without disrupting production seriously. Measures to increase species diversity of grass alleyways and windbreaks and to encourage nesting birds are perhaps the most important. Perhaps the greatest step would be to require a mandatory conservation audit and management plan for the orchard environs. However, although such action is encouraged in IFP, it is not yet a mandatory requirement.

3. Conclusion

IFP has been successful in many European countries and is spreading to other continents. If strictly operated it leads to improvements in horticultural practices and hence to some improvement in external and internal fruit quality. For individual orchards a greater proportion of fruit meets the first quality grade and the overall quality standard is improved. Improvements are variable and difficult to quantify. An important contribution of IFP has been to improve extrinsic fruit quality.

References

1. Cross, J.V. and Dickler, E. *Bulletin IOBC/WPRS* **Vol.17(19),** 1994, 40 pp.
2. Cross, J.V. and Berrie, A.M. *Aspects of Applied Biology* 1994, **37,** 225-236.

MANIPULATING THE QUALITY OF BUTTERCUP SQUASH (*CUCURBITA MAXIMA* CV. DELICA) BY PRE- AND POST-HARVEST MANAGEMENT

W.J. Harvey[1], A. Lush[2], and D.G. Grant[1]

[1]New Zealand Institute for Crop & Food Research Ltd,
Private Bag 4704
Christchurch
New Zealand

[2]Industrial Research Ltd
P.O. Box 2225
Auckland
New Zealand

1. INTRODUCTION

Buttercup squash (*Cucurbita maxima*) is a vegetable fruit weighing between 1.5 and 2.5 kg. It has green striped skin with bright orange flesh. This vegetable is nutritionally outstanding; the skin and flesh are excellent sources of beta-carotene, vitamins A and C, calcium and iron. The flesh has a sweet, nutty flavour with a smooth, dry texture when mature.

Buttercup squash has become a major export to Japan in the past 5 years. In the 1994 season exports exceeded 80,000 tonnes with a value of $70 m. Fruit is shipped by refrigerated sea cargo and takes about two weeks to reach Japan. The major goal for New Zealand exporters has been to improve the consistency of the quality of the fruit arriving there. Early season squash often receive premium prices in the Japanese market by filling a niche in January, when competing countries can only supply low volumes. While Hawthorne[1] showed that fruit harvested early was less likely to rot during transport and distribution, early fruit may be insufficiently mature and thus be of a lower quality when sold. Early season, low quality fruit may result in depression of prices for subsequent shipments.

Previous research[2] has suggested that once a certain level of maturity has been reached, squash will continue to ripen in storage. Furthermore, the quality of the squash is affected by the temperature of storage during shipping[3]. It may be possible to provide a controlled temperature regime either before or during transit to ensure optimum quality of the fruit on arrival in Japan.

Buwalda[4] suggested that Growing Degree Days (GDD's) with a base temperature of 8°C could be used to estimate the optimum time from emergence to harvest. Sharrock & Parkes[5] used days after flowering as a guide. Physical tests such as penetrometer readings of the hardness of the skin within 24 hours of harvest, flesh colour, seed development, Brix value of expressed juice and dry matter of the flesh have all been used as possible indicators of maturity[6]. These physical tests had not been related to sensory quality until our earlier work[7].

Part 1 of the research described here studied the maturation dynamics of buttercup squash at 3 representative growing sites in New Zealand, and during a storage period simulating refrigerated transport to Japan. The aim was to determine time of harvest for optimum sensory quality and minimum incidence of rots, and to define methods for deciding when to harvest. Part 2 of this research involved the evaluation of KIWIFIRM (patent pending) as a non-destructive firmness tester to replace the penetrometer.

In continuing research, reported only briefly here, we are comparing the effect of maturing squash on and off the vine on subsequent quality, particularly sweetness.

2. MATERIALS AND METHODS

2.1 Plant Material

Crops of buttercup squash, cv. Delica, were planted at three sites (Pukekohe, Auckland; Havelock North, Hawke's Bay; and Lincoln, Canterbury) representative of the three most popular commercial squash-growing areas. Crops were planted on 26 November 1992, 16 December, and 3 December 1992 respectively. As female flowers began to appear (from 19 January, 8 February, and 8 February 1993 respectively), each was tagged with the date on which it opened. Twenty fruit were harvested from the tagged flowers at each of 30, 40, 50, 60 and 70 (at Lincoln) days from flowering. Meteorological data for thermal time accumulation (°C days, base temperature 8°C) were collected from emergence to harvest and also from fruit set to harvest.

2.2 Maturity Tests

The following observations were made on the fruit before and after a three week storage period simulating refrigerated (12-14°C) transport:

* skin hardness at harvest (in kg, using 1-12 kg Effegi penetrometer with 4 mm probe)
* total and soluble solids (° Brix)
* flesh colour (using a Hunterlab Labscan reflectance spectrophotometer using the CIElab notation)
* seed development (from the immature jelly-like stage to the formation of a full white cotyledon)
* fructose, glucose, sucrose and starch analyses, which were performed on freeze-dried samples using standard enzymatic methods.

2.3 Sensory Evaluation

The day after the maturity tests were performed, 8 days after harvest for the unstored fruit, and then after a further 21 days at 14° C for the stored fruit, 4 of the 10 fruit from each site and harvest date were selected for tasting by a trained panel of 8 people. Steamed slices were assessed for sweetness, squash flavour, moistness and fibrousness. Panellists were also asked to give an acceptability rating. A 150 mm line scale was used, with descriptive terms as anchor points. Earlier studies, using Japanese nationals to taste individual squash fruit which were then profiled by a trained panel[7], had established the relationship between Japanese preferences and the quality descriptors used.

2.4 Non-Destructive Firmness Tests

A crop of hybrid buttercup squash (*Cucurbita maxima* cv. Delica) was planted in Pukekohe on 4 November 1993. Flowers were tagged with their flowering date between 6 and 28

January 1994 and 10 fruit were then harvested at 21, 28, 35, 42, 49, 56 and 62 days from flowering between 4 February and 11 March 1994. On the day of harvest, the fruit were tested using an Effegi penetrometer and a KIWIFIRM pressure tester to measure skin hardness. Following this, the fruit were assessed using the maturity tests described above to establish the maturity of the fruit independently.

3. RESULTS AND DISCUSSION

3.1 Physical and Sensory Changes

Figure 1 shows how maturity at harvest, and a period of storage, affect acceptability of buttercup squash. Fruit had accumulated sufficient GDD's after 40 days on the vine between flowering and harvest, that the chosen post-harvest treatment removed acceptability differences, all sites achieving excellent acceptability. As expected, the physical parameters that predict acceptability[7] (Total and soluble solids, and colour) and maturity (penetrometer) reflected this behaviour. The observations suggest that after heat accumulation, equivalent in this experiment to 40 days post-flowering, fruit may respond to accumulating heat units similarly on and off the vine. Preliminary results of experiments where fruit of similar maturity are subjected to different post harvest curing regimes tend to confirm this hypothesis - fruit from high latitudes, where cool temperatures slow maturity can be cured to resemble those from the most favoured growing sites.

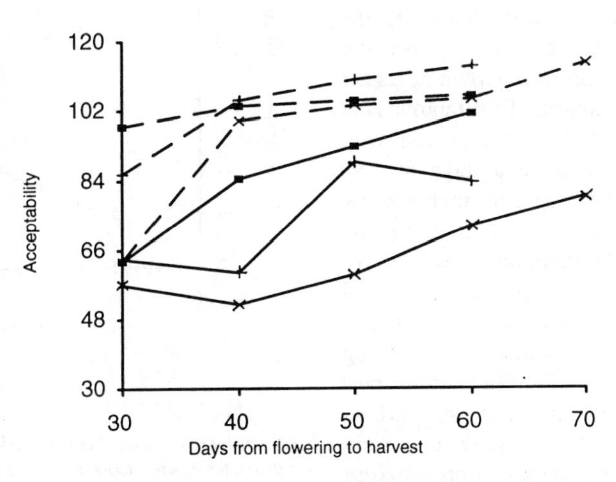

Figure 1:*Variation in squash acceptability with time on the vine, location and storage. Solid lines: immediately post harvest. Dashed lines: after 21 days storage. (■) Pukekohe. (+) Havelock Nth. (X) Lincoln.*

3.2 Non-Destructive Maturity Determination

The results described above are derived from eight panellists, each assessing 4 fruit, and as expected, the level of fruit-to-fruit variation was small. This is due to environmental variation within the crop influencing fruit set the same day. Since fruit set in a crop takes place over a month at higher latitude sites, maturity variation can be greatly magnified. Standard procedures for determining whether a crop is ready to harvest include destructive penetrometer testing of 30 fruit. A mean of 8.5 kg is required, with a minimum of >7 kg. Figure 2 shows that the KIWIFIRM predicts the penetrometer score in the range 6-9 kg with good precision. Log-log transformation, followed by regression gives

$$\text{Penetrometer Score (kg)} = 3.4 \times 10^{-4}(\text{KIWIFIRM})^{2.32}$$

with $R^2 = 0.93$. Since the KIWIFIRM was originally designed for kiwifruit, which are a good deal softer than squash, we are now recalibrating this instrument, and expect to gain further sensitivity in the key range.

4. CONCLUSIONS

Heat accumulation during the transport period can contribute a significant increase in quality to the squash crop, especially when it is grown at higher latitudes. This improvement can be obtained from fruit picked at a stage of crop development that is now considered to be too early. Furthermore, we have shown that there is potential for differentiating between fruit within a crop, using the KIWIFIRM tester pre- or post-harvest, which allows growers and exporters to improve fruit quality and uniformity.

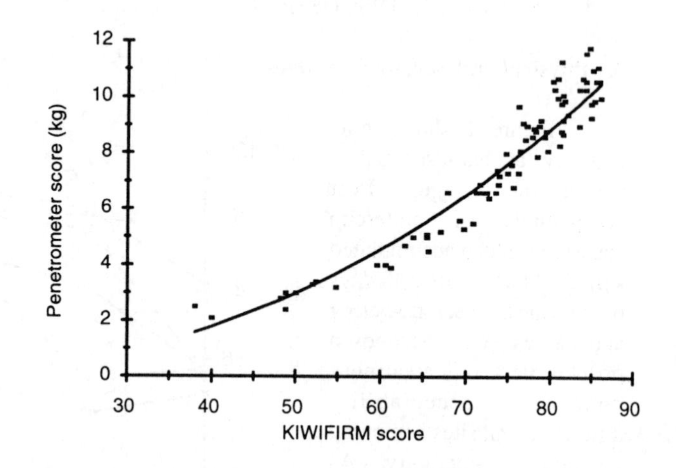

Figure 2 *Relationship between KIWIFIRM pressure tester score and Effeg penetrometer score. Dots are data points and the solid line is the fitte relationship described in the text.*

5. REFERENCES

1. B.T. Hawthorne *N.Z. J. Crop & Horticultural Sci* 1990 **18** 141

2. W.J. Harvey, D.G. Grant, J. Burgmans, J.P. Lammerink and P. Hannan. *N.Z. J. Crop & Horticultural Sci (in preparation)*

3. A. Ngao, T. Indou and H. Dohi *J. Japan. Soc. Hort. Sci.* 1991 **60** 175

4. J.G. Buwalda and R.E. Freeman *Proc. Agron. Soc. N.Z.* 1986 **16** 7

5. K.R. Sharrock and S.L. Parkes *N.Z. J. Crop & Horticultural Sci* 1990 **18** 185

6. D.J. Beever and S.K. Forbes 'Post Harvest Changes in Quality of Buttercup Squash in Different Storage Conditions.' DSIR Fruit and Trees. Auckland, New Zealand, 1991.

7. W.J. Harvey and D.G. Grant *Proc. Agron. Soc. N.Z.* 1992 **22** 25

BIOCHEMICAL APPROACH TO HIGHER ASCORBIC ACID LEVELS IN POTATO BREEDING

G.Ishii,M.Mori and Y.Umemura

Hokkaido National Agricultural Experiment Station
Hitsujigaoka,Sapporo,062 Japan

1 INTRODUCTION

Consumers have become more concerned about relationship between their health and diet especially in food nutrition and safety.[1] They demand potatoes with high vitamin C for table use. A breeding program has started to push up ascorbic acid (AsA) level in potato tuber (*Solanum tuberosum* L.).

A screening for potato varieties with high vitamin C demonstrated that "Kita-akari" was the most rich content of vitamin C among the ten varieties analyzed. This variety was released as a hybrid between the male parent "Tunika" and the female parent "Irish cobbler"in 1988. In contrast, the "Irish cobbler" has a lower lever of vitamin C than "Kita-akari".[2]

To elucidate the carbohydrate metabolic difference in AsA content between them, their activities in biosynthetic enzyme of L-ascorbic acid,AsA contents and sugar composition of tubers were compared from flowering to harvesting.

2 MATERIALS AND METHODS

Two varieties were planted in the open field on early May and harvested periodically during early July to early September. Each 10 g from 5 or 10 tubers was used for analysis.

2.1 Determination of Vitamin C and Sugar

Only L-ascorbic acid was analyzed by HPLC monitoring at 242 nm after extraction of 5% metaphosphoric acid due to the lack of ascorbic acid oxidase in potato tuber.[3]

Each sugar in plant organ was determined by HPLC equipped with RI detector after extraction of 80% ethanol.

2.2 Preparation and Assay of L-Galactono-γ-Lactone Dehydrogenase(EC 1.3.2.3,GLDHase)

The terminal step in the biosynthesis of L-ascorbic acid in potato has been shown to be the oxidation of L-galactono-γ-lactone by an enzyme present in mitochondria, which is called L-galactono-γ-lactone dehydrogenase.
All procedures in extraction and assay of GLDHase followed the method reported by Oba.[4,5] The enzyme from 10 g of tuber tissue was collected from the terminated 500$\times$ g supernant. GLDHase activity was assayed by measuring the L-galactono-γ-lactone-dependent reduction of cytochrome c at 550 nm.

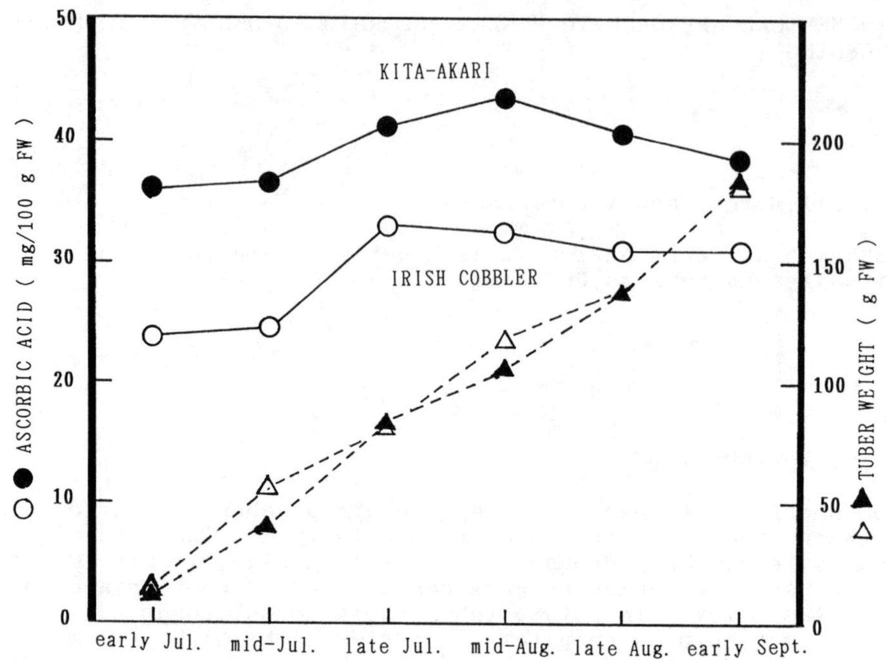

Figure 1 *Changes in ascorbic acid content during the growth of tuber*

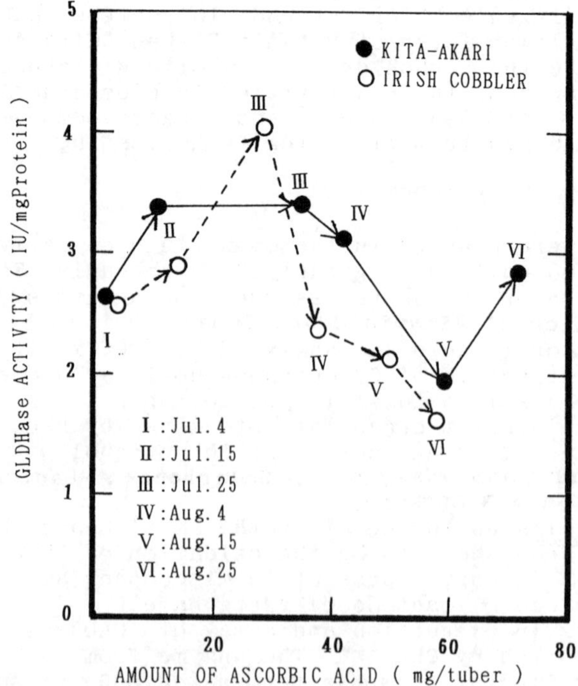

Figure 2 *Changes in GLDHase activity and AsA during the growth of tuber*

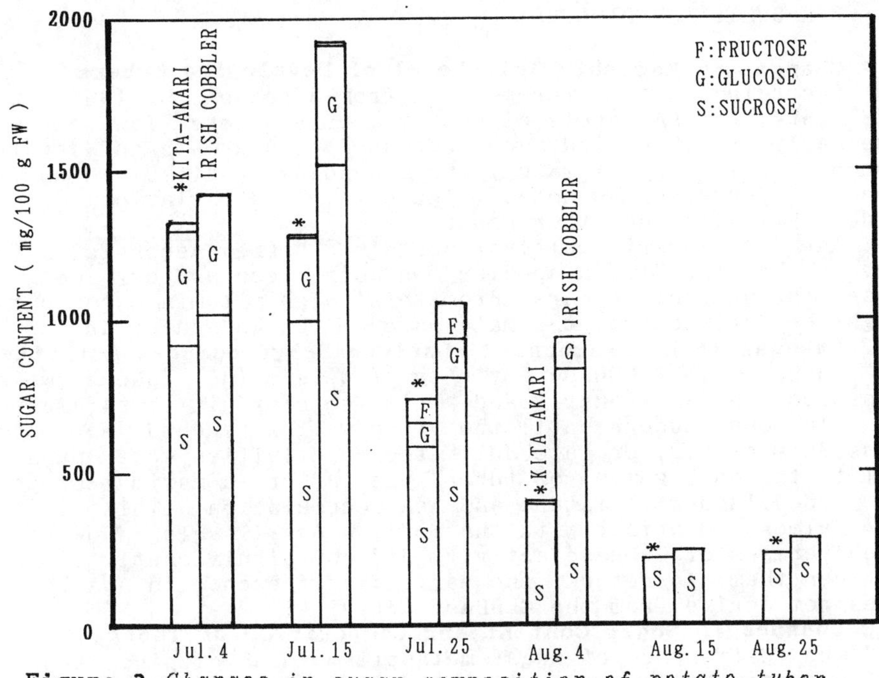

Figure 3 *Changes in sugar composition of potato tuber*

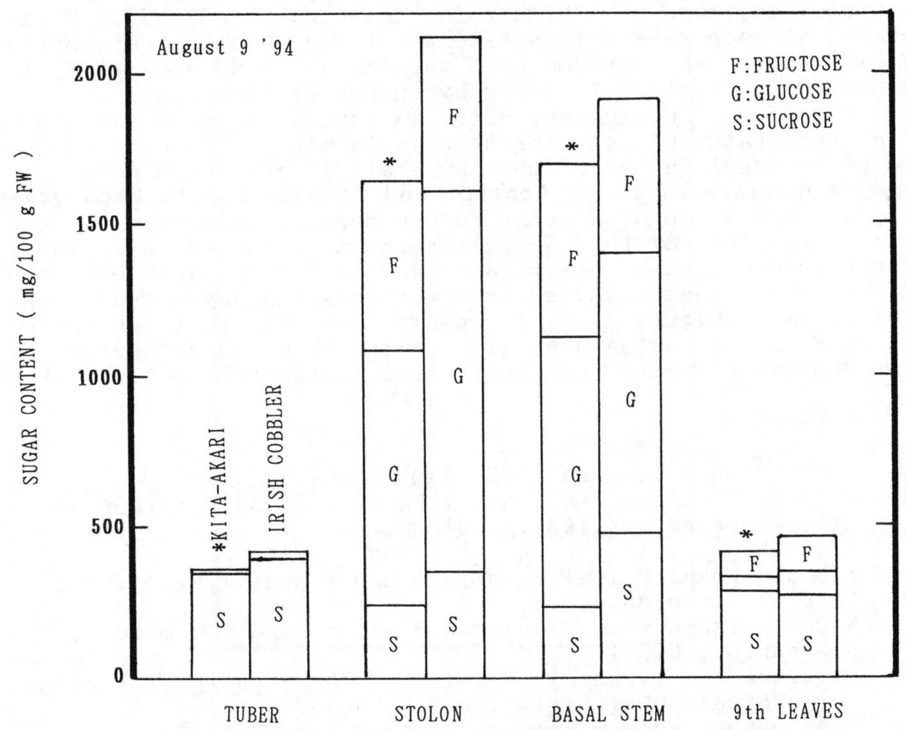

Figure 4 *Varietal difference in sugar composition of organ*

3 RESULTS AND DISCUSSION

3.1 Changes in Ascorbic Acid Level of Developing Tubers

According to the average data from 3 seasons of 1991,1993 and 1994,the AsA levels of both varieties were low during the early growth period, and increased with growth and development until approximately the mid-August, after which a gradual decrease was noted (Figure 1). This variation coincided well with the other paper.[6]

The "Kita-akari" cultivar contained higher ascorbic acid levels than the "Irish cobbler" on both fresh and dry weight bases through the 3 years. From this, high vitamin C contents might be derived from the male parent "Tunika" cultivar.

3.2 Changes in L-Galactono-γ-Lactone Dehydrogenase Activity

Changes in GLDHase activity of developing tubers were measured on "Kita-akari" and "Irish cobbler" in 1994(Figure 2). In young tubers until the fresh weight reached more or less 80 g at mid- or late July, their activities were parallel to the enlargement of tuber, but after a maximum stage, they declined with mature and AsA accumulation. This trend was almost identical with that of the AsA levels. However, the averages of GLDHase activity did not significantly differ between them. Therefore, the varietal difference in AsA level does not derive from the GLDHase activity.

3.3 Changes in Sugar Content and Composition of Tuber

The difference of sugar metabolism in developing tubers was compared between two cultivars(Figure 3). Total sugar and sucrose contents were higher in "Irish cobbler" than "Kita-akari" at each stage. In potato plant, sucrose is transported from leaves as a predominant substance, and converted to glucose and fructose in tuber by enzymatic hydrolysis.

Therefore, D-glucose, which is suggested to be the initial substrate of AsA biosynthesis in plants,[7] does not seem to be a causal factor of the varietal difference in potato.

3.4 Comparison of Sugar Content and Composition in Each Organ

At the date of 2 weeks before normal harvesting, each sugar content of the two cultivars was compared on their tubers,stolons,basal stems and leaves. "Irish cobbler" has more total sugar and sucrose contents in any organ than "Kita-akari"(Figure 4). This result indicates that the varietal difference in AsA content between them can not prove by the GLDHase activity or sugar level in tubers.

References

1. M.Mori and Y.Umemura,1992,JARQ,26,157.
2. M.Mori,T.Maida,G.Ishii and Y.Umemura,Asian Potato Association Proceedings,1991,pp.31-35.
3. K.Oba,Nippon Kasei Gakkaishi,1990,41,715.
4. K.Oba,M.Fukui,Y.Imai,S.Iriyama and K.Nogami,Plant Cell Physiol.,1994,35,473.
5. K.Oba,S.Ishikawa,M.Nishikawa,H.Mizuno and T.Yamamoto, J.Biochem.,1995,117,120.
6. T.Tagawa and Y.Okazawa,Memoirs of the Faculty of Agriculture Hokkaido University,1953,1,240.
7. F.A.Loewus and M.W.Loewus,CRC Crit.Rev.Plant Sci., 1987,5,101.

EFFECTS OF MATURATION ON THE NUTRITIONAL VALUE OF PEA

Periago, M.J.*, Ros, G.*, Martínez, M.C.*, López, G.*, Ortuño, J.* and Rincón, F.**

U.D. Bromatología e Inspección de Alimentos, Faculty of Veterinary, University of Murcia, Murcia 30.071, SPAIN.
Dpt. Bromatología, Faculty of Veterinary, University of Córdoba, Avda. Medina Azahara s/n, Córdoba-14.005, SPAIN.

1.INTRODUCTION

In most developed countries there has been an remarkable increase in the consumption of green legumes, which can be mainly attributed to the availability of frozen legumes, particularly peas[1]. Such legumes, are easier to cook and do not present hard to cook phenomena, as do dry legumes. The chemical composition and nutritive value of pea seeds depend on different factors such as variety, size, climate and soil conditions during their development[2,3]. In general, peas are a rich source of protein, carbohydrates and minerals in the diet[3]. They are rich in iron, calcium and potassium, have a low sodium content and have no cholesterol[4]. However,their low sulphur amino acid content and the presence of antinutritional factors, such as trypsin inhibitor and phytic acid, reduce their protein nutritional value. The aim of this research is to understand the changes in the nutriitional value of peas according to their maturation stage.

2. MATERIAL AND METHODS

2.1 Material

Wrinkled pea seed cultivars Citrina and Warindo were cultivated in Malpica de Tajo (Spain), under the Supervision of Hero, Spain S.A. The peas were classified by diameter and development stage into four commercial sizes: super fine (SF, 4.7-7-5 mm), very fine (VF, 7.8-8.2 mm), fine (FN, 8.3-8.8 mm) and middle (MD, 8.9-10.2 mm). Immediately after harvesting, the pea seeds were freeze-dried, and then ground to a fine flour, which was stored with desiccant until analysis. All samples were analyzed in triplicate for each parameter studied.

2.2 Methods

The proximate composition of peas (moisture, ash, crude protein, total dietary fiber and ash) was determined according to the techniques recommended by the AOAC[5]. Total starch was analyzed using a multienzymatic method described by Englyst et al.[6].To ascertain the nutritional protein value, true protein and free amino acids were determined following the trichloroacetic precipitation and ninhydrin reagent methods, respectively[7]. The non protein nitrogen (NPN) were estimated as the difference between crude protein nitrogen and true protein nitrogen[8]. *In vitro* protein digestibility (IVPD) was carried out using a multienzymatic technique[9]. To ascertain the content of antinutritive factors, trypsin inhibitor activity was measured by the method of della Gatta et al.[10], and phytic acid content was determined as a function of the phosphorus content, after the precipitation as a ferric-phytate and solubilization as sodium-phytate[11].

2.3 Statistical analysis

In the statistical analysis of the data a variance analysis and Tukey's test were applied to ascertain the significance between means.

3.RESULTS AND DISCUSSION

The analysis of variance showed that the chemical composition and nutritive value of peas depend on the pea cultivar and size, except crude protein, which depends only on size (data no show here). The proximate composition of pea seed on a dry matter basis is shown in Table 1.The content of crude protein was significantly ($P<0.05$) higher in peas of Citrina cultivar than Warindo cultivar, with no statistical differences between sizes. Fat increased,while ash and moisture content decreased as pea size increased in both cultivars. The highest ash value was obtained for SF size of the Citrina cultivar probably because it accumulates more minerals in the youngest parts, since these minerals are used for normal development[12]. Total dietary fiber (TDF) decreased as the total starch content and the size of pea increased, since the testa/kernel ratio decreases during pea seed development and starch is stored in the kernel as a physiological reserve of carbohydrates[3].

Table 1 Mean and standard deviation of proximate composition (expressed as g 100 g^{-1} of dry weight) in peas of Citrina and Warindo cultivars classified into four commercial sizes[1].

Variety /Size[2]	Crude Protein	Fat	Ash	TDF	Starch	Moisture
CITRINA						
SF	30.0±0.33^{a}	1.35±0.06^{c}	3.79±0.08^{a}	24.23±0.07^{d}	26.28±0.48^{d}	82.29±0.28^{a}
VF	30.61±1.12^{a}	1.96±0.13^{b}	3.35±0.06^{c}	23.18±0.38^{cd}	29.33±0.15^{c}	77.53±0.84^{b}
FN	29.89±0.85^{a}	2.40±0.28^{b}	3.19±0.05^{d}	22.47±1.14^{d}	31.19±0.81^{b}	75.38±0.68^{bc}
MD	29.32±0.08^{a}	2.31±0.16^{b}	3.14±0.09^{d}	22.75±0.57^{d}	32.84±0.10^{a}	74.29±0.40^{cd}
WARINDO						
SF	26.83±0.87^{bc}	2.31±0.09^{b}	3.51±0.05^{b}	25.36±1.48^{bc}	31.09±0.81^{b}	77.75±0.65^{b}
VF	25.85±0.28^{c}	2.86±0.28^{a}	3.50±0.03^{b}	27.52±0.34^{a}	32.13±0.20^{ab}	72.88±1.30^{cd}
FN	26.23±0.40^{bc}	3.27±0.23^{a}	3.53±0.04^{b}	26.60±0.10^{ab}	32.76±0.64^{a}	71.52±0.80^{de}
MD	27.42±0.05^{b}	3.36±0.25^{a}	3.53±0.02^{b}	26.00±0.75^{ab}	32.14 ±0.05^{a}	70.05±2.56^{e}

[1]Mean ± standard deviation of three determinations expressed as g/100g of dry weight. Different letters in column are significantly different $P<0.05$.

[2]Abbreviations of size (SF: super fine, 4.7-7.5 mm; VF: very fine, 7.6-8.2 mm; FN: fine, 8.3-8.8 mm; MD: middle, 8.8-10.2 mm).

[3]Abbreviations of TDF: total dietary fiber.

Figure 1 shows the content of true protein, free amino acids, NPN, IVPD, trypsin inhibitor activity and phytic acid content in the studied samples. The true protein in peas increased significantly as pea size increased, whereas free amino acids and NPN decreased. This fact is a consequence of the protein synthesis, from free amino acids and NPN, which takes place in the seed kernel during the development of the plant in order to build up a reserve of protein ready for germination. The IVPD increased with the pea size, and was related to the increases of true protein. In general, the digestibility of pea protein could be affected by the presence of antinutritive factor trypsin inhibitor[3] and phytic acid[13]. However, in this study an increase in these antinutritive factors was observed, which was related to the increase of IVPD, and which showed a positive correlation (r=0.92 and r=0.80, respectively).

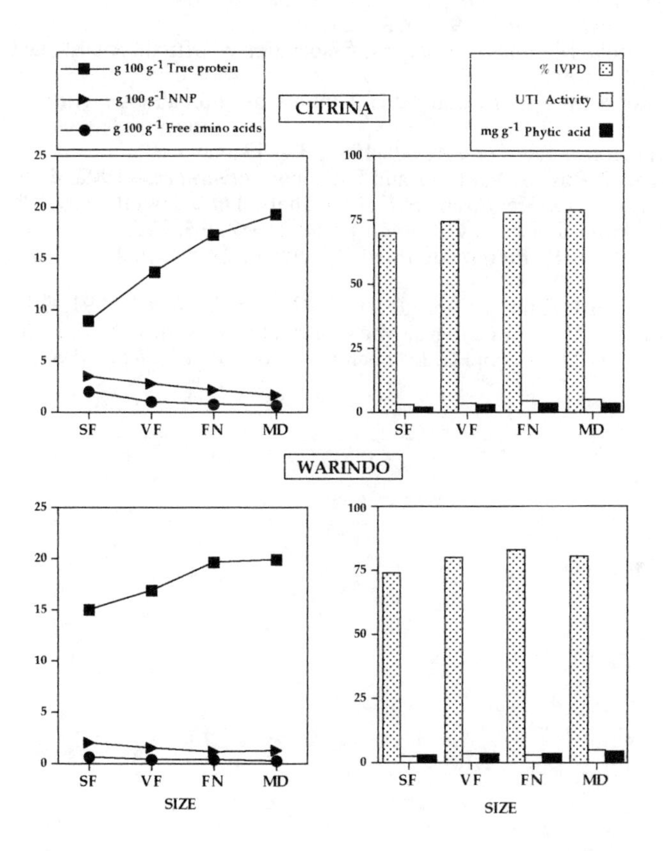

Figure 1. True protein, non protein nitrogen (NPN), free amino acids, *in vitro* protein digestibility (IVPD), trypsin inhibitor activity and phytic acid content, in Citrina and Warindo pea cultivars, classified in four commercial sizes (SF, super fine, 4.7-7.5 mm; VF, very fine, 7.6-8.2 mm; FN, fine, 8.3-8.8 mm, MD, middle, 8.9-10.2 mm).

We concluded that nutritive value of peas was significantly determined by size and cultivar, when peas are studied under the same growing conditions. However, the protein value was higher in mature seeds (FN and MD) than in immature (SF and VF), because the former gave greater true protein and IVPD values. The trypsin inhibitor and phytic acid, which were formed during development, did not affect the IVPD.

References

1. F. Fidanza, The mediterranean diet in health and disease, Van Nostrand Reinhold/AVI, New York, 1991.
2. P. Geervani and U. Devi, *J. Sci. Food Agric.*, 1988, **46**, 243.
3. G.P. Savage and S. Deo, *Nutr. Abst. Rev.*, 1989, **56**, 66.
4. B. Morrow, *Food Technol.*, 1991, **45**, 121.
5. A.O.A.C., Official Methods of Analysis, Association of Official Analytical Chemists, Washington D.C., 1990.
6. H.N. Englyst, S.M. Kingman and J.H. Cummings, *Eur. J. Clin. Nutr.*, 1992, **46**, S33.
7. E.O. Awolumate, *J. Sci. Food Agric.*, 1983, **34**, 1351.
8. M.J. Periago, G. Ros, C. Martínez and F. Rincón, *Alimentaria*, 1992, **4**, 51.
9. L.D. Satterlee, J.G. Kendrich, H.F. Marsahall, D.K. Jewell, R.D. Phillips, G. Sawar and P. Slump, J. *Assoc. Off. Anal. Chem.*, 1982, **65**, 798.
10. C. della Gatta, A.R. Piergovanni and P. Perrino, *Lebensm. Wiss. und-Technol.*, 1988, **21**, 315.
11. S.P. Plaami and J. Kumpulainen, *J. Assoc. Off. Anal. Chem.*, 1991, **74**, 32.
12. J.C. Katyal and N.S. Randahawa, Micronutrientes, O.M.S./F.A.O., Roma, 1986.
13. E. Carnovale, E. Lugaro and G. Lombardi-Boccia, *Cereal Chem.*, 1988, **65**, 114.

EFFECT OF NATURAL AMINO ALCOHOLS ON THE STRESS TOLERANCE AND FOOD QUALITY IN STESSED BARLEY

V. Leinhos and H. Eckert,
H. Bergmann, B. Machelett,

Institute of Nutrition and Environment
The University of Jena/ Germany

1 INTRODUCTION

Unfavourable environments cause changes in the plant metabolism and, depending on the strength of the stress the yield and the food quality in crops decrease. For diminishing the stress response in crops new possibilities are investigated in addition to conventional ways of minimizing the plant stress. Such a new way is the increase of the stress tolerance in plants by using the natural amino alcohol choline and its precursor 2-aminoethanol.[1-4] Furthermore, choline is used as a vitamin-like substance with stress diminishing properties in the animal and human nutrition.[5]

Therefore, we examined in this contribution the influence of plant treatments with the non-ecotoxic compound choline on the formation of stress stimulated components in plants and the yield of protein, some amino acids and grains in stressed barley. The results are discussed in comparison to other crops.

2 MATERIAL AND METHODS

2.1 **Application of Amino Alcohols and Phytohormones**

Choline (as chloride) was applied once in aqueous solution (≤ 0,5 mg/ plant or 1,5 kg/ha) at the shooting stage by spraying or watering. Additionally, in some experiments the phytohormones benzyl aminopurine (BAP) or abscisic acid (ABA) and the choline-precursor 2-aminoethanol (2- AE) were used as reference substances. (Dosages: BAP and ABA = 0,2 mg/plant, split up into several treatments; 2-AE = 0,5 mg/plant, one application).

2.2 **Cultivation of Plants**

Spring barley (*Hordeum vulgare* L. cv. Alexis) was cultivated in Mitscherlich pots (16 plants per pot).[2,3]

After the treatments with amino alcohols or plant hormones the plants were exposed to drought as an "integral stressor" by reducing the soil moisture from 65% of water capacity to 30% water capacity for a week. Afterwards the water loss was replaced (to 65% water capacity). As reference plants were cultivated under sufficient water supply condition and used as control groups.

In long term experiments stressed plant were exposed to 3 periods of drought, as described above. After harvesting the fresh material was lyophilyzed and stored at -20°C. The mature grains were stored in a refrigerator.

2.3 Plant Analysis

Glycine Betaine: Extraction of the plant material with methanol/ chloroform/ water (70:20:10 m/m). Purification by ionexchange-chromatography. Determination by reversed-phase HPLC. Isocratic elution.[3]

Putrescine: Extraction of the plant material with trichloroacetic acid (5%). Synthesis of benzoyl derivatives. Purification with diethylether. Determination by reversed-phase HPLC. Isocratic elution with acetonitrile/water (49/51).[4]

Amino Acids: Hydrolyzation in HCl (6 M) and determination in an amino acid analyzer in a conventional way.

Lignin: Homogenization of the plant material in methanol. Solubilization of the dry residue in 25% (v/v) acetylbromide in acetic acid at 70° C for 30 min. Measuring the absorption of the diluted samples at 280 nm. Quantification by using a standard curve to lignin.[7]

3 RESULTS AND DISCUSSION

3.1 Effect of Choline on the Content of Glycine Betaine

A seasonal water shortage stimulated the formation of the amino acid glycine betaine (trimethylglycine) in barley (Table 1). An additional enrichment of cadmium (Cd) in the soil substrates amplified the betaine formation. After a choline treatment the stress induced formation of glycine betaine (and trigonelline, another betaine) was reduced significantly. The choline precursor 2-AE and the cytokinine BAP affected the betaine synthesis in a comparable extent as choline. However, ABA applied exogenously promoted the betaine formation in a similar way as drought. Thus, choline applied exogenously influenced the stress response in plants different to drought or ABA.

Table 1 *Effect of Choline on the Betaine Content in Barley Plants (Pot Experiments with Different Water Supply. Treatment according to 2.2. Harvesting at Heading Stage)*

Stress Condition	Treatment	Glycine Betaine (mg/g Dry Matter)	Trigonelline (mg/g Dry Matter)
Well-watered	-	4,3	0,5
Drought	-	7,9^{+}	0,7
Drought + 3 ppm Cd	-	9,6^{++}	0,8^{+}
Well-watered	Choline	4,6	-
Drought	Choline	5,2	0,6
Drought + 3 ppm Cd	Choline	7,8^{+}	0,6
Well-watered	BAP	5,0	-
Drought	BAP	5,6	-
Well-watered	ABA	6,5^{+}	-
Drought	ABA	8,7^{+}	-
Well-watered	2-AE	4,6	-
Drought	2-AE	5,5	

Significance: ++ $\alpha \leq 0{,}01$; + $\alpha \leq 0{,}05$; - Not Determined

Moreover, previous experiments with barley showed that the levels of the amino acids proline (a stress indicator) and arginine (precursor of the stress metabolite putrescine) increased in plants cultivated under water shortage, whereas in plants pretreated with choline and subsequently cultivated under drought stress proline and arginine contents similar to that in well watered control plants were found.[8]

3.2 Influence of Choline on the Content of Putrescine and Lignin

The presumption that choline could activate the plant stress tolerance was verified in the same plant material by determining the putrescine content. As shown in Figure 1, drought resulted in an increase of the content of putrescine in barley plants and, similar to the glycine betaine response (Table 1) a plant treatment with choline resulted in a decrease of the putrescine level. The stress-enhanced rise of lignin was also reduced after a plant pretreatment with choline (Figure 1). These changes correlated with alterations in the content of lignin precursors (e. g. ferulic acid, cinnamic acid, coumaric acid; unpublished results).

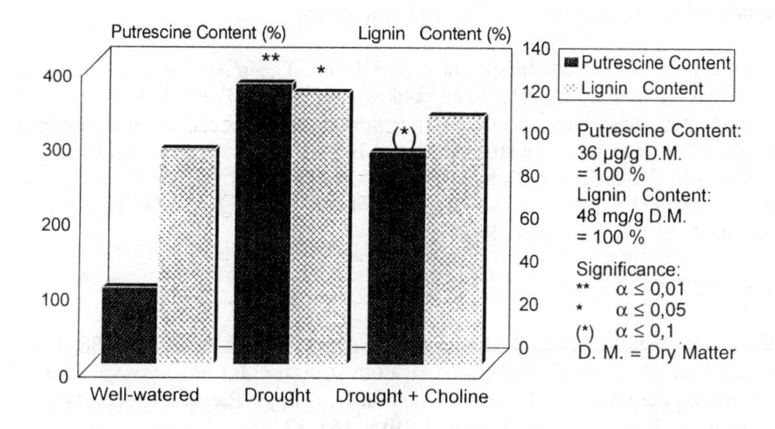

Figure 1 *The Effect of Choline on the Putrescine and Lignin content in Stressed Barley (Pot Experiments, Treatments according to 2.2. Harvesting after Anthesis)*

3.3 The Influence of Choline on the Yield of Protein and some Essential Amino Acids

In 10 pot experiments and 14 field trials the effect of choline on the yield of grains, and grain protein in barley was studied. In favourable environments choline did not influence the yield formation. Under moderate stress an application of plants with choline increased the grain and protein yield up to 10% and under stronger stress conditions the yield-improving effect was yet higher (Table 2).

The yield-stabilizing effect of choline correlated with a higher root surface of the treated plants (up to 40 %).[9]

Table 2 *Effect of Choline on Yield Components in Barley (Cultivar "Alexis") (2 Pot Experiments; Treatment according to 2.2)*

Treatment	Grain Yield (g/pot)	Protein Yield (g/pot)	Yield of Lys + Phe + Val + Met (mg /pot) 1)
Strong Drought	21,2	2,7	427
Moderate Drought	36,3 ++	4,2++	651++
Well-watered	61,8 +++	6,2+++	924+++
Strong Drought and Choline	27,3 +	3,4 +	598 +
Moderate Drought and Choline	38,8 (+)	4,5 (+)	761 (+)

(+) Significant Tendency at $\alpha \leq 0{,}1$ in Relation to "Moderate Drought"
1) Sum of Lysine, Phenylalanine, Valine and Methionine

Because choline and the choline-precursor 2-AE caused a similar improvement of the yield in stressed wheat, rye maize, sugar beet and potato plants in pot experiments as well as in field trials we assume that treatments with the phospholipid components choline and 2-aminoethanol are useful means for improving the stress tolerance in plants. In contrast to ABA the amino alcohols choline and 2-AE do not stimulate the xeromorphogensis and the formation of stress metabolites in plant cells. Therefore, we suggest, that choline and 2-AE may be useful as tools to study the mode of tolerance activation.

References

1. H. Eckert, P. Reissmann and H. Bergmann, *Biochem. Physiol. Pflanzen*, 1988, **183**, 15.
2. H. Bergmann, H. Eckert, C. Weber and D. Roth, *J. Agron & Crop Sci*, 1991, **166**, 117.
3. H. Eckert, H. Bergmann, G. Eckert and H. Müller, *J. Appl. Bot.*, 1992, **66**, 124.
4. I. Horvath and P. R. van Hasselt, *Planta*, 1985, **164**, 83.
5. I. Elmadfa and G. Leitzmann, 'Ernährung des Menschen', Ulmer Stuttgart, 1990, Chapter 3.
6. H. E. Flores and A. W. Galston, *Plant Physiol.*, 1982, **69**, 701.
7. L. M. Lagrimini, *Plant Physiol.* 1991, **96**, 578.
8. H. Bergmann, B. Machelett and V. Leinhos, *Amino Acids* 1994, **7,** 330.
9. B. Lippmann und H. Bergmann, 'Ökophysiologie des Wurzelraumes', Vorträge zur Wissenschaftlichen Arbeitstagung, Schmerwitz-Borkheide, 1994, in press.

STRESS MECHANISMS IN SUGAR BEET

N. A. Clarke, H. M. Hetschkun, and T. H. Thomas

IACR-Broom's Barn
Higham
Bury St Edmunds
Suffolk IP28 6NP

1 INTRODUCTION

Environmental factors such as light, temperature, atmospheric CO_2 concentration, water and the nutritional status of soils, affect the productivity of sugar-beet plants. Optimum conditions for sugar production will be determined by complex interactions that exist between the environment, gene expression and productivity of a particular phenotype. Productivity decreases when environmental conditions depart from this optimum; these conditions are generally regarded as stresses.

Damage by pests and diseases can also be regarded as a form of stress. Environmental and biotic stresses will have complex interacting effects on the sugar-beet crop. The development of cultivars tolerant and/or resistant to these stresses will have obvious benefits to the breeder, grower and processor. This paper describes recent research at Broom's Barn aimed at developing methods to study the stress response of *Beta* material.

2 STRESS TOLERANCE

2.1 Drought Stress

The survival strategy of sugar-beet during drought is to reduce water loss through stomatal closure. Initially the oldest leaves wilt whilst the younger leaves maintain turgidity. Net photosynthetic rate, stomatal conductance and the integrity of photosystem II (PSII) all decrease as leaves age and as stress is applied.[1] Dehydration, high temperatures and strong sunlight are all stresses to plants during drought. Stress tests have been developed that can be used in breeding programmes and have already identified useful cultivars for the analysis of stress tolerance mechanisms.

Water deficiency stress (WDS) was imposed on whole plants by withholding water, and on leaf discs by flotation on mannitol solutions. Stress tolerance of the leaf tissue was estimated from the chlorophyll fluorescence signal, which is a sensitive and early indicator of damage to PSII and the physiology of the plant in general.[2] Several components of the fluorescence induction kinetics were studied but F_v/F_p (the ratio of variable to peak fluorescence) was principally used to quantify the extent of damage. F_v/F_p is indicative of the PSII integrity and is proportional to the quantum yield of photochemistry.

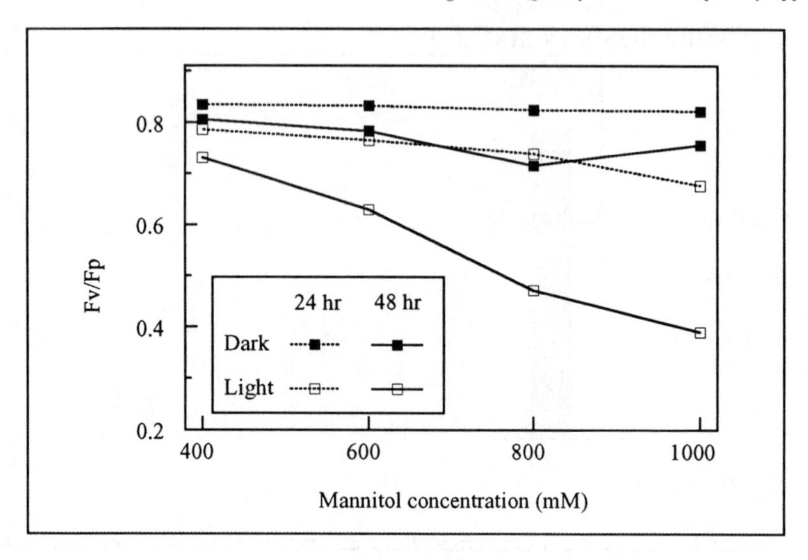

Figure 1 The effect of light and WDS upon PSII integrity.

Experiments with whole plants demonstrated the differential effect of WDS upon five cultivars selected for their extreme responses to stress in national trials.[1] The damage to PSII was least in the cultivar identified as the most stress tolerant in the national trials. Leaf-disc tests have been developed to enable more cultivars to be tested, and to avoid the problem with whole plants of differences in relative growth rate affecting the timing of the onset of stress. The temporal effect of light and increasing WDS upon leaf health is shown in Figure 1. After 48 hrs incubation in the dark the highest WDS reduced F_v/F_p by less than 10%. The addition of light exacerbated the effect of WDS upon leaf health. Thus, after 48 hrs in light the F_v/F_p was reduced by 50% at the highest WDS. Over twenty cultivars from around the world have been screened for WDS-stress tolerance using the leaf-disc test. Results indicate that reductions in F_v/F_p due to stress can vary up to threefold, providing a useful indication of their potential stress tolerance. The stress-tolerant cultivar identified in the national trials again proved to be the most tolerant.

2.2 Heat Stress

Similar experiments have been carried out to investigate the effects of heat shock on sugar-beet plants. Heat stress was imposed either on seedlings and whole plants in incubators, or on leaf discs in microcentrifuge tubes immersed in a water bath. A 4hr heat-shock given to 4d old seedlings indicated that the upper temperature limit (LT_{50}) for sugar beet is approximately 43°C. The LT_{50} was confirmed when the chlorophyll *a* fluorescence signals of heat-shocked plants were used to quantify damage. The effects of pretreating discs at elevated, but not heat-shock, temperatures were examined. A preincubation at 40°C followed by recovery at 21°C was shown to confer some tolerance to a subsequent heat-shock at 45°C (Figure 2). Twenty-five cultivars have also been screened for heat stress tolerance using a modified leaf-disc test. The most tolerant cultivar had 30% less damage than the least tolerant. Yet again the most tolerant cultivar was the same as that identified in the national trials.

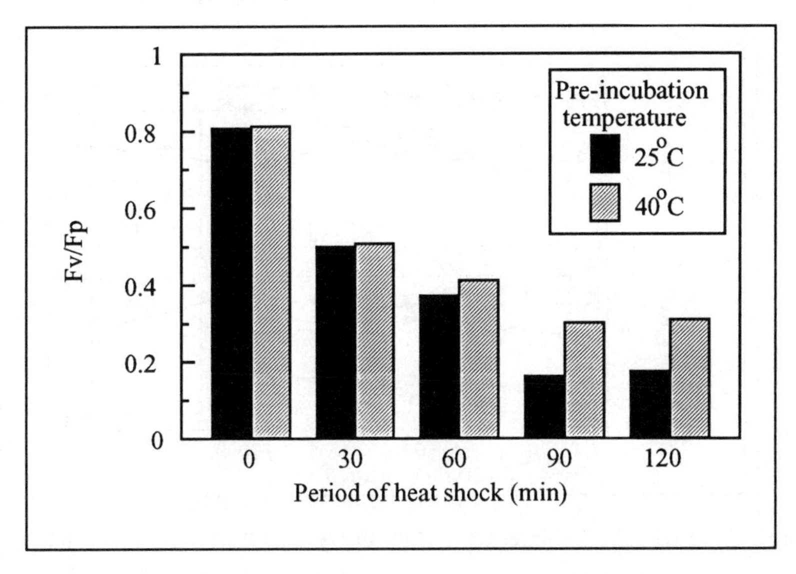

Figure 2 The effect of preincubation and heat-shock upon PSII integrity.

2.3 Virus yellows induced stress

Chloroplast destruction and virus accumulation in the phloem are symptoms of both beet yellows virus (BYV) and beet mild yellowing virus.[3] Field observations have suggested that the yellowing symptoms of virus infection are reduced when poultry manure is applied. This observation has been examined in a series of field experiments at Broom's Barn (1989-92). The results indicate that, whilst poultry manure visibly reduced yellowing, the effects of virus on yield were equally severe. Disruption in phloem transport can be used to explain this phenomenon, but analysis of chlorophyll fluorescence studies provided a further explanation related to chloroplast damage. The ratio of variable to peak fluorescence (F_v/F_p) was again used as an estimate of PSII efficiency. Two months after inoculation with BYV the PSII of the leaves was severely damaged (Figure 3a). The damage followed the classical pattern of infection developing in the leaf tip and spreading down to the base. Poultry manure had no significant effect upon PSII efficiency.

The pool size of electron acceptors on the reducing side of PSII (PSII capacity) can be estimated from the area over the chlorophyll fluorescence curve between F_o and F_p.[2] The effect of virus on PSII capacity showed a similar trend to PSII efficiency (cf. Figure 3a&b). In the virus-infected leaves there was again no significant difference between any of the nitrogen treatments. This suggests that the apparent health of infected plants grown with poultry manure was simply due to elevated chlorophyll synthesis that was not matched by an increased pool of PSII electron acceptors. This observation supports work on peanut plants, which provided evidence that plastoquinone was the probable site of photosynthetic inhibition by peanut green mosaic virus.[4]

In healthy plants, the absence of nitrogen fertiliser caused a significant reduction in PSII capacity compared with the other two fertiliser treatments (Figure 3b). The visible effect of increasing greening of the canopy with nitrogen fertilisation is therefore accompanied by an increase in PSII capacity. However, the extra nitrogen applied in the poultry manure application was not matched with an increase in PSII capacity.

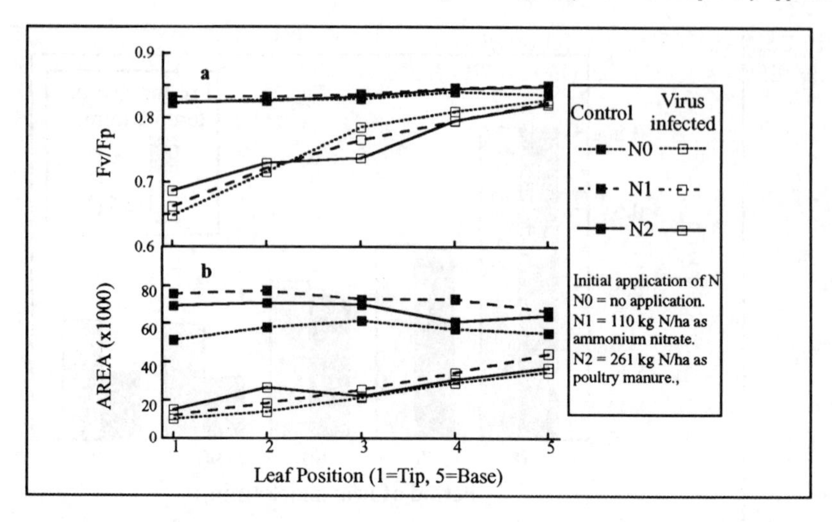

Figure 3 The effect of BYV-infection and N-supply upon PSII integrity(a) and capacity(b).

3 CONCLUSIONS

Chlorophyll fluorescence proved to be a useful technique in assessing damage to plants infected with BYV. PSII electron-acceptor pool size was a measurable symptom of infection, even in plants in which the visible symptoms were masked by excessive nitrogen application.

Sugar-beet crops will be subjected to increased stresses due to global warming. The goal of the plant breeder will be to maximise stress tolerance without reducing yield potential. Chlorophyll fluorescence should be a useful tool for estimating stress tolerance. We believe that selective breeding for high yields and quality is inadvertently reducing stress tolerance. For example, we have evidence that high yielding cultivars tend to have many small stomata whilst more stress-tolerant older cultivars have larger stomata, but at half the frequency. To break such links we need to identify more mechanisms responsible for stress tolerance. Sensible breeding strategies can then be employed in which the goals of high yield, quality and stress tolerance are obtainable.

4 REFERENCES

1. N.A. Clarke, H. Hetschkun, C. Jones, E. Boswell and H. Marfaing, 'Interacting Stresses on Plants in a Changing Climate', M.B. Jackson and C.R. Black, Springer-Verlag, Berlin. 1993, 511.

2. H.A. Bolhar-Nordenkampf, S.P. Long, N.R. Baker, Oquist G., Schreibers U., and Lechner E.G. *Func. Ecol.*, 1989, **3**,497.

3. K. Esau and L.L. Hoefert, *Virol*, 1972, **48,** 724.

4. R.A. Naidu, M. Krishnan, P. Ramanujam, A. Gnanam and M.V. Nayudu, Quoted in 'Biochemistry of Virus-Infected Plants' John Wiley and Sons, 1987, 91.

PRELIMINARY EVALUATION IN THE UK OF AN EMPIRICAL METHOD FOR PRE-HARVEST PREDICTION OF HAGBERG FALLING NUMBER OF WHEAT GRAIN

P.S. Kettlewell[1], B.J. Major[1], G.D. Lunn[2] and R.K. Scott[2]

[1] Crop and Environment Research Centre
Harper Adams Agricultural College
Newport
Shropshire TF10 8NB UK

[2] Department of Agriculture and Horticulture
University of Nottingham
Sutton Bonington Campus
Loughborough
Leicestershire LE12 5RD UK

1 INTRODUCTION

The Hagberg falling number (HFN) test is the commercial method of assessing alpha-amylase activity of wheat grain. Low Hagberg falling number (high alpha-amylase activity) is detrimental to bread loaf quality[2]. There are four possible modes of formation of alpha-amylase in wheat grain which may lead to low Hagberg falling number at harvest[3]. Three of these (retention of pericarp amylase, pre-maturity alpha amylase formation in the absence of sprouting, pre-maturity sprouting) occur before grain has dried to a typical moisture content for combine harvesting. The fourth mode of formation (sprouting after dormancy break) usually occurs when grain is dry enough for combining, but harvest is delayed.

A pre-harvest prediction of Hagberg falling number may help farmers to determine whether Hagberg is already low and this may help decisions on urgency of harvest and marketing. Sampling the standing crop and measuring falling number at regional reference sites has been used in Sweden to achieve this objective without further understanding of the mechanisms involved in alpha-amylase formation[4]. This paper reports a preliminary study in the UK of the relationship between falling number of pre-harvest samples and falling number of harvested grain of eight cultivars at one site in the UK in 1994.

2 MATERIALS AND METHODS

The eight cultivars were selected for a range of sprouting resistance and susceptibility to pre-maturity alpha-amylase formation and were Haven, Hornet, Riband, Pastiche, Recital, Thesee, Soissons and Scipion. Plots were sown in November 1993 at Harper Adams Agricultural College and managed to minimise lodging risk and weed, pest and disease incidence. The experiment was arranged in three randomised blocks with plots 1.65 m x 10 m. About 250 ears were removed from pre-determined locations in each plot at 700 °C days (base temperature 0°C) and 900 °C days after ear emergence. These thermal times corresponded with grain moisture content ranges of 35% to 42% and 14% to 18% and Zadoks Growth Stages of 83 to 85 and 87 to 92 respectively. Ears were air-dried to <15% moisture, threshed mechanically, the grain cleaned over a 2 mm sieve, milled and the Hagberg

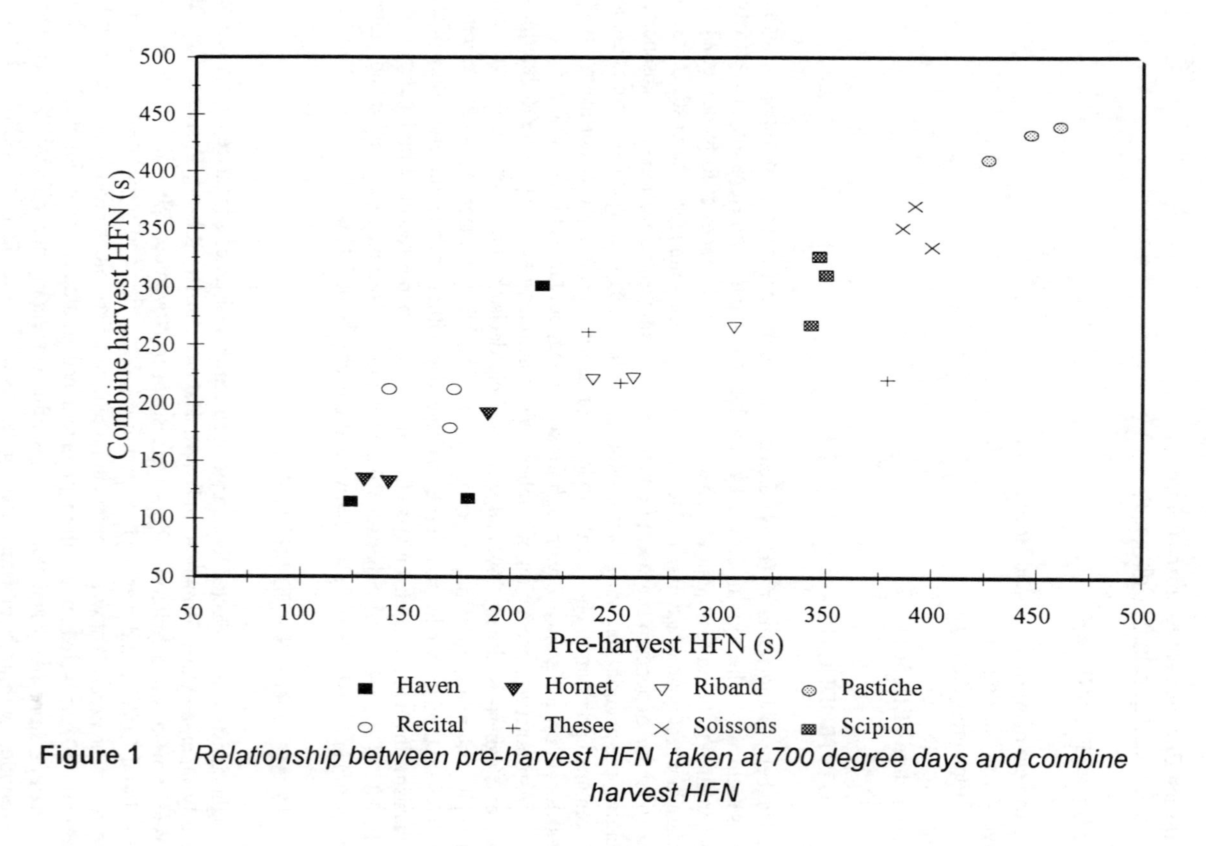

Figure 1 *Relationship between pre-harvest HFN taken at 700 degree days and combine harvest HFN*

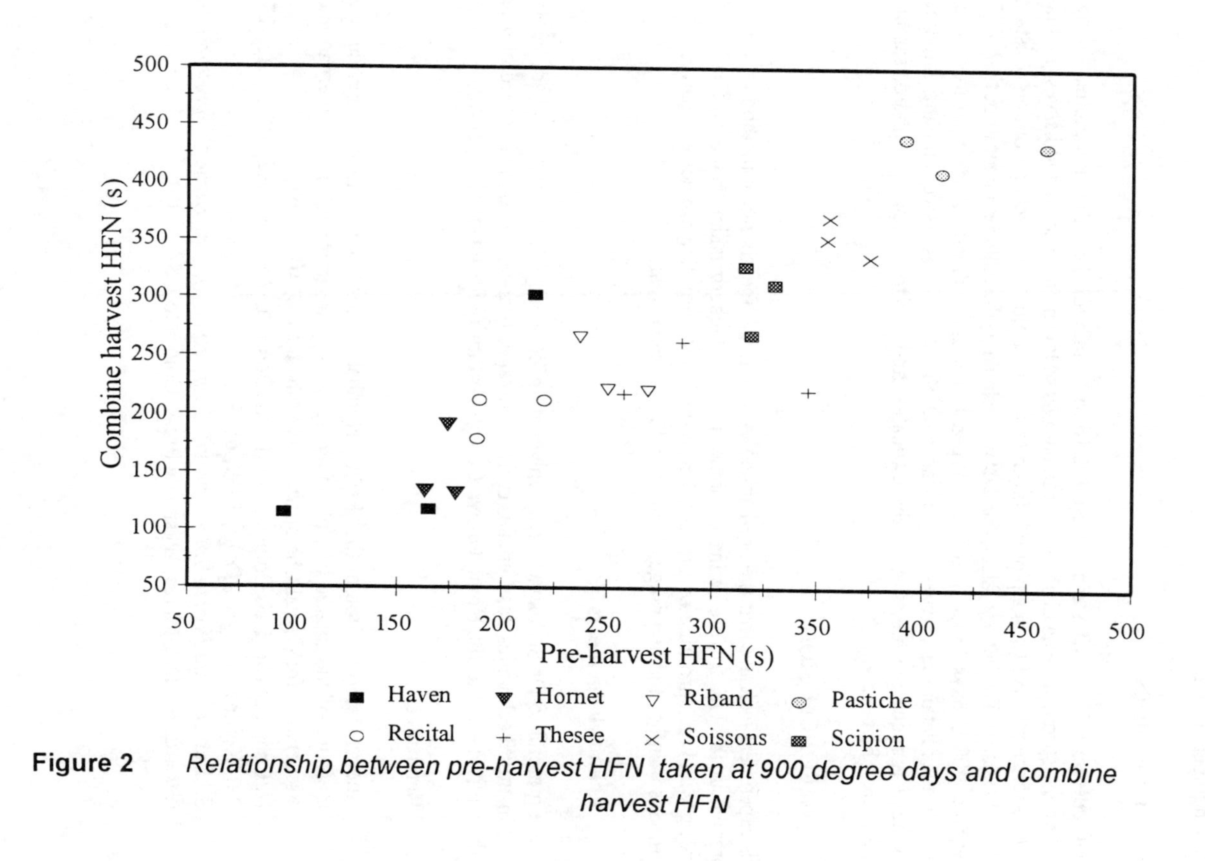

Figure 2 *Relationship between pre-harvest HFN taken at 900 degree days and combine harvest HFN*

falling number determined in duplicate according to the International Cereals Committee method[1]. The plots were combine-harvested on the same day at about 16% moisture (931 to 1094 °C days after ear emergence of the different varieties) and Hagberg falling number determined. Data was analysed by regression of harvest falling number on each pre-harvest falling number separately.

3 RESULTS

Rainfall between 700°C days and harvest was 16 to 40 mm for the different varieties. Some pre-harvest sprouting was observed in the cultivars Haven and Recital and harvest fallling numbers ranged from 115 in a plot of Haven to 439 in a plot of the very sprouting resistant cultivar Pastiche. The harvest falling number was related to the falling number at both 700°C days and 900°C days (Figs. 1 and 2). When block and cultivar effects were removed, however, the relationships were not significant ($0.05<P<0.10$). Excessive sampling variation may have reduced the precision of the estimates, and greater sample size and/or sample number may be needed.

4 CONCLUSION

This experiment indicates that the pre-harvest sampling and measurement method may have potential in the UK for predicting combine harvest Hagberg falling number of cultivars varying widely in sprouting resistance when rainfall and sprouting are not excessive. This method is being tested at a range of sites in a further three years.

Acknowledgements

We are grateful to Dr S. Abeyasekera, University of Reading UK, for statistical advice, P. Gate, Institut Techniques des Céréales et des Fourrages France, for supplying seed of four of the cultivars and to the Home-Grown Cereals Authority for financial support.

References

1. Anon. `Determination of the "Falling Number" according to Hagberg - Perten as a measure of the degree of alpha-amylase activity in grain and flour', International Association for Cereal Chemistry, ICC Standard No. 107, 1968.
2. N. Chamberlain, T.H. Collins and E.E. McDermott, *J.Fd Technol.*, 1981, **16**, 127.
3. P.S. Kettlewell. *Agric. Prog.*, 1989, **65**, 30.
4. S. Larsson, in *Fourth International Symposium on Pre-Harvest Sprouting in Cereals,* ed. D.J. Mares, Westview Press, Colorado USA, 550.

METHODS FOR PRE-HARVEST PREDICTION OF GRAIN WEIGHT IN WINTER WHEAT.

J. E. Macbeth[1], P. S. Kettlewell[1] and R. Sylvester-Bradley[2]

Crop and Environment Research Centre, Harper Adams Agricultural College, Newport, Shropshire TF10 8NB, UK[1].

ADAS Soil and Water Research Centre, Anstey Hall, Maris Lane, Trumpington, Cambridge CB2 2LF, UK[2].

1 INTRODUCTION

Established methods for predicting final grain quality and yield are based on casual observations and experience. This work investigates quantitative on-farm methods for the prediction of some aspects of final grain quality and yield in winter wheat using existing crop physiological knowledge.

During endosperm cell division there is a rapid increase in grain water content which is largely responsible for the grain's volumetric growth at this time. Termination of rapid net water deposition roughly coincides with the end of cell division. Endosperm cell number is positively related to mature grain weight[1]. Cell division may be restricted by spatial limitations[2] relating to water content at the end of cell division, a hypothesis supported by a good relationship between maximum water content and mature grain dry weight[3].

This work investigated the hypotheses that grain water content can be used to predict potential mature grain dry weight, and that discrepancies in achieved grain weights can be explained by source limitation. Naturally varying or artificially manipulated plant source : sink ratios in four experiments with winter wheat gave a range of mature grain dry weights and grain fill. Grain was sampled throughout development to investigate the above hypotheses across a wide range of conditions.

2 MATERIALS AND METHODS

The four experiments were:

Experiment 1 - In 1993 a field of Mercia winter wheat at Harper Adams Agricultural College with naturally varying levels of Gaeumannomyces graminis infection was divided into 10 blocks each of 3 plots to provide a range of mature grain weights resulting from the pathogen infection.

Experiment 2 - In 1993 reference crops of Mercia winter wheat were sown at six sites around the United Kingdom to provide climatic and edaphic variation in grain weight.

Experiment 3 - In 1994 cultivar variation in grain weight was examined using Avital, Maris Huntsman, Mercia, Riband, Rossini, Soissons and Zentos in a field experiment with four randomised blocks. One hundred marked plants per plot were designated controls, whilst one hundred had the top half of the spike removed after the end of endosperm cell division to reduce sink size.

Experiment 4 - A glasshouse experiment was carried out in spring 1994 consisting of three blocks of five treatments of water stress applied at different stages of grain development to create drought-induced variation in grain weight.

Varying numbers of spikes were removed at usually weekly intervals from anthesis to maturity, depending on the experiment. Typically six grains were removed from central spikelets for fresh and dry weight determination. Water content was calculated as the difference between fresh and dry weight. Mature grain dry weight was calculated as the mean of dry weights after the end of dry matter deposition, determined graphically. Shrivelling, defined as pinching or hollowing on the ventral surface of the grain[4], was assessed on a percentage basis using grains from the last four sampling dates to give a crude measure of source limitation.

3 RESULTS AND DISCUSSION.

There was a wide range of mature grain dry weights and shrivelling percentages across and within the experiments. Grain dry matter and water content during development is shown for Riband in Experiment 3 (Figure 1). Different genotypes and experiments showed similar relationships, with an approximate plateau existing for maximum water content from 14-24 days post-anthesis depending upon conditions until grain maturation. Water content at 4 weeks post-anthesis was used as an estimate of maximum water content to keep sampling to a minimum for future on-farm use.

There was a good positive relationship between water content at 4 weeks post-anthesis and mature grain dry weight for all data (Figure 2). Both treatment and control plants in Experiment 3 appeared sink-limited (very little grain shrivelling) so data from this experiment were used to calculate a linear regression line describing the relationship between water content at 4 weeks post-anthesis and potential mature grain dry weight. The equation produced was

$$y = 0.74x + 22.20$$

where y is potential mature grain dry weight and x is water content at 4 weeks post-anthesis. The r^2 value was 0.80. Calculation of potential mature grain dry weight from this equation in the other experiments gave over-estimation of achieved grain weight except for at two sites in Experiment 2 (Figure 3). This was expected as with the exception of these two sites the three experiments all had considerable levels of grain shrivelling. There was a positive correlation with the differences between predicted and observed values and shrivelling percentages for each trial (Figure 4), although the relationship varied between trials. This shows that over-prediction can be partly explained by poor grain fill, but that percentage shrivelled grain may not be a good estimate of source limitation as it does not take into account the degree of shrivelling.

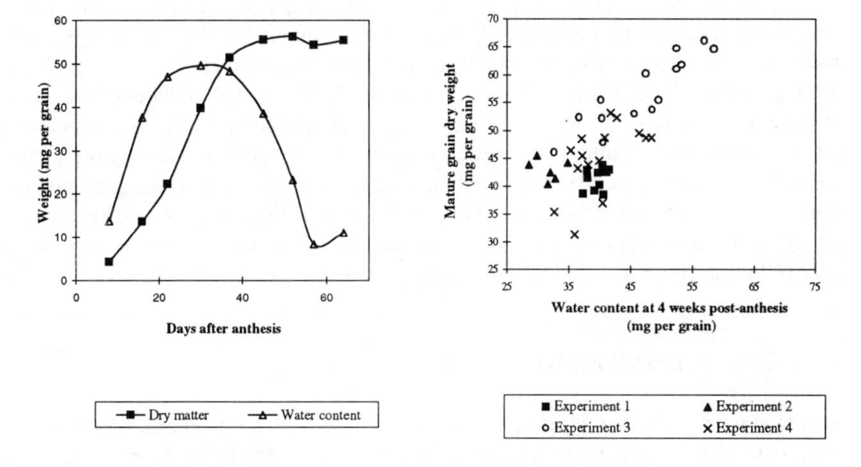

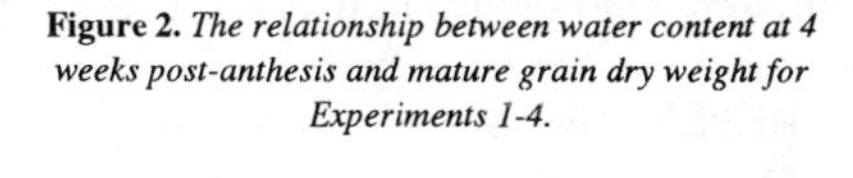

Figure 1. *Dry matter and water content in extracted grains from control spikes of Riband , Experiment 3, from anthesis to harvest.*

Figure 2. *The relationship between water content at 4 weeks post-anthesis and mature grain dry weight for Experiments 1-4.*

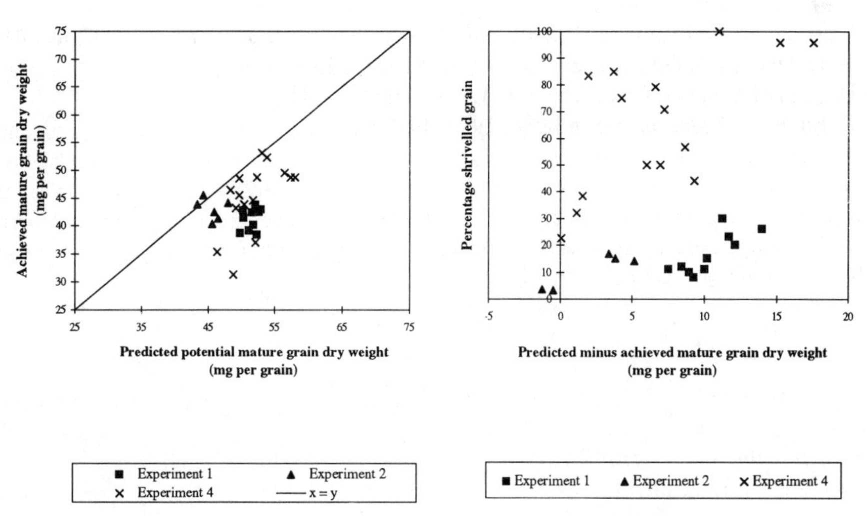

Figure 3. *Predicted potential mature grain dry weight using the equation derived from Experiment 3 against achieved mature grain dry weight for Experiments 1, 2 and 4.*

Figure 4. *The relationship between predicted minus achieved mature grain dry weight and percentage shrivelled grain, using the equation derived from Experiment 3 to predict mature grain weight.*

4 CONCLUSIONS

Inaccuracy and lack of precision in predictions of mature grain dry weight from grain water content at 4 weeks post-anthesis can be partly explained by grain shrivelling, although further investigations are required to clarify this. Potential yield predictions could be made if the remaining yield components of shoots per metre squared and grain number per ear were counted. Initial investigations show that potential grain weight per spike can be predicted from spike water content at 4 weeks post-anthesis, making counts of grain number per spike unnecessary for yield prediction. Calculation of source : sink ratios during grain development may allow predictions of grain shrivelling and deficits in achieved grain weights and yield. This is currently being investigated.

5 ACKNOWLEDGEMENTS

We are grateful for data collected by the Agricultural Development and Advisory Service, University of Nottingham and University of Edinburgh in the H-GCA project number 0044/01/91, and to the H-GCA for their support.

6 REFERENCES

1. I. Sofield, L. T. Evans, M.G. Cook and I.F. Wardlaw, *Aust. J. Pl. Phys.*, 1977, **4**, 799.
2. A. G. Huber and D.F. Grabe, *Crop Sci.*, 1987, **27**, 1252.
3. H. Schnyder and U. Baum, *Euro. J. Agron.*, 1992, **1**, 51.
4. R. Bayles, *J.Natn. Inst. Agric. Bot.*, 1977, **14**, 232.

EFFECTS OF CHANGING NITROGEN, SULPHUR AND POTASSIUM STATUS ON QUALITY CHARACTERISTICS IN SUGAR BEET

J. J. Sexton, J. Bennett, S. B. Reynolds and T. J. Hocking

School of Applied Sciences
University of Wolverhampton
Wolverhampton WV1 1SB

1 INTRODUCTION

There is increasing evidence of sulphur deficiency in agricultural crops in the U.K., partly as a consequence of falling levels of sulphur deposition onto the land.[1] As a macronutrient essential for plant growth, sulphur is required for the synthesis of key amino acids which in turn are needed to produce both structural and functional proteins, including for example, Ribulose bisphosphate carboxylase (Rubisco). In sulphur deficient plants the level of Rubisco falls with a corresponding reduction in photosynthetic activity.[2] Because of this role in the biosynthesis of plant proteins, a close relationship exists between the requirements for sulphur and nitrogen in crop plants.[3,4] Changes in the relative levels of nitrogen and sulphur, as measured by the N/S ratio, can therefore have serious consequences for both crop yield and quality. In sugar beet, when the N/S ratio is increased through high applications of nitrogen or as a result of sulphur deficiency, a higher proportion of assimilated nitrogen may be sequestered into storage pools, in the form of free amino acids and amides (i.e. as alpha amino nitrogen). These compounds are important impurities in sugar beet, adversely affecting the sugar extraction process and reducing the return to the grower.[5]

Earlier studies at Wolverhampton have explored the role that sulphur may play in the growth and quality characteristics of sugar beet.[6] In a field experiment where sugar beet was grown at increasing levels of nitrogen, foliar applications of sulphur reduced the levels of alpha amino nitrogen in the roots at harvest (Table 1, see also ref. 6). At the highest nitrogen level, sulphur applied as late as August reduced the level of these impurities by 35%. Subsequently, hydroponic culture techniques have been used to study the changes in nitrogen metabolism, particularly the synthesis of free amino acids, that occur as the sulphur status is changed. These experiments have been extended to include potassium, a nutrient which, when accumulating in the root, may also affect the sugar extraction process.

2 MATERIALS AND METHODS

Sugar beet plants cv. Primahill were grown in perlite in a heated glasshouse with supplementary lighting. The perlite was watered with a culture solution based on the Long Ashton Formula[7] in which the sulphur (S) concentration was varied by replacing magnesium sulphate with magnesium chloride. A concentration of 1.5 mM S was estimated to provide an adequate supply (described here as normal) of sulphur under these conditions. At intervals, typically six plants per treatment were taken randomly and harvested for measurement of growth and biochemical analysis. Plants were separated into roots and shoots and samples either chopped and frozen prior to amino acid analysis or oven dried below 70° C and milled for sulphur determination. For free amino acids analysis the frozen material was thawed, homogenised, extracted in ethanol and dried extracts derivatised with phenylisothiocyanate prior to separation and identification using a Waters HPLC system as described by Marshall.[8] For the determination of sulphur content, the milled material was analysed by X-ray fluorescence spectroscopy.[9] Chlorophyll content of foliar material was determined spectroscopically following acetone extraction.

3 RESULTS AND DISCUSSION

Measurements from plants harvested 93 days after the start of the sulphur treatments are shown in Table 2. There was a small effect on shoot and root fresh weight as the supply of sulphur was reduced to 1/10th of normal, but the plants adopted a more upright growth habit with proportionately longer petioles and a marked reduction in chlorophyll content. Below 1/10th normal there was a dramatic reduction in growth leading to very stunted plants. There was some evidence of an optimum growth response at 1.5mM S (Table 2a).

Analysis of leaf tissue sulphur revealed a progressive reduction in sulphur content which did not reflect the physiological response (Table 2 b), suggesting a more complex relationship between sulphur content and growth. As the sulphur status was lowered to a critical level, equivalent in these experiments to a supplied concentration of 0.15 mM S, other events were triggered. There was a very marked increase in arginine levels, up to 150 fold, while total free amino acids increased two to four times. The free arginine in the tissue rose from around 2 to over 50 per cent of the total. This change was not progressive and suggests a switch in nitrogen metabolism initiated either by a critical level of sulphur in the tissue, or by a specific change in the N/S ratio, or both.

Table 1 *Effect of foliar sulphur application (10 kg/ha thiovit) on the levels of alpha amino nitrogen at harvest in roots of sugar beet cv. Primahill grown at different levels of nitrogen, applied as a soil dressing. Values are means (n = 10).*

Nitrogen Treatment (kg/ha)	Alpha Amino Nitrogen (mg/100g sugar)		
	Timing of Sulphur Application		
	Nil	June	August
0	67.4	72.2	82.7
87	73.7	89.1	104.9
125	191.1	160.8	99.7
162	176.0	135.9	115.9

Table 2 *Effects of sulphur (S) concentration on (a) shoot fresh weight, root fresh weight and chlorophyll content and (b) sulphur content, arginine content and total free amino acid content of leaves of sugar beet cv. Primahill measured 93 days after start of treatment. Values are means ± S.E. (n = 6).*

(a)

S Supplied (mM)	Shoot f.wt (g)	Root f.wt (g)	Chlorophyll Contents (μg/g)
0	8.5 ± 2.6	3.8 ± 1.3	255 ± 39
0.15	80.6 ± 9.3	78.0 ± 7.4	944 ± 28
1.5	85.1 ± 3.4	98.0 ± 8.7	1397 ± 51
3.0	69.9 ± 6.2	85.3 ± 10.3	1693 ± 69

(b)

S Supplied (mM)	S Content (% w/w)	Total a.a. (μ mol/g.f.wt)	Arginine (μ mol/g.f.wt)	% Arginine
0	0.11	8.31	4.52	54.4
0.15	0.17	4.85	2.49	51.5
1.5	0.36	2.13	0.04	2.0
3.0	0.47	2.11	0.08	3.8

Table 3 *Arginine content in leaves of sugar beet cv. Prima Hill grown in (A) complete culture medium, (B) S deficient culture medium, (C) S deficient culture medium for 72 days followed by progressive transfer to complete culture medium.*

Treatment	Arginine Content (μ mol/g. f.wt) on different days			
	Day 65	Day 72	Day 79	Day 107
A. Complete	0.53	0.45	0.26	0.32
B Deficient	16.04	17.65	21.51	25.42
C Deficient/ Complete	14.20	25.24	4.82	0.01

Table 4 *Effect of potassium concentration on shoot fresh weight and arginine levels in leaves of hydroponically-grown sugar beet cv. Primahill, 75 days after start of treatment.*

	Potassium Concentration (mM)						
	0.034	0.034	1.68	3.37	6.74	13.4	20.2
Shoot f.wt (g)	2.0	27.1	63.4	59.0	60.0	55.1	88.0
Arginine (μmol/g.f.wt)	1.61	0.20	0.05	0.00	0.00	0.00	0.00

This switch is reversible; in a separate hydroponic experiment, plants maintained without any supplied sulphur showed a characteristic arginine response which was reversed when the plants were returned progressively to a complete medium after 72 days (Table 3). In some cases the recovered plants caught up with the plants grown continuously in the complete culture medium.

In an initial experiment where the concentration of potassium supplied in the culture medium was varied over three orders of magnitude, a marked increase in arginine levels was also observed. In this case, however, the major change did not occur until the potassium was lowered to 1/100th of normal, i.e. from 3.37 mM to 0.034 mM. By this stage the plants were severely stunted (Table 4). One possible explanation for this response is that low potassium uptake led to high tissue concentrations of sodium by way of compensation. There is some evidence (D A Fincham, pers. comm.) that high tissue sodium could inhibit the activity of arginase, leading to a rapid accumulation of arginine.

These biochemical events are currently under further investigation in our laboratory. The implications for sugar beet quality, however, are already becoming clear. Sulphur deficiency either directly or via an increase in N/S ratio can lead to large increases in free amino acids contributing to total alpha amino nitrogen impurities before there is a marked effect on growth. These changes reduce the quality of the beet. There may be a threshold level of sulphur in the tissue which leads to a rapid deterioration in quality. Changes in potassium nutrition may contribute to these effects. High potassium may adversely affect quality characteristics directly; low potassium may also have detrimental effects via changes in amino acid metabolism.

Experiments currently in progress indicate that interactions affecting quality may occur between sulphur and potassium levels in the tissue. Further research is necessary to elucidate these relationships, to predict the likely occurrence of nutrient deficiencies in the field and evaluate fully their effects on sugar beet quality.

References

1. I. R. Richards, *Sulphur in Agriculture,* 1990, **14**, 8.
2. S. P. McGrath and A. E. Johnston, *Span,* 1986, **29**, 57.
3. R. G. Coleman, *Soil Sci.,* 1966, **101**, 230.
4. W. Dijkshoorn and A. L. Van Wijk, *Plant and Soil*, 1967, **26**, 129.
5. M. Shore, J. Dutton, B. Houghton and G. Bowler, *British Sugar Beet Review,* 1982, **50**, 54.
6. A. D. E. Martin, PhD Thesis, Wolverhampton Polytechnic, 1990.
7. J. Roberts and D. G. Whitehouse, 'Practical Plant Physiology', Longman, London, 1976.
8. A. Marshall, MPhil Thesis, Wolverhampton Polytechnic, 1991.
9. S. B. Reynolds, A. D. E. Martin, B. Bucknall and B. J. Chambers, *J. Sci. Food Agric.*, 1989, **47**, 327.

PHYSIOLOGICAL PECULIARITIES OF NITROGEN NUTRITION OF PEA PLANTS IN SYMBIOSIS WITH NODULE BACTERIA

E. D. Krugova and N. M. Mandrovskaya

Department of Symbiotic Nitrogen Fixation
Institute of Plant Physiology and Genetics
National Academy of Sciences of Ukraine
Kiev 252127 Ukraine

1 INTRODUCTION

Reduction of protein deficit is one of the major and complex tasks of world agricultural science. The amount of protein synthesized by plants is limited by the quantity of nitrogen available for plants. Leguminous crops are the exception. They form protein assimilating nitrogen from the air at the expense of symbiosis with nodule bacteria. Mineral nitrogen compounds occupy special place in symbiotic relations of leguminous with nodule bacteria.[1,2] It has been shown that leguminous plants even under the most favourable conditions of growing could satisfy their need in nitrogen by 70 to 80% at the expense of symbiotic nitrogen fixation, while under unfavourable conditions this value could fall up to 45%. Leguminous absorb deficit nitrogen from the soil or fertilizers due to the presence of a two-enzyme system of nitrogen reduction - nitrogenase and nitratereductase.[3] It has been found that the function of nitrogenase depends mainly on nitrogen concentration. Taking that fact into consideration, the efficiency of leguminous and rhizobia symbiosis is defined by the peculiarities in nitrogen nutrition of partners within symbiosis. The investigations of the peculiarities in symbiotic functioning under conditions of various mineral and biological supply have been carried out. The application of strains of pea nodule bacteria[4] contrast in their activity, the use of a wide spectrum of pea cultivars of variuos morphotypes[5] as well as rhizobia mutants resistant to increased mineral nitrogen concentrations contributed to this aim. In model experiments the conditions of various degrees of supply by bound nitrogen have been made. Its concentrations in nutrition mixtures were measured. The results of above experiments are presented in this report.

2 MATERIALS AND METHODS

Pea plants (Pisum sativum cv Smaragd) were grown in vegetative experiments (sand culture). The methods of preparation of seeds to sowing and application of nutrient mixtures are described[6] To estimate the intake of mineral nitrogen within the plants the method of isotope indication has been used with the isotope of ^{15}N being applied. Nitrogen (NH_4NO_3) was applied during filling in pots. Experimental variants: I - control (without nitrogen), II - 20 mg, III - 40 mg, IV - 120 mg of mineral nitrogen per kg of substrate. Nitrogen isotope ^{15}N.($NH_4 NO_3$) was abundant with salt by 95.2 at.% ^{15}N. Nitrogen fixating activity, total nitrogen and nitrates were defined by acetylene reduction[7], Kjeldahl method in modification and potentiometer method correspondingly. The share of atmospheric nitrogen was estimated by the formula of isotope dilution[8].

3 RESULTS AND DISCUSSION

Purpose of the paper is to find out peculiarities in the influence of mineral nitrogen on nitrogen fixation, uptake and distribution of biological and mineral nitrogen within plant tissues of pea, grown under conditions of symbiosis of various efficiency. Pea plants had well developed root nodules with high specific activity of nitrogenase (Table 1).

Table 1 Development of Pea Nodules, their Nitrogen Fixating Activity and Nitrogen Content

	Variant	Nitrogenase activity	Nodule	N total
		μ mol C_2H_4 g/nodule h^{-1}	g/plant	% (dr wt)
	I	51.8±2.70	0.969±0.03	4.93±0.06
Flowering	II	44.2±1.60	0.601±0.02	5.08±0.15
	III	30.3±1.45	0.498±0.01	4.35±0.03
Prime condition of beans	I	27.2±1.36	0.812±0.02	4.23±0.03
	II	23.1±0.70	0.867±0.03	4.38±0.06
	III	20.3±1.84	0.901±0.02	4.08±0.08

A small dose of mineral nitrogen (20 mg) did not retard nodule development or inhibit their nitrogenase activity as well. This could be observed within the plants of variant III in the initial period of plant vegetation. However, as nitrogen was being utilized from substrate after plant flowering, nodule mass of these plants has even elevated this index in comparison with control one. The plants of variant IV had no nodules during the whole period of plant development. The nature of nitrogen intake and distribution within plant tissues essentially differed depending on the dose of mineral nitrogen in substrate. In general nitrogen intake in superground mass, roots and beans of plants, grown on high nitrogen dose (variantIV) was more intensive than that of control variant (without mineral nitrogen). The excess of nitrogen absorbed by plants was located in roots and stems as nitrates and it was not utilized for yield development. So, the content of total nitrogen and nitrates in grain within plants of all experimental variants slightly differed (Table 2).

Table 2 Indices of the Efficiency of Symbiotic Nitrogen Fixation and Nitrate Content in Pea Yield

Variant	Superground mass	Beans	N content			NO_3^- content		Protein
			grain	straw	root	grain	straw	grain
	g/pot	Number per/pot	% (dr wt)			mg/kg (fr wt)		%
I	23.09	15.75	3.28	0.61	1.66	251	676	20.50
II	23.10	15.17	3.27	0.66	1.60	249	701	20.44
III	23.95	19.75	3.34	0.81	1.65	256	741	20.88
IV	26.69	18.67	3.41	1.05	2.84	258	2692	21.31

Nitrogen, absorbed from mineral salts and biologically assimilated, was distributed within organs unevenly. According to the degree of abundance by nitrogen isotope ^{15}N, pea organs were arranged in the following decreasing way: roots, superground mass (leaves, stems), green beans, grain, nodules. In nodules symbiotically assimilated nitrogen made up 92 to 98%. Nitrogen, accumulated in pea grain, was also mainly biologically absorbed and amounted to 91 to 99%. Thus, biologically assimilated nitrogen was above all absorbed by root nodules and then it was concentrated in reproductive organs (beans, grain). Nitrogen of mineral compounds was accumulated in vegetative mass (roots and superground mass). The analysis of structural elements in yield has shown that mixed type of nitrogen nutrition, which comprises both biological amd mineral nitrogen, is optimum for pea growing. The application of high doses of mineral nitrogen physiologically seems to be not advantageous since nitrogen retards the growth of root nodules, as it inhibits symbiotic nitrogen fixation or causes disruption of normal nitrate assimilation. Hence, the knowledge of biochemical and physiological processes when molecular and mineral nitrogen reducing, could allow increasing the efficiency of symbiotic nitrogen fixation using genetical potential of legumes. This could make an important contribution to the solution of the problem of protein deficiency.

References

1. P.P.Vavilov, G.S.Posypanov, Leguminous Plants and the Problem of Plant Protein, Rosselkhozizdat, Moscow, 1983, 256.

2. J.H.Silsbury, D.W.Catchpoole, W.Wallace, Austr. J. Plant Physiol., 1986, 13, 257.

3. S.F.Izmaylov, "Nitrogen Metabolism in Plants", Nauka, Moscow, 1986, 320.

4. E.D.Krugova, Physiology and Biochemistry of Cultivated Plants, 1995, 27, 3, 174.

5. E.D.Krugova, D.D.Ostapenko, N.M.Mandrovskaya, Physiology and Biochemistry of Cultivated Plants, 1994, 26, 3, 245.

6. E.D.Krugova, A.S.Tsymbal, O.N.Krymskaya, Physiology and Biochemistry of Cultivated Plants, 1994, 26, 3, 234.

7. R.W.F.Hardy, R.D.Holsten, E.K.Jackson, R.C.Burns, Plant Physiology, 1968, 43, 1185.

8. R.J.Rennie, Canad. J. of Botany, 1982, 60, 6, 856.

JOINT EFFECT OF Ca^{++} AND ETHEPHON ON FORMATION OF CAPSAICINOIDS IN FRUITS OF HOT PEPPER CAPSICUM ANNUUM L.

I. Perucka

Department of General Chemistry
The Agricultural University
Akademicka 15, 20-950 Lublin, Poland

1 INTRODUCTION

Fruits of hot pepper are valuable for pharmaceutical and food industry as they contain specific alkaloids - capsaicinoids which give the fruits their spicy taste, have antiinflammatory effect, and, according to recent research, influence lipid metabolism in high-fat and carbohydrate diet.[1] Red fruits are also a source of natural pigments - carotenoids, ever more often used for colouring food products. In adverse weather conditions hot varieties of pepper may give poor yield of ripe fruits. Hence it seems necessary to apply substances accelerating fruit ripening in field crops.

Ethephon is one of the most frequently used preparations; when applied before harvesting it makes ripening of the fruits uniform.[2] Research carried on recently has shown that prior application of ethephon to hot pepper seedlings not only increases the yield of ripe fruits but also influences growth of the plants.[3] Inhibition of the main shoot and stimulation of lateral stems was noted which gave the plants a bushy shape, resistant to lodging.[3]

2 MATERIALS AND METHODS

The experiments were conducted with the hot pepper variety cv. Bronowicka ostra. When the plants were at the stage of 4-5 leaves they were treated with water solution of Flordimex containing 42% of ethephon. The preparation was applied to leaves in the form of spray in concentrations of 0, 150, 300 and 450 mg of ethephon/l. Ethephon in the concentration of 300 ml/l was sprayed together with 0.1M $CaCl_2$. 10 cm^3 of each solution was applied to 20 plants. The effect of ethephon and Ca^{++} was studied in the fruits at the stage of full ripeness, harvested at three different times: the third decade of August, the second half of September and the first half of October. The mass of fully coloured fruits of control plants and the ones treated with ethephon was determined; the capsaicinoids content was studied. Capsaicinoids were extracted from fresh fruits with acetone-petroleum ether (1:1) until the tissues were completely decoloured. After washing the acetone with water the extract was dried with Na_2SO_4 and evaporated at 35°C. The capsaicinoids content was determined by the modified Jentzsch method[4] after dividing them by the TLC method on plates covered with silica gel(Merck) in composition benzene:ethyl acetate:methanol (75:20:5).

3 RESULTS

The effects of treating hot pepper seedlings with ethephon and $CaCl_2$ on ripening of the fruits are presented in Table 1. The plants sprayed with ethephon had the amount of red fruits increased by 20-30% as compared to controls.

The preparation in the lowest concentrations caused an increase in the mass of ripe fruits as compared to controls mainly in the second harvest period (45% and 20% respectively). The highest ethephon concentration caused an insignificant decrease in the red fruits mass in the first harvest, and in the next ones a 30% increase in the yield as compared to controls. Joint application of ethephon in the concentration 0.2% and of 0.1M $CaCl_2$ had the same final effect as application of the highest concentration. At the same time accelerated fruit maturation was noted. The mass of fruits from the first harvest from plants treated with ethephon and $CaCl_2$ was higher by over 30% than from controls.

Studies on capsaicinoids content in ripe fruits of hot pepper did not reveal a negative effect of ethephon on capsaicinoids formation (Table 2). Only fruits of ethephon-treated plants from the first harvest were characterized by a lower capsaicinoids level than controls. The biggest difference (about 10%) was found after application of the highest concentration of ethephon. A higher alkaloid content was found in dry mass of ripe fruits from ethephon-treated plants calculated per one plant (Table 3). Depending on concentration, the amount of capsaicinoids obtained was by 24% to 35% bigger than in controls.

4 DISCUSSION

In the study it was shown that ethephon applied to hot pepper together with $CaCl_2$ modified the processes connected with fruit maturation. This resulted in an increase of mass of red fruits in all the harvests by 30% as compared to control.

It is supposed that addition of $CaCl_2$ increased the effectiveness of ethephon, as the same increase in mass of ripe fruits was found after the highest ethephon concentration (Table 1). However, with the dose of 450mg/l of ethephon a slight decrease was found in the early crop of fruits which resulted from abscission of fruits and leaves caused by applying big doses of ethephon - 5.000-15.000 ml/l - to hot pepper plants two weeks before harvest.[5]

Another cause of acceleration of fruit maturation could be the increased effectiveness of induction of flowering which was noted by Dass et al. in joint application of $CaCO_3$ and ethephon to pineapples.[6] Addition of $CaCO_3$ probably increased liberation of ethylene from ethephon at a higher pH which gave it easier access to the plant tissues.[7]

The effect of ethephon on formation of capsaicinoids in ripe fruits of hot pepper from the first harvest resulted in a decrease in capsaicinoids content (by less than 10%) as compared to controls when higher concentrations were applied (Table 2). This may be connected with inhibition of the activity of phenylalanine ammonia lyase (PAL) - one of the main enzymes taking part in metabolic changes occurring during maturation and in formation of capsaicinoids.[8]

inhibition of PAL activity caused by ethephon occurred at the final stage of fruit growth and at the beginning of maturation. However, in ripe fruits no significant changes in activity of this enzyme were noted.[8,9] Joint application of ethephon and $CaCl_2$ significantly increased capsaicinoids content in dry mass of ripe fruits harvested from one plant. The biggest amount of capsaicinoids as compared to controls was found in fruits from the second harvest. In conclusion it can be said that application of ethephon to hot pepper seedlings together with $CaCl_2$ can be used in raising pepper in the field as a factor accelerating maturation which increases the mass of red fruits with a higher content of alkaloids typical of pepper.

Table 1 *Joint effect of Ca^{++} and ethephon on mass of red fruits of hot pepper. Uncommonsuperscripts indicate significance ($P<0.05$)*

Treatment mg/*l*	*Harvest term*			*Total*
	I	II	III	
Control	31.3a(100)	111.8a(100)	143.8a(100)	286.9a(100)
Ethephon				
150	33.2a(106)	162.4c(145)	150.1a(104)	345.7b(120)
300	34.7a(111)	134.2b(120)	153.9a(107)	322.8b(112)
450	28.7a(91)	146.3b(131)	193.9b(135)	375.0c(130)
Ethephon+				
$CaCl_2$	40.9b(131)	164.4c(147)	175.0b(122)	380.3c(132)

Table 2 *Capsaicinoids content in hot pepper fruits after treating seedlings with ethephon and Ca^{++}* mg/g d.m.*). Legend as in Table 1.*

Treatment	*Harvest terms*		
mg/l	*I*	*II*	*III*
Control	886.3b(100)	850.6a(100)	726.0a(100)
Ethephon			
150	877.4b (99)	833.6a (98)	704.0a (96)
300	846.1a (95)	856.0a(101)	709.9a (98)
450	812.2a (92)	853.1a(100)	695.7a (96)
Ethephon+ $CaCl_2$	839.0a (95)	842.8a (99)	711.0a (98)

Table 3 *Effect of ethephon and Ca^{++} on capsaicinoids content in red fruits of hot pepper (mg/1 plant d.m.) (Legend as in Table 1)*

	Harvest terms			Total
Treatment mg/l	*I*	*II*	*III*	
Control	3.71b(100)	12.65a(100)	11.98a(100)	27.68a(100)
Ethephon				
150	3.86b(104)	17.60c(139)	12.84a(107)	34.30b(124)
300	3.96b(107)	15.01b(118)	13.44a(112)	32.41b(117)
450	2.33a(100)	16.31b(129)	17.28b(144)	35.92b(135)
Ethephon $+CaCl_2$	4.59c(124)	18.06c(143)	15.35b(128)	38.00c(137)

References

1. U. S. Govindarajan and M. N. Sathyanarayana, *Food Sci. Nutr.* 1991, **29**, 435.
2. A. M. Armitage, *Hort. Science,* 1989, 24, 962
3. I. Perucka, *Pestycydy,* 1994, 2, 1.
4. I. Perucka, *Folia Soc. Sci. Lub.,* 1990, **32**, 13.
5. R. S. Conrad and F. J. Sundstrom, *J. Amer. Soc. Hort. Sci.,* 1987, **112**, 424.
6. H. C. Dass, G. S. Randhava, S. P. Negi, *Scientia Hort.,* 1975, **3**, 231
7. H. L. Warner and A. C. Leopold, *Plant Physiol.,* 1969, **44**, 156.
8. I. Perucka *Proc. IV Conf. EURO FOOD TOX.,* 1994, 2, 448
9. I. Perucka, *Acta Physiol. Plant.* 1995 (in press).

THE INFLUENCE OF STEROID GLYCOSIDES AND GLYCOALKALOIDS ON THE BIOCHEMICAL COMPOSITION OF TOMATO FRUITS

V.A.Bobeica, P.K.Kintia, and G.A.Lupashku

Institute of Genetics
Academy of Sciences
Kishinev 277002
Moldova

Some naturally-occurring compounds of the saponin class isolated from different plants, when used exogenically, have a stimulating effect on seed germination, root formation, growth and development of plants, like phytohormones.[1]

With the aim of more detailed and profound investigation of these properties, we have studied saponins belonging to the group of C-27 steroid glycosides, including spirostanol, spirosolan and furostanol glycosides. The substances were tested on wheat, rice, barley, maize, pepper, tomato, beet and other cultivated plants. Except the above effects, we paid a special attention to the enhancement of plant resistance to phytopathogens, biochemical changes, especially in plant parts used for food. Here are some data on the study of tomato.

Seed presowing treatment with steroid glycosides proved to be the most convenient and effective technique to stimulate seed germination and further growth and development of plants. Phytopathological studies showed that due to seed treatment with steroid glycosides well-developed tomato plants possessed an enhanced resistance to a number of diseases. Thus, the treatment of tomato seeds of the variety Fakel, widely cultivated in the Moldova Republic, with the solutions of a furostanol glycoside, tomatoside, isolated from tomato seeds[2] in the concentration of 10^{-4} - 10^{-3} M during 20-24 hrs prior to their sowing, led to decreased infection rate of adult plants with macrosporiosis (agents Macrosporium solani Ell. et Mart. and Alternaria solani Sor.) by 30-60%, bacterial black spot (agent Xanthomonas vesicatoria Dowson) by 50-60%, wilt diseases (agents Fusarium oxysporum f. Luc. and Verticillium albo-altrum R.) by 40-50%, mosaic disease (agent Tobacco mosaic virus) by 50-60% as compared with the control (plants grown from the seeds treated with distilled water). During the three years when the experiment was repeated, the yield has exceeded the control by 11-37%.

The similar effect has been received when another technique of steroid glycoside application was used, i.e. plant treatment at early ontogenetic stage (2-4 true leaves). Plant spraying with the solutions of a furostanol glycoside, tomatoside, and spirostanol one, α-tomatine[3] in the concentration of 10^{-5} - 10^{-4} M inhibited the development of a number of diseases by approximately 30% and 45%, respectively, comparing with the control (water spraying). In both cases the yield from the plot treated exceeded the control by above 20%.

Table 1. The influence of tomato spraying with steroid glycoside solution (0.005%) on disease development and yield capacity (cv. Ranniy-83, compound consumption 25 g/ha, four sprayings by the fruit formation initiation)

Treatment	Disease development, %			Yield capacity, C/ha
	bacterial black spot	Macrosporiosis	Septoriosis	
Control (water spraying)	31.3	32.5	51.3	571.0
Tomatoside	20.0	20.0	37.5	695.0
α-tomatin	17.5	17.5	42.5	687.0

Accelerated development and increased biomass accumulation in plants grown from steroid glycoside treated seeds, their enhanced resistance to phytopathogens manifest indirectly that metabolic changes are occurring in plants. Tomato plant resistance to a number of fungous and bacterial phytopathogens is known to be related to the presence of C-27 steroid glycoalkaloid, α-tomatine[4] in tomato plant tissues. We have observed the changes of the glycoside content in plant leaves under the effect of seed presowing treatment with furostanol glycosides. It has been shown that α-tomatine content in leaf dry mass of tomato grown from seeds treated with a capsicoside[5] solution in the concentration of 0.1% (hybrid Carlson) at the stage of 6-7 true leaves exceeded the control by 73%. In the treatments with celoside[6] in the concentration of 0.08% the increment over the control reached 100%, in case of purpureagitoside[7] in the concentration of 0.08% by 62%, tomatoside (0.1%) by 65%.

The dry weight of transplant seedlings (6-7 true leaves) of tomato (cv. Fakel) grown from seeds treated with 0.1% solution of capsicoside contained α-tomatine by 44% more than the control. These data agree with the assertion about the role of α-tomatine in the resistance of tomato plants to disease agents and enhanced resistance to glycoside treated plants. They also showed that these kinds of treatment do lead to, at least, qualitative changes in plant me-

Table 2. The influence of the treatment of tomato seeds and plants with steroid glycosides on some biochemical indices of ripe fruits

Variety	Substance & concentration	Treatment technique	Biochemical indices				Yielding capacity, C/ha
			dry residue %	sugar %	total acid content %	ascorbic acid mg/100 g	
Ranniy-83	H_2O (control)	4-fold spraying during the vegetation period	4.40	2.14	0.60	13.68	571
	capsicoside, 0.005%		4.60	2.29	0.57	13.77	698
	-tomatin, 0.005%		4.60	2.29	0.54	15.93	687
Novichok	H_2O (control)	seed treatment	4.66	2.70	0.58	12.76	368
	capsicoside, 0.080%		5.68	2.62	0.62	17.16	527
Novichok	H_2O (control)	plant spraying at the 3-4 true leaf stage	4.71	2.61	0.58	15.40	347
	capsicoside, 0.010%		5.01	2.32	0.71	14.08	528
Novichok	H_2O (control)	spraying at the stage of flowering	5.05	2.70	0.66	15.40	428
	capsicoside, 0.003%		5.82	4.17	0.52	15.40	539

tabolism. From this viewpoint, biochemical changes in tomato fruits are very significant, as they can deteriorate or improve market, eating qualities, as well as influence terms and conditions of storage and transportation.

We also studied qualitative changes of a number of biochemical characteristics of tomato fruits, such as solids mass, total content of sugars and acids, ascorbic acid content and yield capacity (Table 2).

Data received in our experiments show that along with the seed presowing treatment, other techniques of steroid glycoside application have, mainly, a positive effect on the biochemical composition of fruits and yield capacity. In various treatments involving glycoside application, the content of vitamin C exceeded that in the control by 30%, total acidity by 13%, solids by 22%, sugar content by 7% (Table 2).

REFERENCES

1. I.Balansard and F.Pellessier, Compt. Rend. Soc. Chem., 1943, 137, 454.
2. H.Sato and S.Sacamura, Agr. Biol. Chem., 1973, 37, 225.
3. T.D.Fontaine, I.S.Ard and R.M.Ma, J.Amer. Chem. Soc., 1951, 73, 878.
4. J.G.Roddick, Phytochemistry, 1974, 13, 9.
5. R.Tschesche and H.Gutwinski, Chem. Ber., 1975, 108, 265.
6. P.K.Kintia and L.P.Degtiareova, Chim. Prirodn. Soed., 1981, 1, 139.
7. R.Tschesche, A.M.Javellana and G.Wulff, Chem. Ber., 1974, 107, 2828.

THE INFLUENCE OF STEROID GLYCOSIDES FROM SOLANUM MELONGENA L. AND NICOTIANA TABACUM L. SEEDS ON THE YIELD CAPACITY AND QUALITY OF TOMATO FRUITS

S.A.Shvets, P.K.Kintia, and O.N.Gutsu

Institute of Genetics
Academy of Sciences
Kishinev 277002
Moldova

Exogenous natural plant growth regulators are known to manifest their action through the change of the endogenous level of natural hormones, thus allowing the modification of plant growth and development in the direction desired. Steroid glycosides, belonging to saponins, are plant exogenous regulators.[1] As early as in 1940s Balansard et al.[2] showed in the experiments on tomato that presowing treatment of tomato seeds with saponins leads to the enhancement of their germination rate and energy, and during plant development their earlier flowering, as well as abundant harvest by fruit weight and number are observed as compared with the control. The authors ascribe to saponins the role of an activator in the initiation and formation of genuine hormones of growth but do not exclude the possibility of saponins having an effect on plant metabolism a different way, especially when seeds are treated.

We studied the influence of steroid glycosides of the furostan series, isolated from eggplant (Solanum melongena L.) and tobacco (Nicotiana tabacum L.) seeds on the increase of general resistance of tomato plants, yield capacity and biochemical composition of fruits. Melongoside P[3] and nicotianoside E[4] were employed for presowing soaking of seeds. Tomato seeds (cvs. Peremoga and Ranniy-83) were placed into aqueous solutions of glycosides in various concentrations for 24 hrs to sow then in greenhouses. Tomato seeds treated with distilled water and the known growth regulator, ivin,[5] served as control. Transplant seedlings from glycoside treated seeds had a well-developed root system, were more tolerant, after transplanting into field conditions their survival ability was higher than in the control. We have observed acceleration in the phenological stage coming in tomato plants grown from treated seeds, and number of flowering plants in experimental treatments at the early stage of flowering was twice as much as in the control.

Steroid glycosides contribute to the enhancement of tomato plant resistance to a number of diseases, including macrosporiosis (agents Macrosporium solani Ell. et Mart.) and Alternaria solani Sor.), bacterial black spot (agent Xanthomonas vesicatoria Dowson) and wilt diseases (agents

Fusarium oxysporum f. lyc. and Verticillium albo-atrum R.).

Thus, the application of melongoside decreases the rate of infection in tomato plant (cv. Ranniy-83) with bacterial black spot by 28.2%, macrosporiosis by 38.5%, septoriosis by 27% as compared with the control. Similar data have been received for nicotianoside (Table 1).

Table 1. The influence of steroid glycosides on disease development and yield capacity of tomato (cv. Ranniy-83, glycoside concentration for seed soaking 0.05%)

Treatments	Disease development, %			Yield capacity, C/ha		
	bacterial black	macrosporiosis	septoriosis	early		by the end of harvest-
				market	total	
Control (water)	31.3	32.5	51.3	250	338.5	571.5
Melongoside P	22.5	20.0	42.5	351	450.0	652.0
Nicotianoside E	20.9	23.6	39.7	340	404	687

Thus, steroid glycosides, stimulating protective mechanisms of host plants contribute to the enhancement of their overall resistance. It should be noted that glycoside application results in a significantly increased yield capacity of tomato plants, especially in the case of early harvesting.

Along with the enhancement of tomato plant resistance to diseases and yield capacity increase, steroid glycosides increase market quality of fruits increasing their weight and changing biochemical composition (Table 2).

In tomato fruits (cv. Peremoga) from treated plants the number of carbohydrates and total acidity are at the level of control plants, solids contents grows by 4.5% and ascorbic acid content by 16-21%. Nitrate content is nearly twice as lower as in fruits of control plants.

Hence, steroid glycosides contribute to the enhancement of general resistance of tomato plants resulting in their producing capacity increase and improvement of fruit market quality.

Proceeding from this, it is possible to presume that steroid glycosides by penetrating into plant cell perform a role of inducer which interacts with the suppressor and transfers it into an inactive form. Such a form of the suppressor can not interact with the operator gene and, hence, can not prevent from transcription of structural genes. As a result, new RNA matrices and new enzymic systems are synthesizing. Significantly, in addition, new metabolic

Table 2. Yield capacity and biochemical composition of tomato fruits (cv. Peremoga) following seed treatment with glycosides

Treatment	Total yield capacity kg/m^2	% of the control	Mean weight of fruit	Solids %	Sugars %	Ascorbic acid mg%	Total acidity	Nitrates mg%
Control (water)	4.01	100	50	4.72	2.25	12.41	0.62	156.0
Ivin	4.09	102	50	4.51	2.25	11.60	0.59	139.0
Melongoside P	4.88	122	57	4.94	2.26	15.08	0.62	77.7
Nicotianoside E	4.54	113	54	4.93	2.25	14.45	0.61	80.1

ways are realized resulting in the enhancement of plant adaptive potentials. The suggestion is partially confirmed by the works showing that under the action of steroid glycosides in plant cells new proteinous components appear, ribonuclease activity increases by 2.5 times, protein synthesis intensifies, the permeability of membranes for electrolytes and removal rate from plant tissues increase.[6,7]

REFERENCES

1. P.K.Kintia, 2nd Conf. "Plant Growth and Development Regulators", Moscow, 1993, P. 1, 87.
2. I.Balansard and F.Pellissier, Compt. Rend. Soc. Chem., 1943, 137, 454.
3. P.K.Kintia, S.A.Shvets, Phytochemistry, 1985, 24, 1567.
4. Shvets S.A., P.K.Kintia, O.N.Gutsu, 5th Intern. Conf. Chemistry and Biotechnology of Biologically Active Natural Products, Varna, Bulgaria, 1989, 2, 132.
5. Lists of chemical and biological means to protect plants and growth regulators, permitted for application in agriculture and forestry in Moldova, 1994, 2, 374.
6. A.G.Zhakote, P.K.Kintia, G.A.Lupashku etc. 2nd Conf. "Plant Growth and Development Regulators", Moscow, 1993, 1, 91.
7. Balashova I.T., Verderevskaya T.D., Kintia P.K. etc. Preprint, Shtiintsa, Kishinev, 1985, 6.

VARIATION IN THE SAPONIN CONTENT OF DIFFERENT VARIETIES OF AUSTRALIAN-GROWN LUPIN

R.G. Ruiz[1], K.R. Price[1], A.E. Arthur[2], D.S. Petterson[3] and G.R. Fenwick[1]

[1]*Food Molecular Biochemistry Department, Institute of Food Research, Norwich Research Park, Colney, Norwich, NR4 7UA, United Kingdom.*
[2]*John Innes Institute, Norwich Research Park, Colney, Norwich, NR4 7UA, United Kingdom.*
[3]*Western Australia Department of Agriculture, South Perth, WA 6151, Australia.*

1 SUMMARY

Thirty-four samples of *Lupinus angustifolius* were analysed for total saponin content and sapogenol composition. Saponin contents ranged from 379 to 740 mg/kg. Three saponins were detected, two possessing chromatographic properties identical to soyasaponins I and VI, also known as soyasaponin ßg, while the third appeared to possess a novel structure. The changes observed in saponin content were found to be mainly dependant on variety, followed by the site and then the region where grown. No significant effect was found between growing seasons although there was a strong correlation between saponin content and the site grown.

2 INTRODUCTION

Lupin was used as a human food by ancient cultures surrounding the Mediterranean and by those living in the Andean highlands. Although lupin represents an important source of protein for animal and human consumption, the use of this crop was limited due to the presence of toxic alkaloids. This problem has been partially overcome by breeding 'sweet' varieties[1]. Nevertheless, other anti-nutritional factors such as saponins might also interfere with the nutritional use and acceptance of lupin seeds for inclusion in human food.

Saponins are a chemically-complex group of compounds which occur naturally in plants. Their bioactivity has been implicated in disease resistance in the plant and, when consumed, in various antinutritional and toxic effects[2]. Conversely, a beneficial lowering of plasma cholesterol[2] and antiviral activity against human immunodeficiency virus in vitro have in addition been attributed to saponins[3].

Price et al.[4] developed a method for the analysis of saponins which has been used to examine 34 samples of lupin seed from *L. angustifolius* grown in Australia. These samples were chosen to examine differences in saponin content caused by both variety and agronomic variables.

3 MATERIAL AND METHODS

3.1 Material

Seed from 5 cultivars of *L. angustifolius* (Danja, Gungurru, Yorrel, Warrah, Illyarrie) were obtained from crops grown in 1990 and 1991 in areas of Western Australia, Victoria, New South Wales and South Australia and on up to 15 sites within those areas.

3.2 Analysis

Analysis was performed as in Price et al.[4]

3.3 Statistical Methods

The lupin data were subjected to standard two way analyses of variance (without replicates) to determine the statistical significance of differences between the sources of variation being examined.

4 RESULTS

Saponins were detected in all of the lupin seed of *L. angustifolius.* Total saponin content of the varieties grown in the various regions and sites are shown in figures 1-4.

The fast atom bombardment-mass spectra (FAB-MS) showed the presence of 3 saponins in the mixture with molecular weights of 942, 1068 and 1104 respectively. The common aglycone of these was confirmed to be soyasapogenol B by gas chromatography-mass spectrometry. Together with the chromatographic behaviour of these the former saponin has been identified as soyasaponin I and the second one as soyasaponin VI or ßg[5,6]. The third saponin, which appears novel, is the subject of current investigation.

No large inter-regional variations in total saponin content in the 1990 and 1991 growing years were recorded. However, important and significant differences between varieties were found (Figures 1 and 2).

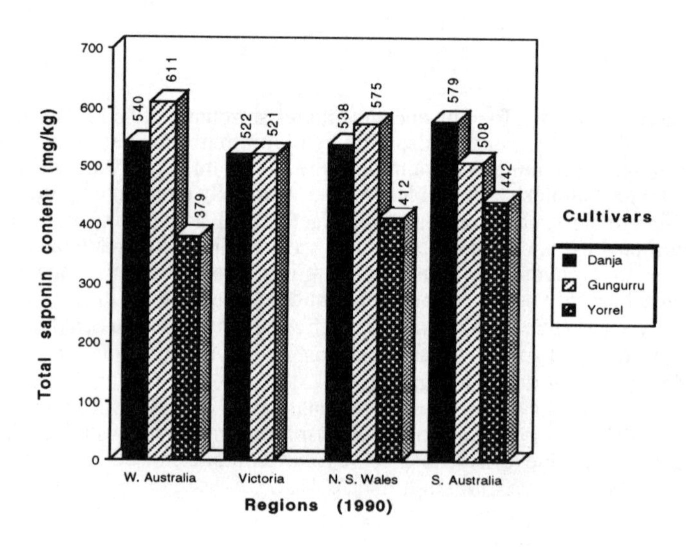

Figure 1 *Total saponin content (mg/kg) - Cultivars vs Regions (1990)*

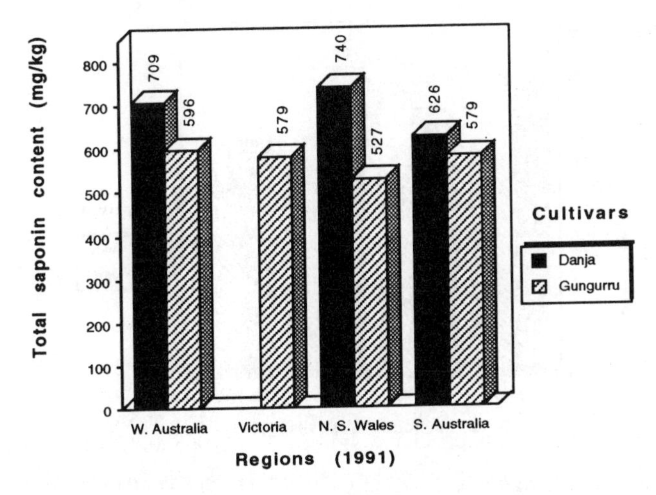

Figure 2 *Total saponin content (mg/kg) - Cultivars vs Regions (1991)*

The range of saponin content in seed grown on sites within S. Australia were larger than those observed between regions. There were also significant differences between varieties (Figures 3 and 4).The cultivar Warrah had a higher saponin content than Illyarrie for all the sites in S. Australia during 1990 (Figure 3). Moreover, for both cultivars, Mannanarie was the site where the highest saponin content was found, followed by Kapunda, Kapinnie and Kangaroo (Figure 3). Similar behaviour was observed for the cultivars Warrah and Yorrel grown in S. Australia during 1991 (Figure 4).

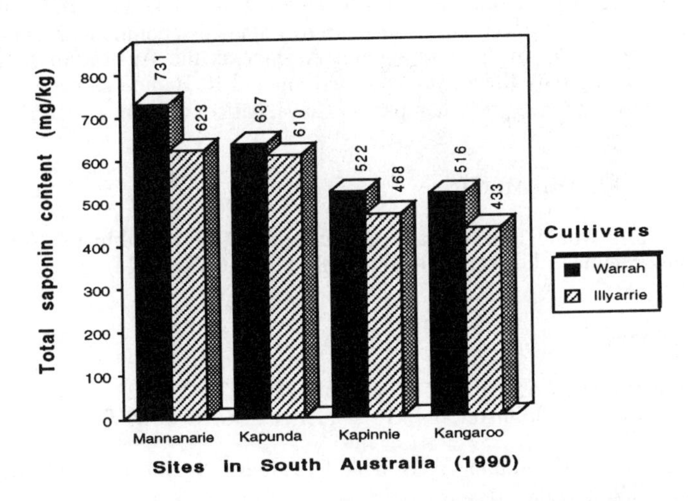

Figure 3 *Total saponin content (mg/kg) - Cultivars vs Sites in South Australia (1990)*

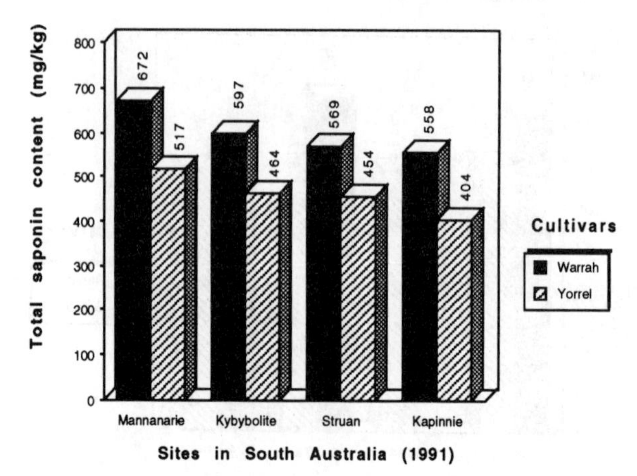

Figure 4 *Total saponin content (mg/kg) - Cultivars vs Sites in South Australia (1991)*

5 DISCUSSION

The total saponin content of Australian *L.angustifolius* was largely dependent on variety, which is in good agreement with Shiraiwa et al[7]. The total saponin was also found to depend on the place where the lupin seed was grown; the most significant differences were shown between sites within the same region. Furthermore, there is a strong correlation between the total saponin content and the site where the lupin seed was grown, which suggests that some specific environmental factors at a particular site could be responsible for higher levels of saponin in the lupin.

The presence of 2 saponins with molecular weights of 942 and 1104 is in agreement with Muzquiz et al[8]. However, the Spanish seed contained saponins with the aglycones of both soyasapogenol B and soyasapogenol A whereas the Australian seed contained saponins possessing only the aglycone soyasapogenol B. Lupin has been shown to be a legume with low saponin content compared to soya, haricot or kidney bean[4].

6 ACKNOWLEDGEMENTS

The authors would like to thank the European Union within the Agro-Industrial Research programme (fellowship contract number: AIR-CT92-5121) for financial support.

References

1. J. S. Gladstones, I.A.B. Field Crop Abstracts, 1970, **23**, 123.
2. K. R. Price, I. T. Johnson and G. R. Fenwick, *CRC Crit. Rev. in Food Sci. and Nutr.*, 1987, **26**, 27.
3. H. Nakashima, K. Okubo, Y. Honda, T. Tamura, S. Matsuda and N. Yamamoto, *AIDS*, 1989, **3**, 655.
4. K.R. Price, C. L. Curl and G.R. Fenwick, *J. Sci. Food Agric.* , 1986, **37**, 186.
5. S. Kudou, M. Tonomura, C. Tsukamoto, T. Uchida, T. Sakade, N. Tanamura and K. Okubo, *Biosci. Biotech. Biochem.*,1993, **57**, 546.
6. G. Massiot, C. Lavaud, M. Benkhaled, L. Le Men-Olivier, *J Nat Prod*, **55**, 1339.
7. M. Shiraiwa, K. Harada and K. Okubo, *Agricol. Biol. Chem.*, 1991, **55**, 323.
8. M. Muzquiz, C. L. Ridout and K. R. Price, *J. Sci. Food Agric.*, 1993, **63**, 47.

INTEGRATING NUTRITION AND AGRICULTURE. THE SOYBEAN STORY IN IGALALAND, NIGERIA.

C. Oguonu[1], N. McNamara[1] and S. Morse[2]

[1]Diocesan Development Services, POB 114, Idah, Kogi State, Nigeria
[2]School of Development Studies, University of East Anglia, Norwich, Norfolk NR4 7TJ.

1 INTRODUCTION

Soybean (*Glycine max*) is not a 'novel' crop in the conventional sense, but there are areas of Africa where the crop can be regarded as 'novel' as the people have little if any experience of growing or eating it. There are relatively few accounts of the problems faced when introducing soybean into such areas, and the aim of this paper is to provide such an account for Igalaland, Kogi State, Nigeria. Particular attention will be paid to the reasoning that lay behind the introduction of the crop and how potential difficulties were overcome.

Soybean has been in Nigeria for almost a century.[1] Indeed, two of the close neighbours of the Igala people, the Tiv and the Idoma of Benue State, have been successfully using soybean for both human and animal consumption for many years.[2,3,4] Soybean has been grown in Igalaland along the border with Idoma, but the Igala people never incorporated the crop into their diet mainly because they did not know how to cook it. Instead it was treated as a cash crop and sold to a processing factory at Taraku, Benue State.

The diet of the Igala is typical of many peoples in West Africa. The starchy staples (cassava, yam, rice, maize, plantain etc.) constitute a major component, but legumes such as cowpea (*Vigna unguiculata*), pigeon pea (*Cajanus cajan*), groundnut (*Hypogea arachis*) and bambarra nut (*Vigna subterranea = Voandzea subterranea*) are also included. The legumes are particularly important as 'snack' food for school children and market people who trade in such 'snacks'. By and large the diet reflects the major crops grown widely in the area, with little or no importation of foodstuffs from outside Igalaland.

The agriculture of the area is based on a mixture of crops from the rainforest and the drier north. The cropping system is primarily a 'slash and burn' rotation, with intercropping as the dominant practice. Soils are generally low in fertility and quite acidic. Approximately 95% of the population are engaged in small-scale farming, the average holding being about 1.9 ha.

The Diocesan Development Services (DDS) is a non-government organisation that has been working in Igalaland since 1971. Its mandate is to help eliminate the causes of hunger, disease and famine, especially among the poorest people. To help achieve this, the DDS operates an integrated rural development programme, with each project organically evolving from the different needs as they arise from the people. Most of the projects have been concerned with agriculture (including a credit scheme) and primary health care, and an extensive nutritional programme evolved from the interface between these two.

2 SOYBEAN AS AN OPTION FOR IMPROVING DIETARY PRACTICES

As the main diet in Igalaland is predominantly based on starchy foods with relatively few vegetables, it was clear to all that the diet of orphans and toddlers about to be weaned was far from ideal. Guinea corn (*Sorghum bicolor*) porridge has traditionally been popular as a weaning food in Nigeria,[3] but cannot be fed to a baby as the protein content is very low. [5] Groundnut and bananas make tasty drinks for children, but are not suitable for babies. Indigenous legumes are also used, but they cannot be processed with existing technology into a milk-like fluid that can be fed to a newborn baby where the mother has died or for some reason cannot feed the baby herself. Dried milk is extremely expensive, and out of the reach of the ordinary villager. In addition, meat is scarce and expensive, and the sheep and goats that are raised are sold mainly for cash.

In view of the above problems with utilising existing foodstuffs for newborn babies and toddlers about to be weaned, the DDS looked elsewhere for potential solutions to broaden the available options. Soybean was an obvious choice, and one that was receiving wide publicity throughout Nigeria in the 1980s.[3] Soybean was being promoted because of its high protein and calcium content,[3,6] and welcomed by people in areas where meat and milk are limited in the diet. This happily coincided with an increasing interest in the crop by the Igalas living along the Idoma borders, whose interest was awakened mainly by the opening of the oil mill at Taraku. With this background, DDS decided to initiate an education programme which would elucidate the use of soybean in nutrition.

Soybean now has several advantages over other legumes. To begin with it can be processed with existing technology into a milk product (soya milk). This is very useful in child welfare programmes. In addition, soybean causes less flatulence than many other grain legumes.[6,8] Also, the residue from the milk production can be the basis of a stew that is highly nutritious, and looks and tastes like the popular egusi soup. The other added attraction is that all other leftovers can be fed to fowl and goats.

Soybean has a number of attractive agronomic attributes besides its ability to fix nitrogen. The indigenous legumes (especially cowpea) are susceptible to a wide range of pests and diseases in the field and during storage, and these are often the major limiting factors in their production.[7] Yield losses are particularly severe before the end of the rains and entry into the dry season where food supply becomes an increasing concern. Initial planting of soybean in Igalaland suggested that it did have the virtue of surviving the local pest and disease complex, a facet noted by many workers in Africa since the 1950s.[2] As soybean is harvested at the beginning of the dry season,[9] its ability to resist the onslaught of intensified attack during that season is yet another bonus.

3 BRINGING SOYBEAN TO THE PEOPLE

Although the soybean crop has many attractions, it does have disadvantages. Foremost among these is the presence of anti-nutritional factors, including trypsin inhibitors, which could be highly toxic.[2,6] That is why people promoting its use as a food had to take great care in the preparation of the dishes, especially the milk.[6] The anti-nutritional factors, especially the trypsin inhibitors, are destroyed by heat.[10] Traditionally the processing took 24 hours, with the beans first having to be soaked and the water changed at least four times before they could be ground into a paste. Very few people had either the time or the water to engage in this lengthy process. Through research based at the International Institute of Tropical Agriculture (IITA),

Ibadan, Oyo State, Nigeria, it was discovered that if the beans were first milled, the anti-nutritional factors were destroyed, and the cooking time could be shortened to a few minutes. Indeed, milk could be made immediately after milling.

A second disadvantage is the bad smell and taste that can surround foodstuff made solely from soybean seeds. These problems are also largely due to the presence of the anti-nutritionl factors, and can also be resolved by milling.

The DDS began a coordinated campaign of soybean promotion in 1989. As well as having its own programme, it formed a network with other organisations like hospitals, clinics and women's resource centres. The new method of processing (milling followed by cooking) was incorporated in 1991. A week was spent in each village demonstrating the preparation and uses of the soybean in diets. When these visits coincided with the planting season (late July to early August), seeds would be distributed, especially to women. Since snacks form an important source of income for women, they are taught the multiple uses of soybean as a means of generating income.

A second component of the programme is the provision of a revolving loan fund for grinding mills. Soybean can be processed by the same method as is commonly employed for other grain crops. Women are very keen on these mills, and special provision is made for them. Currently over a hundred people benefit from the fund.

In parallel with the processing and nutritional dimensions, soybean was incorporated into the DDS on-farm research (OFR) programme as the Igala farmers have little experience of growing the crop. Initial studies in 1987 suggested that the particular strain of *Rhizobium* required by the soybean was not sufficiently common in Igala soils. Soybean crops typically looked sickly, and the pods often had no seed. Soybeans are unusual among grain legumes in having a requirement for special strains of *Rhizobium*.[2] IITA solved this problem by releasing varieties that were not so strain specific.[4]

The model followed in the OFR programme was the maize/pigeon pea intercrop, a well-known system in Igalaland. Results to date suggest that soybean can be readily and successfully intercropped with maize. Yields of soybean from the intercrop are often about 80 to 90% that of a sole crop, closely parallelling the plant population used in the intercrop. Maize yields from the system are also reasonable (55 to 65% that of a sole crop). Although the yield output of the maize/soybean system is higher than with equivalent areas of sole crops, it is interesting that Igala farmers prefer sole cropping soyabean. A survey conducted in 1994 found that 72% of farmers growing soybean planted it as a sole crop, with only 22% preferring an intercrop system. The reason is linked to the mode of harvesting, as the farmers prefer to uproot the plant and thrash it whole. Intercropping with maize makes this uprooting more difficult.

4. CONCLUSIONS FROM THE DDS SOYBEAN PROGRAMME

Soybean answers a very specific need in child welfare programmes, namely the provision of milk for infants. In addition, its versatility as a food is highly acceptable and its value as a cash crop is becoming increasingly attractive to women as news of its nutritional and culinary characteristics spreads.

From the agronomic viewpoint, it is clear that once established in the field soybean presents very few problems to the farmer. Farmer feedback via several methods suggests that the crop is rapidly increasing in popularity.

However, DDS has learnt some lessons from its experience with soybean. To begin with, DDS had to adapt the planting recommendations provided by IITA[9] as dry weather is required at harvest. The IITA advice was to plant before the middle of July, but the best results in Igalaland were obtained from planting in early August. Unfortunately, the initial problem with planting dates did result in the loss of some varieties. In retrospect, more frequent visits to Tiv farmers would have been invaluable during the early experimental period.

An additional problem has been the relatively rapid loss of viability that soybean seeds undergo during storage. Again, this is a well-known problem,[2] but as yet there is no complete solution, especially as high temperature and humidity, the conditions that result in the rapid loss of viability,[4] are 'endemic' to Igalaland. This is an area of concern that could be addressed by plant breeders.

The fact that there is a possibility of selling the soybeans as a cash crop to the mill at Taraku can often mean that not enough seed is kept for meeting household requirements. However, this is not perceived as being a major problem as the price offered at Taraku is not guaranteed.

The success of the DDS soybean programme further suggests that the Igalas are a highly adaptive people, quite willing and able to take on new ideas if they fit their traditional systems and taste in food. Provided new ideas offer an advantage the people respond, but time frames cannot be imposed. The fact that the DDS programme was integrated offered a substantial advantage as it allowed all aspects of soybean utilisation to be covered. In addition, DDS worked with the people in 'fine tuning' the results of 'on station' scientific research. The OFR approach is in vogue with many development agencies for field crop production, and maybe should be extended to cover post-harvest aspects of crop utilisation as well.

References

1. F. O. C. Ezedinma, *Proc. Ag. Soc. Nig.*, 1964, **3**, 13.
2. W. R. Stanton, 'Grain legumes in Africa', Food and Agriculture Organization of the United Nations, Rome, 1966.
3. AERLS, 'Soybeans in the Nigerian diet', Extension Bulletin no. 21, Agricultural Extension and Research Liason Service, Samaru-Zaria, Nigeria, 1982.
4. D. A. Shannon, 'Prospects for commercial soybean production in Nigeria: a compilation of agronomic and economic information for prospective soybean growers and processors', International Institute of Tropical Agriculture, Ibadan, Nigeria, 1983.
5. O. U. Eka and J. K. Edijala, *Nig. J. Sci.*, 1976, **6**, 157.
6. H. O. Ogundipe, 'Soybean utilization in Nigeria', International Institute of Tropical Agriculture, Ibadan, Nigeria, 1987.
7. S. R. Singh, 'Grain legume entomology', International Institute of Tropical Agriculture, Ibadan, Nigeria, 1977.
8. A. K. Smith and S. J. Circle (eds.), 'Soybeans: Chemistry and Technology. Volume 1. Proteins', Avi Publishing Corporation, Westport, 1972.
9. IITA, 'How to grow soybeans in Nigeria', International Institute of Tropical Agriculture, Ibadan, Nigeria, 1987.
10. I. E. Leiner and M. L. Kadade, 'Protease inhibitors', in 'Toxic Constituents of Plant Foodstuffs', I. E. Leiner (ed.), Academic Press, New York, 1969.

Section 3: Quality Improvement by Manipulation of Protein, Starch and Lipid

MANIPULATION OF WHEAT PROTEIN QUALITY

P.R. Shewry[1], A.S. Tatham[1], J. Greenfield[1], N.G. Halford[1], F. Barro[2], P. Barcelo[2] and P. Lazzeri[2]

[1]IACR-Long Ashton Research Station, Department of Agricultural Sciences, University of Bristol, Long Ashton, Bristol BS18 9AF
[2]IACR-Rothamsted, Harpenden, Herts AL5 2JQ

1 INTRODUCTION

Proteins account for about 10 to 15% of the dry weight of the mature cereal grain, and storage proteins for about half of the total. This contrasts with the higher levels of protein present in legumes such as soybean (up to 40% dry weight). Nevertheless cereals form the most important source of dietary protein for mankind and his livestock, with storage proteins determining the quality and end use properties of the grain. Much of the barley, wheat and maize grown in the developed world is used for livestock feed, and the nutritional quality for monogastic livestock (e.g. pigs and poultry) is limited by the low levels of lysine (all three species), threonine (wheat, barley) and tryptophan (maize) in the prolamin storage proteins. As a result it is necessary to supplement cereal-based diets with sources of these amino acids, such as soybean or fish meal.[1] In the case of wheat the prolamins form a visco-elastic network in doughs (called gluten) which is crucial for the manufacture of breads, other baked goods, pasta and noodles.[2] If the level of gluten is low, or it is insufficiently elastic, it is necessary to fortify the breadmaking grist with either strong wheats imported from N. America or gluten isolated using an industrial separation process. The prolamin storage proteins also affect the quality of barley for malting, brewing and distilling. However, in this case, the effects may be both negative (in limiting starch modification) or positive (in providing hydrophobic peptides that contribute to foam stability and cling). These are discussed in detail by Lazzeri and Shewry.[3]

The economic and technological importance of cereal storage proteins has provided an important stimulus for their study. In addition, they have been an attractive model system for molecular biologists interested in the mechanisms of gene regulation. The results of these studies have provided a valuable insight into the molecular basis for various quality traits and have indicated strategies for improvement through the application of molecular genetics. In this article we focus on one aspect of cereal protein quality: the structure of wheat gluten and its relationship to breadmaking quality, and discuss the application of molecular genetics to explore its functional properties and produce improved varieties of wheat.

2 WHEAT GLUTEN PROTEINS AND BREADMAKING QUALITY

Wheat gluten consists essentially of the prolamin storage proteins which are synthesised in the developing starchy endosperm and deposited in discrete structures called protein bodies. These bodies are disrupted as the cells become distended with starch, and the protein forms a matrix in the dry mature grain. Milling, wetting and kneading result in the formation of continuous gluten network within the dough which traps carbon dioxide produced by fermentation, leading to the light porous crumb structure that is characteristic of leavened bread. Although the gluten proteins form a network in dough, they can be isolated in a form which is substantially pure by washing to remove starch and water-soluble components.

Gluten consists of a mixture of over 50 individual proteins, which are classically divided into two groups. The gliadins are readily soluble in aq. alcohols and are monomeric. In contrast, the glutenins are not generally soluble in aq. alcohols and consist of subunits assembled into polymers stabilized by inter-chain disulphide bonds. This classification has proved to be very durable because these groups also have functional significance: the gliadins are responsible mainly for gluten viscosity and the glutenins for elasticity. Wheats grown in the UK and Western Europe often produce gluten which is not sufficiently elastic for breadmaking and the glutenins have consequently been an important topic for research. One group of glutenin proteins, called the high molecular weight (HMW) subunits, appear to be particularly important in this respect.

3 HMW SUBUNITS OF GLUTENIN: QUANTITATIVE AND QUALITATIVE EFFECTS ON QUALITY

Evidence for a role of the HMW subunits in gluten elasticity comes from two sources. Firstly, they are only present in high M_r (above about $1x10^6$) polymers, the amounts of which are correlated with quality.[4] Secondly, genetical studies have shown that allelic variation in the number of HMW subunits (3, 4 or 5 in bread wheat) and their mobilities on SDS-PAGE are correlated with differences in breadmaking quality.[5]

All bread wheats contain six HMW subunit genes, present at single loci on chromosomes 1A, 1B and 1D. Each locus consists of two genes encoding subunits termed x-type and y-type, but specific gene silencing means that 1Ay subunits are never present in bread wheats, and that 1Ax and/or 1By subunits are only present in some cultivars. Quantitative analyses have shown that this gene silencing results in an effect on the total amount of HMW subunit protein, with each expressed gene accounting, on average, for about 2% of the total extractable proteins.[6] Consequently HMW subunits account for about 8% of the extractable protein in cultivars with four expressed genes and 10% in cultivars with five expressed genes. These results indicate that it may be possible to improve the breadmaking quality of wheat by increasing the number of expressed genes for HMW subunits.

However, variation in breadmaking quality is also associated with allelic variation (observed as differences in mobility on SDS-PAGE) in the HMW subunits encoded by the genes on chromosomes 1B and 1D,[5] which is presumably due to qualitative differences in the properties of the proteins themselves, rather than quantitative differences in subunit amount.[7]

Detailed physico-chemical studies of purified HMW subunits and the isolation and sequence analysis of the corresponding genes have enabled us to develop a model for HMW subunit structure.[7] The mature proteins consist of 627 to 827 amino acid residues, with M_r of 67,500 to 88,100. Their sequences can be divided into three structural domains, with short N-terminal (81 to 104 residues) and C-terminal (42 residues) domains flanking a long central repetitive domain (Figure 1). The latter consists of tandem and interspersed blocks of three, six and nine residues, and appears to adopt a loose spiral structure based on regularly repeated β-reverse turns. In contrast, the N- and C-terminal domains are rich in α-helix and are probably compactly folded. Between four and seven cysteine residues are present, with three or five in the N-terminal domain, one in the C-terminal domain and, in some subunits only, one in the repetitive domain.

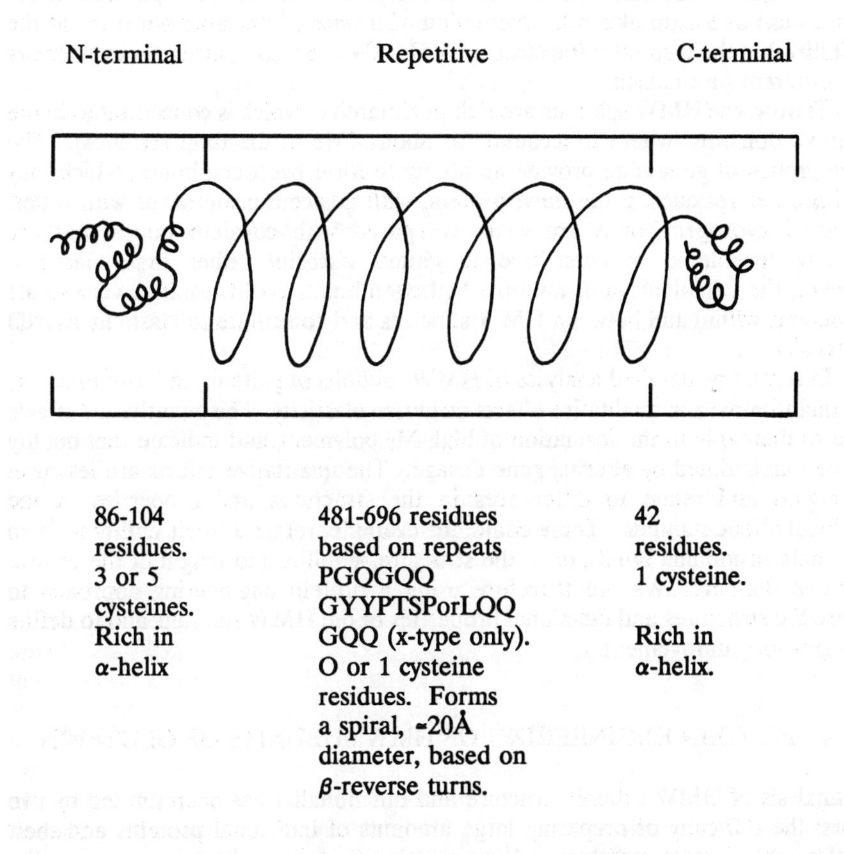

Figure 1 *Summary of HMW Subunit Structure*

Three aspects of HMW subunit structure could be relevant to their role in forming an elastic network in gluten. The first is the number and distribution of cysteine residues. These are potential sites for cross-links, and the degree of cross-linking could affect the bulk elastic properties, as in rubbers. In relation to this it is perhaps significant that the only major difference noted between allelic pairs of subunits associated with good breadmaking quality (1Dx5 + 1Dy10) and poor breadmaking quality (1Dx2 + 1Dy12) is the presence of an additional cysteine residue in subunit 1Dx5.[7] This could lead to the formation of more highly cross-linked and, hence, more elastic, glutenin polymers.

The second aspect of HMW subunit structure is the β-spiral conformation adopted by the repetitive central domain.[7] This appears to be based on regularly repeated β-reverse turns and may be similar to the structure adopted by a synthetic polypentapeptide based on a repetitive sequence present in elastin, an elastomeric protein of mammalian connective tissues.[8] The synthetic polypentapeptide also exhibits intrinsic elasticity, with an elastic modulus similar to that of elastin itself.[8] We proposed, therefore, that the β-spiral formed by the HMW subunit may also be intrinsically elastic, with a molecular mechanism similar to that of elastin.[9] Although nuclear magnetic resonance spectroscopy subsequently showed that a purified HMW subunit was not elastin-like in its interaction with water,[10] this does not rule out the possibility that the β-spiral spiral structure of HMW subunits contributes to elasticity via a different mechanism.

Thirdly, the HMW subunits are rich in glutamine, which is concentrated in the repetitive domains (where it accounts for about 40% of the total residues). The amide groups of glutamine provide an ability to form hydrogen bonds, which may be with other residues in the same protein, with adjacent proteins, or with water. Individual hydrogen bonds are weak compared with covalent bonds and are generally considered to contribute to gluten viscosity rather than elasticity. However, the combination of multiple hydrogen bonds could result in very stable interactions within and between HMW subunits and contribute to elasticity as well as viscosity.

In summary, detailed analyses of HMW subunits of glutenin indicate that they have quantitative and qualitative effects on gluten elasticity. The quantitative effects relate to their role in the formation of high M_r polymers, and indicate that quality may be manipulated by altering gene dosage. The qualitative effects are less well understood and relate to differences in the structures and properties of the individual allelic subunits. These could, for example, relate to their ability to form inter-chain disulphide bonds, or to the structure, stability and length of the central repetitive domain. We are therefore using a protein engineering approach to explore the structures and functional properties of the HMW subunits and to define strategies for improvement.

4 PROTEIN ENGINEERING OF HMW SUBUNITS OF GLUTENIN

The analysis of HMW subunit structure and functionality has been limited by two factors: the difficulty of preparing large amounts of individual proteins and their complex multidomain structures. We are using protein engineering to overcome these problems.

Galili[11] showed that HMW subunits could be expressed at high levels in *E.coli*, using the pET3d vector system. We are therefore using a related vector to express an M_r 57,000 fragment corresponding to residues 103 to 643 of subunit 1Dx5 (see Figure 2). This fragment is wholly repetitive and contains no cysteine residues. We have therefore also expressed a mutant form with single cysteine residues close to the N- and C-termini. Circular dichroism and Fourier-transform infra-red spectroscopy of the repetitive fragment have confirmed that it is rich in β-reverse turns. This conformation was proposed previously on the basis of structure prediction[9] and spectroscopy of short synthetic peptides,[12] but could not be demonstrated by spectroscopic analysis of the whole protein due to the presence of α-helix rich N- and C-terminal domains. These β-turns appear to form a loose spiral, as observed by scanning tunnelling microscopy,[13] and the availability of large amounts of the repetitive peptide may allow details of this structure to be obtained using X-ray crystallography or 2-D nuclear magnetic resonance spectroscopy.

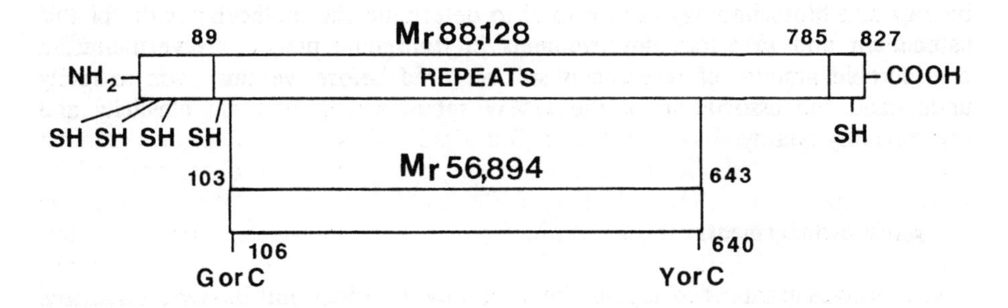

Figure 2 *Schematic Structure of HMW Subunit 1Dx5 and of the Repetitive Fragment Expressed in E.coli*

The ability to produce large amounts of the repetitive peptide and to control the pattern of cross-links via the insertion of cysteine residues will also allow us to explore the relative contributions of the repeat structure itself and the number and distribution of cross-links to gluten elasticity, using standard rheological procedures.

5 MANIPULATION OF GLUTEN QUALITY IN TRANSGENIC WHEAT

Protein engineering studies of the type described above are essential in order to determine the precise structures of the HMW subunits and their role in gluten structure and functionality. However, the fact that HMW subunits have quantitative effects on quality means that this knowledge is not an essential prerequisite for attempts to improve quality by genetic engineering. We have therefore established a project to manipulate the number and properties of the HMW subunits of wheat using transformation mediated by particle bombardment. As recipient lines we selected an isogenic series produced by crossing lines with null (i.e. silent) alleles at the HMW subunit loci on chromosomes 1A, 1B and 1D. This series consists of lines with zero, one (1Ax1), two (1Dx5 + 1Dy10 or 1Bx17 + 1By18), three (1Dx5 +

1Dy10, 1Ax1 or 1Bx17 + 1By18, 1Ax), four (1Dx5 + 1Dy10, 1Bx17 + 1By18) or five (1Dx5 + 1Dy10, 1Bx17 + 1By18, 1Ax1) expressed subunits.[14] Genes are available for three of these subunits (1Ax1, 1Dx5, 1Dy10)[6,15,16] and are being used to transform the various lines, with expression controlled by their own promoters. The first transgenic plants from this project are currently being analyzed.

6 CONCLUSIONS

We have discussed the HMW subunits in some detail as they form the best documented example of the association of a complex quality parameter with a group of grain components. They also provide an example of how a multidisciplinary approach, combining genetics and plant breeding, biochemistry, chemistry, molecular biology and biotechnology can be used to determine the molecular basis for this association and to effect improvements in transgenic plants. Nevertheless, a considerable amount of research is still required before we can claim to really understand the association of the HMW subunits with glutenin elasticity and breadmaking quality.

Acknowledgements

IACR receives grant-aided support from the Biotechnology and Biological Sciences Research Council of the United Kingdom.

References

1. S.W.J. Bright and P.R. Shewry, *CRC Crit. Rev. Pl. Sci.* 1983 **1**, 49.
2. B.J. Miflin, J.M. Field, and P.R. Shewry 'Seed Proteins' Eds. J. Daussant, J. Mosse, and J. Vaughan. Academic Press, London, 1983 pp. 255.
3. P. Lazzeri and P.R. Shewry 1993, *Biotech. & Genet. Eng. Rev.* Ed. M. Tombs Intercept Ltd., U.K. p.79.
4. J.M. Field, P.R. Shewry and B.J. Miflin, *J. Sci. Food Agric.* 1983 **34**, 370.
5. P.I. Payne, *Ann. Rev. Pl. Phys.* 1987 **38**, 141.
6. N.G. Halford, J.M. Field, H. Blair, P. Urwin, K. Moore, L. Robert, R. Thompson, R.B. Flavell, A.S. Tatham and P.R. Shewry, *Theor. Appl. Genet.* 1992 **83**, 373.
7. P.R. Shewry, N.G. Halford and A.S. Tatham, *J. Cer. Sci.* 1992 **15**, 105.
8. D.W. Urry, C.M. Venkatachalam, M.M. Long and K.U. Prasad 'Conformations in Biology' Eds R. Srinvason and R.M. Sarma. Adenine Press, New York, USA, 1983 pp. 11.
9. A.S. Tatham, P.R. Shewry and B.J. Miflin, *FEBS Lett.* 1984 **177**, 205.
10. P.S. Belton, I.J. Colquhoun, J.M. Field, A. Grant, P.R. Shewry and A.S. Tatham, *J. Cer. Sci.* 1994 **19**, 115.
11. G. Galili, *Proc. Natl. Acad. Sci. (USA)* 1989 **86**, 7756.
12. A.S. Tatham, A.F. Drake and P.R. Shewry, *J. Cer. Sci.* 1990 **11**, 189.

13. M.J. Miles, H.J. Carr, T. McMaster, P.S. Belton, V.J. Morris, J.M. Field, P.R. Shewry and A.S. Tatham, *Proc. Natl. Acad. Sci (U.S.A.)* 1991 **88**, 68.
14. G.J. Lawrence, F. Macritchie and C.W. Wrigley, *J. Cer. Sci.* 1988 7, 109.
15. R.B. Flavell, A.P. Goldsbrough, L.S. Robert, D. Schnick and R.D. Thompson, *Bio/Technology* 1989 7, 1281.
16. O.D. Anderson, F.C. Greene, R.E. Yip, N.G. Halford, P.R. Shewry and J.-M. Malpica-Romero, *Nucleic Acids Res.* 1989 17, 461-462.

PATTERNS OF HORDEIN SYNTHESIS AND DEPOSITION IN RELATION TO BARLEY MALTING QUALITY

P.R. Shewry[1], J. Broadhead[1], R. Fido[1], G. Down[1], I. Cantrell[2], D. Griggs[2], D. Smith[3], C. Brennan[4], M. Amor[4], N. Harris[4] and A.S. Tatham[1]

[1]IACR-Long Ashton Research Station, Department of Agricultural Sciences, University of Bristol, Long Ashton, Bristol BS18 9AF, UK
[2]Pauls Malt Limited, Bury St. Edmonds, Suffolk IP32 7AD, UK
[3]PBI Cambridge, Maris Lane, Trumpington, Cambridge CB2 2LQ, UK
[4]Department of Biological Sciences, University of Durham, Science Laboratories, South Road, Durham DH1 3LE, UK

1 INTRODUCTION

Hordein storage proteins comprise about half of the total nitrogen in the mature barley grain, the amount increasing disproportionately with fertilization.[1] Hordein consists of four groups of polypeptides, called B, C, D and γ-hordeins, which differ in their structures, properties and contributions to the total fraction.[2]

It has long been known that high levels of grain protein are disadvantageous for malting quality, due to a negative correlation with hot water extract, and that this effect varies with variety.[3] More recent studies indicate that the relationship between hordein and malting quality may be more complex, with both negative and positive effects.[4]

Baxter and Wainwright[5] suggested that malting quality was related to the proportion of hordein that was present in high M_r polymers stabilized by disulphide bonds, and Shewry *et al.*[6] provided evidence in support of this. These polymers can be extracted directly from milled grain with 1.5% (w/v) sodium dodecylsulphate and Smith and Lister[7] showed a clear inverse correlation between the weight of gel proteins and malting quality. It is possible that this gel forms an impermeable layer around the starch granules, hindering digestion during malting and brewing.

More recent work indicates that not only the properties of hordein, but also its distribution within the grain may be important. Thus, Millet *et al.*[8] compared the distribution of monomeric and polymeric hordeins in fractions prepared by the sequential pearling of single good and poor quality cultivars. They showed that a higher proportion of hordein was present in the peripheral layers of the poor quality line, and that more of this was present in polymers. These results have not, however, been confirmed by analysis of a wider range of germplasm.

Finally, hordeins may also contribute positively to the brewing performance of barley, by providing hydrophobic peptides involved in foam stability and cling[9].

2 RESULTS AND DISCUSSION

We are re-evaluating the relationship between hordein and malting quality, using a combination of fine structural, immunocytochemical and biochemical analyses. Pearling fractions prepared from mature barley grains are being analyzed to determine the amount, composition and properties (i.e. polymer distribution) of hordein present, while developing and malted seeds of the same lines are being subjected to both biochemical analyses and electron microscopy.

Smith and Lister[7] demonstrated that D hordein was a major component of gel protein, and we are therefore comparing a series of near isogenic lines differing in the presence or absence of D hordein. Preliminary analyses of six such pairs of lines confirmed that the absence of D hordein resulted in a lower level of gel protein in all cases, with mean values of 0.15g gel protein/g meal in the lines lacking D hordein, and 0.39g gel protein/g meal in the D hordein+lines. Similarly, preliminary analyses of pearling fractions have shown cultivar-dependant differences in the spatial distribution of gel protein within the grain,[10] although the relationship to quality is not clear.

The fine structural studies are also showing interesting results. For example, scanning electron microscopy of freeze-fractured mature grains of good and poor malting quality cultivars of both spring and winter types showed clear variation in the extent of starch/protein adhesion. The poor quality cultivars yielded fracture planes across the starch granule/protein interface, indicating a high degree of starch/protein binding. These cultivars also had a higher level of grain hardness, as determined by the milling energy procedure of Allison *et al.*[11]

This combination of approaches should yield more information on the precise relationships between the amount, composition and distribution of hordein and the patterns of endosperm modification during malting.

Acknowledgements

This project is supported by BBSRC (AFRC) under the Collaboration with Industry Scheme. We are grateful to S. Swanston (SCRI, Dundee) for carrying out the milling energy analyses.

IACR receives grant-aided support from the Biotechnology and Biological Sciences Research Council of the United Kingdom.

References

1. M.A. Kirkman, P.R. Shewry and B.J. Miflin, *J. Sci. Food Agric.*, 1982 **33**, 115.
2. P.R. Shewry, 'Barley: chemistry and technology' eds. J. McGregor and R. Bhatty. American Association of Cereal Chemists, St. Paul Minnesota, U.S.A. 1992.
3. L.R. Bishop, *J. Inst. Brew.* 1930, **36**, 421.
4. P. Lazzeri and P.R. Shewry 'Biotechnology & Genetic Engineering Reviews' ed. M. Tombs. Intercept Ltd., U.K. 1993, p.79.
5. E.D. Baxter and T. Wainwright, *J. Am. Soc. Brew. Chem.* 1979, **37**, 8.
6. P.R. Shewry, M.S. Wolfe, S.E. Slater, S. Parmar, A.J. Faulks and B.J. Miflin, 'Barley Genetics IV' Edinburgh University Press. U.K. 1981 pp. 596.
7. D.B. Smith and P.R. Lister, *J. Cer. Sci.* 1983, **1**, 229.
8. M.-O. Millet, A. Montemboult and J.-C. Autran, *Sciences des Alimentations* 1991, **11**, 155.

9. J.A. Kauffman, E.N.C. Mills, G.M. Brett, R.J. Fido, A.S. Tatham, P.R. Shewry, A. Onishi, M. Proudlove and M.R.A. Morgan, *J. Sci. Food Agric.*. 1994 **66**, 345.
10. P.R. Shewry, J. Price, A.S. Tatham, D.B. Smith, I.C. Cantrell, J.T. Davies and N. Harris, *Aspects of Applied Biology*, 1993, **36**, 43.
11. M.J. Allison, I. Cowe and R. McHale, *J. Inst. Brew.*, 1976 **82**, 166.

THE SIGNIFICANCE OF PLANT LIPOXYGENASES TO THE AGRIFOOD INDUSTRY

R. Casey, C. Domoney, C. Forster
John Innes Centre, Colney Lane, Norwich NR4 7UH

D. Robinson and Z. Wu
Procter Department of Food Science, University of Leeds, Leeds LS2 9JT

1 INTRODUCTION

Lipoxygenases (LOX) catalyse the hydroperoxidation in plant tissues of polyunsaturated fatty acids (PUFAs), including linoleic (18:2) and linolenic acids (18:3), the predominant PUFAs of legume seeds, and arachidonic acid (20:4). The positional and stereo-specificity of LOX vary according to the source of the enzyme and the reaction conditions. LOX have been identified in a wide range of plant organs[1]; many of the flavour and taste volatile compounds of fruits and vegetables, including off-flavours, are thought to derive from the action of LOX. The ability of certain LOX to catalyse co-oxidation processes, including chlorophyll bleaching and protein disulphide formation, is significant to the food and breadmaking industries.

In addition to the above phenomena, all of which are of direct relevance to the food industry, LOX play a role in plant productivity under stress conditions, through their involvement in responses to drought, wounding and pathogen attack.

2 FLAVOUR FORMATION

2.1 Off-Flavour Formation in Pea Seeds and Soybean Seed Products

It has been recognized for some time that a specific isoform of soybean seed LOX, LOX-2, is responsible, in conjunction with hydroperoxide lyase, for the formation of *n*-hexanal from linoleic acid. Hexanal is responsible for the unacceptable "grassy-beany" flavour of soybean products. Genetic removal of LOX-2 decreases *n*-hexanal production and increases consumer acceptability of such products[2]. Fresh-frozen peas form rancid off-flavours during storage unless they are heat-treated (blanched) before freezing. Pea seeds contain two major LOX isoforms that are very similar in sequence to LOX-2 and -3 from soybean[3,4]; pea genetic variants that lack LOX-2 produce less *n*-hexanal than wild-type lines[5,6] and may be useful to the fresh-frozen vining pea industry. The LOX-2 null phenotype is being introgressed into a vining pea background for evaluation.

2.2 Desirable Tastes and Flavours

Many of the tastes and flavour volatiles of fruits and vegetables, in particular aldehydes, alcohols and other short-chain carbon compounds, have been attributed to the initial action of LOX on PUFAs. These include hexanal and cis-3-hexanal from tomato[7,8], and the volatile components of bananas[9], peppers[10] and cucumbers[11], for instance. Variants with altered amounts of LOX activity[12] may be significant in the production of flavour volatiles, but their significance is difficult to assess in widely differing genetic backgrounds, where other variables may also influence flavour. Genetic engineering affords the opportunity to increase or decrease LOX amounts in specific organs of particular plants; it is, for instance, possible to reduce the amount of LOX in potato tubers by "antisense" technology and evaluate the consequences of this for flavour production.

3 CO-OXIDATION PROCESSES IN FOODS

Some LOX can catalyse co-oxidations in the presence of PUFAs; the most significant of these are the bleaching of pigments and the oxidation of thiol groups in proteins, although the chemistry of both processes is uncertain. Co-oxidation potential varies between different LOX; wheat seed LOX, for instance, are much less efficient at co-oxidation than those of legume seeds[13]. Addition of legume LOX, in the form of soybean flour 'improver', is commonly part of the breadmaking process; it leads to improved mixing tolerance[14], a reduced rate of staling, and a whiter crumb. The first is thought to be a consequence of the formation of disulphide bonds in wheat gluten proteins and the last a result of cooxidative bleaching of pigments. Biotechnology may contribute to breadmaking in a number of ways, including the genetic engineering of breadmaking yeast; a genetically modified strain of bakers' yeast that expresses pea seed LOX may prove useful in this context[15].

4 SPECIFICITY OF HYDROPEROXIDATION

Different plant organs contain different LOX with different positional specificities with respect to their PUFA substrates.[16] This inherent positional specificity can be modulated by the reaction conditions and the nature of the substrate. Studies on mammalian lipoxygenases have indicated that the sequence of amino acids within the active site can influence positional specificity, probably by affecting the degree to which the PUFA can enter the substrate pocket, thus determining which particular double bond comes into proximity with the active site iron.[17]

The site of hydroperoxidation is significant to the subsequent formation of taste and flavour volatiles through the action of hydroperoxide lyases. Studies with pea mutants that lack one or other of the two major seed lipoxyenases suggest that the two enzymes differ in positional specificity with respect to hydroperoxide formation from linoleic aid, with LOX-2 showing a preference for the 13-, and LOX-3 the 9-, position.[5] This is probably the basis of the decrease in the production of *n*-hexanal in pea lines that lack LOX-2[6].

5 MULTIPLE EFFECTS OF ALTERING LOX ACIVITY

Genetics and biotechnology represent powerful tools for the alteration of the amounts of LOX in plant foods. By overexpressing, or down-regulating, the amounts of specific LOX in specific plant organs, it is possible to produce a range of novel plant forms and thus influence the production of taste volatiles in fresh, frozen and cooked foods. Altering LOX activities can also have effects on physical properties of food, including colour, texture and rheology, through co-oxidation processes. It is, however, important to remember that LOX play a role in normal plant development, and some changes to LOX activities may be deleterious to plant health or yield.

6 LIPOXYGENASES AND PLANT PHYSIOLOGY

There is evidence to support a role for LOX in plants' responses to wounding, pathogen attack and water deficit.

6.1 Wounding

It has long been recognized that LOX activity increases on wounding of plant tissues, but only recently has evidence been provided for an induction of LOX gene expression in response to wounding [18]. Several physiologically active compounds are thought to arise from the initial action of LOX on PUFAs, including traumatic acid ("wound hormone") and jasmonates; the latter are potent activators of plant defence genes, including protease inhibitors. Thus, the induction of the synthesis of phytoprotectant inhibitors in leaves is thought to be mediated through the so-called "octadecanoid signalling pathway" that includes the hydroperoxidation of linolenic acid by LOX as an initial step.[19]

6.2 Pathogen attack

LOX gene expression is induced in the vegetative parts of plants as a response to bacterial or fungal attack.[20,21] One consequence of the increase in LOX activity is the formation of short-chain aldehydes, some of which can reach bacteriostatic levels.[20] Although such an apparent direct response has not been observed for fungal pathogens, there are clear differences in induction of LOX gene expression and enzyme activity, between plants that exhibit sensitive and resistant responses, that may be significant to plants' defence against fungal attack.

6.3 Drought responses

LOX is believed to be on the biosynthetic pathway to abscisic acid,[22,23] a plant growth regulator that is produced in response to drought and is significant to plants' ability to control water deficit through the closure of stomata. The site of action of LOX is thought to be the co-oxidative cleavage of the carotenoid, xiolanthin, to the abscisic acid precursor xanthoxin; it is possible that such activity is membrane-bound, given the chemical nature of carotenoids.

Thus, LOX may play significant roles in plants' health and ability to withstand stress.

7 CONCLUSIONS

LOX are important to the food industry through the production of desirable and undesirable tastes and flavours, and through the effect of co-oxidation processes on the physical properties and appearance of foods. Manipulation of the amounts of LOX, by genetic or biotechnological means, can have significant effects on such phenomena and could be of considerable value to the food industry. It will be necessary to be selective in making these changes; LOX activity may be undesirable in certain plant organs or tissues, but vital to plants' health and survival in others. Such selectivity requires a greater understanding of the regulation of the many different LOX genes in plants and of the role of the various types of LOX in wounding, pathogen and drought responses. Given such understanding, it should be possible to manipulate the amounts of LOX to the benefit of the food industry without deleteriously affecting plant yield.

References

1. A. Pinsky, S. Grossman and M. Trop, *J. Food Sci.,* 1971, **36**, 571.
2. C.S. Davies, S.S. Nielsen and N.C. Nielsen, *J. Am. Oil Chem. Soc.*, 1987, **64**, 1428.
3. P. M. Ealing and R. Casey, *Biochem. J.*, 1988, **253**, 915.
4. P. M. Ealing and R. Casey, *Biochem. J.*, 1989, **264**, 929.
5. Z. Wu, D. S. Robinson, C. Domoney and R. Casey, *J. Agric. Food Chem.*, 1995, 337.
6. Z. Wu, D. S. Robinson, R. Casey and C. Domoney, in preparation.
7. T. Galliard and J. A. Matthew, *Phytochemistry,* 1977, **16**, 339.
8. T. Galliard, J. A. Matthew, A.J. Wright and M. J. Fishwick, *J. Sci. Fd. Agric.,* 1977, **28**, 863.
9. R. Tressl and F. Drawert, *J. Agric. Fd. Chem.*, 1973, **21**, 560.
10. M. I. Minguez-Mosquera, M. Jaren-Galan and J. Garrido-Fernandez, *Phytochemistry*, 1993, **32**, 1103.
11. T. Galliard and D. R. Phillips, *Biochim. Biophys. Acta,* 1976. **431**, 278.
12. S. F. Vaughn and E.C. Lulai, *Plant Sci.*, 1992, **84,** 91.
13. H. W. Gardner, *Adv. Cereal Sci. & Technol.*, 1992, **9**, 161.
14. P. J. Frazier, *Bakers Digest*, 1979, **53**, 8.
15. B. Knust and D. von Wettstein, *Appl. Microbiol. Biotechnol.,* 1992, **37**, 342.
16. D. S. Robinson, Z. Wu, C. Domoney and R. Casey, *Food Chem.*, 1995, in press.
17. D. L. Sloane, R. Leung, C.S. Craik and E. Sigal, *Nature*, 1991, **354,** 149.
18. A. Geerts, D. Feltkamp and S. Rosahl, *Plant Physiol.,* 1994, **105**, 269.
19. E. E. Farmer and C. A. Ryan, *The Plant Cell,* 1992, **4**, 129.
20. K. P. C. Croft, F. Jüttner and A. J. Slusarenko, *Plant Physiol.,* 1993, **101,** 13.
21. Y-L. Peng, Y. Shirano, H. Ohta, T. Hibino, K. Tanaka and D. Shibata, *J. Biol. Chem.*, 1994, **269**, 3755.
22. A. D. Parry and R. Horgan, *Phytochemistry,* 1991, **30**, 815.
23. R.A. Creelman, E. Bell and J. E. Mullett, *Plant Physiol.*, 1992, **99,** 1258.

STUDY OF TECHNOFUNCTIONAL PROPERTIES OF ALBUMIN FRACTIONS OF PEA VARIETIES

B. Czukor, I. Gajzágó-Schuster, Zs. Cserhalmi,

CFRI - H-1022 Budapest, Hungary

1 INTRODUCTION

Pea protein is an important component of animal feeds and due to protein shortage its significance has grown as a potential raw material in human food too. Therefore, this protein source has been studied from many aspects for years. Major interest has centred on globulin fractions in spite of the advantageous properties of albumin. The amino acid composition of pea albumin is more favourable from the nutrition point of view due to a high content of essential amino acids, sulfur containing amino acids included.

The purpose of the present study was to determine the albumin content of pea varieties cultivated in Hungary and some chemical and functional properties of albumins originated from different pea varieties.

2 MATERIALS AND METHODS

2.1 Materials

Pea seed varieties (Hunor, Junak, Türkiz, UM-1073, UM-1095) was obtained from the Research Station of Vegetable Crops Research Institute, Újfehértó.

2.2 Methods

The albumin fraction of pea varieties was extracted by the method of GWIAZDA et al.[1]

Emulsifying activity index (EAI) and the emulsion stability index (ESI) was measured by the method of Kato et al.[2] after slight modification. Protein surface hydrophobicity was fluorometrically determined according to the method of KESHAVARZE and NAKAI[3] after slight modification.

The sulphydryl and disulphide groups were determined by the method VOUTSINAS and NAKAI[4].

3 RESULTS AND DISCUSSIONS

3.1 The albumin content of pea varieties

The albumin content of 5 pea varieties as a percent of buffer soluble protein is summarized in Table 1.

Table 1

Variety	Albumin content %
Hunor	19,2
Junak	17,0
Türkiz	16,1
UM-1073	17,2
UM-1095	19,7

The albumin content was the highest (19,2 - 19,7 %) in Hunor and UM-1095, two pea varieties had 17% albumin (Junak, UM-1073) and the lowest value (16,1%)was measured in Türkiz.

3.2 Sulfhydryl and disulfide groups of albumins

The sulfhydryl (free and buried SH) and total sulfhydryl (SH + reduced SS) groups of albumins can be seen in Table 2.

Table 2

Variety	SH (μM/g)		SH + r.SS (μM/g)	
	x	s	x	s
Hunor	11,40	0,32	115,85	8,35
Junak	8,17	0,12	73,90	2,50
Türkiz	9,89	0,22	133,75	7,06
UM-1073	9,37	0,57	105.65	10,7
UM-1097	13,40	0,47	137,79	13,9

As the Table 2 shows the Junak albumin contained the lowest SH and SS groups. The highest values were measured in UM-1095 albumin.

3.3 The functional properties of pea albumins

3.3.1 Protein surface hydrophobicity. The protein surface hydrophobicity of albumins are listed in Table 3.

Table 3

Variety	Surface hydrophobicity S_o
Hunor	980
Junak	730
Türkiz	1160
UM-1073	1140
UM-1095	820

Two pea variety albumin (Türkiz, UM-1073) had high protein surface hydrophobicity, the lowest was measured in Junak albumin.

3.3.2 Emulsification property of pea albumins. On the basis of conductivity measurement the emulsifying activity and stability index of albumins can be seen in Table 4.

Table 4

Variety	EAI (μS)		ESI (min)	
	x	s	x	s
Hunor	15,5	0,86	13,70	0,60
Junak	20,2	0,75	5,90	0,25
Türkiz	17,5	0,86	9,25	0,25
UM-1073	21,0	3,00	11,00	1,00
UM-1095	13,0	1,73	5,70	0,82

On the basis of the results listed in Table 4 it seem that the emulsification properties of albumins isolated from different varieties differ in their emulsifying activity index. Two varieties had high emulsifying activity index (20,2 - 21,0 μS) and one variety (UM-1095) had low EAI (13,0 μS).

The albumins differ in their emulsifying stability too. The two albumins having high EAI differ in the emulsion stability.

In the case of low EAI the emulsion stability was low too (UM-1095).

Summarizing the results, the examined pea varieties differed in their albumin, SH and SS groups content. Differences were measured in protein surface hydrophobicity and emulsifying properties, but any relationship has not been found between the measured properties.

On the basis of results of chemical and functional properties it can be said that this characterictics are cultivar dependent.

Further investigations are needed for establishing relationship between the chemical and technofunctional characteristics of different pea varieties.

References

1. S.Gwiazda, K.D. Schwenke and A. Rutkowski, *Die Nahrung, 1980, 24, 939*
2. A.Kato, T.Fujishige, N.Matsudomi and K.Kobayashi, *J.Food Sci., 1985, 50, 56*
3. E. Keshavarze and S. Nakai, *Biochem. Biophys. Acta, 1979, 576, 269*
4. L.P. Voutsinas, S. Nakai and V.R. Harwalkar, *Can. Inst. Food Sci. Technol. J., 1983, 16, 185*

IMPRUVEMENT OF LUCERNE AMINO ACID COMPOSITION BY BIOLOGICAL NITROGEN FIXATION

M.M.Nichik, S.Ya.Kots, E.P.Starchenkov

Department of symbiotic nitrogen fixation
Institute of Plant
Physiology and Genetics, Natl.
Acad. of Sci. of Ukraine
31/17 Vasilkovskaya str. Kiev,
252022, Ukraine.

1 INTRODUCTION

Lucerne is one of the perennial legumes which is of a paramount importance for forage crop production.[1] The quality of lucerne forage crop depends on the content easy to assimilate protein and amino acids, particularly, irreplaceable ones, the composition of which, primarily is defined by a variety and plant growth phase.[2,3] Unlike other plants amino acid composition of lucerne shoot mass essentially depends on the intensity of nitrogen fixing process in root nodules, that is connected with the efficiency of Rhizobium strains.[4]

The present paper seeks to study the influence of Rhizobium meliloti strains of different efficiency on the qualitative and quantitative composition of bound and free amino acids in lucerne leaves.

2 MATERIALS AND METHODS

The field small plot experiments with lucerne (cvs Zaikievich and Yaroslavna) were carried out on dark grey podzol soil (pH 6,3, humus content 2,36) on the background of $N_{30}P_{60}K_{60}$. The experiments were replicated 3 times, the area of recorded plot was 4 square meters. The applied agrotechnics was that agreed for uncovered lucerne crops. Before sowing the seeds were inoculated by eight new Rhizobium meliloti strains of various efficiency. Control variants were inoculated by local races of Rhizobium meliloti. To carry out the analyses the plants were selected by lucerne growth stages. Hay cutting was made at the stage of budding - the initiation of flowering. The quantity of total nitrogen of lucerne shoot mass was determined by Kjeldal's method.[5] The bound amino acid content was detected in acidic hydrolizates of dry leaves on amino acid Biotronik LC 6001 analyzer.[6] The amount of free amino acids was defined in fixed by ethanol plant material on AAAT 339 analyzer.[7]

3 RESULTS

The clearest differences between the strains of Rhizobium in respect of their influence on the green mass crop and protein content have been revealed within the plants of the first year. The greatest harvest of green mass of lucerne cv Zaikievich has been received with inoculation by strain M4 (reliable gain to control is 29 %). Using inoculation by M6 strain maximum protein yield (11,6 c/ha) has been obtained (Table 1).

Table 1. *The inoculation effect of Rhizobium meliloti on green mass crops, nitrogen and bound amino acids content in lucerne leaves.*

Strains	Green mass		Nitrogen	Bound amino acids, mg/g	
	c/ha	% to control	in air dry mass, %	The sum of irreplaceable	Total content
Control	231	100	2.46	95.60	228.37
425a	228	98.7	2.55	105.00	264.38
435a	267	115.5	2.61	98.67	243.78
404b	246	106.5	2.49	99.00	248.30
412b	258	111.5	3.02	117.56	296.43
M3	269	116.4	2.55	115.48	288.34
M4	299	129.4	2.59	117.08	294.95
M6	272	117.7	2.99	112.65	283.73
M7	243	105.2	2.54	112.30	282.38
LSD 05	30.2				

The plants inoculated by M4 and 412b strains considerably exceeded control in respect of plant protein yield. It has been shown that inoculation of lucerne seeds by 5 from 8 tested Rhizobium meliloti strains favours both the increase of yield and protein harvest by 10 to 30 %. Our investigations show that the improvement of qualitative composition of green mass occurs when inoculating lucerne by nodule bacteria. The greatest number of bound amino acids including irreplaceable are found in lucerne leaves (cv Zaikievich) when inoculating by M3, 412b and M4 strains. The analysis of amino acids within 2 years of plant vegetation shows that in lucerne leaves of both cultivars 18 bound amino acids are permanently detected. Among them amides of aspartic and glutamic acids, phenylalanine, alanine, arginine, leucine and lysine predominate in all variants of the experiments.

One should note that quantitative composition of bound amino acids somewhat varies depending on a strain, stage, plant growth and cultivar. So, when inoculating by M4, M6, 435a and 425a strains the amount of phenylalanine in lucerne (cv Zaikievich) is increased by 13 to 20 % in comparison with noninoculated control. Leucine content is increasing by 7 to 14 % when inoculating by M4 and 425a strains correspondingly. Within the process of plant vegetation the number of bound amino acids is increasing during the transition from the stage of stemming to budding. Its content in the plants of the second year of growing is 1,2 to 1,5 times higher. In this period the content of asparagine, glutamine and alanine is increasing by 40 to 50%. It causes the increase of the amount of some irreplaceable amino acids (lysine, arginine, histidine).

Qualitative and quantitative composition of free amino acids, which content in lucerne leaves is by 3-4 orders lower, than that of bound, is an evidence for the direction and intensity of metabolic process in a plant cell (Table 2). Inoculation by nodule bacteria of tested strains essentially changes quantitative relationship of free amino asids in leaves.

So, inoculation by M6 strain and byM3 and 425a strains increases lysin content 6 and 4 times correspondingly in comparison with noninoculated control. Maximum content of free amino acids, including irreplaceable ones in the leaves of lucerne (cv Zaikievich and cv Yaroslavna), is revealed when inoculating by M6 strain for the first and second years of growing. A comparative analysis of data on the stages of plant growth shows that the total number of free amino acids is 2 to 10 increased by the second year of vegetation period within both cultivars depending on a strain. It has been established that in

Table 2. *Free amino acids in the leaves of lucerne (cv Zaikievich) when inoculating by various R. meliloti strains (the second year of vegetation period - the stage of budding-initiation of flowering)*, mg/kg

	Strains of nodule bacteria							
Amino acids	control	425a	435a	412b	M6	M4	M3	M7
Arg	5	6	2	5	12	8	9	6
Val	14	31	15	17	33	18	24	21
I le	7	13	4	10	15	15	17	10
Leu	7	12	3	13	22	17	23	11
Lys	3	13	1	5	18	7	12	5
Met	6	21	7	5	11	6	12	15
Thr	12	18	8	12	27	17	19	22
Phe	7	9	5	5	15	12	16	14
The sum of irreplaceable	61	123	45	72	153	100	132	104
Asparagine	105	117	66	159	182	181	30	104
Asp	60	40	40	55	93	120	53	41
Glutamine	1	8	3	5	11	7	2	5
Glu	89	82	89	138	146	192	124	132
γ-aminobutyric	20	25	21	32	51	50	48	36
Tyr	10	13	10	14	16	15	24	22
Cys	20	50	30	20	30	60	50	50
Total content	502	784	442	631	939	932	691	672

the process of development within lucerne leaves of both cultivars 26 free amino acids are found. Some of them (histidine, tryptophane and γ-aminobutyric acid) are revealed not at all stages of growth.

It shoud be noted that at the stage of budding in the leaves of lucerne (cv Zaikevich and cv Yaroslavna) the content of giutamine is lower by 1 to 2 orders than that of asparagine. It is an evidence for the worsening of carbohydrate exchange in plant leaves.

Thus, when inoculating lucerne (cvs Zaikievich and Yaroslavna) by active strains of Rhizobium not only the yield of green mass and protein harvest is increased by 10 to 30 %, but the quality of forage protein is as well improved due to the increase of the content of some bound amino acids including irreplaceable ones. Maximum amino acid content in leaves of lucerne of both cultivars is found at the stage of budding of the second year of plant vegetation. That is why hay cutting is recommended to carry out exactly in this period.

Reference

1. E.P.Starchenkov, S.Ya. Kots, Biological nitrogen in agriculture and role of lucerne in addition of its resources in the soil. "Physiology and biochemistry of cultivated plants", Kiev, 1992, 24, 4, 325.
2. Simovic Milutin, Sadrzaj sirovih proteina w nadzemnim organima lucerke zavisno od fasa razva ja poothosima "Poljopr. i sum.", 1988, 34, 1.93.
3. M.M.Nichik, N.V.Peterson, S.Ya.Kots et al. Bound and free amino acids of lucerne leaves when inoculating by new Rhizobium strains "Phys. and biochemistry of cultivated plants", Kiev, 1991,23,5, 439.
4. N.V.Peterson, T.A.Chernomurdina, Amino acid composition of nodules and leaves within legume plants when inoculating by nodule bacteria of various efficiency "Microbiological Journ,"Kiev.1987, 49, 2, 23.
5. B.P.Pleshkov, " Practical course on plant biochemistry." Kolos, Moscow, 1985, 3, 255.
6. V.G.Ryadchichov " The improvement of cereal plants and their estimation." Kolos, Moscow, 1988, 368.
7. A.S.Timoshenko, O.N.Kasuto, E.E. Gordeeva Quantitative estimation of putrescin in plant material in the presence of free amino acids, "Plant Physiology,"Moscow, 1980, 80, 6, 1308.

MANIPULATION OF STARCH COMPOSITION AND QUALITY IN PEA SEEDS

C. L. Hedley[a], T. Ya. Bogracheva[b], J. R. Lloyd[a] and T. L. Wang[a]

[a] John Innes Centre
Norwich Research Park
Colney
Norwich NR4 7UH

[b] The Institute of Biochemical Physics
RAS
Moscow
Russia

1 INTRODUCTION

Starch comprises more than 30% of the average diet in the UK on a dry weight basis. As well as being a major component of natural foodstuff, such as bread, it is also used widely in the food processing industry because of its ability under certain circumstances to form pastes and gels. Most starch used for processing within the UK and the EU comes from North America as maize starch. There is also a substantial market in the EU for potato starch, although this is used mainly in the non-food sector.[1]

At present much of the starch used in food is hydrolysed to produce glucose, fructose, maltose and syrups for drinks and confectionary. Most of the remainder is chemically modified to suit specific uses (Table 1) by reducing problems such as retrogradation, or reassociation, of amylose or amylopectin in aqueous solutions, variable peak viscosity and the lack of long term stability.[2] Many of these processes utilise chemicals which are toxic, demanding expensive toxicity testing of the resulting food product. They also hold a risk of environmental pollution.

There is an advantage, therefore, in producing a wider range of natural starches which may replace or reduce the need for such processes. At present the main sources of starch are from cereals and potatoes but very few industries in Western Europe have considered grain legumes for this purpose.

1.1 Cereal Starch

There are a number of reasons why maize starch is popular with food processors. Together with other diploid cereals, there is a wide range of genetic diversity available within maize which has increased the potential uses and hence the market. In particular, genes have been identified which give a range of amylopectin in the starches from 20% to 75%. Furthermore, following extraction of the starch from maize meal the byproducts, oil and protein, provide additional commercially valuable material.

The main problem with utilising maize starch is the difficulty of extracting the starch from the grain. There is the initial problem of separating the germ from the grain and the secondary problem of separating the protein from the starch. In addition to these technical problems, all maize starches used in the UK have to be imported.

Table 1.

Some important properties and uses of modified starches

Modified starch	*Treatment (example)*	*Advantages over natural starch*	*Examples of use in food*
Pregelatinized	heat/moisture	cold-water soluble	pie fillings, 'instant products' coatings
Acid-thinned	acid	low hot paste viscosity high gel viscosity	gums, jellies
Oxidized	hypochlorite	increased clarity, reduced set-back	gravy/sauce thickeners, jellies
Hydroxyalkyl ethers	propylene oxide	increased clarity, increased stability reduced set-back	salad dressings, pie fillings instant foods
Esterified	acetic anhydride	increased clarity, forms films, fibres	frozen foods
Monophosphates	phosphoric acid	increased stability to freeze/thaw cycles	frozen foods, infant formulae
Cross-linked, eg di-starch phosphate	phosphorus oxychloride	increased stability to heat, pH, shear and freeze-thaw cycles	wide range of canned and frozen foods

Information derived from Starch: Properties and Potential. Ed. T. Galliard. Pub. John Wiley and Sons, Chichester, New York, Brisbane, Toronto, Singapore, 1987.

The possibility of using wheat as a home grown alternative to maize suffers from a lack of the genetic diversity found in maize, plus all of the problems of extraction associated with the other cereals. In addition, wheat starch has two populations of starch granules which differ in size. Like maize, wheat starch also contains significant amounts of lipid (ca. 1%), which adversely affects the functional properties of the starch and is prone to oxidation giving 'off flavours'.[3]

1.2 Potato Starch

It has been suggested that potato starch would be preferred to maize starch if it was available in sufficient quantities. This is because of its superior functional properties due in part to its very low level of lipid. It has excellent pasting, film making and binding characteristics and it retrogrades very slowly.[2]

Although a significant potato starch industry has developed within The Netherlands, there are serious economic problems with the wider establishment of a starch industry based on this species. Unlike maize, the byproducts have little commercial value and there is less available genetic diversity for widening the uses.

The main problems with utilising potatoes for starch production, however, are associated with harvesting, transporting and storage of the tubers, which are composed of about 80% water and only about 17% starch. The result of these problems is that it is only possible to extract and process starch from potatoes for four months of the year. There would be a financial penalty, therefore, of maintaining an extraction plant which is only used for part of the year.[1]

1.3 Grain Legume Starch

There is a need within the UK and the EU for a home based starch industry to provide raw material to food processors. For the reasons given above the existing crops, in particular maize and potato, both have specific economic and technical problems. There is a suggestion, in particular from countries in Eastern Europe and the States of the Former Soviet Union, that the main contenders to fill this role are the grain legumes and specifically the pea (*Pisum sativum).*

The pea crop will grow and yield well in the temperate climate of Western Europe and in recent years it has become a significant crop in terms of acreage, particularly in France. At present the main use for pea seed is in the production of high protein animal feedstuff and a range of new varieties have been developed for this purpose. Although peas are thought of as a protein crop, starch is the major storage product, contributing about 50% of the seed dry weight - twice that of the protein.

Within this short presentation we will discuss some of the particular characteristics of legume starches, and pea starch especially, that distinguish them from cereal and potato starches. We will demonstrate the range of variation for starch content and composition that can be generated at present within the pea and finally discuss the potential for producing new starches in pea seeds according to nutritional and industrial requirements.

2. GENERAL CHARACTERISTICS OF PEA STARCH

All starches, including those from peas and other legumes, are composed of two types of molecule, amylose and amylopectin. Amylose is an essentially linear polymer composed of glucose monomers linked by α-(1-4) glycosidic linkages, with an average molecular weight of up to 2×10^6 g mol^{-1}. Amylopectin, unlike amylose, is a highly branched macromolecule consisting of linear α(1-4) linked glucose chains with α-1-6) branch points after every 20-25 glucose residues. The two types of molecule are characterised by their iodine binding behaviour under standard conditions; amylose binding ca. 20% w/w and amylopectin less than ca. 1% iodine.[4] This definition is confused by some amylopectins having longer than normal chains which enhances the iodine binding capacity giving an overestimate for the apparent amylose content of starch.[5]

2.1. Starch granule structure

All native starches contain semi-crystalline regions caused by the ordered packing of adjacent branches of amylopectin molecules. The structure of legume starches, including peas, differs from that of cereal and potato starches, however, in the way that these crystalline regions are constructed and packed within the starch grain. This difference can be demonstrated by the patterns produced using wide angle X-ray diffraction (Fig 1).

Maize and potato are characterised by having 'A'- and 'B'-type patterns respectively, whereas legume starches are characterised by a 'C'-type pattern. It is believed that starches with a 'C'-type pattern are composed of a mixture of 'A' and 'B'-type polymorphs, or crystalline structures.

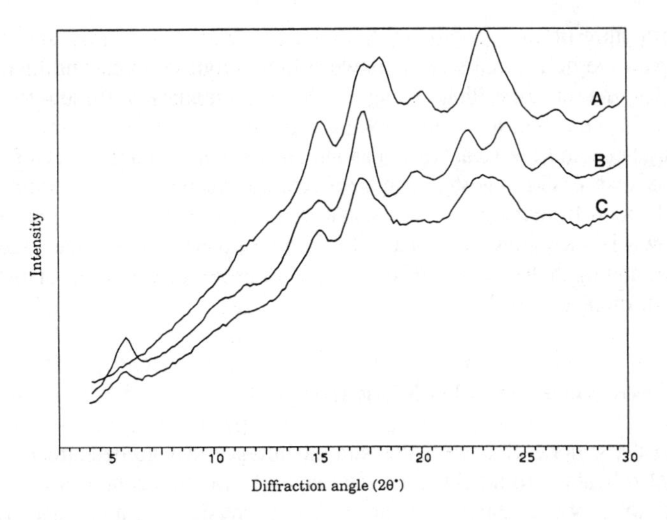

Figure 1 *WIde angle X-ray diffraction patterns for A-type (A; Maize); B-type (B; Potato) and C-type (C; Pea) starches.*

2.2. Functional Properties

The structure of starches is related to the functional properties which they demonstrate during processing and when present in different foods. Starch-containing foods are usually cooked by heating in the presence of water. During this process starches exhibit the useful property of thickening and gelling aqueous solutions. The functional properties of the starches are usually described by their viscosity, swelling and solubility behaviour under heating, followed by cooling and storage.

In general, legume starches are characterised by a higher level of solubilisation and restricted swelling of the grains, compared with cereal and potato starches[6,7,8]. The consistency of pastes produced by legume starches is usually comparable with that found for other starches, although there are some reports that the viscosity of heated and cooled pea starch suspensions is very high[6,7,8,9] , which could be very useful for the food industry. Another useful property of legume starches is that their heated suspensions have a stable viscosity during storage under high temperature[6,7,8,9].

When pea starches are modified the functional properties are changed; oxidation giving hard gels and drying suspensions at high temperatures give a high and stable level of swelling in cold water (Bogracheva and Davidova; unpublished data).

2.3. Nutritional Properties

In general starches are degraded to a large extent in the gastro-intestinal tract. The starch in cooked foods, however, is known to retrograde in storage. This is due mainly to the agregation of amylose resulting in a starch fraction which is resistant to α-amylase. Such starches are known as resistant starch[10] and is generally unavaliable as a carbohydrate for digestion. This characteristic of resistance to digestion makes resistant starch analagous to dietary fibre and may result in its inclusion in the analysis of total dietary fibre of some foods[11].

Comparative nutritional studies between high amylose maize and high amylose peas (containing the mutant *r* allele) using rats has shown that that these two 'resistant' starches differ in the extent to which they are digested. The maize starch appeared to be partialy digested and the remainder was almost totally fermented. The pea starch, however, was less easily digested and also more resistant to fermentation[12]. The sinificance of this difference between resistant maize and pea starch is not known. It has been shown in rats, however, that there is no evidence for resistant starch from either source having a beneficial effect on the loss to faeces of protein and fats compared with dietary fibre[13].

3. MANIPULATION OF PEA STARCH

In addition to the general characteristics and advantages of legume starches discussed above, peas also benefit from having a wide range of available genetic variation which can be used to produce starches with different physico-chemical and functional properties. They are also easy to manipulate genetically because the starch is produced in the diploid embryo rather than in a triploid endosperm, as in cereals. In addition, the pea is an inbreeding species which means that the stabilisation and maintenance of naturally occurring or induced variation is relatively easy.

3.1. Mutants Affected in Starch Content and Composition.

Genes at two loci, *r* and *rb*, in peas have been known for many years to affect the content and composition of starch within the seed.[14,15] Genes at both loci are known to encode key enzymes within the starch biosynthetic pathway; the *R* gene for starch branching enzyme I and the *Rb* gene for ADPglucose pyrophosphorylase.[15,17]

A mutant gene at the *r* locus reduces the starch content from about 55% to about 35% of the embryo dry weight and changes the proportion of amylose in the starch from about 35% to about 65%. As well as affecting the content and composition of the starch, the presence of a mutation at the *r* locus also changes the shape of the starch grains in the mature seed from simple and oval to compound in appearance[18] (Plate 1).

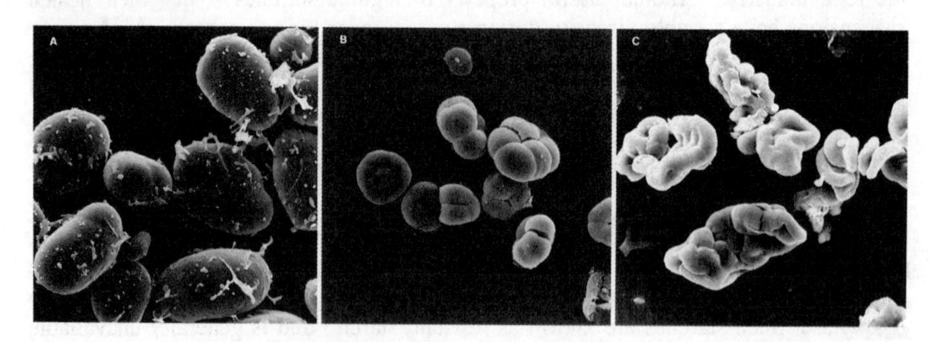

Plate 1. *Starch granules from wild type (A) and r (B), rug-5 (C) mutants.*

The presence of a mutant gene at the *rb* locus, while decreasing the starch content by a similar amount to the *r* gene, reduces the proportion of amylose in the starch to about 20%[19]. Unlike genes at the *r* locus, genes at the *rb* locus do not appear to affect the shape of the starch grain in the dry seed.

The presence of mutant genes at both the *r* and *rb* loci reduces the starch content to about 25% of the seed dry weight and gives an amylose content in the starch of about 50%. Starch grains from the dry seeds of this double mutant have a similar compound appearance to those found when only the *r* mutation is present[19].

3.2. Structure and Properties of 'Mutant' Starches

The inclusion of mutations at either or both the *r* and *rb* loci gives rise to four pea genotypes (*RRRbRb, rrRbRb, RRrbrb* and *rrrbrb*) which differ for the content and composition of starch in the dry seed. Following a backcrossing programme, lines which are near-isogenic except for genes at these two loci have been produced.[18,19] These lines have been used to study the structure and physico-chemical properties of the starches produced.

3.2.1. Starch Structure A comparative study of the constituent chain profiles from the amylopectin component of the four starches has revealed that a mutation at the *rb* locus has no apparent affect on the structure of the starch.[20,21] A mutation at the *r* locus, however, results in an increase in the amylose content of the starch and a change in the structure of the amylopectin.[20,21] The structure of the amylopectin has been studied following debranching by isoamylase and separation of the branches using ion exchange and size exclusion chromatography[20,21] (Fig 2). Starch from the *r* mutant has a range of branch sizes which is wider than in lines with the wild-type allele. In addition, there is a suggestion that the presence of the *r* mutation results in the production of a class of molecules which are intermediate in structure between amylose and amylopectin.[22]

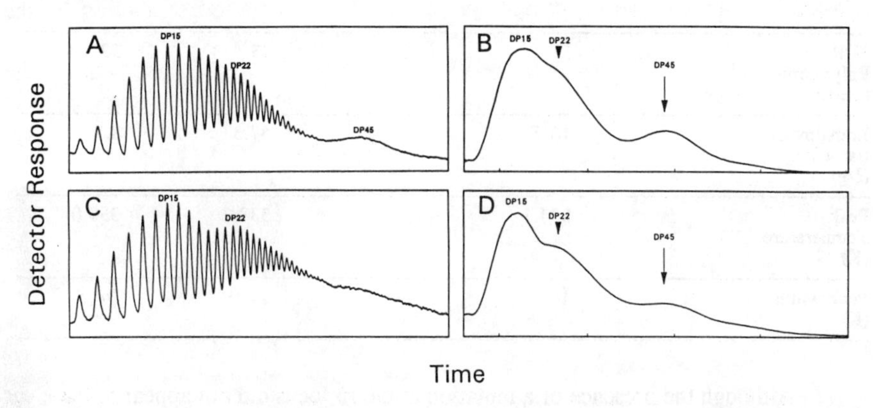

Figure 2. *Ion exchange (A and C) and size exclusion (B and D) chromatography of amylopectin branches obtained from wild-type (A and B) and rrRbRb (C and D) starches.*

3.2.2. Physico-chemical Properties As was mentioned earlier, legume starches can be distinguished by having a 'C'-type structure which differs from that found in either maize ('A'-type) or potato ('B'-type) when studied by wide angled X-ray crystallography. The molecular order of the starch can be further studied by following the process of gelatinisation during heating in excess water using Differential Scanning Calorimetry (DSC).

It has been reported that during this process wild-type pea starches give single gelatinisation endotherms[23]. This single peak has been shown to become double in a range of wild-type starches when the gelatinisation is carried out in salt solutions[24,8]. It has been proposed that this is due to the 'A' and 'B' polymorphs, which make up the 'C'-type starch, being in different independent cooperative structures within the starch grain[24,8].

When the starches from the *r* and *rb* near-isogenic lines were studied using X-ray diffraction they were all found to have 'C'-type diffraction patterns but different contents of 'A' and 'B' polymorphs[25] (Table 2). The presence of a mutation at either locus increased the 'B' polymorph content in the starches, the effect being much greater when the mutation was at the *r* locus which greatly reduces the crystallinity. In addition, the starch from the *rrRbRb* line did not show the usual single peak during gelatinisation in excess water. The inability of the *rrRbRb* starch to undergo the cooperative transition which gives rise to the gelatinisation peak is correlated with the very high proportion of amylose and observed defects in the crystalline structure of this starch.

Table 2. *Characteristics of starches obtained from wide angle X-ray diffraction and DSC analysis.*

	RRRbRb	rrRbRb	RRrbrb	rrrbrb
% Crystallinity of starch	27	16	30	22
% B Polymorph content	37	73	43	92
Enthalpy of transition (J/g)	16.7	-	57.3	3.8
Peak Temperature (K)	334.1	-	338.2	350.0
Peak width (K)	16	-	14	24

Although the presence of a mutation at the *rb* locus did not appear to have such a dramatic effect on the crystalline structure of the starch, the *RRrbrb* starch did have a greatly increased enthalpy of gelatinisation in excess water compared with the wild-type starch[25] (Table 2). It is suggested that this is connected with specific charge interactions between the molecules in the starch granule.

DSC and X-ray studies of starch from the *rrrbrb* line indicated that there was an interaction between the two genes to produce a starch with properties which reflected the effect of both mutations. This starch had a relatively wide gelatinisation peak in excess water, with low enthalpy and a very high peak temperature of transition[25] (Table 2).

4. DEVELOPMENT OF NEW PEA STARCHES

Although it is now known that genes at the *r* and *rb* loci encode enzymes affecting specific steps in the starch biosynthetic pathway, they were first identified and their inheritance determined by following the effect of mutations at these loci on the phenotype of the seed[14,15]. The dry seed of mutants which have either or both recessive alleles at the *r* or *rb* loci have a wrinkled appearance, compared with the round seeded wild-type. It is generally accepted that this is due to an increase in water content of developing mutant seed, attributed to a rise in the sucrose level and a decrease in starch synthesis induced by the mutation[26].

4.1. Selection of New Wrinkled Seeded Mutants

The phenotypic effect on seed shape associated with mutations at the *r* and *rb* loci has been used as a primary criterion to select new mutants affecting starch content and composition, following a chemical mutagenesis programme carried out in 1987[27,28,29].

The primary selection based on seed shape was followed by a secondary selection based on chemical analysis of the starch content and composition of the mutant seed. From this programme we have isolated about 30 new mutants which have wrinkled seed and changes in the content and composition of the starch, compared with the wild-type round seeded parent line[27].

Genetic analysis of all the mutant lines has demonstrated that a number of these new mutants are allelic to the original r and rb mutant alleles,[30] although differing in the severity of their effect on starch and amylose contents[14,15,31,32] (Table 3). In addition, mutations have been identified at three new loci, now given the gene symbols rug-3, rug-4 and rug-5[30] (Table 3).

Table 3. *Effect of mutations at 6 independent loci on starch and amylose contents.*

	Starch (% dry weight)	Amylose (% starch)
Wild Type	50	35
r	27-36	60-75
rb	30-37	23-32
rug-3	1-12	0-1
rug-4	38-43	31-33
rug-5	29-36	43-52
lam	39-49	4-10

Mutants at the rug-3 locus have starch contents which range from 1 to 12% and amylose contents between 0 and 10%. Mutants at the rug-4 locus are similar in starch content and composition to those at the *rb* locus; the starch contents ranging from 38 to 43% with amylose contents between 31 and 33%[14,28].

The presence of mutations at the *rug-5* locus reduces the starch content to 29-36% and increases the amylose content to 43-52%. In addition, starch granules extracted from *rug-5* mutant seeds have a unique, irregular, appearance[20] (Plate 1) and the structure of amylopectin isolated from the starch also appears to be unique with a proportion of very long side chains following debranching[20].

4.2. Selection of Round Seeded Starch Mutants

The primary selection criterion applied to identify new starch mutants was a change in seed shape, which has invariably been associated with a decrease in the starch content of the seed[15]. It is evident that such a selection system, therefore, will not identify mutants which have an altered starch composition but show no reduction in starch content. Such mutants are known to exist in other species, for example the starch of the *waxy* mutant in maize has a very low amylose level but the seed has a normal starch content.

Such low amylose starches give a red-brown colour with iodine rather than the indigo colour associated with starch containing the more usual 30% amylose. A screening method was devised, therefore, which was based on a colour difference in the starches between wild-type and mutant[33].

The screen produced five mutant lines, all of which had starches which stained red-brown with iodine. The seeds from these lines were similar in shape to the wild-type and the starch content was not significantly reduced. Following chemical analysis, the amylose content has been shown to be very low. The starch grains appear normal in shape but on staining with iodine they attain a blue core with a pale periphery[33]. The basis of this iodine staining is unclear but may be due to an effect of the gene on the amylopectin chain length.

Genetic analysis has shown that the new mutants are all monogenic recessives and reside at a common locus. The new mutants have been assigned the symbol *lam* (*l*ow *am*ylose)[34].

5. FUTURE DEVELOPMENTS

To date six loci (*r, rb, rug-3, rug-4, rug-5* and *lam*) have been identified which affect the content and composition of starch in pea seeds. In addition, several alleles, which vary in the degree of their affect, have been identified at each of these loci. It has already been shown that mutations at two of these loci (*r* and *rb*) dramatically affect the structure and physico-chemical properties of the starch. With the variation available, it seems very likely, therefore, that a wide range of natural starches can be produced by manipulating the genes at these loci[35,36]. In addition, it has been shown that the starch from the *rrrbrb* double mutant has different properties to that produced by either of the single mutants at these two loci[25]. There is the additional possibility, therefore, of increasing the starch variation further by producing a range of such double or even triple mutants.

There is an essential requirement at present to fully study the structure and physico-chemical properties of the mutant starches and to relate these to the functional properties. With this information and with the available genetic variation it should be possible to produce pea starches which are designed for specific uses within the food industry[14,35,36,37].

References

1. T. Galliard, 'Starch: Properties and Potential', Wiley and Sons, Chichester, New York, Brisbane, Toronto, Singapore, 1987.
2. G. F. Visser and E. Jacobsen, Tibtech, 1993, **11**, 63.
3. W. R. Morrison and H. Gadan, *Journal of Cereal Science*, 1987, **5**, 263.
4. T. R. Noel, S. G. Ring and M. A. Whittam, 'Modern Methods of Plant Analysis', Springer-Varlag, Heidelberg, 1992.
5. W. R. Morrison, 'Modern Methods of Plant Analysis', Springer-Varlag, 1992.
6. J. L. Doublier, *Journal of Cereal Science*, 1987, **5**, 247.
7. R. Hoover and T. Vasanthan, *Carbohydrate Research* 1994, **252**. 33.
8. N. I. Davydova, S. P. Leontiev, Ya. V. Genin, A. Yu. Sasov and T. Ya. Bogracheva, *Carbohydrate Polymers*, 1995, (in press).
9. R. Stute, *Starch/Stärke*, 1990, **42**, 178.
10. H.N. Englyst, H.S. Wiggins and J.H. Cummings, Analyst, 1982, **107**, 307.
11. H.N. Englyst, H. Trowell, D.A.T. Southgate and J.H. Cummings, American Journal of Clinical Nutrition, 1987, **46**, 873.
12. R.M. Faulks, S. Southon and G. Livesey. British Journal of Nutrition, 1989, **61**, 291.
13. G. Livesey, I.R. Davies, J.C. Brown, R.M. Faulks and S. Southon, British Journal of Nutrition, 1990, **63**, 467.
14. T. L. Wang and C. L. Hedley, 'Peas, Genetics, Molecular Biology and Biotechnology', CAB International Press, Wallingford, 1993.
15. T. L. Wang and C. L. Hedley, *Seed Science Research*, 1991, **1**, 3.
16. A. M. Smith and K. Denyer, *New Phytol.*, 1992, **122**, 21.
17. M. K. Bhattacharyya, A. M. Smith, T. H. N. Ellis, C. L. Hedley and C. Martin, *Cell*, 1990, **60**, 115.
18. C. L. Hedley, C. Smith, M. J. Ambrose, S. Cook and T. L. Wang, *Annals of Botany*, 1986, **58**, 371.
19. C. L. Hedley, J. R. Lloyd, M. J. Ambrose and T. L. Wang, *Annals of Botany*, 1994, **74**, 365.
20. J. R. Lloyd, PhD Thesis, University of East Anglia, 1995.
21. V. J. Bull, C. L. Hedley, J. R, Lloyd, S. G. Ring and T. L. Wang, *Carbohydrate Polymers*, 1995, (in press).
22. P. Colonna and C. Mercier, *Carbohydrate Research*, 1984, **126**, 233.
23. A. Buleon, H. Bizot, M. M. Delage and B. Pontoire, *Carbohydrate Polymers*, 1987, **7**, 461.
24. T. Ya. Bogracheva, S. P. Leotiev and Ya. V. Genin, *Carbohydrate Polymers*, 1994, **25**, 227.
25. T. Ya. Bogracheva, N. I. Davydova, Ya. V. Genin and C. L. Hedley, *J. Experimental Botany*, 1995, (in press).
26. T. L. Wang, C. M. Smith, S. K. Cook, M. J. Ambrose and C. L. Hedley, *Annals of Botany*, 1987, **59**, 73.
27. T. L. Wang, A. Hadavizideh, A. Harwood, T. J. Welham, W. A. Harwood, R.

Faulks and C. L. Hedley, *Plant Breeding*, 1990, **105**, 311.
28. T. Wang, M. Macleod, S. Johnson, A. Jones, L. Barber, C. Martin and C. Hedley, 'Proceedings of the IV International Workshop on Seeds', ASFIS, Paris, 1993.
29. C. L. Hedley, M. Macleod, S. Johnson, A. Jones, L. Barber and T. L. Wang, 'Proceedings of the 1st European Conference on Grain Legumes', Angers, 1992.
30. T. L. Wang and C. L. Hedley, *Pisum Genetics*, 1993, **25**, 64.
31. M. R. Macleod, C. L. Hedley, C. R. Martin and T. L. Wang, *Aspects of Applied Biology*, 1991, **27**, 263.
32. M. Macleod, C. L. Hedley, C. R. Martin, D. A. Jones, S. Johnson, A. Bakhsh, L. M. Barber and T. L. Wang, 'Proceedings of the 1st European Conference on Grain Legumes', Angers, 1992.
33. K. Denyer, L. R. Barber, R. Burton, C. L. Hedley, C. M. Hylton, S. Johnson, D. A. Jones, J. Marshall, A. M. Smith, H. Tatge, K. Tomlinson and T. L. Wang, *Plant Cell and Environment*, 1995, (in press).
34. T. L. Wang, L. R. Barber, R. Burton, C. L. Hedley, C. M. Hylton, S. Johnson, D. A. Jones, J. Marshall, A. M. Smith, H. Tatge, K. Tomlinson and K. Denyer, *Pisum Genetics*, 1994, **26**, 39
35. C. L. Hedley and T. L. Wang, 'Proceedings of the 1st European Conference on Grain Legumes', Angers, 1992.
36. C. L. Hedley and T. L. Wang, *Aspects of Applied Biology*, 1991, **27**, 205.
37. C. L. Hedley and T. L. Wang, *Agrofood Industry Hi-Tech*, 1994, 14.

EVOLUTION OF NON-STARCH POLYSACCHARIDES AND RESISTANT STARCH CONTENT DURING PEA SEED DEVELOPMENT

Periago, M.J.* and Englyst, H.N.**

* U.D. de Bromatología e Inspección de Alimentos. Faculty of Veterinary. University of Murcia. 30.071-Murcia, SPAIN.
** MRC Dunn Clinical Nutrition Centre, Hills Road, Cambridge CB2 2DH, UK.

1. INTRODUCTION

The beneficial effects associated with legumes consumption, which are related to the slow rate of starch digestion and the high resistant starch and dietary fiber content have been reported by several authors[1,2]. In addition, the content of non-starch polysaccharides in legumes is high, which inhibits swelling and dispersion of starch during processing[3]. The resistant starch content and non-starch polysaccharides of plant foods play an important role in the colonic fermentative process since they are fermented by colonic bacteria, and form short chain fatty acids, which have beneficial physiological effects in humans. The aim of this study is to investigate the evolution of the non starch polysaccharides content and resistant fraction of starch in peas during seed development.

2. MATERIAL AND METHODS

2.1 Material

Wrinkled green peas (Citrina and Warindo cultivars) were classified into four commercial sizes (superfine: SF, 4.7-7.5 mm; very fine: VF, 7.6-8.2 mm; fine: FN, 8.3-8.8 mm and middle: MD, 8.9-10.2 mm) according to seed diameter and development.

2.2 Methods

All analyses were carried out with dry samples after freeze-drying the pea seeds. The non-starch polysaccharides (NSP) content was studied using an enzymatic-gravimetric and GLC technique described by Englyst et al.[4] Resistant starch (RS) and total starch (TS) were determined by controlled enzymic hydrolysis and measurement of the released glucose, as decribed by Englyst et al.[5].

3. REULTS AND DISCUSSION

Figure 1 shows the total, soluble and insoluble non-starch polysaccharides (NSPt, NSPs and NSPi, respectively), and the TS and RS content of in the studied samples. Table 1 give the cellulose and uronic acids content, and neutral sugar composition of NPSt, NSPs and NSPi residues of peas. The NPSt, NSPs and NSPi

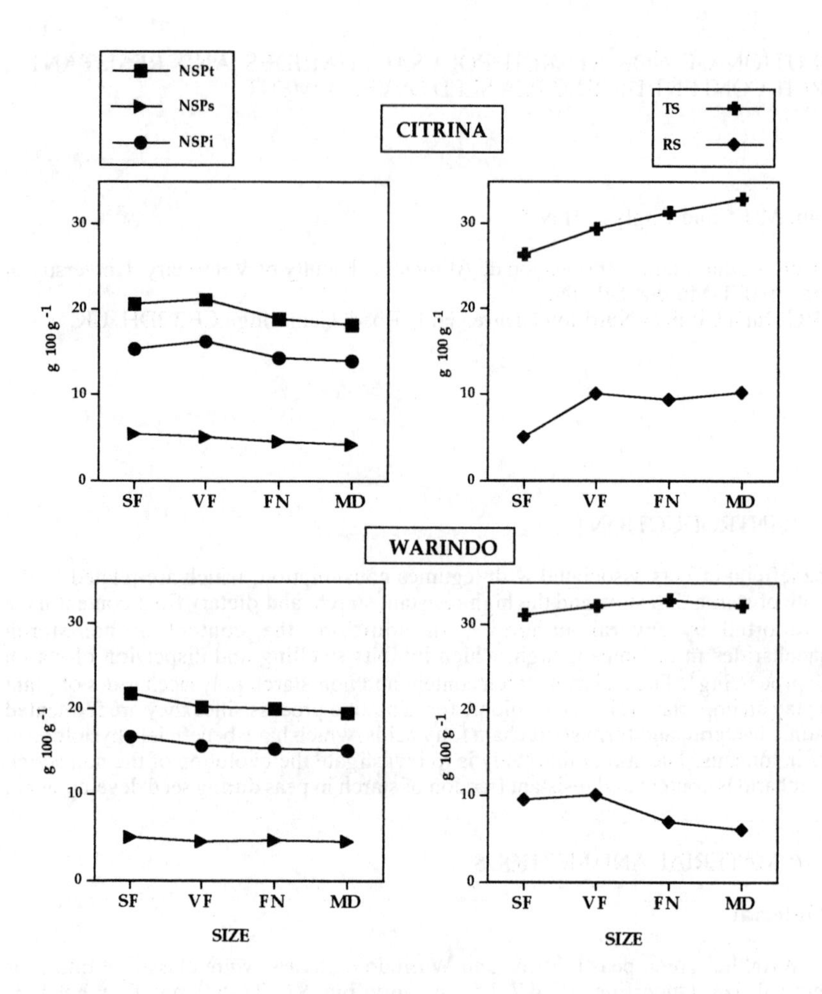

Figure 1 Evolution of total, soluble and insoluble non-starch polysaccharides (NSPt, NSPs and NSPi), total starch (TS) and resistant starch (RS), in Citrina and Warindo pea cultivars, classified in four commercial sizes (Abbreviations of size = SF, super fine, 4.7-7.5 mm; VF, very fine, 7.6-8.2 mm; FN, fine, 8.3-8.8 mm; MD, middle, 8.9-10.2 mm.).

content decreased in both pea cultivars during development (Figure 1). The due to a significant fall in cellulose and uronic acids (Table 1). As regards for neutral sugars, the arabinose and xylose content increased during pea seed development, whereas the glucose content decreased (Table 1). As the peas matured there was a modification of the non-cellulosic polysaccharides, hemicellulose (mainly arabinose and xylose) and pectin (which is mainly uronic acid). In general, the non-cellulosic polysaccharide content of peas is constant during seed development since hemicellulose increases and pectin decreases (Table 1).

TS increased as pea size increased, which is related to the decrease in NSPt (Figure 1). Since starch is stored in the kernel as a physiological reserve of carbohydrates[6]. However, in Warindo cultivar the TS increase was not significant between sizes. RS in Citrina cultivar increased as pea size increased, the SF size showing the lowest value. On the other hand, in Warindo cultivar, RS decreased in FN and MD sizes, because this variety presented a higher maturity index and the natural amylolysis process would probably have hydrolyzed the RS in the seed.

Table 1. Mean values of cellulose, neutral sugar and uronic acid contents of soluble, insoluble and total non-starch polysaccharides residues in four sizes of peas of Warindo and Citrina cultivars.

			Neutral sugars (g 100 g $^{-1}$)							
Size[1]	Residue	Cel[2]	Rha	Fuc	Ara	Xyl	Man	Gal	Gluc	Ur Ac
SF_W	soluble	t[3]	0.46	0.07	1.23	0.23	0.10	0.43	t	2.43
	insoluble	9.30	0.40	0.03	2.37	0.77	0.10	0.30	1.60	1.40
	total	9.30	0.86	0.10	3.60	1.00	0.20	0.73	1.60	3.83
VF_W	soluble	t	0.40	0.07	1.33	0.20	0.07	0.40	t	2.03
	insoluble	8.67	0.40	0.03	2.97	0.93	0.10	0.30	1.00	1.27
	total	8.67	0.80	0.10	4.3	1.13	0.17	0.70	1.00	3.30
FN_W	soluble	t	0.40	0.08	1.55	0.23	0.08	0.50	t	1.85
	insoluble	8.33	0.40	0.02	3.20	0.98	0.10	0.30	0.68	1.30
	total	8.33	0.80	0.10	4.75	1.20	0.18	0.80	0.68	3.15
MD_W	soluble	t	0.50	t	1.40	0.20	0.10	0.40	t	1.75
	insoluble	7.90	0.40	0.10	3.10	0.95	0.10	0.35	0.60	1.20
	total	7.90	0.90	0.10	4.50	1.15	0.20	0.75	0.60	2.95
SF_C	soluble	t[3]	0.53	t	1.13	0.30	0.06	0.60	t	2.70
	insoluble	10.16	0.60	0.10	1.20	0.50	0.13	0.30	1.06	1.53
	total	10.16	1.13	0.10	2.33	0.80	0.20	0.90	1.06	4.23
VF_C	soluble	t	0.70	0.10	1.10	0.26	0.06	0.44	t	2.26
	insoluble	11.03	0.43	t	1.46	0.70	0.10	0.26	0.76	1.33
	total	11.03	1.13	0.10	2.56	0.96	0.16	0.70	0.76	3.60
FN_C	soluble	t	0.43	0.10	1.16	0.16	t	0.36	t	2.23
	insoluble	9.16	0.43	t	1.66	0.90	0.10	0.30	0.50	1.20
	total	9.16	1.00	0.10	2.83	1.06	0.10	0.66	0.50	3.43
MD_C	soluble	t	0.43	0.10	1.43	0.16	0.03	0.40	t	2.03
	insoluble	8.90	0.43	t	1.53	1.00	0.10	0.30	0.60	1.20
	total	8.90	0.86	0.10	2.96	1.16	0.13	0.70	0.60	3.23

[1] Abbreviations of size = SF, super fine, 4.7-7.5 mm; VF, very fine, 7.6-8.2 mm; FN, fine, 8.3-8.8 mm; MD, middle, 8.9-10.2 mm. Subscript W and C means Warindo and Citrina cultivars, respectively.
2 Abbreviations of parameters = Cel, cellulose; Ram, Rhamnose; Fuc, fucose; Ara , arabinose; Xyl, xylose; Man, manose; Gal, galactose; Gluc, glucose; Ur Ac, uronic acids.

We concluded that NSP content and RS of peas depend on the development stage of the seed, since NSP decreased and RS increased during development. Although, RS decreased at the beginning of maturation process.

References

1. D.J.A. Jenkins, T.M.S. Wolever, R. Taylor, H.M. Baker, H. Fielder, J.M. Baldwin, A.C. Bowling, H.C. Neewman, A.L. Jenkins and D.V. Goff, *Am. J. Clin. Nutr.*, 1981, **34**, 362.
2. S.A. Truswell, *Eur. J. Cli. Nutr.*, 1992, **46**, S91.
3. P. Würsch, S. del Vedovo and B. Koellreuter, *Am. J. Clin. Nutr.*, 1986, **43**, 25.
4. H.N. Englyst, M.E. Quigley and G.J. Hudson, *Analyst,* 1994, **119**, 1497.
5. H.N. Englyst, S. M. Kingman and J.H. Cummings, *Eur. J. Cli. Nutr.*, 1992, **46**, S33.
6. D.P. Savage and S. Deo, *Nutr. Abst. Rev.*, 1989, **55**, 66.

CHARACTERISTICS OF PEA VARIETIES ACCORDING TO GALACTOOLIGOSACCHARIDES CONTENT

F. Kvasnička, R. Ahmadová-Vavrousová, M. Mrskoš[2], J. Velíšek[1] and P. Kadlec

Institute of Chemical Technology
Department of Carbohydrate Chemistry and Technology
1 -Department of Food Chemistry and Food Analysis
Prague, Czech Republic
2 - State Institute For Agriculture Supervision And Testing, SKZÚZ
Brno, Czech Republic

1 INTRODUCTION

The oligosaccharides raffinose, stachyose, verbascose and ajugose (RFO = Raffinose Family Oligosaccharides) that are common in legume seeds are thought to be the major producers of flatulence when those foods are consumed. Owing to the absence of α-galactosidase enzyme capable of hydrolysing the α-1,6-galactosidic linkage, these oligosaccharides accumulate in the lower intestine and undergo anaerobic fermentation by bacteria. They may produce diarrhoea, flatus gas and their attendant discomfort [1]. Fibre polysaccharides and indigestible starch have been also associated with flatulence [2]. On the other hand, the volatile fatty acids that are produced when carbohydrates are fermented in the colon have been suggested to have positive effects on cholesterol and carbohydrate metabolism. The decrease in faecal pH, due to production of acids, has also been suggested to be preventive factor against cancer in the large bowel [3]. The benefits of oligosaccharide ingestion arise from increased population of indigenous bifidobacteria in the colon which, by their antagonistic effect, suppresses the activity of putrefactive bacteria and reduce the formation of toxic fermentation products [4].

All these factors of flatulence were described in detail in foreign literature, but only from point of view of searching the causes of this phenomenon but not as to the usage for research of existing varieties or screening of a great number of genotypes in pea breeding programme as well as older varieties. These are often used as parental components in the formation of new variety or in searching for genetic dependence for a possible way of breeding.

The objective of this work is finding of pea varieties with extreme RFO levels (both low and high levels). Such pea varieties will be further analyse by gel electrophoresis with the aim to find "working gene" which controls RFO synthesis. Finding a suitable genotype, working gene respectively, would remove the up to now existing and deep-rooted prejudices which for greater percentage of population be caused by natural dietetic problems. Finding it could lead to the renaissance of this valuable food.

2 EXPERIMENTAL

Experimental work was divided into following steps:
1) finding out of RFO distribution in pea plant
2) monitoring of RFO in pea germplasm collection
3) statistical evaluation of the obtained data
4) selection of suitable pea varieties for further experiments

2.1 Material and methods

2.1.1 Pea samples. A set of 250 ground pea seeds of variety CH 589 (breeding company Selgen) was used for finding out of the distribution of RFO in pea plant. The RFO monitoring was done using a world-wide pea germplasm collection representing 1200 samples of variety *Pisum* (two taxonomic subgroups - Pisum sativum var. sativum ssp. hortense (L01) and Pisum sativum var. speciosum ssp. arvense (L02)). This pea collection was provided from genetic resources of Agritec, Ltd., Šumperk, Czech Republic. A database of morphological characteristics descriptors (34 descriptors) of pea varieties was also provided from Agritec, Ltd. [5].

2.1.2 HPLC analysis. Two different chromatography methods [6] were employed for RFO determination in 80% ethanolic pea extracts - RP-HPLC (reversed phase - silica C18) and IMP HPLC (Ion Moderated - cation exchanger in Ca form). The RP HPLC method was used for finding out of distribution of RFO in pea plant and the IMP HPLC for RFO monitoring analysis.

2.1.3 Statistical evaluation. Statistical techniques employed were Simple Descriptive Statistics, Least Squares Analysis and Discriminant Analysis which were performed on an IBM compatible computer using software SPSS/PS+ V3.1 purchased from SPSS GmbH, München, Germany.

3 RESULTS

The finding out of RFO distribution in pea plant was made prior to start of RFO monitoring of pea collection. The aim of this analysis is to determine variations of RFO content in pea seeds that come from different plants of the same variety, different branches and different pods. For this purpose a 250 pea sample was analysed on RFO content (fifty plants, each plant possesses of up to 7 branches and each branch has up to 3 pods). The obtained data were statistically evaluated by Least Square Analysis. On the basis of the obtained results it can be stated that there is not statistically significant differences of RFO levels between seeds come from different plants and/or from different parts of plant. So that a pea meal obtained from several pea seeds of certain variety can be taken as representative sample of such variety.

Up to now 700 samples of different pea variety (666 samples of Pisum sativum var. sativum ssp. hortense and 34 samples of Pisum sativum var. speciosum ssp. arvense) were analysed. The results of descriptive statistics of the obtained data are summarise in Table 1. Pea varieties with extreme saccharide content are listed in Table 2.

The pea varieties were divided into two groups - group 1 and group 2 that comprised the pea varieties with RFO content lower than average content (4.86 %) and higher, respectively. By discriminant analysis of such data set (totally 454 pea samples of both subgroups L01 and L02) was finding the relationship between morphological descriptors and RFO content, i.e., the order of descriptor significance for classification of pea varieties into the above mentioned groups. Some samples and/or descriptors were removed from database due to missing values. Results of discriminant analysis are summarised in Table 3. Percentage of "grouped" cases correctly classified using all 29 descriptors was 76 (L01), 83 (L02) and 76 (L01 + L02), respectively.

Table 1 *Descriptive statistics of analysed samples*

Parameter	*VERBASCOSE (g/100 g)*	*STACHYOSE (g/100 g)*	*RAFFINOSE (g/100 g)*	*SUCROSE (g/100 g)*	*RFO* (g/100 g)*
Average	2.47	1.86	0.53	2.48	4.86
Minimum	0.65	0.79	0.16	0.63	3.29
Maximum	5.77	4.97	1.60	6.92	9.97

* sum of Verbascose, Stachyose and Raffinose content

Table 2 *Pea varieties with extreme saccharide content*

Variety	*Parentage*	*Description*	*Saccharide content (g/100 g)*
Frostar (L01)	unknown	green-seeded market pea, wrinkled, early	9.97 - RFO
Lowadis (L01)	unknown	green-seeded market pea, wrinkled, early	9.84 - RFO
Kelvedon Wonder (L01)	unknown	green-seeded market pea, wrinkled, early	9.70 - RFO
Lowador (L01)	unknown	green-seeded market pea, wrinkled, early	9.65 - RFO
Lorina (L01)	unknown	green-seeded market pea, wrinkled, early	9.78 - RFO 5.77 - Verb
WAW 111 12 (L01)	unknown	green-seeded market pea, wrinkled, early	7.15 - RFO 4.97 - Stach
HS 8282 (L01)	Dětenice 11 * Usatyj	yellow-seeded field pea, round and smooth, semi-early vegetation period	5.71 - RFO 1.31 - Raff
Kujawski Pozny (L01)	unknown	yellow-seeded field pea, round, late	3.29 - RFO
Zeiners Tiefgrüne (L01)	unknown	green-seeded field pea, round	3.34 - RFO
Pisum abyssinicum (L02)	unknown	colour-flowered field pea, round, late	3.89 - RFO 0.65 - Verb
Hallorengold (L01)	unknown	yellow seeded field pea, round, long stem	3.94 - RFO 0.79 - Stach
Superba (L01)	unknown	yellow-seeded field pea, round	4.32 - RFO 0.16 - Raff

Table 3 *Results of Discriminant Analysis*

Variety subgroup	*No of cases / No of descriptors*	*The first four morphological descriptors***
L01	300/29	Leaflet colour (0.61) Seed shape (0.6) Stem internode length under first productive node (-0.33) Stem length to first productive node (-0.32)
L02	12/29	Flower vexillum apex (0.75) Seed colour (0.61) Leaflet colour (- 0.47) Shape of leaf tendrils (0.37)
L01 + L02	312/29	Leaflet colour (0.63) Seed shape (0.59) Stem internode length under first productive node (-0.34) Stem length to first productive node (-0.33)

** the value in parenthesis means the correlations between discriminating variables (descriptors) and canonical discriminant function

4 CONCLUSIONS

Close correlation as to shape of seed suggests that the most high RFO content can be expected - and our results show it - according to the genetics of seed shape and in ascending order of increasing RFO content from RRRbRb toward the rrRbRb, RRrbrb and rrrbrb genes (affecting the wrinkling of the seed, that is from field pea varieties toward the market pea varieties), actually from round and smooth seeded varieties to the wrinkled ones. High RFO content can be seen in the market pea varieties such as Lowadis, Kelvedon Wonder, Lorina and WAW 111 12 with wrinkled seed. This fact might be interesting under aspect of starch content [7]. The higher RFO content, the lower starch content [8]. It could mean that increasing RFO content replaces decreased starch content as another storage product content. Other two significant characters - stem length to the first fertile node and length of internode under the first fertile node - must be treated as a group which is influenced by complex of genes (*le, lk, lo, long-1, long-2, ls, la, cel, coe, coh, cona, cot, cres, cry, fr/fru and h*) modifying internodes length, leaflets size and colour. The lower RFO content, the later flowering time. Late varieties usually have longer stem to the first fertile node and higher number of nodes, too. In most cases lateness goes hand in hand with longer internode under the first fertile node. Following varieties with significantly lower RFO content are rather late and have longer stem: Kujawski Pozny (Pozny = Late), Zeiners Tiefgrüne, Ukosnyj, Pisum abyssinicum, Hallorengold, Superba. It is possible there is strong link between RFO content and the above mentioned group of genes, but probably there will another working gene responsible for it. We do not see any linkage between RFO content and flower vexillum apex now. The significance of seed colour in connection with RFO content must be tested and studied in whole range of varieties, so that results would be known later.

In connection with screening RFO content of the above mentioned *Pisum* varieties it is necessary to say it is the first work of a such kind and scope in unprospected field for the time being. The results would have direct impact to further understanding of genetical background of oligosaccharides problems itself as they display extremities on both sides which can influence an intended selection of suitable parental components in breeding process.

Acknowledgements

The authors gratefully acknowledge support from the Grant Agency of the Czech Republic (grant No 509/93/2385) and supply of world-wide assortment of pea varieties from genetic resources of Agritec, Ltd., Šumperk, Czech Republic.

References

1 - S. E. Fleming, *J. Food Sci,* 1981, **46**, 794.
2 - D. Marthinsen, S. E. Fleming, *J. Nutr.*, 1982, **112**, 1133.
3 - M. Siljeström, I. Björk, *Food Chemistry,* 1990, **38**, 145
4 - H. Tomomatsu, *Food Technol.*, 1994, **48**, 61
5 - A. Pavelková, J. Moravec, D. Hájek, I. Bareš, J. Sehnalová: *Descriptor List, genus Pisum L.*, Research Institute of Crop Production, Prague, 1986
6 - F. Kvasnička, R. Ahmadová-Vavrousová, J. Frias, K. R. Price and P. Kadlec, *HPLC Determination of Raffinose Family Oligosaccharides in Pea*, in preparation for publishing
7 - T. L. Wang and C. Hedley, *Seed Sci Research*, 1991, **1**, 3
8 - J. Cerning-Béroard and A. Filiatre-Verel, *Lebensm.-Wiss. u. -Technol.*, 1979, **12**, 273

THE EFFECT OF ANNEALING ON THE PHYSICOCHEMICAL PROPERTIES OF LEGUME STARCHES.

R. Hoover and H. Manuel
Department of Biochemistry
Memorial University of Newfoundland
St. John's, Newfoundland
Canada A1B 3X9

1. INTRODUCTION

Annealing is a process whereby a material is held at a temperature somewhat lower than its melting temperature, which permits modest molecular reorganization to occur and a more organized structure of lower free energy to form.[1] Annealing of starches has been studied at various starch: water ratios (1:1, 1:3, 1:5) and at temperatures ranging from 50 to 75°C[2-4]. The physical aim of annealing is to approach the glass transition temperature (Tg) which thereby enhances molecular mobility without at the same time triggering gelatinization[1]. A great deal is known about the effect of annealing on gelatinization parameters, granular swelling, enzyme digestibility, X-ray diffraction intensities, paste viscosities and amylose leaching of cereal and tuber starches. However, there is a dearth of information on the physicochemical properties of annealed legume starches. The present investigation was undertaken to determine the effect of annealing (75% moisture, 55°C, 24h) on swelling factor, amylose leaching, enzyme hydrolysis, gelatinization and retrogradation properties of legume starches.

2. MATERIALS

Legume seeds (green arrow pea, black bean, pinto bean, express field pea and eston lentil) were obtained from the Department of Crop Science, University of Saskatchewan. Starches were isolated and purified by the method outlined in an earlier publication.[9]

3. METHODS

Standard AACC methods[5] were used for the determination of moisture, ash and nitrogen. Amylose content was determined by the method of Chrastil.[6] Lipids (unbound, bound and total), swelling factor, amylose leaching, enzyme hydrolysis, x-ray diffraction and differential scanning calorimetry were performed according to procedures outlined in an earlier publication.[4] Gels (50% w/v) were prepared as described by Krusi and Neukom.[7] Gel powders were prepared from stored gels according to the procedure of Roulet et al.[8] Starches were annealed[4] at 55°C for 72 h at a moisture content of 75%.

4. RESULTS AND DISCUSSION

The proximate analysis of the starches are presented in Table I. All starches were of very high purity (< 0.13% nitrogen). The x-ray diffraction spacings and intensities of the major peaks of native starches changed only slightly on annealing. However, in all starches d-spacings remained unchanged. This implies that only marginal changes in crystallinity occurs on annealing. In all starches the swelling factor (SF) and amylose leaching decreased on annealing (Table 2). At 95°C, the decrease in SF on annealing in green arrow pea, black bean, pinto bean, express field pea and eston lentil were 76.8, 32.2, 31.9, 36.7 and 32.7% respectively, while at the same temperature, decreases in AML for the above starches were 13.2, 2.2, 2.3, 1.9 and 1.5% respectively. The decrease in SF and AML suggests interactions may have occurred between the starch components (amylopectin - amylopectin, amylose-amylopectin, amylose-amylose) during annealing. Of these interactions, those involving amylose chains may have been the main causative factor, since the extent of the reduction in SF and AML was more pronounced in green arrow pea (85% amylose) and was fairly similar in the other three starches (37.8-44.7% amylose).

Annealing increased the gelatinization transition temperatures (black bean > pinto bean > eston lentil > express field pea) and the enthalpy (ΔH) of gelatinization (black bean > eston lentil > express field pea > pinto bean) [Table 3]. However, in all starches the gelatinization temperature range decreased on annealing (black bean > pinto bean ~ eston lentil > express field pea). The difference in ΔH between native and annealed starches, suggests that the amount of double helices that unravel and melt during gelatinization is higher in annealed (indicated by higher ΔH values) than in native starches. This suggests that the double helical content of legume starch granules increases on annealing. Formation of double helices during annealing has been shown to result mainly from interactions between the outer branches of amylopectin.[4] The increase in gelatinization transition temperatures on annealing has been shown to result mainly from interactions that occur between amylose and the outer branches of amylopectin.[4] The above results suggest that the outer branches of amylopectin are probably longer in black bean than in the other starches. The decrease in the gelatinization temperature range on annealing is due to a closer packing of the pre-existing crystallites within the granule.

The susceptibility of starch granules to hydrolysis by porcine pancreatic α-amylase increased on annealing (black bean > pinto bean > express field pea > eston lentil > green arrow pea). Franco et al[9] postulated that α-amylase initially hydrolyses the amorphous regions of the granule. Thus, the increase in hydrolysis, is probably indicative of starch chain realignment (within the amorphous regions) on annealing.

The thermal characteristics and x-ray diffraction intensities of native and annealed starch gels (50% w/v) stored at 40°C for 4 and 7 days are presented respectively, in Tables 5 and 6. The retrogradation enthalpy of native and

Table 1. *Chemical composition of native starches*[a]

Characteristics	Composition (%)				
	Green arrow pea	Black bean	Pinto bean	Express field pea	Eston lentil
Moisture	6.4 ± 0.0	6.3 ± 0.1	7.1 ± 0.7	8.5 ± 0.7	9.0 ± 0.1
Ash	0.09 ± 0.01	0.18 ± 0.03	0.06 ± 0.03	0.06 ± 0.02	0.11 ± 0.02
Nitrogen	0.13 ± 0.01	0.02 ± 0.01	0.05 ± 0.01	0.06 ± 0.01	0.06 ± 0.01
Lipid					
Solvent extracted:					
chloroform-methanol	0.07 ± 0.01	0.04 ± 0.01	0.02 ± 0.01	0.02 ± 0.01	0.04 ± 0.01
n-Propanol-water	0.15 ± 0.02	0.16 ± 0.03	0.16 ± 0.02	0.10 ± 0.01	0.09 ± 0.01
Acid hydrolyzed[d]	0.22 ± 0.03	0.20 ± 0.02	0.18 ± 0.03	0.12 ± 0.02	0.13 ± 0.02
Amylose content					
Apparent[e]	67.5 ± 0.7	35.7 ± 1.2	37.9 ± 0.1	39.8 ± 0.8	35.6 ± 1.1
Total[f]	85.6 ± 0.9	41.2 ± 1.1	44.7 ± 2.0	43.2 ± 0.5	37.8 ± 0.6
Amylose complexed with native lipid[g]	21.1	13.3	15.2	7.8	5.8
Starch damage	2.4 ± 0.2	0.7 ± 0.1	0.2 ± 0.1	0.4 ± 0.1	1.8 ± 0.5

[a]All data reported on dry basis and represent the mean of three determinations.
[b]Lipids extracted from native starch by chloroform-methanol 2:1 (v/v) at 25°C (mainly unbound lipids).
[c]Lipids extracted by hot n-propanol-water 3:1 (v/v) from the residue left after chloroform-methanol extraction (mainly bound lipid).
[d]Lipids obtained by acid hydrolysis (24% HCl) of the native starch (total lipids).
[e]Apparent amylose was determined by iodine binding without removal of free and bound lipids.
[f]Total amylose was determined by iodine binding after removal of free and bound lipids.
[g]Total amylose - Apparent amylose x 100

Table 2. *The swelling factor and amylose leaching of legume starches at 95°C*[a].

Starch source and treatment	Swelling Factor	Amylose Leaching (%)
Green arrow pea		
Native	6.9 ± 0.1	35.6 ± 0.6
annealed	1.6 ± 0.2	22.4 ± 0.1
Black bean		
Native	28.2 ± 0.1	32.6 ± 0.4
annealed	19.1 ± 0.1	30.4 ± 0.7
Pinto bean		
Native	19.1 ± 0.3	33.4 ± 0.4
annealed	13.0 ± 0.1	31.1 ± 0.5
Express field pea		
Native	21.5 ± 0.2	44.8 ± 0.7
annealed	13.6 ± 0.2	42.9 ± 0.1
Eston lentil		
Native	27.5 ± 0.3	25.4 ± 1.3
annealed	18.5 ± 0.1	

[a]All data represent the mean of three determinations.

Table 3. *Thermal characteristics of native and annealed starches.*[a]

Starch source and treatment	Transition Temperatures (°C)[b]			ΔT	ΔH[c] (J/g)
	T_o	T_p	T_c	(T_c-T_o)	
Black Bean					
Native	58.5 ± 0.5	67.0 ± 1.9	74.2 ± 1.5	15.7	8.4 ± 0.5
Annealed[d]	69.6 ± 0.8	74.0 ± 1.3	80.3 ± 1.7	10.7	12.2 ± 0.1
Pinto Bean					
Native	59.0 ± 0.5	68.0 ± 0.5	73.0 ± 1.2	14.0	9.2 ± 0.3
Annealed	69.4 ± 0.0	73.5 ± 0.1	79.0 ± 0.2	9.6	9.6 ± 0.1
Express Field Pea					
Native	58.7 ± 1.3	69.3 ± 2.1	72.0 ± 0.5	13.3	9.6 ± 0.6
Annealed	64.6 ± 1.2	73.3 ± 0.2	76.2 ± 1.1	11.6	11.6 ± 0.2
Eston Lentil					
Native	58.6 ± 1.2	65.4 ± 1.9	70.9 ± 0.3	12.3	8.0 ± 0.4
Annealed	66.9 ± 4.3	70.0 ± 2.9	74.9 ± 3.5	8.0	10.2 ± 0.2

[a]All data represent the mean of three determinations.

[b]T_o, T_p and T_c represent the onset, peak and conclusion transition temperatures, respectively; starch:water was 1:3. T_o, T_p and T_c of green arrow pea were too high to be recorded on the DSC.

[c]Enthalpy.

[d]Starches were heat treated at 55°C for 24 hours at a starch to water ratio of 1:3.

Table 4. *Thermal characteristics of retrograded starches gels.*[a]

Starch source and treatment	Storage time at 40°C (days)	Transition temperatures[b,c] (°C)			Enthalpy (ΔH_R J/g)[b,d,e]
		T_o	T_p	T_c	
Black Bean					
Native	4	40.3	42.8	64.8	10.5
	7	40.3	44.8	61.5	12.2
Annealed[d]	4	41.0	44.7	63.0	13.5
	7	43.0	56.0	66.0	20.5
Pinto Bean					
Native	4	41.1	57.3	67.0	5.0
	7	39.2	52.8	65.5	5.9
Annealed	4	40.6	47.3	58.5	5.0
	7	40.5	43.0	58.3	6.0
Express Field Pea					
Native	4	45.6	56.7	84.0	3.0
	7	45.9	55.7	79.6	3.5
Annealed	4	42.6	49.1	63.4	3.4
	7	40.1	51.5	65.1	4.9
Eston Lentil					
Native	4	42.6	46.3	56.5	1.5
	7	43.9	46.2	63.0	2.0
Annealed	4	40.8	48.7	65.1	1.8
	7	42.8	45.7	60.6	2.5

[a]Starch:water, 50:50 (w/w dry basis) the gels were converted to a powder[8] prior to examination by DSC (starch:water 1:3).
[b]All data represent the means of 3 determinations.
[c]Average standard deviation, 0.5°C.
[d]Average standard deviation 0.5 J/g.
[e]Enthalpy of retrogradation.

Table 5. *X-ray diffraction spacings and intensities of retrograded starch gels.*[a]

Starch source and treatment	Storage Time at 40°C (Days)	Moisture Content (%)	Interplanar spacing (d) in Å with intensities (CPS)				
Green Arrow Pea							
Native	4	9.8	5.16 (612)		4.56 (418)	3.94 (434)	3.77 (440)
	7	10.3	5.19 (913)	5.11 (830)	4.50 (462)	3.93 (615)	3.77 (450)
Annealed[b]	4	9.8	5.22 (627)		4.44 (434)	4.03 (467)	3.91 (372)
	7	11.5	5.21 (675)		4.54 (395)	4.00 (473)	3.86 (414)
Black Bean							
Native	4	9.8	5.10 (741)		4.40 (439)	4.03 (461)	3.87 (429)
	7	9.6	5.10 (786)		4.38 (63.0)	4.02 (469)	3.96 (496)
Annealed[b]	4	9.8	5.26 (810)	4.70 (318)	4.48 (466.5)	4.37 (490)	4.01 (440)
	7	9.8	5.19 (1563)	4.71 (1012)	4.50 (1017)	4.33 (984)	3.95 (1058)
Pinto Bean							
Native	4	9.8	5.10 (776)	3.95 (473)	3.75 (375)		
	7	9.7	5.09 (779)	3.94 (501)	3.74 (521)		
Annealed[b]	4	9.8	5.19 (796)	3.99 (486)	3.76 (382)		
	7	9.8	5.11 (804)	3.95 (521)	3.74 (467)		
Express Field Pea							
Native	4	9.8	5.11 (700)	4.71 (327)	4.33 (309)	4.04 (443)	3.79 (359)
	7	9.5	5.16 (765)	4.78 (330)	4.44 (341)	4.01 (471)	3.76 (349)
Annealed[b]	4	9.8	5.16 (702)		4.13 (388)	4.03 (450)	3.70 (285)
	7	9.8	5.17 (724)		4.15 (426)	4.00 (455)	3.76 (401)
Eston Lentil							
Native	4	9.8	5.10 (701)	3.96 (451)	3.85 (442)	3.69 (342)	
	7	10.1	5.12 (734)	3.97 (465)	3.86 (484)	3.70 (363)	
Annealed[b]	4	9.8	5.19 (724)	4.57 (669)	4.20 (713)	4.05 (700)	3.83 (528)
	7	9.8	5.15 (747)	4.58 (668)	4.16 (773)	4.08 (763)	3.96 (584)

[a]Starch:water, 50:50 (w/w) dry basis. The gels were converted to a powder prior to examination by x-ray diffraction.
[b]Counts per second.

annealed starches increased on storage. The extent of this increase was more pronounced in annealed starches (black bean > express field pea > eston lentil > pinto bean) [Table 4]. The x-ray intensities increased only marginally during storage of native starch gels. In annealed starches, the increase in x-ray intensities were very pronounced in black bean starch, but were only marginal in the other starches (Table 5). The above results suggest that on annealing, the degree of separation between the outer branches of adjacent amylopectin chain clusters are reduced to a greater extent in black bean than in the other starches. Consequently, during gel storage, the formation and lateral association of double helices involving amylopectin chains, would be easier and much stronger in black bean starch. This would then explain the higher rate of retrogradation of black bean starch.

5. CONCLUSION

The results showed that in all starches, SF, AML, and the gelatinization transition temperature range decreased on annealing, whereas increases were observed for gelatinization transition temperatures, gelatinization enthalpy and rate of retrogradation and enzyme hydrolysis. X-ray diffraction patterns remained unchanged and x-ray intensities changed only marginally in all starches. These changes in physicochemical properties were due to increased interaction between starch components and to starch chain realignment on annealing.

REFERENCES

1. Blanshard, J.M.V. Starch granule structure and function: A physicochemical approach. In Starch: properties and potential, (T. Galliard ed.), John Wiley & Son, New York, 1987, p. 16.
2. Knutson, C.A., *Cereal Chem.*, 1990, **67**, 376.
3. Krueger, B.R., Walker, C.E., Knutson, C.A. and Inglett, G.E., *Cereal Chem.*, 1987, **64**, 187.
4. Hoover, R. and Vasanthan, T., *J. Food Biochem.*, 1994, **17**, 303.
5. American Association of Cereal Chemists. Approved Methods of the AACC, 1984. 8th ed. St. Paul, MN.
6. Chrastil, J., *Carbohydr. Res.*, 1987, **159**, 154.
7. Krusi, V.M. and Neukom, H. *Stärke,* 1984, **132**, 83.
8. Ph. Roulet, Macinnes, W.P., Wursch, P., Sanchez, R.M. and Raemy, A., *Food hydrocolloids*, 1988, **2**, 281.
9. Franco, C.M.L., Petro, S.J., DoR, Ciacco, C.F. and Tavares, D.Q., *Stärke*, 1988, **40**, 29.

WHAT IS THE POTENTIAL FOR IMPROVING THE QUALITY OF OIL IN DRIED PEAS?

D.A. Jones, L.M. Barber and C.L. Hedley
Applied Genetics Department
John Innes Centre
Norwich Research Park
Norwich NR4 7UH UK

INTRODUCTION

We screened a wide selection of pea lines from the John Innes Gene Bank and found considerable variation in the fatty acid composition of the seed storage oil[1]. In addition we have identified a series of lines, resulting from a chemical mutagenesis programme[2,3], which have novel seed compositions; no starch, higher than normal protein content and 10% oil. Using this material we have been able to identify lines with a modified fatty acid composition similar to that of oilseed rape and olive oil. Generally the quality of plant derived oils is superior to those from animal sources. The saturated fatty acid component is less than 20%, whilst mono-desaturated fatty acids can be as high as 70% of the total. Although pea has a much lower oil content (1-2%) to that of soybean (20-25%) they both have similar fatty acid profiles.

MATERIALS AND METHODS

The pea lines used in this work were produced and the lipids and fatty acids were extracted and analysed by gas chromatography, using the methods described[2,4-7,11].

RESULTS AND DISCUSSION

Fatty acid composition
The five fatty acids commonly found in most plant lipids are; palmitic acid (16:0), stearic acid (18:0), oleic acid (18:1), linoleic acid (18:2) and linolenic acid (18:3). In pea these account for over 99% of the total lipid content. From a dietary point of view, a lipid composition containing predominately mono-desaturated fatty acids, i.e. oleic acid (18:1), is considered the best [the so called 'Mediterranean Diet'. Oleic acid takes its name from olives (*Olea europea*) where the oil extracted from the fruits contains about 70% oleic acid]. High dietary consumption of saturated fatty acids, mostly derived from animal fats and dairy products, can lead to clogging of the arteries and heart disease. Both linoleic and linolenic acids are essential in the diet to ensure healthy growth and development, and

a consensus of views suggest that a recommended daily intake would be around 10g/day for linoleic and 1g/day for linolenic acids[8]. The fatty acid composition of pea is acceptable, but it could be altered to be more like that of olive oil, which would further enhance its value as a food source. We have produced a wrinkled seeded cross which has 4% total lipid with 50% oleic and 25% linoleic acids, a reversal of the normal wild type composition. This new composition would be 'healthier' and add value to the pea crop. Table 1 shows the profiles of several major commercial sources of plant oils.

One of the main problems with extracted oils is their susceptibility to oxidation, leading to the production of breakdown products which impart unpleasant flavours and odours to the oil as well as altering its properties. The more desaturated the fatty acids are in the oil, the more likely the oil is to degrade. This can be controlled by the addition of antioxidants, some of which are natural products and are found in the seeds, but by selecting lines which are low in linolenic and linoleic acids the storage and keeping properties of the oil can be increased. Pea oil has relatively low amounts of linolenic acid (about 10%), but there is variation which ranges from 2.5 to 14.5%. A similar situation exists in soybean where breeders have developed low 18:3 varieties[9,10].

Oil content

Fats and oil form an essential part of a balanced diet, although an excess can cause growth and health problems for both humans and animals. Fat is an important source of the energy component in food and its presence also helps to make the food more palatable. It is generally accepted that a good diet should contain 25-30% of the total calories consumed as fat[8].

Oil is one of the bulk storage materials found in many seeds. It is mainly a mixture of triacylglycerols (triglycerides) and is stored in discrete organelles (oil bodies, as storage oil) within the seed cells. There are other lipids found in the cell and most of these are present within the membrane systems (membrane lipids). It is possible by using solvents of different polarities to extract the different lipid components from a sample of pea flour. Non-polar solvents such as n-hexane, 2,2,4-trimethylpentane (iso-octane) and petroleum ether, for example, tend to selectively dissolve and extract the oils, while more polar solvents such as water saturated butan-1-ol will dissolve and extract the membrane lipids[11].

We have found that genes at three *rugosus* loci; *r*, *rb* and *rug-3*, acting through a pleiotropic effect increase the lipid content of the seed[2,5] at the expense of starch content. Figure 1 shows the effects of these loci on the total lipid and that there is a 'real' increase in lipid content on a weight per seed basis.

Normal round pea varieties (wild type) have a total lipid content of 2%, (about 1% of storage oil and 1% membrane lipid). As the oil content is increased there is a smaller increase in membrane lipid. The oil is stored in oil bodies and these are bound by a single unit membrane, therefore, an increase in membrane lipid is to be expected. Further measurements show that there is a real increase in oil content as the weight of oil extracted from the round genotype; *RRRbRb* ++ (4 mg/seed) increases to about 28 mg/seed in the ++ ++ *rug-3 rug-3* line, (Table 2). The average weight of seed is in the range 200-250mg.

Recently Somerville and colleagues[12] have cloned most of the genes encoding for the enzymes concerned with the lipid biosynthetic pathway in *Arabidopsis thaliana*, a close relative of oilseed rape. Using these clones it is now possible to produce genetically transformed rape plants, producing different fatty acids to order, depending upon end use.

Table 1.

	Total Lipid	Palmitic 16:0	Stearic 18:0	Oleic 18:1	Linoleic 18:2	Linolenic 18:3
Olive	70.0	13.7	2.5	71.1	10.0	0.6
Corn	55.0	12.2	2.2	27.5	57.0	0.9
Oilseed Rape	45.0	3.9	1.9	64.1	18.7	9.2
Sunflower	40.0	6.8	4.7	18.6	68.2	0.5
Soybean	20.0	11.0	4.0	23.4	53.2	6.0
Pea	2.0	14.1	3.5	24.0	50.6	8.6

% dry weight analysed as Fatty Acid Methyl Esters [FAMES]

Table 2.

Genotype	++ ++ ++	rr ++ ++	++ rbrb ++	rr rbrb ++	++ ++ rug3rug3
Oil	1.0	2.2	3.2	3.5	7.6
Membrane	1.1	1.5	1.5	2.3	2.5
TOTAL %	2.1	3.7	4.7	5.8	10.1

% dry weight analysed as Fatty Acid Methyl Esters [FAMES]

Figure 1.

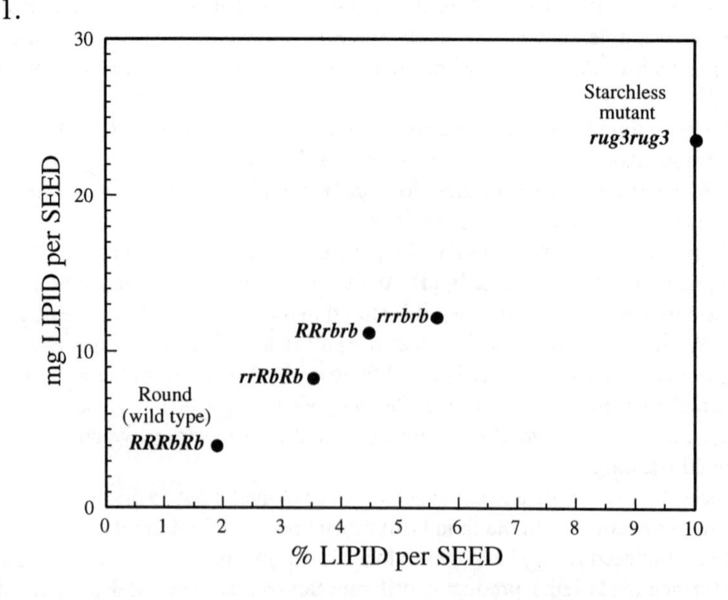

The points are the means of five single seed analyses (triplicate analysis on each seed). The values are presented on a dry weight basis.

These clones could be introduced into any plant species that can be genetically transformed. Reliable pea transformation systems are still being developed, but should be available soon, which would allow further and perhaps more controlled manipulation of both the composition and the amount of oil produced in the seed.

CONCLUSIONS

Over the past 40 years we have seen the emergence of oilseed rape, which is now a major source of oil and protein. The oil content of the seed has been increased slightly during this period, but the major advance has been an overall increase in yield, and the development of vigorous and stable lines. Similar improvements have been made to the other major oilseed crops, maize and soybean. The oil from traditional crops is in great demand for uses outside food, in the chemical industry. There is a need for alternative oilseed crops to make up the shortfall in demand. Good farming practice requires crop rotation and legumes have long been recognised as being beneficial to the soil (through the symbiotic relationship with the bacteria in the root nodules), requiring less fertilisers to be used. If the same level of resources and effort that have been applied to the more traditional oilseed crops, were put to improving the pea crop, then the full potential of the pea could be realised.

The John Innes Centre is supported via a grant-in-aid from the BBSRC. This work was also carried out with support from MAFF. Lorraine Barber was supported by the EU ECLAIR P.E.A. programme, contract AGRE CT90 0048.

REFERENCES

1. D.A. Jones, L.M. Barber, A.E. Arthur and C.L. Hedley, *Plant Breeding*, 1995, **114**, In Press.
2. T.L. Wang, A. Hadavizideh, A. Harwood, T.J. Welham, W.A. Harwood, R. Faulks, and C.L. Hedley, *Plant Breeding*, 1990, **105**, 311-320.
3. T.L. Wang and C.L. Hedley, *Seed Science Research*, 1991, **1**, 3-14.
4. C.L. Hedley, C.M. Smith, M.J. Ambrose, S. Cook and T.L. Wang, *Annals of Botany,* 1986, **58**, 371-379.
5. C.L. Hedley, J.R. Lloyd, M.J. Ambrose and T.L. Wang, *Annals of Botany*, 1994, **74**, 365-371.
6. D.A. Jones, A.E. Arthur, H.M. Adams, D.T. Coxon, T.L. Wang and C.L. Hedley, *Plant Breeding*, 1990, **104**(2), 144-151.
7. D.A. Jones, *Journal of Biological Education,* 1994, **28**(3), 154-156.
8. F.D. Gunstone, J.L. Harwood and F.B. Padley, 'The Lipid Handbook', Publishers; Chapman and Hall, 1986, ch 12, 527.
9. B.D. Rennie, J. Zilka, M.M. Cramer and W.D. Beversdorf, *Crop Science*, 1988, **28**, 655-657.
10. B.D. Rennie and J.W. Tanner, *Crop Science*, 1991, **31**, 297-301.
11. W.W. Christie, 'Gas Chromatography and lipids', Publishers: The Oily Press, Ayr, Scotland,1992, ch 2, 11-41.
12. C. Somerville and J. Browse, *Science*, 1991, **252**, 80-86.

FATTY ACIDS IN PROTEIN CONCENTRATES OBTAINED FROM ALFALFA JUICE WITH CHEMICALS AS PROTEASE INHIBITORS

B.M. Baraniak

Department of General Chemistry
University of Agriculture, Lublin
Poland

1 INTRODUCTION

Concentrates obtained from the juices of green plant parts, besides proteins are also a rich source of carotenoids i.e., biologically active dyes. At the time of propagation of low-fatty foodstuffs they can replace fat in human nutrition because they contain about 10 % of this constituent and a half of it constitutes non-saturated fatty acids.

The content of particular fatty acids, similarly as chemical composition of concentrates, depends on the kind of raw material used as well as on every stage of the technological process (degree and method of green mass maceration, method of precipitation and drying of preparation and the conditions of its storage). The aim of the study is to compare the content of fatty acids in the flocculated concentrates from alfalfa juice at the presence of selected chemical compounds as proteolytic enzyme inhibitors.

2 MATERIALS and METHODS

2.1 Materials

Alfalfa juice (*var. Kleszczewska*) was obtained by initial milling the green fodder in the screw press, then pressing the obtained pulp using the pug mill, getting 570 g of juice from 1 kg of plants. The following substances were added to the juice: Ag^{+}, Hg^{++} (0.5 mM/dm^3), acetic aldehyde (1 ml/dm^3), simazine (1 g/dm^3), filtrate obtained from fractionating the proteins of alfalfa juice at temperature (55°C) and by adding the flocculant (Magnafloc LT 26) in the volume ratio 1:10, and juice pressed out of the buckwheat biomass in the volume ratio of 1:4. From this juices and from juice with out any additions (the control test) the chloroplastic concentrates were precipitated by the Magnafloc LT 26 anionic flocculant, or, the amount of 300 mg/dm^3 of juice, alkalized by 2M NaOH to pH=7.5.

The concentrates were centrifuged (4000 rpm), twice washed out with distilled water, dried in the temperature of 50°C and after milling stored at 6°C.

2.2 Analytical Methods

The content of total protein (N total x 6.25) was determined after Kjeldahl, with an automatic Kjel-Foss apparatus. The reducing sugars were analysed with DNS (1) in neutralized filtrates obtained as a result of hydrolysis of concentrates with 6M HCl. The fat was extracted with hexane and the level of particular fatty acids has been determined with gas chromatography (2).

3 RESULTS

Fats, besides carbohydrates being the main energetic component of food, also provide organisms with vitamins soluble in it (A, D, E and K) and indispensable non-saturated fatty acids (NSFA). The source of the latter are plant fats and they cannot be, contrary to fats of animal origin, substituted with other nutritive components. Animal organisms are not able to synthesise necessary non-saturated fatty acids. Linoleic acid, the precursor of arachidonic acid, i.e., the principal one for animal organisms, and linolenic acid are synthesised by plants through dehydrogenation of oleic acid due to aerobic or anaerobic conversion. Non-saturated fatty acids consist main component of fatty fraction of protein concentrates coagulated from juices of the plants, green part. Their contents can exceed 70% of the total amount of fat present in the concentrate (3), and it is determined by both the species of the plant used (4, 5, 6) and the conditions of technological process performance. During the processing of protein concentrates a significant role is played by inactivation of proteolytic enzymes, because the loss of proteins due to its enzymatic hydrolysis can reach 50%. The introduction of selected chemicals into alfalfa juice as protease inhibitors caused the differentiation of the level of fats, proteins and reducing sugars in flocculated preparations (Table 1).

During the coagulation of proteins from juice, sugars form complexes with glycolipids originating from photosynthesising membranes, the main fats occurring in chloroplastic concentrates. The contact of solvent with fat molecules is only possible after disruption of the formed complexes and the applied solvents must be able to form hydrogen bonds with sugars. The differences in the amount of extracted fats (with the same solvent and constant conditions) obtained in this study, as well as the amount of sugars obtained allow us to assume that the complexes formed are the most stable in the case of addition of buckwheat juice (the highest level of sugars at the lowest amount of fats), and the least stable in the case of simazine.

The presence of chemical substances in alfalfa juice caused some changes in the content of saturated and non-saturated fatty acids in fatty fraction (Table 2 and 3), as compared to the control -i.e. flocculated concentrate from the juice without any additives.

Table 1 *Effect of chosen substances on fat, protein and reducing sugars content in protein concentrates obtained from alfalfa juice*

Substance	*Fat* %	*Total protein* %	*Reducing sugars* %
Ag^{+}	7.4	41.6	9.0
Hg^{++}	6.6	42.0	9.0
Acetic aldehyde	8.0	34.4	8.5
Filtrate from fractionation by:			
Flocculant	6.6	37.2	6.5
Temperature	9.2	41.3	10.0
Simazine	9.3	37.4	6.5
Buckwheat juice	6.4	37.4	12.0
Control	9.0	38.8	11.5

Table 2 ***The content of saturated fatty acids (%) in fatty fraction of protein concentrates obtained from alfalfa juice with chemicals***

Substance	*14:0*	*16:0*	*18:0*	*20:0*
Ag^{+}	2.0	34.5	4.9	-
Hg^{++}	1.0	33.3	6.6	1.7
Acetic aldehyde	2.2	34.5	4.6	0.9
Filtrate from fractionation by:				
Flocculant	2.7	32.4	5.5	3.4
Temperature	1.8	31.5	4.6	4.1
Simazine	2.4	31.8	3.2	8.5
Buckwheat juice	1.8	33.4	6.6	4.5
Control	2.6	40.4	5.5	-

Table 3 ***The content of non-saturated fatty acids (%) in fatty fraction of protein concentrates obtained from alfalfa juice with chemicals***

Substance	16:1	18:1	18:2	18:3
Ag^{+}	2.9	5.4	19.0	31.4
Hg^{++}	1.6	5.0	23.3	29.0
Acetic aldehyde	2.4	2.7	17.3	35.4
Filtrate from fractionation by:				
Flocculant	2.8	4.4	18.4	30.4
Temperature	2.0	4.4	19.8	31.8
Simazine	2.9	2.7	17.4	31.3
Buckwheat juice	2.9	3.9	18.5	28.2
Control	3.0	2.9	19.4	26.2

In all the cases an increase in the content of non-saturated acids has been stated, the highest being after the application of silver and mercury ions as well as acetic aldehyde. The most significant differences were observed in the level of palmitic and linolenic acid.

The latter one and its metabolites, belonging to the family n-3, show lower biological activity than the family n-6, where the remaining NSFA belong (linoleic and arachidonic acid and their metabolites). The requirements of any organisms for indispensable non-saturated fatty acids is discussible. In general, it is assumed that their dose should constitute from 3% to maximum 10 % of the total energetic value supplied with food. Non-saturated fatty acids play an important role in the prevention and treatment of sclerosis - they decrease the level of cholesterol and triacylglycerols and prevent against the appearance of vascular thrombus. They also have structural value because they are permanent components of phospholipids of cell and mitochondric membranes. They are also precursors of prostagliadynes, i.e. modulators of hormonal activity. Therefore, it is important that components rich in non-saturated fatty acids would not be eliminated from the diets of both humans and animals. The application of protein concentrates in nutrition allow to supply NSFA together with a greater amount of proteins and carotenoides - precursors of vitamin A.

References

1. G. E. Miller, Anal.Chem., 1959, **31**, 426.
2. I. H. Lima, T. Richardson. M. A. Stahmann, J. Agric. Food Chem., 1965, **13**, 143.
3. R. Carlsson, E. M. Wendy Clarke, Qual. Plant., Plant Foods Human Nutr., 1983, **33**, 127.
4. S. Nagy, H. E. Nordby, L. Telek, J. Agric. Food Chem., 1978, **26**, 701.
5. A. Rosas Romero, A. C. Diaz, Acta Cient. Venezolana, 1983, **34**, 72.
6. A. Rosas Romero, M. E. Vivas, A. T. de Rodriguez, Acta Cient. Venezolana, 1984, **35**, 424.

OLEIC/LINOLEIC RATIO IMPROVEMENT IN PEANUT OIL FROM CORDOBA, ARGENTINA

N. R. Grosso[1,3], A. L. Lamarque[2], J. A. Zygadlo[2], D. M. Maestri[2], C. A. Guzmán[2,3], and E. H. Giandana[4]

[1] Química Biológica, Facultad de Ciencias Agropecuarias (UNC), CC 509, 5000 Córdoba, República Argentina.
[2] Química Orgánica, FCEFyN (UNC).
[3] Centro de Excelencia de Productos y Procesos de la Provincia de Córdoba (CEPROCOR).
[4] Instituto Nacional de Tecnología Agropecuaria (INTA).

1 INTRODUCTION

Traditionally in the U.S., Runner market types have been predominatly utilized for the peanut butter trade, and oil composition likewise plays an important role in the manufacturing of this end-use product (1).

Runner-type peanut accounted for over 80% of the total production area in Argentina. The main cultivar utilized in Córdoba is Florman bred in Argentina from Florunner (2).

Differences in fatty acid composition of peanut due to the effect of the geographic location were observed. Higher temperatures during the last weeks before harvest resulted in higher oleic to linoleic (O/L) ratio (3).

The oleic to linoleic ratio and iodine value are important indicators of oil stability and quality (1). The objective of this study was to know the better harvest time and to determine the fatty acid variation of Runner-peanut among geographic localities from Córdoba, Argentina.

2 MATERIALS AND METHODS

All seed samples of peanut were obtained from Córdoba (Argentina). Localities studied: North Reducción (NRe), South Reducción (SRe), South General Cabrera (SGC), North General Cabrera (NGC), General Cabrera (GC), West General Cabrera (WGC), East General Cabrera (EGC), General Deheza (GD), South General Deheza (SGD), Río Cuarto (RC), Rincón (Ri), Las Higueras (LH), Gigena (Gi) and West Manfredi (WM). The cultivar used in this work was Florman from the 1991, 1992 and 1993 crop years. Fatty acid variation in different harvest times was determined from Florman peanut cultivated in Instituto Nacional de Tecnología Agropecuaria (INTA) of Manfredi (1992 and 1993 crop years). Three harvest times was established: early harvest time (EHT) (141 days after sowing), normal harvest time (NHT) (164 days after sowing), and late harvest time (LHT) (177 days after sowing). The material was provided by Georgalos S. A. and INTA of Manfredi, Córdoba.

Sound and mature seeds were milled and oil was extracted for 16 h with petroleum ether (boiling range 30-60°C) in a

Soxhlet apparatus. The extracted oils were dried over anhydrous sodium sulfate and the solvent removed under reduced pressure in a rotary film evaporator. The fatty acid methyl esters of total lipids were prepared by transmethylation with a 3% solution of sulfuric acid in methanol, as previously described (4). Methyl esters were analysed on a Shimadzu GC-R1A gas chromatograph equipped with flame ionization detector (FID). Capillary column AT-WAX Superox II (30m x 0.25mm i. d.) was used. Column temperature was programmed from 180 °C (held for 1 min.) to 240 °C at 4 °C min^{-1}. The injector temperature was 250 °C. The carrier gas (Nitrogen) had a flow rate of 10 mL min^{-1}. A standard fatty acid methyl ester mixture was run in order to use retention times in identifying sample peaks. Fatty acid levels were reported as a relative proportions of the total composition. Iodine values were calculated from fatty acid percentages (5) using the formula: (% oleic x 0.8601) + (% linoleic x 1.7321) + (% eicosenoic x 0.7854).

T-test (least significant differences) was used for statistical mean separations (1).

3 RESULTS AND DISCUSSION

Palmitic (16:0), stearic (18:0), oleic (18:1), linoleic (18:2), arachidic (20:0), eicosenoic (20:1), behenic (22:0), and lignoceric (24:0) acids were detected (Table 1). Percentage means were all within the ranges recently reported by Branch et al. (1), except for oleic and linoleic acids wich appear to be lower and higher, respectively. Significant differences were found within the fatty acid profile among 15 geographic localities studied.

Higher oleic to linoleic ratios and lower iodine values would suggest better oil stability and longer shelf-life (1, 3). Accordingly, Runner peanut from Reducción zone (north and south) followed by Las Higueras locality had higher oleic acid contents and correspondingly the best O/L ratios and iodine values, respectively (Table 1). These zones are located southeast of cultive area and have soils more sandy and higher precipitations than the other localities (6). Here, the peanut is early sown and reaped. Therefore, the temperature during seed maduration is higher. Holaday and Pearson (3) found that higher temperatures during the last weeks before harvest resulted in higher O/L ratio. These environmental factors could favor the O/L ratios and iodine values in these localities.

The peanut of EHT had higher oleic acid content correspondingly the best O/L ratio and iodine value (Table 2. Probably due to a higher temperature during the last weeks before harvest than in NHT and LHT.

Fatty acid composition, iodine value and O/L ratio of Florman peanut from Córdoba, Argentina could vary in the different production locations and harvest times.

Table 1 *Average percentages (1991-1993) of fatty acid composition of Florman peanut from 14 geographic localities (Loc) of Córdoba, Argentina.*

	Fatty acid composition (%)*									
Loc	*16:0*	*18:0*	*18:1*	*18:2*	*20:0*	*20:1*	*22:0*	*24:0*	*O/L*	*IV*
NRe	9.2c	1.6a	49.0d	33.4g	1.2b	1.6ad	2.5ac	1.4ab	1.46d	101a
SRe	9.9a	2.3f	45.8c	35.3h	1.1bc	1.8b	2.5ac	1.4ab	1.30c	102a
SGC	9.9a	1.7ab	44.0a	39.1ab	0.9a	1.8b	1.9ef	0.8f	1.12a	107b
NGC	9.9a	1.7ab	44.1a	38.6a	0.9a	1.6ad	2.1bde	1.1cde	1.14a	106b
GC	9.8a	1.7ab	45.2b	37.0c	0.9a	1.8b	2.3ab	1.3abc	1.22b	104c
WGC	10.2b	1.8bc	44.1a	38.6a	0.9a	1.5d	2.0def	1.0def	1.14a	106bc
EGC	10.0ab	1.3e	44.0a	39.5b	0.7d	1.8b	1.8f	0.9ef	1.11a	108b
GD	10.2b	1.9cd	45.2b	36.7c	1.0ac	1.6ad	2.3ab	1.2bcd	1.23b	104c
SGD	9.8a	1.4e	43.8a	38.1ad	0.9a	1.8b	2.7c	1.5a	1.15a	105c
RC	9.9a	1.8bc	44.0a	37.8d	0.9a	1.8b	2.4a	1.3abc	1.16a	104c
Ri	10.2b	1.8bc	40.8e	40.6e	0.7d	2.1c	2.5ac	1.3abc	1.00e	106c
LH	9.9a	1.9cd	45.8c	36.1f	1.0ac	1.7ab	2.4a	1.2bcd	1.26bc	103ac
Gi	9.8a	1.9c	41.7f	40.6e	0.7d	2.2c	2.3ab	1.2bcd	1.02e	108b
WM	9.3d	1.8bc	44.1a	37.9d	0.9a	1.8b	2.4a	1.3abc	1.16a	105bc

* Percentages within each column followed by the same letter do not differ significantly at P = 0,05.

Table 2 *Average percentages (1992 and 1993) of fatty acid composition of Florman peanut in different harvest times (HT).*

	Fatty acid composition (%)*									
HT	*16:0*	*18:0*	*18:1*	*18:2*	*20:0*	*20:1*	*22:0*	*24:0*	*O/L*	*IV*
EHT	9.5a	1.7b	43.4a	37.1b	1.1b	2.0a	2.8a	1.5a	1.17b	104b
NHT	9.3a	1.5a	43.5a	38.8a	0.9a	2.1a	2.5a	1.3a	1.12a	107a
LHT	9.6a	1.5a	43.8a	38.5a	0.8a	1.6b	2.6a	1.4a	1.13a	107a

* Percentages within each column followed by the same letter do not differ significantly at P = 0,05.

References

1. W. C. Branch, T. Nakayama and M. S. Chinnan, *J. Am. Oil Chem. Soc.*, 1990, **67**, 591.
2. J. R. Pietrarelli, E. H. Giandana y R. Sanchez. *Oleico*, 1985, **31**, 39.
3. C. E. Holaday and J. L. Pearson, *J. Food Sci.*, 1974, **39**, 1206.
4. M. D. Jellum and R. E. Worthington, *Crop Sci.*, 1966, **6**, 251.
5. I. B. Hashin, P. E. Koehler, R. R. Eitenmiller and C. K. Kvien, *Peanut Sci.*, 1993, **20**, 21.
6. R. A. Miatello, *Geografía Física de la Provincia de Córdoba*, Boldt, Córdoba, Argentina, 1979.

Section 4: Texture and Cell Wall Components

BEANS, FIBRE, HEALTH AND GAS

C.L.A. Leakey[1] and C.A.J. Harbach[2]

[1]Peas & Beans Ltd.
Girton, Cambridge

[2]M-Scan,
Sunninghill, Ascot, Berkshire

1 INTRODUCTION

Grain legumes or pulses have always bean known to provoke gas. Pythagoras suggested that people should refrain from eating them and their use by nuns was banned by St Augustine on grounds that "involuntary bodily functions, including flatulence, signify man's Fall from Grace"[1].

Flatulence is a perfectly natural phenomenon, but one of many of these, which, in modern domesticated societies and more particularly those in which people who are required to work in close quarters, normally seek to control. There are serious environmental issues in the atmosphere of work places. Sometimes flatulence may be well beyond the norms and be associated with irritable bowel syndrome[2], or with more serious pathological conditions of the gut[3,4]. It has been suggested moreover that attempting to restrain flatulence may contribute to diverticulitis[5]. The dietary fibre hypothesis[6], and the refinement of what is meant by dietary fibre[7] and how it should be measured[8] and that it ought to be increasingly used in human diets have had a revolutionary effect on catalysing work on nutrition and health and the development of what are beginning to be called nutraceuticals[9] as well as influencing choices among natural foodstuffs.

Many different sources of dietary fibre, among normal foodstuffs, have been promoted because of their probable effects on a range of medical conditions. Specifically for grain legumes, perceived benefits include lowering risks of atherosclerosis and ischaemic heart disease through modifying circulating lipoproteins[10] reducing the risk of onset of Type 2 diabetes[11], and reducing the risks of colon cancer[12]. Must all of these potential benefits from dietary fibre be won at a cost of increased flatulence? Will increased flatulence tend to lower the compliance with advice to improve fibre intake?

White[13] in a survey in California found that 24% of households either did not consume beans or restrained consumption because of the associated flatulence. This may mean that there is a very large potential untapped market if the problem could be overcome by breeding or by food technology. The development of a galactosidase enzyme as a food additive in the USA[14] would not have occurred unless a significant problem was recognised as being worth addressing commercially. Are there ways in which the benefits to health from higher fibre diets can be achieved without an increase in the social embarrassments and nuisance of flatulence, let alone contributions to the greenhouse gas emissions?

Murphy[15] reported genetic variation in flatulence factors and suggested breeding for reduced flatulence. DeLumen[16] has suggested bio-technological, genetic engineering approaches by which this might be achieved. Price *et al*[17] also suggest that breeding against flatulence could be useful. We have been working for some years towards this end[18-20].

2 FLATULENCE AND INDIGESTION

2.1 Flatus as an Index of Digestibility

It is probable that most flatus gas arises from bacterial fermentation of substrates that reach the lower gut having passed the stomach and small gut without having been digested and absorbed[21]. To the extent that fermentation is the source of gases as well as other medically-significant metabolic products such as short chain fatty acids, flatus volume and chemical composition may be a useful index of indigestibility and be a source of information about microbial metabolism in the lower gut. Whilst gas volume actually present within the GI tract might have even greater value[22] it is also not an index so easy to obtain outside a clinical setting, and thus less potentially useful.

Indirect measurements via respired hydrogen[23] of which up to 13% of fermentation produced hydrogen may be lost[24] are more genteel for nutritionists but may be less informative and also normally require a clinical environment. Direct measurement of flatus gas volumes and its collection for chemical analysis has been the subject of several studies but a new instrument and analytical techniques are claimed as substantial improvements[25] and open the way for data collection in ambulant, non-clinical, as well as in clinical settings.

2.1.1 Carbohydrates digestibility. Bean starches have a long history of study[26,27]. Amylose and amylopectin have been assayed in a number of different types of *Phaseolus* beans[26-28] and technical starch fractions been isolated[29]. "Resistant starch" has been identified as contributing to dietary fibre[30] and heteropolysaccharide been isolated in a fairly refine state[31,32]. There are many studies on glycoprotein inhibitors of alpha amylases present in a wide range of varieties of beans[33,34]. The major inhibitor seems to be the same or similar in all and to be highly resistant to destruction in acid and to proteolysis. What is not reported is its stability during normal cooking of beans but it is suggested as being possibly anti-nutritional *in vivo*.

2.1.2 Protein digestibility. There is a wealth of literature on protein digestibility with regard to the proteins themselves, but also on the presence of protease inhibitors. The trypsin inhibitors, important in domestic livestock nutrition[35] may be rather unimportant in cooked pulses as they are easily heat labile. The less well known heat stable protease inhibitors[36,37] deserve far more attention than they have received.

2.2 Changes to Gut Functionality and Other Aspects of Anti-Nutrition

There has been a great deal of work on the effects of lectins as anti-nutrients in domestic and experimental animals though very much less in human subjects in which experimental study with toxic substances is more inhibited by morality. Kik[38] is an important source of information which bridges the veterinary-medical gulf. Sgarbieri's massive monographic review[39] of antinutrients in *Phaseolus* beans is a major source of information from the very significant Latin-American literature as well as covering the North American literature well. De Muelenaere's work on S. Africa drew attention to the real risks from residual lectins[40]. More recently there is continuing work on lectins at Wageningen[41].

2.2.1 Survival of anti-nutrients through cooking or processing. The recognition of the toxicity of inadequately cooked red kidney beans was prominently publicised following the report of Noah *et al*[42]. Earlier reports in other countries and notably in Tanzania and in early post war Germany in Berlin[43], should be kept in mind whenever there are proposals to use beans in unusual ways that might risk insufficient lectin destruction. Since some varieties of beans may be very much higher in lectin content than others and residual quantities of these high initial levels may resist even normal cooking at altitudes well above sea level[44]. There needs to be continuing vigilance. Lectins of other food crops also should be more widely studied.

2.2.2 Indigestible foods and malabsorption. The nature of the foods themselves, or abuses to normal small intestinal absorbtion of nutrients, may either, or both, result in non-absorbed carbohydrate and or proteinaceous nutrients reaching the lower gut, ileum, caecum and colon, and there being "fermented" by resident bacteria. Gases (inorganic and organic volatiles) among other chemicals will result. The study of these volatiles may be capable of throwing useful light on fermentative metabolism. In our view this is an interesting reason for capturing and analysing flatus gases to see whether a semi-invasive capture of a new bio-product may be useful in "marking" different conditions as well as by providing flatulence indices for different foods and for different food-subject combinations in food tolerances studies.

3 SCOPE FOR PLANT BREEDING

3.1 Scope Through Genetic Variation

3.1.1 Is there useful genetic variation in chemical composition? There appears to be, but too little has yet been done to make use of this in breeding for better quality for consumers. There is documented variation in beans in composition of testa flavonoids. These include the flavonol glucosides, the proanthocyanidins and the anthocyanins[45]. The two former of these are likely to be bio-active. Lectin variation is so great that there is a real possibility that in some circumstances and with some varieties there may be residual lectin activity in the consumption even of cooked beans. There are known differences in the raffinose oligosaccharides and in the hetero-polysaccharides between genera and species and every likelihood that such differences also exist within *Phaseolus vulgaris*. Little is known of the within-species variation in protein and in non-protein sulphur amino acids whose metabolism may lead to mercaptanoid production in the colon.

Bressani long ago pointed out differences in digestibility in Central American beans associated with differences in the "tannins" of these testas[46]. Our own work has been largely empirical. We have modified the composition of various flavonoids by conventional breeding methods and then tested the flatulence index of the resulting beans which in chemical composition most closely corresponds with those bean classes folk-reputed to be non-flatulent in Chile. This has led us simultaneously into flatulence technology and an unusual aspect of plant breeding.

3.2 Plant Breeding in Practice

3.2.1 Flavonoids. By making an appropriate cross between parents carrying between them the appropriate recessive Mendelian genes (Figure 1) it has been possible to re-synthesise new Mantecas beans containing no detectable tannins[47], from non-yellow parents.

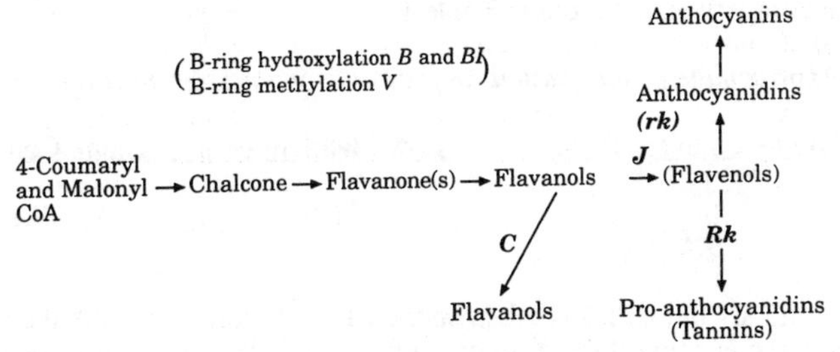

Figure 1 - Outline scheme showing the probable metabolic pathways in testa flavonoids controlled by Mendelian major "colour" genes **C**, **J** and **Rk**. (Based on Leakey [18])

Early maturing, erect, determinate Manteca beans crossed with an indeterminate true Coscorron from Chile can then yield early maturing determinate Coscorrons with a simple segregation only involving the patterning and not the colour genes. This breeding programme provides a range of new material of novel and striking appearance, whose organoleptic and nutritional value can and should be explored.

3.2.2 Destruction of Lectins. We have not bred for reduced lectins *per se* in red-seeded beans but have addressed the problem through breeding for size and shape so that it may be easier in cooking or processing to destroy the lectins through better heat penetration.

4 PRELIMINARY RESULTS AND FURTHER WORK NEEDED

Because of the potential increasing importance of flatulence in ageing populations and the constraint that flatulence already places on the consumption of beans as demonstrated in the only known study of this factor[13] we have developed an improved method of capturing and assessing flatus gas from ambulant human subjects that is sufficiently user-friendly as to find possible non-clinical as well as clinical diagnostic use[25]. In Figure 2 we present the date from a four replicate single subject trial of beans as test meals compared with muesli.

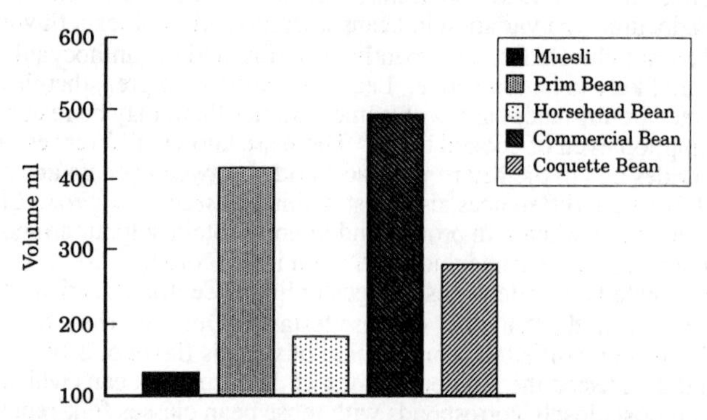

Figure 2 - Flatus produced following ingestion of various meals. Mean of 4 replicate tests of volumes of flatus gas collected during the 5-8 hours following the test meal of the food indicated taken at breakfast time. Fisher's F value 4.999. 5% value 3.26, 1% value 5.41.

In Figure 3 we show gas chromatographic data with marked differences in peaks corresponding to acetone and 2-propanol in the collected flatus gas between flatulent and non-flatulent beans. The two strongly differing peaks correspond to acetone and 2-propanol and the comparison is set out in Table 1.

Table 1 *Approximate concentrations in flatus gas in micrograms/litre*

Compound	Highly Flatulent Bean	Low Flatulent Bean	Blank Control
Acetone	11	1.5	0.34
2-Propanol	9.4	1.3	-

These are preliminary but indicative data on the basis of which we think the topic sufficiently interesting to suggest that further studies should be made. Specifically these metabolites tend to suggest that difference may exist in a sugar termination in a carbon with two attached methyl groups. Such a compound might be a non-toxic analogue of phaseolunatin or even this compound itself, which is known to produce acetone as one of its metabolic products[48]. It is has been recorded at acutely sub-toxic levels in *P. vulgaris*[49], though higher by an order of magnitude than those previously reported[50]. The Chilean work however has never been confirmed.

It is too long already since Price *et al.*[18] reviewed flatulence and suggested food scientists should take the subject more seriously. Perhaps the time has come.

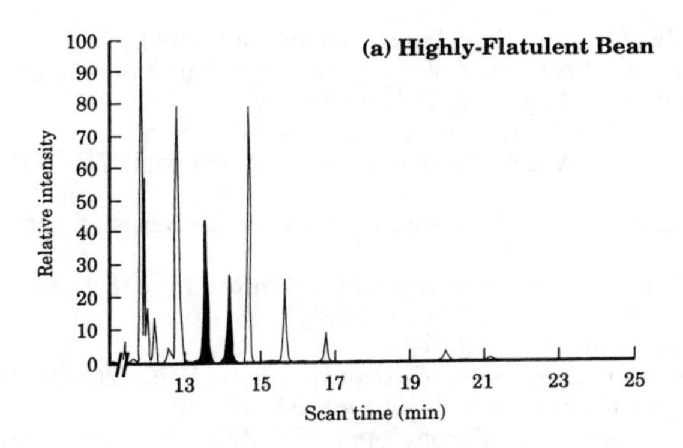

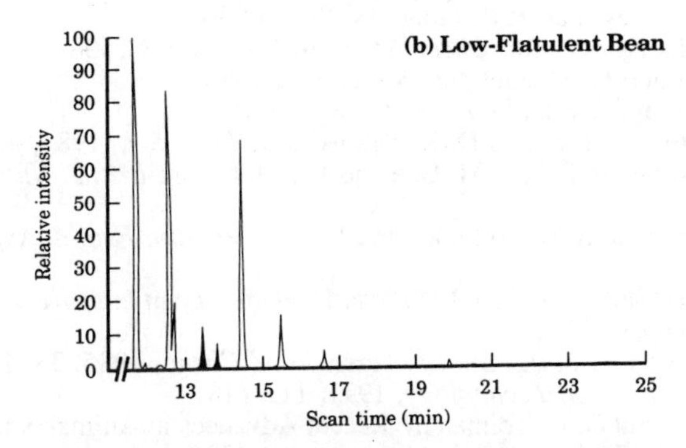

Figure 3 - Thermal desorbtion chromatograms from flatus gases. The two shaded peaks appearing between 13-15 minutes correspond to acetone and 2-propanol. The areas under the peaks give the data in Table 1 above.

References

1 T. Bolin and R. Stanton in 'Wind Breaks - Coming to terms with wind', Margaret Gee Publishing, NSW Australia, 1993.

2 J. Hunter in 'Irritable Bowel Syndrome', Chapter 17 pp 203-219, Read, N. W., ed., Blackwell Scientific Publications, 1991.

3 J. L.A. Roth in 'Bockhus' Gastroenterology', Chapter 9 pp 142-166, Berk, J. E. et al eds., W. B. Saunders, 1985.

4 W. S. Haubricht in 'Bockhus' Gastroenterology', Chapter 135, Berk, J. E. et al eds., W. B. Saunders, 1985.

5 G. Wynn-Jones, *Lancet*, 1975, **2**, 211-212.

6 H. Trowell, D. Burkitt and K. Heaton, eds., 'Dietary ibre, fibre-depleted foods and disease', Academic Press, London, 1985.

7 R. Selvendran, B. J. H.Stevens and M. S. DuPont, *Adv. Fd. Res.*, 1987, **31**, 117-209.

8 N. H. Englyst and J. H. Cummings, *J. Assoc. Off. Anal. Chem.*, 1988, **71**, 808-814.

9 A. Fernandez-Quintela *et al*, this conference.

10 D. J. A. Jenkins *et al., Am. J. Clin. Nutrit.*, 1983, **38**, 567-573.

11 P. R. Ellis, 'Functional Foods - status, opportunities and research in the UK', Seminar, Westminster, UK 1995.

12 J. Cummings in 'Cancer of the large bowel', Trowell, H. *et al*, eds., 1985, 161-189.

13 E. D. White, Report to 10th Dry Bean Research Conference, Davis, USA, Aug 1970.

14 J. E. Brody, *The New York Times*, October 4th, 1990.
15 E. L. Murphy, 'Proceedings of a symposium on Nutritional Improvement of Food Legumes by Breeding', July 1972, pp272-276.
16 B. DeLumen, *Food Structure*, 1994, **11**, 34-46.
17 K. R. Price, J. Lewis, G. M. Wyatt and G. R. Fenwick, *Die Nahrung*, 1988, **32**, 609-626.
18 C. L. A. Leakey, 'Genetic resources of Phaseolus beans', Gepts, P., ed., Kluwer, 1988.
19 C. L. A. Leakey, *Bean Improvement Cooperative Report*, 1992.
20 C. L. A. Leakey, *Grain Legumes*, 1994, **5**, 18-19.
21 G. Grimble, *Gut*, 1989, **30**, 6-13.
22 M. D. Levitt and J. H. Bond, *Gastroenterology*, 1970, **59**, 921-929.
23 D. Calloway, *Gasteroenterology*, 1966, **51**, 383-389.
24 D. Calloway and E. L. Murphy, *Ann. N.Y. Acad. Sci.*, 1968, **150**, 82-95.
25 C. L. A. Leakey, Patent Pending UK No 9409360.6.
26 M. Eichelberger, *J. Am. Chem. Soc.*, 1922, **44**, 1407-1408.
27 R. Hoover and H. Manuel, this conference, 1995.
28 S. E. Fleming, *J. Food Sci.*, 1981, **46**, 794-798.
29 S. K. Sarthe, V. Iyer and D. K. Salunkhe, *J. Food Sci.*, 1981, **46**, 617.
30 S. J. Fairweather-Tait, J. M. Gee and I. T. Johnson, *Brit. J. Nutr.*, 1983, **49**, 303-312.
31 S. W. Appelbaum, U. Tadmoor and H. Podoler, *Ent. Exp. & Appl.*, 1970, **13**, 61-70.
32 C. B. Pena-Valdivia and M. L. Ortega-Delgardo, *Bean Improvement Cooperative Annual Report*, 1983.
33 F. M. Lalojo and F. F. Filho, *J. Agric. Food Chem.*, 1985, **33**, 133-138.
34 S. S. Nielsen, *Food Technology*, 1991, 112-118.
35 J. Huisman and G. H. Tolman in 'Recent Advances in Animal Nutrition', Garnsworthy, P. C. et al, eds., Butterworth, 1992.
36 D. Seidl, M. Jaffe and W. G. Jaffe, *J. Agric. Food Chem.*, 1969, **17**, 1218-1221.
37 V. C. Sgarbieri, E. M. W. Clarke and A. Pusztai, *J. Sci. Food Agric.*, 1982, **33**, 881-889.
38 M. J. L. Kik, Thesis University of Utrecht, 132pp, 1991.
39 V. C. Sgarbieri, W*orld Review of Nutrit. Diet*, 1989, **60**, 133-198.
40 J. H. E. De Muelenaere, *Nature*, 1964, **201**, 1021.
41 J. Huisman, A. F. B. van der Poel, J. M. Mouwen and E. J. van Weerden, *Brit. J. Nutr.*, 1990, **64**, 755-764.
42 N. D. Noah, A. E. Bender, G. B. Reai and R. J. Gilber, *British Medical Journal*, July 1980.
43 R. Korte, *Ecol. Food Nutr.*, 1972, **1**, 303-307.
44 J. H. E. De Muelenaere, *Nature*, 1965, **206**, 827-828.
45 W. J. Feenstra, *Proc. Koninkl. Neder. Akad. Wetternschappen Ser C*, 1959, **62**, 119-130.
46 R. Bressani and L. G. Elias, *UN Food and Nutrition Bulletin*, 1979, **1**, 23-34.
47 J. C. Harborne, private communication.
48 I. E. Leiner in 'Advances in Legume Science', Summerfield, R. J. and Bunting, A. H., eds., R. B. G. Kew, 1980, 157-177.
49 R. Palma, V. and C. Ciuidad, *Agricultura Tecnica*, 1972, **32**, 122-127.
50 R. D. Montgomery in 'Toxic constituents of plant foodstuffs', Liener, I. E., ed., 1969.

THE TEXTURE OF PROCESSED POTATOES; FROM GENES TO TEXTURE

Netty van Marle, Elvis Biekman, Monique Ebbelaar, Kees Recourt, Dougan Yuksel and Cees van Dijk

Agrotechnological Research Institute (ATO-DLO)
Bornsesteeg 59
6708 PD WAGENINGEN, The Netherlands.

1. INTRODUCTION

Potatoes constitute an integral part of the European food pattern. They have a valuable contribution to the diet on basis of their contents such as starch, fibres and vitamins (B1 and C)[1]. However, it has to be kept in mind that the glycoalkaloids are negative impact compounds with regard to human health[2].
For the consumer as well as for the potato processing industry the textural behaviour of the processed potato is an important quality attribute. The key determinants responsible for the texture of cooked potatoes are the cell walls in combination with the remaining interactions between adjacent cells within the tissue. In addition the gelatinized starch may also contribute to the final texture. In our studies two extremes with respect to textural behaviour after cooking were studied in depth; variety Irene representing a mealy and variety Nicola representing a firm potato.

2. GENERAL CONSIDERATIONS

The major determinant of texture in plant based-foods are the plant cell walls[1]. At the molecular level the key determinants of texture are the chemical nature of the cell wall and the interrelations between the component polymers; the cellulose-xyloglucan network embedded in a pectin matrix. During growth and storage the cell wall polymers and there cross-links are modified by enzymatic, during processing both by enzymatic and chemical reactions. At the cellular level the key determinant of texture is the tissue archestructure which comprises cell size, cell wall thickness, cell-cell adhesion. Textural properties will be measured at the organ level by rheological and/or sensory analysis. All these aspects have to be taken into consideration in order to be able to relate characteristics of the raw material with those of processed food product.

3. SPECIFIC CONSIDERATIONS

3.1 Molecular biological and biochemical aspects underlying texture

Since the pectin matrix acts as glue encapsulating microfibrils, the contribution of enzymic action on the pectin matrix and the resulting effects on the final texture of processed potatoes is of importance. Previously, it has been shown that pectolytic enzymes can affect textural aspects of potato. More specific, pectyl methylesterase (PE: EC 3.1.1.11) is believed to be activated at temperatures between 60 and 70 °C and demethylate pectin allowing calcium ions to crosslink pectic chains thereby increasing the resistance of the tissue to further thermal degradation[4]. In addition, it is generally accepted that demethylated pectin is less susceptible to chemical breakdown at higher temperatures.

To investigate the role of pectinesterases in determining potato texture, a biochemical and molecular biological study was initiated. PE activity was present in all vegetative plant tissues and ranged between 50 nkat/mg protein for roots and 20 nkat/ mg protein for leaves. For tubers and microtubers of cvs Irene and Nicola, enzymatic activities were significantly lower and appeared approx 5 nkat/mg protein. Initial purification studies show that tubers and sprouts contain at least two PE isoenzymes. At a molecular biological level, PE-specific oligonucleotides were designed and used to isolate potato-specific pectin esterase cDNAs. By using sprout mRNA as a template for the PCR reaction, two cDNA clones of 670 bp were isolated which were tentatively named PE1 and PE2. In comparison to a fruit specific tomato pectin esterase, the homology of PE1 and PE2 appeared approx 73% and 91% respectively. Both cDNAs are abundantly expressed during sprout development but only PE1 mRNA was detectable during tuber and microtuber development. Currently, the role of both PE genes is investigated using the antisense approach.

3.2 Chemical aspects underlying texture

Cell wall material was isolated from non-cooked and cooked tissue of the varieties Nicola and Irene. The chemical composition of the cell wall material was analyzed. It was shown that, corrected for the surface area of the averaged cell size, the non-mealy cooking cultivar Nicola had significantly less cell wall material as compared to the mealy cooking cultivar Irene. No significant differences could be observed in the overall chemical composition of cell wall material for both cultivars[5].

3.3 Histological aspects underlying texture

A TEM analysis of the cell walls of non-cooked tissue of cv. Irene indicated that this variety seems to have thicker and more dense cell walls than cv. Nicola[5]. This would be in agreement with the observed difference in the amount of cell wall material per unit area, between the two cultivars.

A cryo-SEM[6] evaluation of the fracture planes of cooked samples of the firm variety Nicola showed the presence of large intercellular contacts, with flat cells and cell surfaces showing folds and cracks. Cooked samples from mealy potatoes (cv Irene) showed almost no intercellular contacts and exhibited round cells and smooth surfaces. This suggests that either the way and/or the intensity by which adjacent cells are "glued" together is different for both cultivars.

In vitro cultivation of both cultivars showed that the micro-tubers exhibit the same cooking behaviour with regard to the histological aspects as compared with the field grown potatoes.

3.4 Rheological aspects underlying texture

The softening of potato tissue of the varieties Nicola and Irene was studied by measuring the decrease in firmness of the potato disks (16 mm in diameter and 4 mm thick) in time during steam cooking, making use of an universal testing machine in its shear press mode. The decrease in firmness could for both varieties been described by an exponential decay curve. The rate constants of the softening process for the varieties Nicola and Irene are respectively $(4.7).10^{-3}$ sec^{-1} and $(7.3).10^{-3}$ sec^{-1}; the initial compression values 349 N, respectively 471 N and the calculated rest values for firmness at t-->∞ are for both varieties 0 N. Despite of the obvious differences in textural behaviour, no statistical significant differences could be observed on basis of compression measurements.

3.5 Analytical sensory aspects and the texture of potatoes

An in depth analytical sensory study was performed for three consecutive years to characterize the texture of cooked potatoes (ten varieties; including Nicola and Irene) on basis of 12 descriptors of which 8 are based on aspects related to mouthfeel. A Principal Component Analysis (PCA) of the data showed that two principal components explained over 95% of the variance. The first component showed the difference between varieties for mealy and smooth characteristics (explained part; 86%); the second component explained firm and moist characteristics (explained part; 10%). Next to this it was shown that independent of the variety an increase in tuber size ran parallel with an increase in mealy and firm characteristics.

References

1 C.M.J. Vencken and G. Ebbenhorst, *VMT*, 1990, **23**, 9
2 W.M.J. van Gelder, PhD Thesis, Wageningen, The Netherlands, 1989
3 R.R. Selvendran, *J. Cell Sci.Suppl.*, 1985, **22**, 51
4 L.G. Bartolomo LG and J.E.Hoff, *J. Agric. Food Chem* 1972, **20**, 266
5 J.T. van Marle, T.Stolle-Smits, C.van Dijk, A.G.J. Voragen and K.Recourt, *J.Agric.Food Chem.*, submitted
6 J.T. van Marle, A.C.M. Clerx and A. Boekestein *Food Structure*,1992, **11**, 209

MODIFICATION TO THE FIBRE MATRIX ASSOCIATED WITH THE PRODUCTION OF PHYTATE-REDUCED WHEAT BRAN

C. N. Jayarajah, J. A. Robertson and R. R. Selvendran

Department of Food Molecular Biochemistry,
Institute of Food Research,
Norwich Research Park, Colney,
Norwich NR4 7UA

INTRODUCTION

Wheat bran is a rich source of dietary fibre with widespread use as a supplement in food products. The elevated concentration of phytate in 'fibre-enriched' wheat bran has implications for the bioavailability of minerals and proteins in foods[1,2]. Dephytinisation can be achieved during the fermentation of bread dough and may involve yeast phytase activity as well as endogenous phytase in flour[3,4]. Endogenous phytase is activated during grain germination and in bran preparations has been used to reduce phytate levels. During grain germination polysaccharide-degrading enzymes are also activated and thus there is the potential for both degrading the fibre matrix and phytate[5,6]. The objective of this study was to determine whether endogenous phytase activity could be optimised to reduce phytate concentration and also whether activation of phytase was associated with an activation of polysaccharide degrading enzymes.

METHODS

Wheat bran samples were supplied by Sofalia, Ennezat, France as fibre-enriched products, 250μm mean particle size. Fibre analysis was as non-starch polysaccharides (NSP) recovered[7] after dephytinisation treatments with quantification of neutral sugars as their alditol acetates[8].

Dephytinisation treatments involved either leaching using excess water or incubating with limited moisture.

Leached: samples incubated in 20 vol water, (pH 5.1); 30 min, room temperature, with shaking and recovered by centrifugation and freeze dried.

Limited moisture: samples incubated with known ratios of acidified water (pH 5.1) at 55 C for set time periods and recovered by freeze drying.

Exogenous phytase activity: samples incubated under limited moisture conditions (25 C) using an actively fermenting yeast culture.

Phytate was estimated from the difference between total phosphate and inorganic phosphate[9] ie $\mathbf{P_{phy} = P_t - P_i}$ where: **P_{phy} = phytate phosphate; P_t = total phosphate and P_i = inorganic phosphate**. Preliminary experiments confirmed that phytate was the major source of phosphate in bran (~90%) and that inorganic phosphate represented less than 5% of total phosphate. Hence dephytinisation was monitored from the loss of total phosphate from bran or the relative increase in inorganic phosphate.

RESULTS AND DISCUSSION

The untreated bran had a phytate content of 37.9mg/g (Table 1). Leaching treatment removed the bulk of the phytate to leave a 'phytate free' fibre preparation but the treatment also removed around 30% of the original bran (recovery 71.5%). The loss in dry weight was ascribed to the removal of protein, residual starch and mineral rather than to a change in NSP. The increase in supernatant P_i and decrease in P_{phy} indicated a significant degradation of phytate during leaching, consistent with endogenous phytase activity. Incubation under limited moisture conditions also resulted in a significant dephytinisation (~90% conversion of P_{phy} to P_i) but the exogenous source of phytase had no effect on the extent of dephytinisation.

The pH used (5.1) was chosen to optimise endogenous phytase activity and is lower than the pH of fermenting dough (~pH 6). Under conditions used for dough fermentation then exogenous phytase may contribute to dephytinisation. However, the results indicate that endogenous phytase activity can lead to substantial reduction in phytate concentration.

Table 1 *Effects of Dephytinisation Treatments on the Distribution of Phosphate in Wheat Bran Samples*

Treatment	P_t	P_i	P_{phy}	*Phytate*
		(mg/g)		
Untreated	11.3	0.6	10.7	37.9
Leached bran	0.8	0.2	0.6	2.2
- supernatant	8.8	1.9	6.9	24.4
Endo.Enzyme	11.6	10.2	1.4	5.0
Endo.+Exo. Enzyme	11.8	10.3	1.5	5.3

Endo. enzyme = endogenous treatment after 18h incubation; 2:1 ratio water:bran
Endo. + Exo. enzyme = treatment as for endo. enzyme + yeast culture.

A moisture ratio of 2:1 achieved around 50% dephytinisation with a 6 hour incubation (Fig. 1a), which was only slightly greater than that achieved at a moisture ratio of 1:1. At a moisture ratio of 2:1 the dephytinisation progressed more rapidly in the initial 2 hours of incubation. Thus at a moisture ratio of 1:1 the rate of dephytinisation was slower and 4.5

hours were required to achieve a dephytinisation level corresponding to 1.5 hours at a

Figure 1 *Effect of Moisture Content and Incubation time on Endogenous Phytase Activity*

a) Effect of moisture content (6 hours incubation)

b) Effect of time at different moisture contents

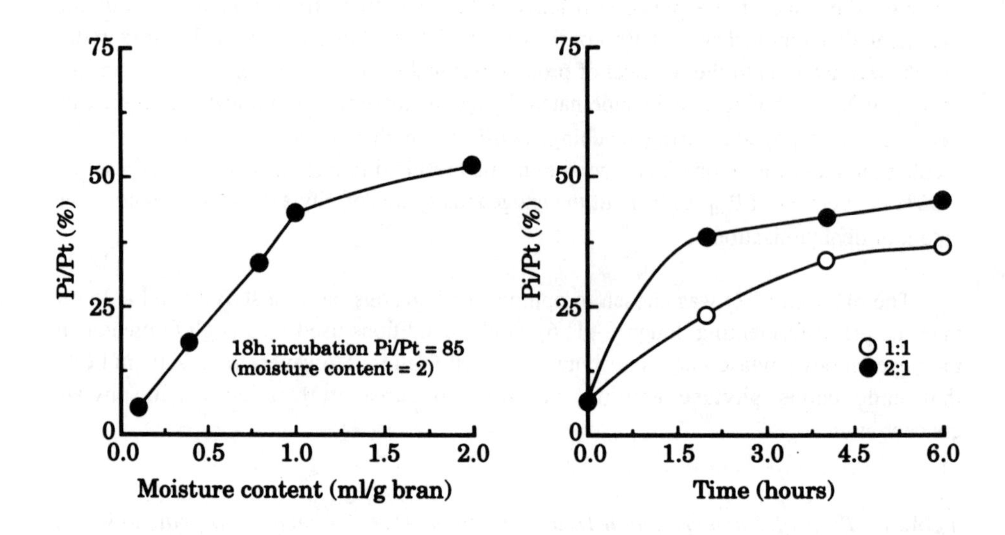

moisture ratio of 2:1. That differences in rates of dephytinisation can be achieved at different moisture contents and that there is a distinction between phytate readily susceptible to degradation and that requiring a prolonged incubation provides a convenient mechanism to manipulate the progression of dephytinisation to 'optimise' phytate. This provides a convenient mechanism for the preparation of reduced phytate bran.

Compared to the untreated bran (Table 2) recovery of NSP was similar after both dephytinisation treatments (~43%) and cellulose content and the ratio of X:A was typical of wheat bran[10]. A slight but consistent increase in X:A in dephytinised samples indicated there may be some arabinosidase activity but this could at most account for a reduction of only 5% of total arabinose. The treatments were also without effect on fibre solubility. Although soluble NSP was lower in leached samples the leaching had resulted in a loss of soluble material, including NSP. Even allowing for the loss of all soluble NSP through leaching then total soluble NSP through leaching would remain only 2.7% of bran.

Table 2 *Effects of Dephytinisation Treatments on the Fibre/Non-stach Polysaccharide Content of Wheat Bran Samples*

Treatment	*NSP*	*Insoluble (% bran)*	*Soluble*	*Cellulose (% NSP)*	*X/A*
Untreated	44.0	42.1	1.9	23.6	1.8
Leached	42.3	41.5	0.8	22.5	1.9
Endo. Enzyme	42.9	40.6	2.3	23.9	1.9

Samples correspond to treatments in Table1. Cellulose estimated as glucose released by Saeman hydrolysis less glucose released by 1M acid hydrolysis. X/A = ratio of xylose: arabinose in NSP residues.

CONCLUSION

A dephytinisation of wheat bran can be achieved at relatively low moisture content through activation of endogenous phytase. The activity can be sufficient to achieve a 50% dephytinisation within 1.5 hours and at the hydration levels required there was no apparent loss or modification of the fibre matrix.

REFERENCES

1. E.M. Widdowson and R.A. McCance, *Lancet*, 1942, **1**, 588.
2. P. Ryden and R.R. Selvendran, 'Encyclopedia of Food Science, Food Technology and Nutrition', [eds R. Macrae, RK. Robinson & MJ. Sadler], Academic Press, 1992, p.3582
3. B.F. Harland, and J. Harland, *Cereal Chem.*, 1980, **58**, 226.
4. G. Giovanelli, and R. Polo, *Ital. J. Food Sci.*, 1994, **1**, 71.
5. F.G. Peers, *Biochem. J.*, 1953, **53**, 102.
6. E. Mellanby, 'A Story of Nutritional Research', Williams and Wilkins, Baltimore, 1950, Chapter 8, p. 248.
7. R.M. Faulks and S.B. Timms, *Food Chem.*, 1985, **17**, 273.
8. R.R. Selvendran, J.F. March and S.G. Ring, *Anal. Biochem.*, 1979, **96**, 282.
9. A.O.A.C. 'Official Methods of Analysis of the AOAC' Association of Official Analytical Chemists, Arlington, Virginia, 1990, Method 986.11. p.980.
10. R.R. Selvendran, B.J.H. Stevens and M.S. DuPont, *Adv. Food Res.*, 1987, **31,** 117

ACKNOWLEDGEMENTS

Funding for the work, as part of the EEC FLAIR Programme, (Contract AGRF-CT90-0038C), is gratefully acknowledged.

THE POTENTIAL FOR THE IMPROVEMENT OF QUALITY IN FOOD SYSTEMS THROUGH INCORPORATION OF DEFINED FIBRES OBTAINED FROM PROCESSED VEGETABLE BY-PRODUCTS

A. Femenia[1], A.-C. Lefebvre[2], J.-Y. Thebaudin[2], J.A. Robertson[1] and C.-M. Bourgeois[2]

1. Department of Food Molecular Biochemistry, Institute of Food Research, Norwich Research Park, Colney, Norwich NR4 7UA, UK. 2. ADRIA, 6 rue de l'Université, 29334 Quimper, France

1 INTRODUCTION

High fibre diets have been recommended to maintain or improve health but there has been less consideration of the organoleptic properties of these diets. There is a tendency, however, for high fibre diets to be less acceptable to the consumer and hence a need to consider ways to incorporate fibre into foods and which will be acceptable to the consumer. This has led to the concept that as well as amount of fibre the "type of fibre" should also be considered[1]. Certainly the physiological response to fibre is related not only to the amount present and properties of the constituent polysaccharides, but also to the structural features of the plant cell walls[2,3]. Allied to this, food processing by-products can be rich sources of fibre, at present discarded but with potential as ingredients to improve industrial technology. Such products are often derived from the more mature parts of the plant and it is known that cell wall polysaccharides are susceptible to modification during growth as well as during processing[4]. Hence there is a need to evaluate fibre sources for chemical and structural modification in food systems and how these changes may influence their behaviour in foods. However, it is more difficult to predict the consequences of structural changes in foods.

In this investigation the objective was to characterise structural features of cauliflower fibre in different tissues; how these were affected by processing (dehydration) and the consequences on the physico-chemical properties of the fibre. On the basis of this evaluation selected fibre preparations were incorporated into food systems to determine the main physical and sensorial characteristics resulting.

2 MATERIALS AND METHODS

1.1 Materials

Fresh cauliflower (*Brassica oleracea* L. var. *botrytis*) plants supplied by Sofalia (Ennezat, France) were separated into the florets or curd and stem. Samples were analysed as the fresh material and after dehydration. Dehydration was performed either at 40°C or 75°C.

1.2 Methods

1.2.1 Fibre analysis. Fibre was determined as soluble, insoluble and total non-starch polysaccharides (NSP)[5,6].

1.2.2 Functional properties. Functional properties[4] measured included hydration (water holding, retention, swelling), fat adsorption capacity (FAC) and textural behaviour.

1.2.3 Incorporation into food systems. Selected preparations were incorporated into a range of food systems, including meat products, emulsions, sauces and delicatessen/bakery products. Incorporation levels were 2% w/w, either as an addition or substitution.

1.2.4 Evaluation of food systems. The analyses performed varied depending on the type of preparation. Technological yield, textural properties and colour were the major parameters determined. A sensory evaluation was conducted to assess the acceptability of the experimental products and for comparison with control formulations.

3 RESULTS AND DISCUSSION

3.1 Fibre analysis

The NSP composition of the different samples revealed important differences between mature and immature tissues (Table 1). Whilst florets and upper stem were similar and contained mainly pectic polysaccharides and cellulose, the lower stem was enriched in hemicelluloses (acidic xylans) and cellulose associated with the lignified tissues. Pectic substances were also present in the lower stem but in amounts lower than the upper stem.

Fibre analysis revealed that only minor changes either in the amount or the composition occurred during drying (Table 1). However, a decrease in the degree of methyl-esterification (DME) occurred when samples were dried at temperatures above 40°C .

Fibre analysis also revealed an apparent decrease in water solubility with increasing temperatures. This effect was evident using non-degradative methods of fibre analysis[6] and cell wall fractionation[7] but was not revealed by methods which predisposed fibre to solubi-

Table 1 *NSP composition of fresh and processed cauliflower tissues*

	%NSP[6] (fresh weight)	%NSP as cellulose	hemicellul.	pectin	DME (%)	Solub NSP[6] (%)
Floret						
fresh	2.05	34.7	8.5	56.8	59	9.8
dried 40°C	2.07	37.5	9.6	52.9	57	9.3
dried 75°C	2.02	38.1	10.3	51.6	46	5.8
Upper stem						
fresh	2.85	36.6	10.6	52.8	65	9.9
dried 40°C	2.81	37.9	11.1	51.0	61	9.3
dried 75°C	2.79	38.4	11.4	50.2	53	6.1
Lower stem						
fresh	6.02	51.9	21.8	26.3	38	2.2

lisation[5]. Therefore processing can result in structural changes either in the mature or immature cell walls, but further changes can also be induced with more severe processing.

3.2 Functional properties

Functional properties measured for the different preparations (Table 2) showed that at higher temperature of drying, hydration properties were all reduced. However, floret and stem tissue had similar hydration properties. FAC was much lower than hydration properties and the effects of temperature were less marked. Textural capacities differed between floret and stem samples, with values for stem being higher. This was consistent with the 'toughening' of more mature tissues. Similarly the higher values for texture after heating were consistent with the modification of the food matrix during cooking.

No major compositional differences were detected between NSP composition of the dried samples which indicated that the functional properties were more dependant on the structural arrangement of the polysaccharides than on the overall NSP composition.

3.3 Incorporation into food systems

The samples which exhibited higher functional properties (Table 3) were selected for incorporation into model food systems, chosen to include meat-based products, fat emulsions, delicatessen/bakery products and sauces.

3.3.1 Meat products. The incorporation of fibre supplements resulted in a significant increase in the technological yield. The addition of stem fibres also increased the firmness of the beef burgers. The organoleptic properties were considered acceptable and confirmed the potential of these fibres to be incorporated into meat products.

3.3.2 Fat emulsions. Fat emulsions are considered to be intermediate products, hence no sensorial analysis was performed on these preparations. In terms of texture and technological yield, there were no significant differences between the control and the experimental formulations. These fibre preparations can be considered neutral and therefore have the potential to be incorporated in foods but without modifying texture.

3.3.3 Delicatessen/Bakery products. Fibres were incorporated into vegetable paté (carrot) and quiche "Lorraine". The organoleptic characteristics of the experimental paté formulations were considered acceptable, but these products underwent slight discoloration. The quiche "Lorraine" formulations had a "very agreeable" taste and a texture similar to the control. Thus, the quiche can be considered a suitable product for the addition of fibres.

Table 2 *Functional properties of fibre-enriched preparations*

Preparations	SW (ml/g)	WHC (g/g)	WRC (g/g)	FAC (g/g)	Texture (raw) (N)	Texture (cooked) (N)
Floret 40°C	16.9	6.4	12.8	1.3	8.9	10.2
Floret 75°C	4.2	1.9	5.7	0.9	0.6	2.2
Upper Stem 40°C	17.5	7.3	13.4	2.1	17.1	27.3
Upper Stem 75°C	8.7	6.0	9.1	1.2	13.2	17.1
Flor + Stem 40°C	17.0	8.9	18.2	1.4	8.7	14.0

Table 3 *Features and organoleptic properties of the experimental food systems*

Food products	Main features	Organoleptic evaluation
Beef-burgers	Improve technological yield and firmness	Taste and flavour acceptable
Forcemeat	Improve technological yield	Acceptable
Emulsion	Neutral, no differences to the control	Not evaluated (intermediate product)
Vegetable paté	Decrease in firmness and adherence	Discoloration, no acceptable
Quiche "Lorraine"	Texture similar to the control	Very good taste and appearance
Tomato sauce	Granular texture	Colour slightly modified but acceptable
Béchamel sauce	Slight increase in viscosity	Acceptable

3.3.4 Sauces. Fibres were incorporated into béchamel and tomato sauces. The experimental formulations of tomato sauce had a granular texture. The béchamel sauce with added fibres, despite a slight increase in viscosity, was acceptable to the sensory panel.

4 CONCLUSIONS

Fibre supplements from cauliflower tissues are susceptible to modification during processing and this can affect their functional properties. Changes in functional properties are a function of matrix structure rather than polysaccharide composition. The ability to control functional properties is important if fibre supplements are to be successfully incorporated and/or substituted into refined foods. However, a greater understanding of the structure and functional behaviour of fibre in foods is required before attempting predictive modelling of the effects of fibre supplementation. Nevertheless, this study has provided the basis for the potential use of standardised food systems with well defined fibres, and for the subsequent assessment of their nutritional response.

References

1. R.R. Selvendran and J.A. Robertson, 'Physico-chemical Properties of Foods and Effect of Processing on Micronutrients Availability', EEC Brussels, 1994, p. 11
2. N.-G. Asp, 'New Developments in Dietary Fiber', Plenum Press, N.Y., 1990, p. 227
3. R.R. Selvendran and J.A. Robertson, 'Dietary Fibre: Chemical and Biological Aspects', The Royal Society of Chemistry, Cambridge, 1990, p. 27
4. J.-F. Thibault, M. Lahaye and F. Guillon, 'Dietary Fibre - A Component of Food', Springer Verlag, London, 1992, p. 21
5. H.N. Englyst, M.E. Quigley, G.J. Hudson and J.H. Cummings, *Analyst*, 1992, **117**, 1707
6. R.M. Faulks and S.B. Timms, *Food Chem.*, 1985, **17**, 273
7. R.J. Redgwell and R.R. Selvendran, *Carbohydr. Res.*, 1986, **157**, 183

Acknowledgements.

Funding for the work, as part of the EU FLAIR Programme (Contract AGRF-0038C), is gratefully acknowledged.

CONTENT AND COMPOSITION OF NON-CELLULOSIC POLYSACCHARIDES OF ALLOPLASMIC RYE COMPARED TO PARENTAL SPECIES - TETRAPLOID TRITICALE AND RYE

M. Cyran[1], M. Rakowska[1] and B. Łapiński[2]

[1] Department of Nutritional Evaluation of Plant Materials
[2] Department of Genetics
Institute of Plant Breeding
Radzików, P.O. Box 1019, 00-950 Warsaw, Poland

1 INTRODUCTION

The non-cellulosic polysaccharides (NCP), derived from plant cell walls are the main constituent of dietary fiber (DF). These polysaccharides, together with cellulose, lignin, protein, polyphenolics and minerals, form a mixture, which is resistant to hydrolysis by the digestive enzymes of monogastrics[1]. In spite of relatively small contribution to the total weight of grain, their unique properties, especially a high water-binding capacity[2] influence the baking quality of rye products[3]. Addition of soluble, DF polysaccharides to human diet, diminishes plasma cholesterol level, the risk of colon cancer and heart disease as well as improves the tolerance of glucose[4]. On the other hand, DF constituents, particularly soluble fraction, are a factor limiting availability of other food components[5], due to their ability to form highly viscous solution.

There are attempts at rye improvement by crossing it with tetraploid (4x) triticale and backcrossing to rye (4x triticale passage method). Therefore, content and composition of NCP, viscosity, protein and lysine contents, as the indicators of quality were studied in lines of alloplasmic rye, compared to parental forms - 4x triticale and two lines of rye with high and low soluble NCP content.

2 MATERIAL AND METHODS

One line of 4x triticale (TS-2), two lines of rye, R1 with low soluble NCP content and R2 with high soluble NCP content, six lines of alloplasmic rye from crossing TS-2 x R1 and six lines from crossing TS-2 x R2 were used. All cereals were grown under the same soil and climatic conditions.

Soluble and insoluble DF components were isolated by gravimetric method based on enzymatic digestion of starch and protein as described by Asp et al.[6]. The DF fractions were hydrolyzed in 1N trifluoroacetic acid (TFA) at 125°C for 1 h and, directly from hydrolysates, aldononitrile acetate derivatives of neutral sugars were obtained, according to McGinnis[7]. The samples were analyzed by gas chromatography.

Protein content was determined by the Kjeldahl method using a Kjeltec System. Lysine content was assayed according to Mason et al.[8] using a Beckman Amino Acid Analyser.

Suspension viscosity of ground grains (1:3 w/v) in 0.2 M sodium phosphate buffer

Table 1 *Content and Composition of Soluble Non-cellulosic Polysaccharides in Alloplasmic Rye and in Parental Species, Tetraploid Triticale (TS-2) and Rye with Low and High Content of Soluble Arabinoxylan (R1 and R2, respectively)*

	Arabinoxylan	*Mannose*	*Glucose*	*Galactose*
		% of dry matter		
Tetraploid triticale TS-2	1.86	0.15	0.30	0.15
Alloplasmic rye (TS-2 x R1), n=6	2.31 ± 0.36	0.20 ± 0.06	0.69 ± 0.07	0.17 ± 0.02
Rye R1	2.74	0.22	0.77	0.18
Alloplasmic rye (TS-2 x R2), n=6	2.94 ± 0.61	0.21 ± 0.05	1.20 ± 0.50	0.17 ± 0.04
Rye R2	3.98	0.36	2.33	0.29

The values for alloplasmic rye represent the mean values ± standard deviation

(pH 5.6) was measured with Rheotest-2 at constant temperature (30°C).

3 RESULTS

In lines of alloplasmic rye (TS-2 x R1) the mean content of soluble arabinoxylan (2.32% of dry matter) was lower, than that of rye R1 (2.74%) and higher, than the respective value of 4x triticale (1.86%, Table 1). Similarly, in lines of alloplasmic rye from the second crossing (TS-2 x R2) the mean content of soluble arabinoxylan (2.94%) was notably lower, than in rye R2 (3.98%).

Contents of soluble glucose found in lines of alloplasmic rye (TS-2 x R1) and rye R1 (0.69% and 0.77%) were practically the same, while in lines from crossing (TS-2 x R2) almost twofold higher level of soluble glucose was found (1.20%), but it was lower, than in the parental form of rye - R2 (2.33%).

The minor constituents of NCP, mannose and galactose were present in small amount. Only in rye R2 higher contents of both sugars have been observed, whereas there were no significant differences in the rest of samples. It is worthy to note, that among the breeding material from both crosses the content of soluble NCP was an intermediate in four lines, close to 4x triticale in one of them and also close to parental forms of rye in one of lines.

Composition of soluble NCP fractions of alloplasmic ryes was an intermediate between sugar spectrum observed in rye parental forms. Arabinoxylan constituted 68%, 66% of NCP in lines from crosses TS-2 x R1, TS-2 x R2 and 70 %, 57% in rye R1 and R2, respectively, whereas 75% in 4x triticale. Glucose polymers constituted 12 % in 4x triticale, 20% and 21% in rye R1 and in lines of alloplasmic rye (TS-2 x R1),

Table 2 *Content and Composition of Insoluble Non-celullosic Polysaccharides in Alloplasmic Rye and in Parental Species, Tetraploid Triticale (TS-2) and Rye with Low and High Content of Soluble Arabinoxylan (R1 and R2, respectively)*

	Arabinoxylan	*Mannose*	*Glucose*	*Galactose*
		% of dry matter		
Tetraploid triticale TS-2	5.26	0.29	1.31	0.29
Alloplasmic rye (TS-2 x R1), n=6	6.57 ± 0.64	0.22 ± 0.06	1.77 ± 0.11	0.39 ± 0.06
Rye R1	7.28	0.24	1.80	0.33
Alloplasmic rye (TS-2 x R2), n=6	6.49 ± 0.34	0.21 ± 0.04	2.20 ± 0.21	0.32 ± 0.04
Rye R2	8.06	0.26	2.45	0.49

The values for alloplasmic rye represent the mean values ± standard deviation

respectively, while 26% and 34% in lines (TS-2 x R2) and rye R2.

There were no significant differences in the mean contents of insoluble arabinoxylan (Table 2) between lines of alloplasmic rye from both crosses (6.57% and 6.49%), although these values were notably lower, than in parental ryes (7.28% and 8.06% in R1 and R2, respectively). Similar amount of insoluble, non-cellulosic glucose was found in lines of

Table 3 *Protein, Lysine contents and Viscosity of Buffer Suspension of Ground Grain in Alloplasmic Rye and in Parental Species, Tetraploid Triticale and Rye*

	Protein % of dry matter	Lysine g/16g N	Viscosity mPa·s
Tetraploid triticale TS-2	20.2	4.49	343
Alloplasmic rye (TS-2 x R1), n=6	16.5 ± 1.0	3.74 ± 0.10	1708 ± 474
Rye R1	18.2	3.39	2728
Alloplasmic rye (TS-2 x R2), n=6	15.4 ± 1.0	3.82 ± 0.16	3526 ± 336
Rye R2	16.1	3.70	4284

The values for alloplasmic rye represent the mean values ± standard deviation alloplasmic rye (1.77%, 2.20%) and in rye parental forms (1.80%, 2.45%).

Composition of insoluble NCP fractions was similar, where lower variability of constituent sugars has been observed.

The samples of alloplasmic rye shown evidently lower buffer suspension viscosity compared to the respective value of parental ryes (Table 3). However, twofold lower viscosity had samples of alloplasmic rye derived from crossing with rye R1 (1708 mPa·s), than these from crossing with rye R2 (3526 mPa·s).

High lysine content (3.74 and 3.82g/16g N) as well as high protein content (16.5% and 15.4%) were found in samples of alloplasmic rye, similarly to the high values of these components in the parental forms.

4 CONCLUSIONS

Lines of alloplasmic rye obtained by crossing 4x triticale with rye demonstrate notably lower level of soluble as well as insoluble NCP compared to the parental forms of rye.

The amount of NCP in the parental rye infulences content of soluble NCP fraction in the hybrid progenies, while no impact of this factor on content of insoluble fraction has been observed.

It is to be expected, that lower viscosity and high lysine and protein contents found in these lines will result in better nutritional quality of the grain.

References

1. H. Trowell, D. A. T. Southgate, T. M. S. Wolever, A. R. Leeds, M. A. Gassul and D.J.A Jenkins, *Lancet*, 1976, **1**, 967.
2. F. Meuser and P. Suckow, *Spec. Publ. R. Soc. Chem.*, 1986, **56**, 42.
3. J. A. Delcour, S. Vanhamel and R. C. Hoseney, *Cereal Chem.*, 1991, **68**, 72.
4. K. L. Roehring, *Food Hydrocolloids*, 1988, **2**, 1.
5. M. Rakowska, R. Kupiec and K. Rybka, *Pol. J. Food Nutr. Sci.*, 1992, **1**, 103.
6. N. G. Asp, C. G. Johansson, H. Hallmer and M. Siljestrom, *J. Agric. Food Chem.*, 1983, **31**, 476.
7. G. D. McGinnis, *Carbohydr. Res.*, 1982, **108**, 284.
8. V. C. Mason, S. B. Andersen and M. Z. Rudeno, *Tierphysiol. Tierernähr. u. Futtermittelkde.*, 1980, **43**, 146.

A STUDY OF THE EFFECTS OF WATER CONTENT ON THE MECHANICAL TEXTURE OF BREAKFAST CEREAL FLAKES

D.M.R. Georget and A.C. Smith,

Institute of Food Research,
Norwich Research Park,
Colney,
Norwich, NR4 7UA

1 INTRODUCTION

Commercial production of breakfast cereals produces large variations in size, shape and density of individual pieces. Breakfast cereals are complex foods and in order to understand their textural characteristics it is important to study their mechanical properties[1,2]. Mechanical properties of multiple flakes depend on packing, geometry and structure of flakes and composition, particularly water content of the flake matrix.

In this study breakfast cereal flakes have been compacted in the pressure range 100 Pa to 85 MPa. At lower pressures (LP) the bulk density of the flakes may be related to the pressure and this relationship is sensitive to a combination of the geometry, structure and matrix properties of the flakes. At higher pressures (HP) attrition of the flakes occurs such that the matrix properties dominate the response.

Tests using simple deformations give data in engineering units but are not always easily applicable to small or irregular-shaped samples. Tests of single flakes were carried out using a pin loading the flakes centrally. Mechanical properties depend on the structure and composition of the flakes. The test may be analysed using classical mechanics to provide stresses, strains and elastic moduli on the assumption that the flake approximates a concentrically- supported disc. This approximation was made more exact by flattening flakes using a hydration - drying step.

Breakfast cereals are composed of biopolymers, principally starch and gluten. The mechanical properties of these polymers are determined to a large extent by their glass transitions (T_g)[3]. Solely composition effects are investigated by measuring the mechanical behaviour of flakes which were milled and reconstituted as bar-shaped test pieces. Dynamic mechanical thermal analysis (DMTA) together with impact failure were used to compare the response of multiple component systems with that published for simpler one and two component materials which were also in the form of hot-pressed bars.

2 EXPERIMENTAL

Wheat breakfast cereal flakes were produced commercially and had an initial water content of 3 to 4 % (w.w.b). Flakes were approximately disc-shaped with typical diameters in the range 10 to 20 mm. The flakes were conditioned to different water contents using saturated salt solutions. They were subjected to compaction of multiple flakes[4], indentation of single flakes[5] and deformation of pressed bars of ground flakes[6].

3 RESULTS AND DISCUSSION

The relationship between apparent density and the logarithm of applied pressure for flakes at one water content is shown in Figure 1. The test is able to differentiate samples of different geometry, structure and composition produced by different processing methods. The Heckel deformation stress was obtained from the relationship between the relative density and the axial pressure (Figure 2). In the earlier study of potato starch[7], the decrease in deformation stress with water content was discussed in relation to the large changes in elastic modulus.

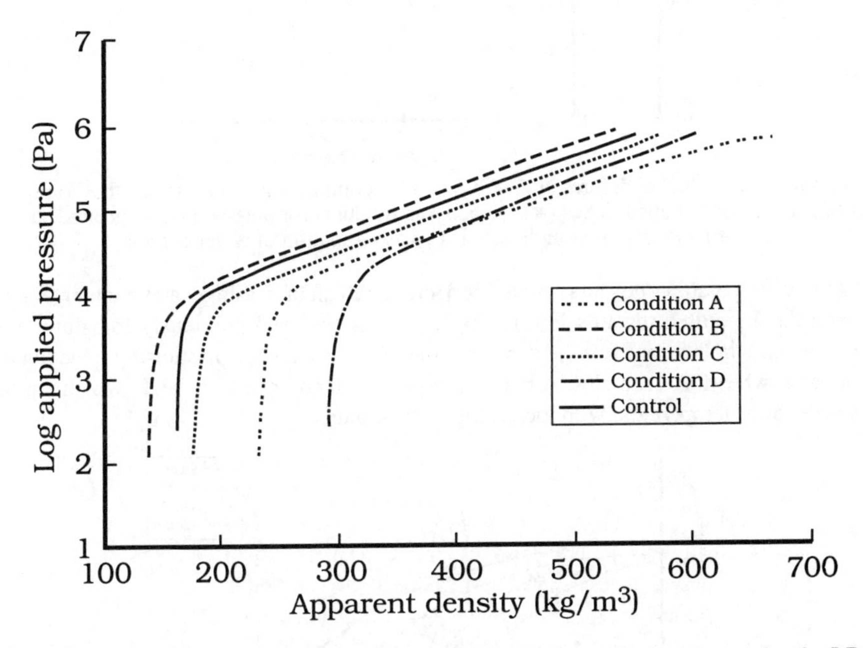

Figure 1: Apparent compact density as a function of log (applied pressure) for the LP experiments for different compositions

Figure 2 shows the Young's modulus for the as-supplied flakes and as a function of water content for flattened flakes. It also shows the variation of the bending modulus of moulded bars with water content. Apart from the thermal history necessitated by the moulding treatment, the flattened flakes and moulded bars differ in particle size and porosity, at a water content of 18 % (w.w.b) the densities of a flake and moulded disc were found to be 750 and 1450 kg m^{-3} respectively. The modulus falls with the inclusion of air according to a method of mixtures or density scaling law. An estimate of the critical energy release rate was made based on the energy to break for different notch sizes[6]. The present study shows that the addition of water made the samples harder to break or tougher (Figure 2).

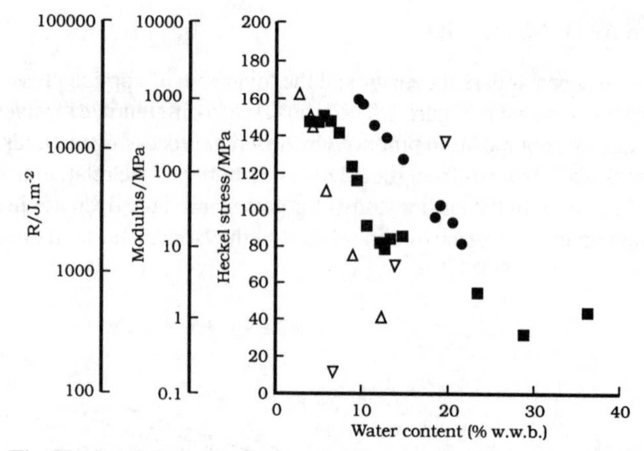

Figure 2: The Heckel deformation stress (Δ) for compacting wheat flakes, the Young's modulus, E, for flattened flakes (■) the bending modulus for pressed bars of ground flakes (●) and fracture toughness, R (∇) as a function of water content.

At low temperature the glassy modulus decreased with increasing water content (Figure 3) and the T_g shifted towards lower temperature as observed previously for gluten and amylopectin [8,9]. This confirms the expected role of water having a plasticising effect on the fabricated wheatflake samples. A maximum peak in tanδ occurs at the T_g which shifted towards lower temperature with increasing water content.

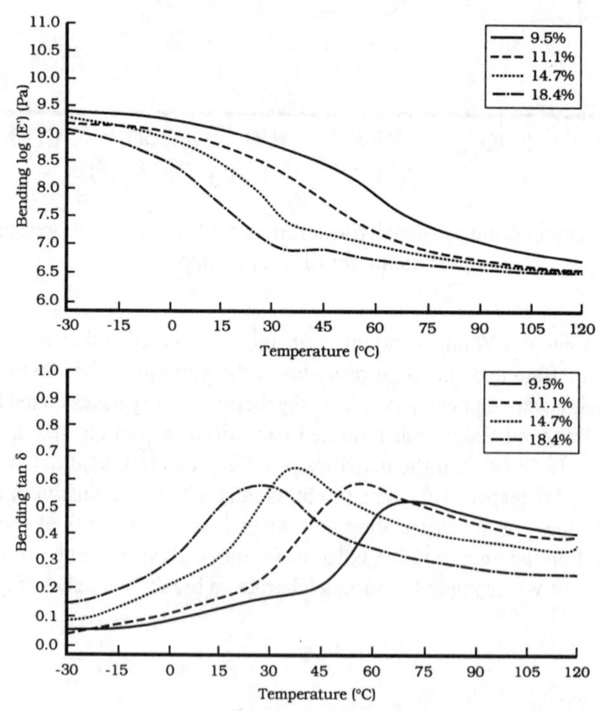

Figure 3: DMTA Log E' and tanδ for the samples for different water contents as a function of temperature.

4 CONCLUSION

The stiffness of hot-pressed test pieces of ground wheat flakes, the stiffness of flattened flakes and the Heckel deformation stress all fell with increasing water content, whereas the fracture toughness increased. The changes in mechanical properties can be related to shifts of the glass transition temperature. The general features of the gluten and amylopectin systems appear to be exhibited by the more complex samples. Whereas at low water contents the samples are glassy and only small elastic deformations occur, at the higher water contents the particles become increasingly rubbery and allow extensive deformation.

This work was supported by a MAFF - DTI LINK grant under the Food Processing Sciences Programme involving Weetabix Ltd, APV Baker, Campden Food and Drink Research Association, University of Nottingham and Novo Nordisk (UK) Ltd.

REFERENCES

1. J. Loh and W. Mannell, in 'Dough Rheology and Baked Product Texture', H. Faridi and J.M. Faubion (eds.), Van Nostrand Reinhold, New York, 1990, p. 405.
2. F. Sauvageot and G. Blond. J. Text. Studies, 1991, **22**, 423.
3. L. Slade and H. Levine (1993). In 'The Glassy State in Foods', J .M. V. Blanshard and P. J. Lillford (eds.), Nottingham University Press, Nottingham, UK, p. 35.
4. D. M. R. Georget, R. Parker and A. C. Smith, Pow. Tech., 1994, **81**, 189.
5. D. M. R. Georget, R. Parker and A. C. Smith, J. Text. Studies , 1995, **26**, 161.
6. D. M. R. Georget and A .C. Smith. Carbohydr. Poly. In press.
7. A.-L. Ollett, A .R. Kirby, R. Parker and A. C. Smith, Pow. Tech., 1993, **75**, 59.
8. Kalichevsky, M.T., Jaroszkiewicz, E.M., Ablett, S., Blanshard, J.M.V. and Lillford, P.J. Carbohydr. Polym., 1992, **18**, 78.
9. Kalichevsky, M.T., Jaroszkiewicz, E.M. and Blanshard, J.M.V. Int. J. Biol. Macromol., 1992, **14**, 257.

Connection between the rheological properties and the sensory characteristics of raw and preserved cucumbers

János Zana – Péter László – Erzsébet Kristóf

Department of Physics and Control
University of Horticulture and Food Industry
Budapest, Hungary

1. INTRODUCTION

Among the properties related to the quality of foods there is one which has a higher importance, the sensory value that expresses of the delight of its consuption. The food qualifying tests tend in one hand to evaluate food analytics properties, in the other hand to express the consumers' claims. Today it is clear that both evaluating methods are important. This time we searched for the connection between the two qualifying methods. In this report we give results of a partial field.

2. MATERIALS AND METHODS

The National Institute for Agricultural Quality Control (Hungary) gave us samples of five cucumber varieties: Express ZKI, Levina RS, Inge RS, Minerva SG and Potomac ASG as fresh as processed for testing. The varieties are predominantly female except of the Inge which is all female. The Levina and the Potomac are spined and the others are smooth skined varieties.

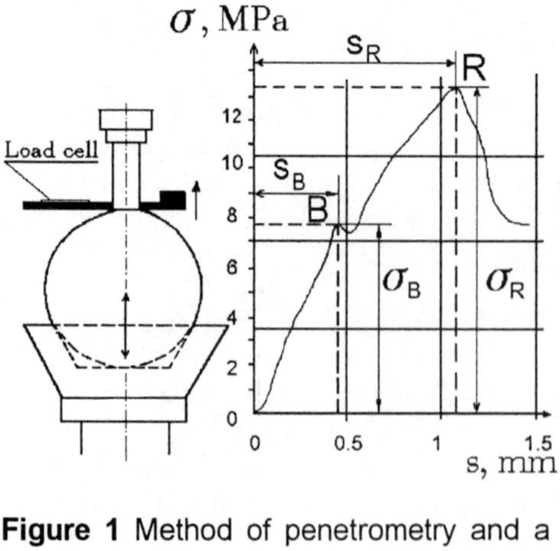

There was a double aim in research. We had to elaborate a method for estimating the quality of the preserved product with the help of the physical properties. The other aim was to find among the ranges of the properties the best connection possible with the sensory evaluation.

Our measuring method is the penetrometry and is usually implemented by Instron-type machines (Figure 1). We take a stress-deformation graph on the fruit using a 6 mm diameter plunger. The typical points of the graph are the bioyield point and the rupture point.

Figure 1 Method of penetrometry and a sample rheogram

The composed data storing and

computing program can give more properties: e.g. the deformations belonged to the typical stresses, the deformation energy, the most important moduli (Young's modulus).

The places characterising the cucumber cell structure are the stem, the calyx and with the skin removed, the flesh. We discovered in the past years that we can get most information about the fresh cucumber at the calyx end and at a 10 mm slice crosswise in the middle of the fruit. At the processed cucumbers we did measurements similarly at the calyx end and on the peeled flesh.So we tested the same tissue type for the comparison.

The preliminary tests prooved that we could not use the properties which were directly evaluated while measuring e.g. the rupture stress for determining the information on the structure or consistency of the fruit flesh. It is required to determine new physical properties using the results of the measuring. We constructed for the present research the ratio of the rupture and the yield stresses:

$$\mu = \frac{\sigma_r}{\sigma_y}, \text{ where}$$

$\sigma_r = \frac{F_r}{A}$ (F_r the force causes the rupture, A is the surface of the plunger),

$\sigma_y = \frac{F_y}{A}$ (F_y is the force at the bioyield point).

We took it as a variable which expresses the homogeneity of the fruit flesh. The results showed variety dependency. We can state that the variances generally belong to different groups.

It is well known about the normal distribution 95 % of the heap of the results can be found inside a range determined by the average and two times of the standard error. The width of the interval which contains the heap of the results is a function of the standard error. We involved in the researches the stress ratio considered it as a variable from normal distribution, estimated the width of the interval belonging to the 95 % probability level.

3. THE MEASURED RESULTS AND THEIR EVALUATION

We could see after the first year penetrometric tests that there was no significant difference in the stress ratio of the varieties Inge and the Levina (according to their flesh structure) but we could find 140 mJ deformation work at the Levina (computed up to the rupture point) and 100 mJ only at the Inge which ment a significant difference. This means that the Levina has better characteristics for preservation purposes (it is called *cracking*).

Next year we compared the Levina with the Minerva, testing the stress ratio. We searched how many percents of the results can be found inside a determined interval. We compared these figures with the points given at the sensory evaluation (Table 1).

This table prooves the hypothesis that the measuring results of the Levina belong into a narrower interval than those of the Minerva, related to the same probability level. Minerva has lower sensory value than the Levina. This proves that Levina is better for making canned products considering the cracking of flesh..

Table 1 Connection between stress ratio and the sensory points (second year).

Variety	Stress ratio as variable σr/σy=μ	Percent of results belonging into the interval		Average sensory points (max 9 pts.)
		Fresh, %	Preserved, %	
Levina	1.2...3.5	52		
	2...6		56	7.4
Minerva	1.2...3.5	97		
	2...6		89	6

We compared the Levina the Express, the Inge and the Potomac in the third year. We computed at testing the fruit flesh, how wide is the interval of the 95 % of the stress ratio values (Table 2).

Table 2
Connection between the sensory points and the stress ratio in the third year.

Variety	Size of the interval of the stress ratio		Average values of sensory evaluation (max. 9 points)
	Fresh	Canned product	
Express	5.63	2.31	5.81
Inge	5.24	2.65	5.82
Levina	4.14	2.17	5.71
Potomac	2.13	1.79	6.20

Table 2 gives evidences on the experiences of the previous years. The width of the interval at the Potomac is the narrowest and its sensory points are the best. We got similar sensory values at similarly wide intervals (Express, Inge). We could not find differences between the Inge and the Levina in this research year. It has to be mentioned that the stress ratio values at the samples we had got in the recent years there is no difference at Inge (3.44) and Levina (3.40).
We prepared the density functions for any of the varieties upon the average and the standard error values. (Figure 2 and 3)

4. RESUME

The penetrometry at cucumber can be used for evaluating the connection between the sensory values and the physical properties.
The stress ratio is in close connection with the sensory evaluation results.
The quality of the preserved product can be estimated by testing the fresh cucumber.

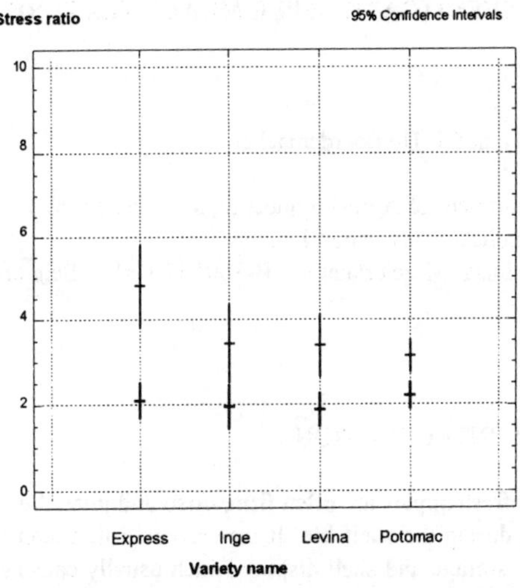

Figure 2 Average and standard error of the stress ratio measured on cucumber flesh.

5. REFERENCES

P. LÁSZLÓ – J. NYÉKI – J. ZANA: Researches on the rheological properties of fruit flesh. *Journal of Food Physics.* Budapest 1989 Vol. LIII. 41...44. p.

P. LÁSZLÓ – P. SASS – J. ZANA: Penetrometric properties of apple varieties in Hungary. POSTHARVEST' 93. Kecskemét. *Acta Horticulturae* 1994. No. 368. Vol. I. 333..343. p.

E. KRISTÓF – P. LÁSZLÓ – J. ZANA: Rheological properties of cucumber varieties related to the sensory evaluation. *Journal of Food Physics.* 1994. Vol.1. 55.-59.p. 1st ISFP

ZS. FÜSTÖS – A. OMBÓDI – J. ZANA: Investigation of Onion's Agrophysical Parameters. *Journal of Food Physics.* 1994.Vol. 2. 28-31.p. 1st ISFP

J. ZANA – P. LÁSZLÓ – E. KRISTÓFNÉ: Rheological and sensory evaluation properties of cucumber. *The Post-Harvest Treatment of Fruit and Vegetables.* 1994. October 19-22, Oosterbeek, Netherland. 87. p.

Nuri N. MOHSENIN: Physical properties of plant and animal materials. Gordon and Breach Publishers. New York 1984.

Acknowlidgement. We have to thank our second form students, Miss Beatrix Kiss and Miss Ágota Ivancsó for helping us at the research work

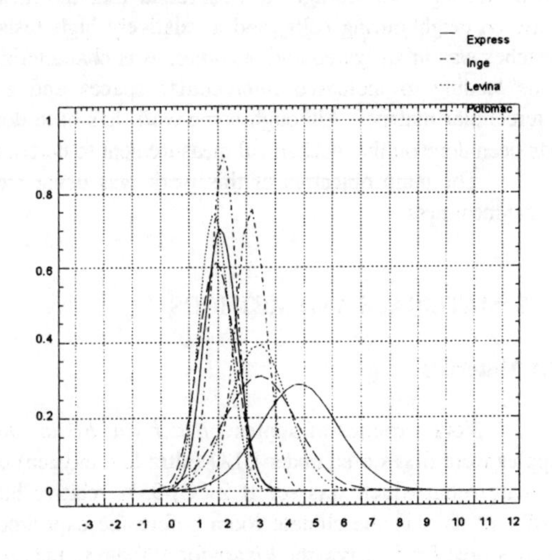

Figure 3 Density function at varieties. Preserved left, fresh cucumbers right.

INVESTIGATE APPLE MEALY TEXTURE USING INSTRUMENTAL METHODS

K. Tu and J. De Baerdemaeker

Department of Agro-Engineering and Economics
Katholieke Universiteit Leuven
Kardinaal Mercierlaan 92, B-3001 Heverlee, Belgium

1 INTRODUCTION

The fresh apples are often firm, crisp and juicy, but the texture can deteriorate during cold storage and during the shelf life. It was reported that later picked apples tend to lose more water during cold storage and shelf display which usually cause significant wilting, softening, shriveling, and a dry (mealy) taste[3].

Szczesniak[8] suggested the texture profile technique to identify the key texture attributes of food. The mealy texture was classified as[5]: possessing the texture property manifested by the presence of components of different degrees of firmness and toughness.

In the literature the physiological changes of *Braeburn* apple were studied as they became mealy during cool storage. It was found that the mealiness was associated with low adhesion between neighbouring cells, and a relatively high resistance to cell rupture[3]. Chilling injury in peaches was investigated and mealiness was characterized by separation of mesocarp parenchyma cells leading to increased intercellular spaces and accumulation of pectic substances in the intercellular matrix[7]. Although some work has been done on the physiological aspect, little work has been done on the mechanical measurement to determine apple mealiness.

The main objective of this work was using mechanical and analytical methods to study apple mealiness.

2 MATERIALS AND METHODS

2.1 Materials

Four varieties of apples : *Braeburn, Elstar, Jonagold* and *Teser T219* were tested. The apples were first stored under ULO (ultra low oxygen) conditions and then moved to the controlled climate room which was set at 2 °C, 95% relative humidity (RH). The apples were stored for different days in the climate room before the experiment began. The *Braeburn* and *Teser T219* were stored for 55 days, the *Elstar* for 100 days, and *Jonagold* for 22 days.

2.2 Methods

The apple texture was determined just after cool storage at 2°C, 95% RH and the tests were repeated after 7 days storage under 20°C, 65% RH room conditions.

2.2.1 Non-destructive Measurements. Twenty apples of each variety were taken out of cool storage chamber and put in room for some time to get equilibrium temperature with the environment. The non-destructive acoustic impulse response measurement was first carried out.

The set up was the same as used by Chen and De Baerdemaeker[2]. The apple flesh firmness can be indicated by stiffness factor expressed as $f^2m^{2/3}$ (f: peak frequency,Hz; m: mass of the apple, Kg).

2.2.2 Destructive Measurements. The destructive measurements were performed following the non-destructive measurements. The apple texture was determined with Texture Profile Analysis (TPA) and the tensile test using the Universal Testing machine System (UTS Test System GmbH).

Texture profile analysis (TPA) is the procedure to describe food properties by compressing bite-size pieces of food using mechanical device and analyzing the force-time curve which indicating the simulated mastication[9]. A typical two-cycle compression TPA curve during compression is shown in Figure 1. The parameters of hardness (the peak force measured during the first compression cycle), adhesive force (peak negative force during the up stroke of the first compression), cohesiveness (is the ratio of A_2/A_1) were determined for 4 apple varieties. Bourne and Comstock[1] studied the compression degree effect on texture profile parameters. In this research, 50 mm/min compression rate and 80% compression deformation were chosen based on some previous experiments.

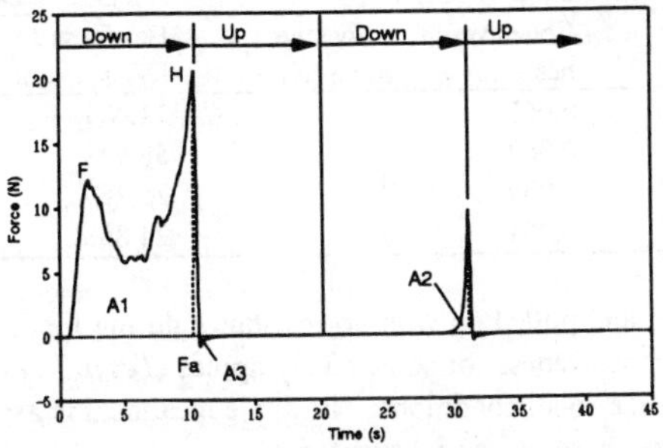

Figure 1 *Texture Profile Analysis (F: Fracturability, H: Hardness, A1: Area under H, A2: Positive area under second compression, A3: Adhesiveness,* F_a*: Adhesive force)*

The tensile test was carried out using a ring-shaped sample subjected to radial loading as described by Verlinden and De Baerdemaeker[10]. The force-deformation curve was recorded. The UTS cross head moving speed was 50 mm/min.

2.2.3 Analytical Methods.

The pH value, internal air space (IAS), juiciness, soluble solids concentration (SSC), colour were determined just after cool storage and 7 days later under shelf conditions for four apple varieties.

pH value: pH of each apple variety was measured at 20°C.

Internal air space: The internal air space of whole fruit can be calculated according to the formula (1). The specific gravity (SG) of apple was determined in the same way as described by Hatfield and Knee[4]. The specific gravity of apple juice was the average of estimations using a pycnometer.

$$IAS(\%) = 1 - \frac{SG\ fruit}{SG\ juice} \times 100 \qquad (1)$$

Juiciness: In this experiment, the apple juiciness was determined by the weight gain of filter paper (Whatman) during a single cycle compression similar to Lee and Patel[6]. The apple juiciness is expressed in terms of % expressible fluid.

SSC: The juice expressed during compression was collected directly onto a hand-operated refractometer (0% to 85%; Zeiss, Germany) to measure SSC.

Colour: The apple background colour was determined using the colour chart suggested by the research committee on storage (Belgium) which assigned the greenest fruit score 1 and the yellowest background score 8.

3 RESULTS AND DISCUSSION

The result of acoustic impulse measurement of apple firmness is correlated well with the destructive measured hardness. The three parameters of the Texture profile analysis measured in the experiment can be seen from Table 1. It is obvious that the hardness decreased for all apple varieties except for the *Elstar* apple. This indicates that the apples became softer after one week under room conditions. The *Elstar* apple has been stored for more than 3 months and the texture seems not decrease any more.

The raw apple normally shows no adhesiveness. As the apple ripened, the adhesiveness may appear, especially for the variety of *Teser T219*. Harker and Hallett[3] conclude that mealy apple cells are difficult to rupture while non-mealy apple cells are relatively easy to rupture and release their contents during chewing. For the apples just after cool storage, the UTS cross head

Table 1 *Results of the Texture Profile Analysis*

Apple	After cool storage (2°C)			One week later at (20°C)		
Variety	Hardness (N)	Cohesive-ness	Adhesive force (N)	Hardness (N)	Cohesive-ness	Adhesive force (N)
Teser T219	23.475	0.067	0	21.071	0.037	0.4 ~ 0.6
Braeburn	54.733	0.067	0	50.918	0.056	0.1 ~ 0.3
Elstar	24.078	0.039	0	25.169	0.041	0.1 ~ 0.3
Jonagold	26.674	0.051	0	24.888	0.047	0.1 ~ 0.3

compressed the sample and pulled up with no resistance during the TPA test. After one week, the TPA showed obvious adhesiveness for some mealy apples (*Teser T219*). The cohesiveness of most apples decreased under the room conditions. The apple mealiness is associated with the decrease of cohesiveness and a small increase of adhesive force.

Table 2 *Changes of the apple color, pH value and SSC*

Apple	After cool storage (2°C)			One week later at (20°C)		
Variety	colour	PH	SSC (%)	colour	PH	SSC (%)
Teser T219	7.0±0.2	3.54±0.05	12.9±0.1	8	3.82±0.05	13.4±0.2
Braeburn	7.0±0.2	3.55±0.05	12.0±0.2	8	3.72±0.05	12.2±0.3
Elstar	> 8	3.50±0.05	13.1±0.1	>8	3.88±0.05	14.1±0.2
Jonagold	7.0±0.5	3.50±0.05	11.8±0.3	8	3.69±0.05	12.0±0.3

The apple SSC, apple colour, and pH value are determined as shown in Table 2. The SSC value increased after one week room storage. The apples were all described ripen after one week room storage based on the colour. The pH value increased for all the apples. It is clear that the IAS (Table 3) will increase as the apple become ripening and mealy. This observation indicates that

Table 3 *Expressible fluid (Juiciness) and IAS of the apples*

Apple	After cool storage (2°C)		One week later at (20°C)	
Variety	Expressible fluid (%)	IAS(%)	Expressible fluid (%)	IAS(%)
Teser T219	25.4±5.8	20.2±0.2	25.1±3.7	23.2±0.2
Elstar	23.2±2.5	14.2±0.2	17.2±3.0	16.5±0.2
Braeburn	28.8±3.8	20.7±0.2	18.9±2.4	24.8±0.3
Jonagold	26.6±1.8	19.4±0.2	21.1±0.9	19.8±0.3

mature apples have a large volume of air space and likely cause smaller cell wall surface area. Other researches[3,4,11] have also demonstrated that the large internal air spaces were associated with poor apple texture (mealiness). It should be mentioned that the IAS are different for different varieties of apples. If we take *Elstar* apple as a reference mealy apple that has been stored for a long time, it may suggest that the apple is mealy as the IAS of apple is more than 21%. But the *Braeburn* apple may taste mealy as its IAS reached 16-17%.

The percentage of expressible fluid directly linked to the crisp and juicy apple texture. Table 3 shows that the apple has less expressible juice after one week under shelf life conditions. It suggests that the apple is not juicy as the percentage of expressible fluid less than 20% during these experimental conditions and the apple may be turning to poor texture.

4 CONCLUSIONS

The apple firmness decreases under shelf life conditions. The relative fresh non-mealy apple will not show adhesiveness during texture profile analysis but the ripened, mealy apple may show some adhesiveness.

Internal air space seems to be a good indicator of apple mealiness. Although the IAS for different apple varieties may be different. The higher IAS (>=21%) normally corresponds to poor, mealy texture.

Juiciness of apple is directly indicated by expressible juiciness. The apple is considered not juicy as the expressible juiciness is less than 20%.

More work still needs to be done to identify mealiness accurately and reliably with mechanical methods.

References

1. Bourne M. C. and Comstock S. H. 1981. Effect of compression on texture profile parameters. *J Texture Studies* **12**, 201-216.
2. Chen H. and De Baerdemaeker J. 1993. Effect of apple shape on acoustic measurements of firmness. *Journal of Agricultural Engineering Research,* **56**, 253-266.
3. Harker F.R. and Hallett I.C. 1992. Physiological changes associated with development of mealiness of apple fruit during cool storage. *HortScience*, **27(12)**, 1291-1294.
4. Hatfield S.G.S. and Knee M. 1988. Effect of water loss on apples in storage. *Int. J. Food science and Technology* **23**, 575-583.
5. Jowitt, R. 1974. The terminology of food texture. *J Texture Studies* **5**, 351-358.
6. Lee C. M. and Patel K. M.,1984. Analysis of juiciness of commercial frankfurters. *J. Texture Studies* **15**, 67-73.
7. Luza J. G., Van Gorsel R., Polito V.S. and Kader A.A. 1992. Chilling injury in peaches: a cytochemical and ultrastructural cell wall study. *J AMER Soc Hort Sci* **117(1)**, 114-118
8. Szczesniak, A.S. 1973. Instrumental methods of texture measurements. In: *Texture measurements of foods*, ed Kramer A. and Szczesniak, A.S., D. Reidel, Dordrecht, Holland, pp 71-108.
9. Szczesniak, A.S. 1975. General Foods texture profile revised - ten year perespective. *J. Texture Studies* **6**, 5-17.
10. Verlinden Bert E. and De Baerdemaeker J., 1994. Development and testing of a tensile method for measuring the mechanical properties of carrot tissue during cooking. *XII C.I.G.R. world congress and AgEng '94 conference on agricultural engineering, 29 Aug.-1 Sept., Milan, Italy, 1994.*
11. Vincent, J.F.V. 1989. Relationship between density and stiffness of apple flesh. *J. Sci. Food Agr.* **47**, 443-462.

MODELLING THE MECHANICAL AND HISTOLOGICAL PROPERTIES OF VEGETABLE TISSUE DURING BLANCHING AND COOKING TREATMENTS

B. E. Verlinden*, T. de Barsy§, J. De Baerdemaeker* and R. Deltour§

* Department of Agro-Engineering and Economics, Katholieke Universiteit Leuven, Kardinaal Mercierlaan 92, 3001 Heverlee, Belgium
§ Département de Botanique, Service de Morphologie végétale, Université de Liège, Boulevard du Rectorat Bât. B22, Sart Tilman, B4000 Liège, Belgium

1 INTRODUCTION

Quality aspects that are affected by the blanching and cooking process are texture, colour and nutritional value.[1,2] Lee et al. and Bourne investigated the effect of blanching on the texture of canned carrots and green beans.[3,4] In the case of potato processing, blanching is used to inactivate peroxidase, to improve the texture, the colour and to some extend the flavour of the final product. Blanching has a firming effect which gives resistance to physical breakdown and sloughing during further processing which affects the final texture. The effect of blanching at low temperatures for longer times on the firming of vegetables has been acknowledged for a long time. The biochemical mechanisms which cause this firming are under investigation.[5-7]

In this work a kinetic model describing the texture changes during blanching and cooking treatments is presented. It can simulate the firming effect of blanching treatments. The model structure is based on the biochemical processes in the cell wall during processing. The rupture force and the rupture mode of the tissue are taken as a measure of texture. These experimental data are used to identify the model parameters.

2 MATERIALS AND METHODS

Carrots of the variety 'Amsterdamse bak' and potatoes of the variety 'Irene' were used for this work. Ring shaped samples with inner diameter 10 mm, outer diameter 20 mm and a height of 6 mm for the carrot samples and a height of 10 mm for the potato samples were cut. These samples were subjected to a blanching-cooking process in de-ionised water. After certain times the samples are taken out of the water and the texture is measured.

The texture changes of the samples during the blanching-cooking process are measured by the changes of the mechanical rupture properties in a tensile test using a universal testing machine and by means of histological investigation of the resulting rupture surfaces on the ruptured samples. A detailed description of the ring tensile test can be found elsewhere.[8]

For the histological investigation, each ring sample was cut with a razor blade at 4-5 mm of the broken zone on the whole ring thickness. Samples were coloured in Ruthenium Red, a special colorant for plant walls and embedded in ice. Sections of 70 μm thick were made with an ice Reichert microtome from the inner diameter of the sample up to the outer diameter and mounted in glycerol. Cells along the whole rupture zone were categorised either as broken through the cell wall or separated along the middle lamella by observation with a light photo-microscope. The percentage of cell wall breaks was calculated as the ratio of the number of cells in the cell wall break category and the total number of observed cells times 100.

3 MODELLING

3.1 Model Equations

A kinetic model to describe the changes of the texture of vegetables during blanching and cooking processes has been developed. The model structure is based on the knowledge of the biochemical changes that occur in the cell wall during processing that can be found in literature.

The model consist of three differential equations which describe how three major state variables change in time as a function of the state of the system and of the processing temperature θ. A first state variable is the texture component 1, T_1 which can be broken down by two mechanisms. A first mechanism is a chemical reaction, modelled by a first order degradation with rate constant k_1. The second mechanism is an enzyme reaction. The enzyme with concentration E converts the texture component 1, T_1, into texture component 2, T_2. This is modelled by a second order reaction with rate constant k_2.

$$\frac{dT_1}{dt} = -k_1 . T_1 - k_2 . T_1 . E \quad \text{with } T_1(t=0) = T_1^0 \tag{1}$$

The second texture component is formed out of texture component 1 and is not susceptible to chemical degradation.

$$\frac{dT_2}{dt} = k_2 . T_1 . E \qquad \text{with } T_2(t=0) = T_2^0 \tag{2}$$

A third state variable is the enzyme concentration E. The enzyme is denaturated by heat. This is modelled by a first order degradation.

$$\frac{dE}{dt} = -k_e . E \qquad \text{with } E(t=0) = E^0 = 1 \tag{3}$$

To model the temperature dependence of the reaction rate constants k_i the Arrhenius model is used.

$$k_i(\theta) = k_i^{ref} . \exp\left(\frac{Ea_i}{R}\left(\frac{1}{\theta} - \frac{1}{373}\right)\right) \quad \text{with } i = 1,2,e \tag{4}$$

k_i^{ref} are the reference rate constants at the reference temperature of 373 K and Ea_i the activation energies of the reactions. R is the universal gas constant.

The output of this model, the quantity of interest being the measured texture, is modelled as the sum of texture component 1, texture component 2 and a component for the turgor tension T_{turg}.

$$T = T_{turg} + T_1 + T_2 \tag{5}$$

The term T_{turg} is actually a fourth state variable, however the change of the turgor tension in time is modelled only in a crude way because the lack of information about this process. The turgor component is non-zero for the raw material (process time = 0) and becomes zero for process times > 0.

3.2 Simulation and Parameter Estimation

The model is fitted to the experimental data using a non-linear regression procedure to obtain the model parameters. The kinetic model is integrated with a variable order, variable-step Gear method, NAG integration routine D02EBF. The sum of squares of the differences between model values and experimental values at every time-temperature combination are calculated and minimised using the NAG minimisation routine E04FDF. When the sum of squares has reached its minimum value the NAG routine E04YCF is used to calculate the variance covariance matrix of the estimated model parameters to calculate their confidence intervals.

4 RESULTS

4.1 Carrot Tissue During a Cooking Process at Three Constant Temperatures

The carrot samples were cooked at 73, 85 or 99°C for times ranging from 2 minutes to 25 minutes. It was assumed that at these temperatures the enzyme concentration E drops fast because of denaturation and the model reduces to equations (1),(4) and (5). The results of the rupture force measurements and the percentage cell wall breaks are plotted in Figure 1. The model is fitted to the rupture force data and the resulting estimates of the model parameters are listed in Table 1.

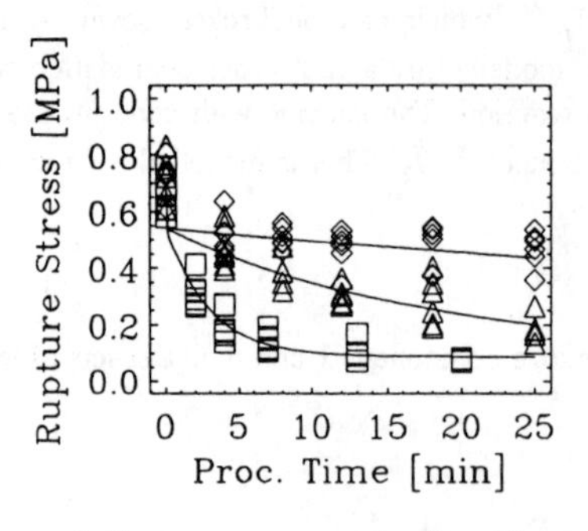

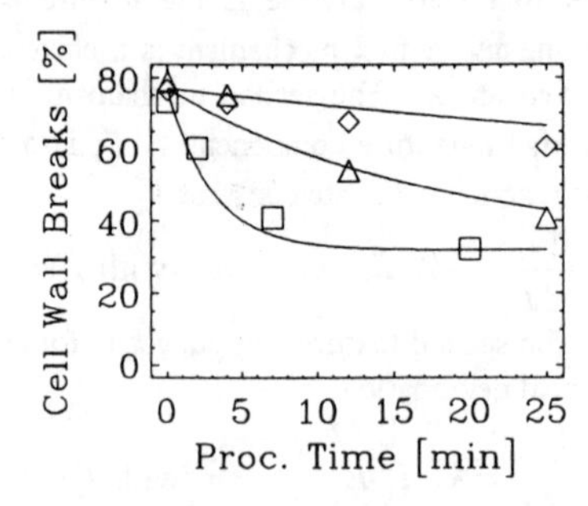

Figure 1 *Rupture stress and cell wall breaks of the carrot samples as a function of processing time. The symbols denote measured values at the different cooking temperatures □ :99°C, Δ:85°C and ◊: 73°C, the lines are the fitted model.*

4.2 Potato Tissue During a Blanching and Subsequent Cooking Process

The potato samples were first treated with a blanching process at 55, 65 or 75°C for 15 or 30 minutes. After the blanching step the samples were cooled down in ice water for 10 minutes and then cooked at 100°C for 7 or 15 minutes. For each blanching temperature the results of the rupture force measurement and the percentage cell wall breaks are plotted against the process time in Figure 2. The estimates of the model parameters can be found in Table 1. The same model parameter values were also able to describe the histological data for the carrot as well as the potato data. Only the initial values had to be fitted again because they have a different scale.

Table 1 *Model Parameter Estimates with 95% Confidence Intervals*

Model	Carrots		Potatoes	
parameters	Rupture force	Cell Wall Breaks	Rupture force	Cell Wall Breaks
T_{turg}^{0}	(0.16 ± 0.04)MPa	0 %	(3 ± 3) N	(6 ± 12) %
T_{1}^{0}	(0.45 ± 0.04)MPa	(42 ± 7) %	(33.1 ± 0.4) N	(63.8 ± 0.1)%
T_{2}^{0}	(0.09 ± 0.01)MPa	(34 ± 6) %	(0.94 ± 0.09) N	(5.1 ± 0.1) %
k_{1}^{ref} [1/min]	0.38 ± 0.05		0.27 ± 0.03	
Ea_1 [kJ/mole]	140 ± 20		150 ± 6	
k_{2}^{ref} [1/min]	-		0.053 ± 0.005	
Ea_2 [kJ/mole]	-		45 ± 2	
k_{e}^{ref} [1/min]	-		30 ± 3	
Ea_e [kJ/mole]	-		270 ± 10	

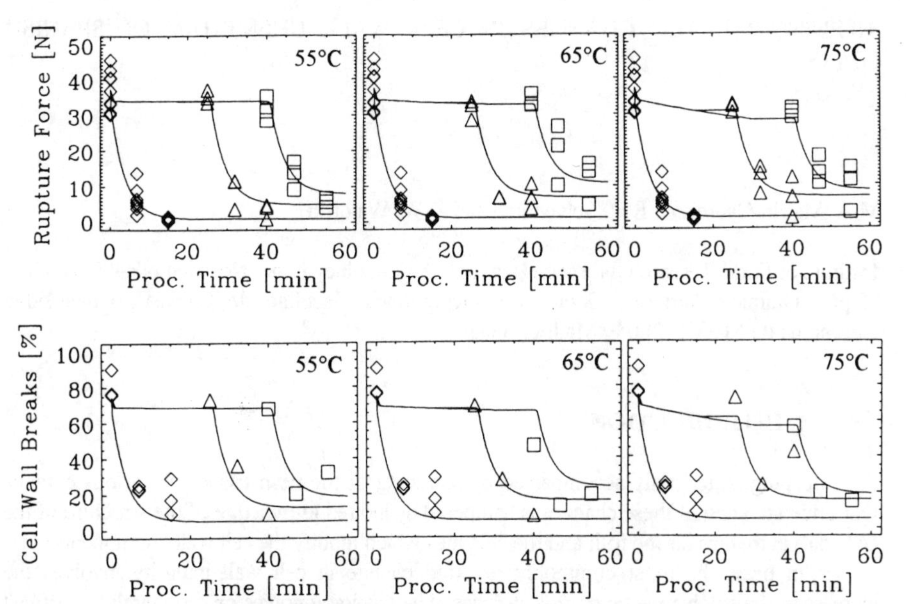

Figure 2 *Rupture stress and cell wall breaks of the potato samples as a function of processing time. The symbols denote measured values with different blanching times before cooking at 100°C.* □*: 30 min,* Δ*:15 min and* ◊*:0 min, the lines are the fitted model. The blanching temperatures are denoted on the graphs.*

5 CONCLUSIONS

A kinetic model was presented that is able to describe the texture changes in a normal cooking process as well as the firming effect encountered in a combined blanching and cooking process. The model structure is based on biochemical knowledge of the cell wall processes during processing.

The model parameter estimates derived from the rupture force data were also able to predict the mode of tissue rupture expressed as the percentage cell wall breaks in the rupture area.

Acknowledgements

This study has been performed as part of AIR-project AIR1-CT92-0278. The authors wish to thank the European Communities for the financial support.

References

1. P. Varoquaux, F. Varoquaux and L. Tichit, *Journal of Food Technology*, 1986, **21**,401.
2. S. Mirza and I.D. Morton, *Journal of the Science of Food and Agriculture*, 1977, **28**, 1035.
3. C.Y. Lee, M.C. Bourne and J.P. Van Buren, *Journal of Food Science*, 1979, **44**(2), 615.
4. M.C. Bourne, *Journal of Food Science*, 1987, **52**(3), 667.
5. L.G. Bartolome and J.E. Hoff, *Journal of Agriculture and Food Chemistry*, 1972, **20**(2), 266.
6. A. Wu and W.H. Chang, International Journal of Food Science and Technology, 1990, **25**, 558.
7. A.Andersson, V. Gekas, I. Lind, F. Oliveira and R. Oste, *Critical Reviews in Food Science and Nutrition*, 1994, **34**(3), 229.
8. B.E. Verlinden and J. De Baerdemaeker, AgEng paper No. 94-G-055, 1994.

RIPENING-RELATED CHANGES IN CELL WALL CHEMISTRY OF SPANISH PEARS

M.A. Martín-Cabrejas*, R.R. Selvendran and K.W. Waldron

Institute of Food Research, Norwich Research Park, Colney Lane, Norwich NR4 7UA, UK.
*Dpto. Química Agrícola, Geología y Geoquímica, Facultad de Ciencias. Universidad Autónoma de Madrid, 28049 Madrid, Spain

1. INTRODUCTION

Fruit ripening is generally accompained by softening of the flesh tissues to various extents. Our understanding of these changes is hampered by limited knowledge of the structure of the cell walls in mature unripe fruit and the enzymes which modify the cell wall components (1).

In pear, the most commonly reported change in cell wall integrity involves the increase in the soluble pectic polysaccharides. The change involves on an overall loss of cell wall arabinose and uronic acids (3,4). However, in those studies which have attempted to fractionate the cell wall have confined themselves to extracting pectic polymers by water and chelating agents, or have used methods of cell wall extractions including hot water and strong alkali, which will have solubilised many of the constituent pectic polysaccharides through β-eliminative degradation (5), masking many of the ripening-related in the solubility of cell wall polymers.

Indeed, studies have indicated that in addition to cell wall degradation, some cell wall synthesis continues during the ripening process (2). This poses the possibility that new cell wall interpolymeric cross-links may be formed during ripening, thereby modifying some of the effects of wall degrading enzymes. No evidence has been provided for the synthesis of secondary cell walls during ripening. In the present paper, we have reported studies of cell walls of "Blanquilla" Spanish pears during ripening, and the changes occur in the fractions of the cell wall from which they are been released by making use of modern methods of plant cell wall sequential extraction (6).

2. MATERIALS AND METHODS.

2.1. Plant material

Mature green unripe Spanish pears (*Pyrus communis* L. cv. Blanquilla) were obtained from the central fruit market of Madrid (MercaMadrid). Fruits were selected for uniformity in size and were free from blemishes. Fruits (samples of 25) were used immediately in their unripe state or after being allowed to ripen at 18°C in the dark for 1 week. Whole pears were peeled and the ovary and central vascular tissue excised. The fruit flesh was sliced and diced, frozen immediately in liquid N_2, and stored at -40°C until required.

2.2. Preparation of alcohol-insoluble residue and its fractionation

The alcohol-insoluble residue (AIR) was obtained by immersing the frozen pear in boiling ethanol (85%) homogenising with a Waring Blender and then boiling for 5 min. The material was filtered and resuspended in absolute alcohol (twice), acetone (twice), and then air-dried after which it was stored in a sealed container at room temperature. The fractionation method is based on that developed for cell wall material of onion bulbs (7). AIR was sequentially extracted with water; chelating agent (CDTA-Na salt) and alkali (KOH) which were carried out with O_2-free solutions under argon. After each extraction, the solubilised polymers were separated from the acidified residue after centrifugation. The alkali extracts were acidified to pH 5 with HOAc and all extracts were filtered through glass fibre filter paper GFC (Whatman). The cellulose-rich residue remaining after the final extraction was also acidified to pH 5. The filtered extracts and residue were dialysed exhaustively, concentrated and lyophilised.

Sugars were released from AIR and sub-fractions by dispersing in 72% in H_2SO_4 followed by dilution to 1M and hydrolysing for 2.5h at 100°C (Saeman hydrolysis). Where necessary, cell wall glucose which did not originate from cellulose was estimated after hydrolysing the samples in 1M H_2SO_4 for 2.5h at 100°C. Neutral sugars were reduced, acetylated and then quantified by gas chromatography as described (6). Total uronic acid was determined colorimetrically by the method of Blumenkrantz and Asboe-Hansen (8). All analyses were performed in triplicate.

3. RESULTS AND DISCUSSION

The values for carbohydrate composition obtained in this study are shown in Figure 1. The AIR of unripe pears was rich in xylose and glucose which, together, comprised approximately half of the cell wall carbohydrate.

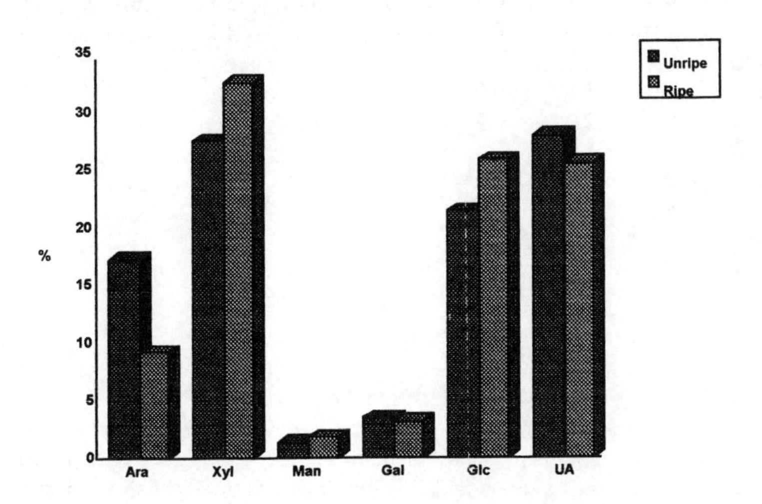

Figure1 *Carbohydrate composition of AIRs from unripe and ripe pear*

The remainder of the AIR carbohydrate comprised pectic polysaccharides as indicated by the levels of arabinose and uronic acid. The small quantities of deoxisugars are not shown. The level of xylose and glucose increase. This was accompained by, and largely due to continued maturation of lignified sclereids (4). The cross-linking between these and the parenchyma cells has a profound effect on cell-cell adhesion and therefore texture during ripening, reducing early over-softening in this variety of pears.

Cell walls were sequentially fractionated in order to gain information on the polymers in which the ripening-related changes occurred. The extraction procedure was designed to minimise β-eliminative degradation of pectins and to solubilise the cell polymers in as close to their native form as possible. Data are shown in Table 1.

Table 1 *Carbohydrate composition of polymers extracted from cell walls of pears*

Sample	Carbohydrate composition (μg/mg original AIR)						
	Ara	Xyl	Man	Gal	Glc	UA	Total
Water							
Unripe	33.1	11.5	2.1	11.5	7.5	84.1	149.8
Ripe	38.3	11.2	1.0	9.0	6.2	103.2	168.9
CDTA							
Unripe	3.0	1.0	0.3	0.7	1.0	14.1	20.1
Ripe	6.1	2.0	0.4	1.1	1.0	27.0	37.6
Na_2CO_3							
Unripe	31.0	2.5	0.1	3.1	1.0	52.2	89.9
Ripe	5.1	1.5	0.2	1.0	1.0	14.1	22.9
0.5 M KOH							
Unripe	4.0	6.1	0.2	1.0	3.0	2.0	16.3
Ripe	1.0	3.0	0.2	1.1	2.1	2.0	9.5
1 M KOH							
Unripe	1.0	1.1	0.0	0.2	0.5	1.5	4.3
Ripe	2.0	2.0	0.2	1.0	1.0	1.0	7.2
4 M KOH							
Unripe	5.0	65.2	4.0	8.1	25.3	17.0	124.6
Ripe	5.1	70.0	6.1	9.0	26.1	35.7	152.0
Residue							
Unripe	11.0	24.5	2.0	3.1	43.5	25.0	109.1
Ripe	3.1	24.5	1.0	1.0	38.7	19.1	87.2

Water and CDTA fractions were rich in pectic polysaccharides, not cross-linked into the cell wall and also held in the wall by Ca^{2+}, respectively. These polysaccharides exhibited heterogeneity in the ratios of neutral to acidic sugars. In addition, these polymers contained small but significant quantities of xylose, indicative of pectin-xylan complexes as identified in olive pulp cell walls (6). The KOH extracted polymers were rich in hemicelluloses, particularly xylans, as indicated by the presence of xylose, especially in the 4M KOH extraction. But also, the extracted polymers contained some pectic polysaccharides consistent with the presence of pectic-xylan complexes.

The cellulose-rich residue was rich in cellulosic glucose and xylose, probably originating from the sclereids which could be visualised in the residue by light microscopy, staining sections 1-2 μm thich with 1% toluidine blue. A significant amount of insoluble (cross-linked) pectic material was also present in the residue.

The ripening-related loss of arabinose and uronic acid from the wall, and the increase in water-soluble pectic polymers was concomitant with the very large decrease in Na_2CO_3-soluble polymers (Table 1) and also, to a lesser extent, residue-insoluble pectic polymers. This is similar to the changes in cell walls of ripening kiwi fruit (9) and suggests that the Na_2CO_3-soluble polymers may have some role in cell-cell adhesion which decreased during ripening. These results would not have been found without the use of minimally-degradative methods of extraction.

4. CONCLUSIONS

This study demonstrates that ripening in Spanish pears is accompanied by: (1) an increase in soluble pectic polysaccharides and a corresponding decrease in Na_2CO_3-soluble polymers; (2) an increase in xylans associated with continued development of sclereids.

5. REFERENCES

1. C. Brady, *Ann. Rev. Plant Physiol.*, 1987, **38**, 155.
2. E. Mitcham, K. Gross and T.Ng, *Plant Physiol.*, 1989, **89**, 477.
3. A. Dick and J. Labavitch, *Plant Physiol.*, 1989, **89**, 1394.
4. M.A. Martín-Cabrejas, K. Waldron, R.R. Selvendran, M.L. Parker and G. Moates, *Physiol. Plant.*, 1994, **91**, 671.
5. M. Jermyn and F. Isherwood, *Biochem. J.*, 1956, **64**,12
6. M. Coimbra, K.Waldron and R.R. Selvendran, *Carbohydr. Res.*, 1994, **252**, 245.
7. R. Redgwell and R.R. Selvendran, *Carbohydr. Res.*, 1986, **157**,183.
8. N. Blumenkrantz and G. Asboe-Hansen, *Anal. Biochem.*, 1973, **54**, 484.
9. R. Redgell, L.D. Melton and D.J. Brash, *Phytochem.*, 1991, **209**, 191.

CHINESE WATER CHESTNUT: THE ROLE OF CELL WALL PHENOLIC ACIDS IN THE THERMAL STABILITY OF TEXTURE

K.W. Waldron, M.L. Parker, A. Parr and A. Ng

Institute of Food Research
Norwich Research Park
Colney
Norwich NR4 7UA
UK

1 INTRODUCTION

The texture of plant tissues are dependent mainly on the cell wall[1]. Application of heat, for example during cooking, will usually induce tissue softening in most fruits and vegetables. This results from dissolution of the middle lamella (pectic) polysaccharides[2], which leads to an increase in cell separation. The solubilisation in these polymers is thought to result from a combination of depolymerisation by ß-eliminative cleavage and chelation of divalent cations by endogenous organic acids.

Chinese water chestnut (CWC) is the corm of a sedge that is commonly used in oriental food. The fleshy, edible portion consists, predominantly, of thin-walled, starch-rich storage parenchyma similar to that in potato tissue. Unlike potato however, CWC is able to maintain a crunchy texture after prolonged heat treatments such as canning[1]. This is because such treatments fail to induce cell separation[3].

In order to elucidate the basis for the thermal stability of texture in CWC, a number of studies have been performed to compare the cell walls of CWC with other physiologically similar storage tissues such as potato parenchyma[3,4,5]. These have concentrated mainly on aspects of histology and cell wall carbohydrate chemistry. Virtually no attention has been given to other cell wall components such as glycoproteins and phenolics.

In this paper, we report the presence of significant amounts of simple phenolics in the cell walls of CWC and their possible contribution to the thermal stability of texture in this vegetable.

2 MATERIALS AND METHODS

2.1 Plant Material

Fresh and canned CWC (*Eleocharis dulcis*) were purchased from a local supplier.

2.2 Microscopy

2.2.1. Autofluorescence. This was examined with a HBO 50 W mercury arc lamp and an exciter and barrier filter combination with transmission of 340-380 nm and >430 nm respectively. Hand-cut sections were mounted either in 0.1M sodium acetate, pH 4.5, or in 20mM NH_4OH (pH 10).

2.2.2. *Scanning electron microscopy.* This was carried out as described[6]

2.3 Chemical and biochemical methods

2.3.1 Preparation of cell wall material (CWM). CWM was prepared from fresh CWC tissue essentially by the method of Redgwell and Selvendran[7] except that starch was removed by washing the tissue-SDS homogenate with cold water followed by sieving through 75μm nylon mesh. Several washes effectively removed the intracellular starch.

2.3.2. Extraction and identification of phenolic acids. Phenolic acids were extracted from CWM with 0.1M NaOH (1h, 25°C) under N_2 (O_2-free) by the method of Hartley and Morrison[8]. After filtration and addition of *trans*-cinnamic acid (internal standard), the filtrate was acidified and partitioned into ethyl acetate. This was reduced under N_2. The sample was dissolved in 500ml litre^{-1} methanol and analysed by HPLC.

2.3.3. Sequential extraction of CWC tissue. Sections of CWC tissue (10x10x1mm approximately) were extracted sequentially with continual stirring in solvents normally used for extracting isolated cell walls[9]. In addition, sections were treated with purified, specific endoxylanase[10] and trifluoroacetic acid (TFA).

2.3.4. Vortex-induced cell separation (VICS). After an extraction/treatment of the cell wall, ease of cell separation was assessed by placing two sections in a 10 ml screw-capped tube with 3 ml water, vortexing for 1 minute, and shaking vigorously 10 times. The degree of disruption was assessed visually and assigned a score on a scale of 0-5 with 0 representing intact tissue and 5 representing total disruption, mostly single, separated cells.

3 RESULTS AND DISCUSSION

3.1 Microscopic analysis

Fracturing of CWC tissues, both fresh and cooked, involved cell wall breakage, not cell separation (Figure 1). Under UV light and alkaline conditions only, the cell walls of CWC parenchyma emitted a bright green/yellow autofluorescence (Figure 2). This is consistent with the presence of cinnamic acid derivatives.

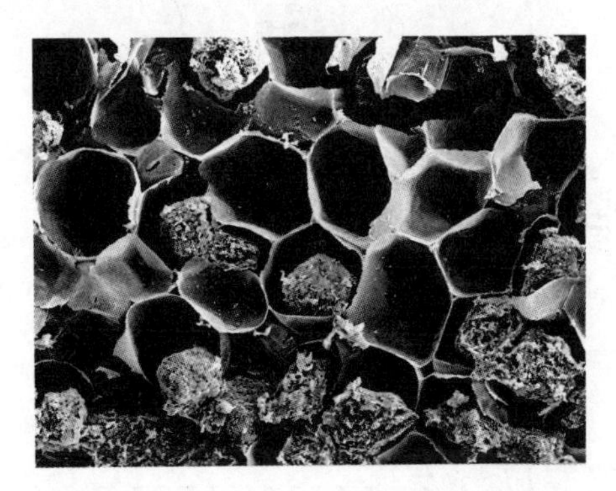

Figure 1 - SEM of ruptured, cooked CWC tissues, exhibiting cell breakage and intracellular contents. Scale: ——— 100μm.

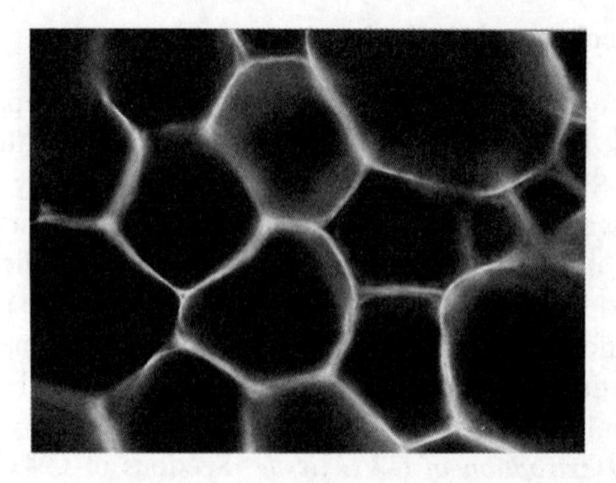

Figure 2 - Fluorescence micrograph of CWC tissue (pH 10).
Scale: ———— 100μm.

3.2 Cell separation

Sections of CWC tissue were subjected to a variety of extraction procedures designed to cleave cell wall chemical bonds.

3.2.1 Hot TFA. Treatment of CWC tissue sections (fresh or canned) in hot (100°C) TFA at 1M resulted in total VICS and loss of pH-dependant autofluorescence (PDA) within 30 min (Figure 3). In 0.1 and 0.05M hot TFA, total VICS was also induced within 55 and 75 min respectively, but PDA was retained (Figure 4). In 0.01M hot TFA, no significant VICS occurred within 90 min.

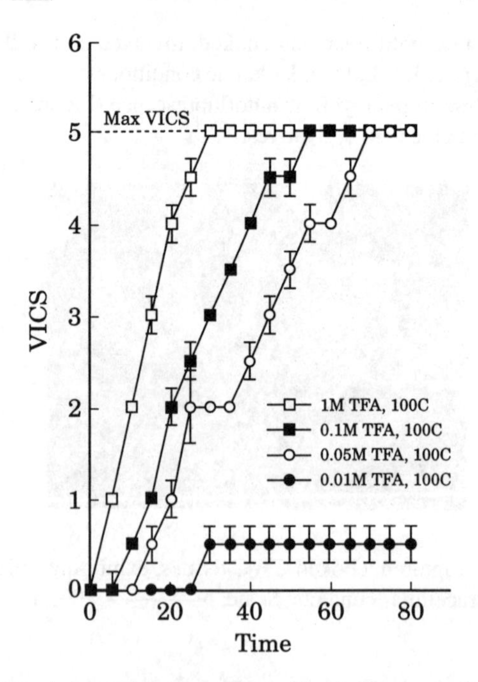

Figure 3 - Graph demonstrating effect of hot trifluoroacetic acid on VICS.

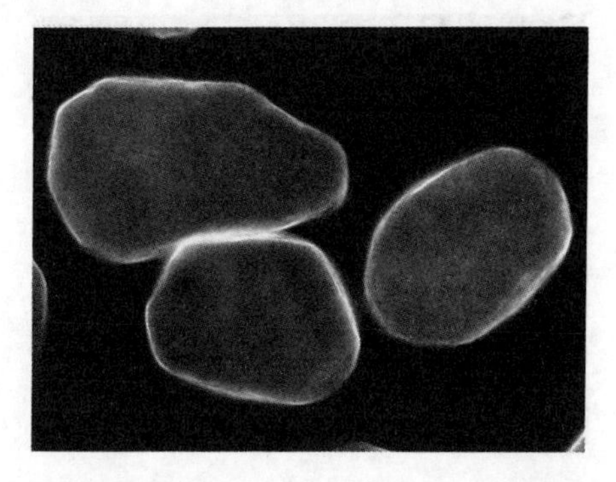

Figure 4 - Fluorescence micrograph showing dilute, hot-acid-separated cells that retain their fluorescence. Scale: ———— 100μm.

3.2.2. Sequential extraction in chelating agents and alkali. Fresh or canned CWC tissue sections failed to undergo VICS after extraction in cold CDTA followed by cold, dilute alkali (Na_2CO_3, 50mM and KOH, 50mM). However, VICS could be induced by subsequent extraction in stronger alkali, and this was accompanied by a corresponding decrease in PDA (Figure 5). Treatment in alkali generally induced a yellowing of the tissues, indicating the presence of simple phenolic compounds. Interestingly, extraction in hot, dilute alkali (Na_2CO_3, 50 mM, 100°C) induced total VICS of fresh CWC tissues within 30 min. This was accompanied by a loss of most PDA, except in certain locations within the cell wall (Figure 6). During early stages of alkali-induced VICS, fracturing the tissue resulted in rupture of cells deep within the tissues where alkali had failed to penetrate, and where PDA was unchanged.

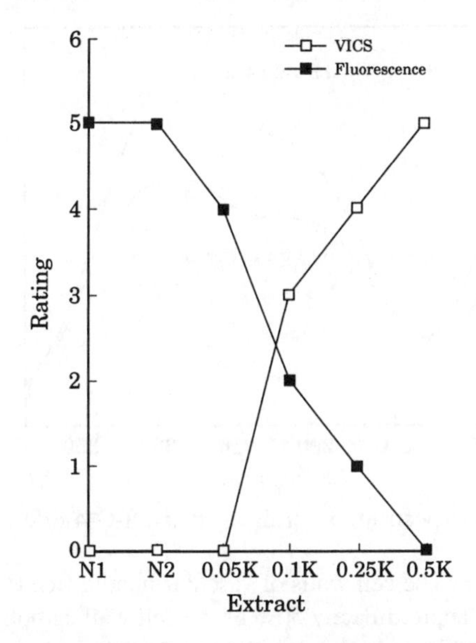

Figure 5 - Graph showing induction of VICS and corresponding decrease in PDA during extraction of CWC tissues in dilute alkali for 16h

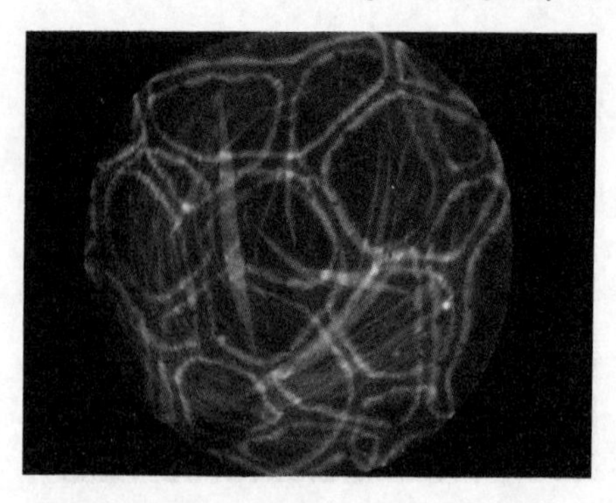

Figure 6 - Fluorescence micrograph of a single cell of CWC after separation in hot, dilute alkali. The fluorescence is localised to the perimeters of areas of the cell surface in contact with adjacent cells. Scale: ——— 50µm.

3.2.3. treatment with cell-wall-degrading enzymes. Extraction of tissue sections with endoxylanase induced total VICS with little change in the PDA.

3.3 Chemical analysis

3.3.1. Phenolic composition. CWM was prepared from fresh CWC and analysed for phenolic composition. Lignin was identified by staining tissue sections with phloroglucinol/HCl and was present in small quantities only in the vascular bundles which were a minor component. The phenolic acid component was analysed as described above. At least 75% of the phenolics extracted consisted of ferulic acid of which over 30% was in the form of ferulic acid dimers. These gave similar UV spectra to ferulic acid (Figure 7).

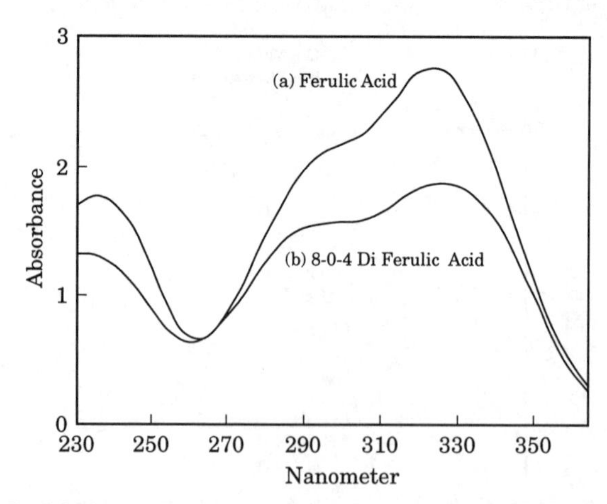

Figure 7 - Diode array spectrum of ferulic acid and 8-O-4-diferulic acid

3.3.2. Cell wall composition. The cell walls of vascular-bundle-free CWC parenchyma tissues were rich in cellulose (approximately 50% of the cell wall carbohydrate), and also contained large amounts of arabinose and xylose (results not shown), consistent with the presence of arabinoxylan hemicelluloses[5].

3 DISCUSSION

The coincidental decrease in PDA during alkaline-induction of VICS (Figure 5) suggests that ferulic acid-containing moieties have the same extraction characteristics as those involved in cell-cell adhesion. It is possible that the ferulic acid dimers play a key role since they have been shown to cross-link polysaccharides (usually arabinose-containing polymers[11]) and have been implicated in cell-cell adhesion[12]. The alkaline conditions which induce VICS would be expected to de-esterify these diferulic acid cross-links.

The involvement of arabinose-containing polymers is indicated by the induction of VICS in 0.05M TFA, conditions which will hydrolyse furanoside linkages which are common to arabinose-containing polysaccharides[13] but not pyranoside linkages which require much stronger acid. Their attachment to xylan hemicelluloses (possibly arabinoxylans) is also indicated by the induction of VICS with a pure, specific endoxylanase.

4 CONCLUSIONS

The results are consistent with the involvement of ferulic-acid-containing arabinoxylan hemicelluloses in the thermal stability of texture in Chinese Water Chestnut.

5 ACKNOWLEDGEMENTS
The authors wish to thank the UK Office of Science and Technology for funding via the BBSRC.

6 REFERENCES

1. D. M. Klockeman, R. Pressey and J. J. Jen, *J. Food Biochem.*, 1991, **15**, 317.
2. J. P. S. Van-Buren, *J. Text. Stud.*, 1979, **10**, 1.
3. J. Loh, W. M. Breene and E. A. Davis, *J. Text. Stud.*, 1982, **13**, 325.
4. J. Loh and W. M. Breene, *J. Text. Stud.*, 1981, **12**, 457.
5. R. Pressey, *J. Food. Biochem.*, 1993, **17**, 85
6. M. L. Parker and K. W. Waldron, *J. Sci. Food Agric.*, 1995, **68**, 337.
7. R. Redgwell and R. R. Selvendran, *Carbohydr. Res.*, 1986, **157**, 183.
8. R. D. Hartley and W. H. Morrison, *J. Sci. Food. Agric.*, 1991, **55**, 365.
9. K. W. Waldron and R. R. Selvendran, *Phytochem.*, 1992, **31**, 1931.
10. M. A. Coimbra, N. M. Rigby, R. R. Selvendran and K. W. Waldron, *Carbohydr. Polyms.*, 1995, *(in press)*.
11. N. Shibuya, *Phytochem.*, 1984, **23**, 2233.

SPECIFIC RELEASE OF PHENOLICS FROM FOODSTUFFS USING MICROBIAL ESTERASES

G. Williamson, C.B. Faulds, P. A. Kroon and B. Bartolomé

Food Molecular Biochemistry Department, Institute of Food Research, Norwich Research Park, Colney, Norwich, NR4 7UA, UK.

1 ABSTRACT

We have characterised ferulic acid esterases from *Aspergillus niger*, *Streptomyces olivochromogenes* and *Pseudomonas fluorescens*, which release phenolic compounds from plant cell walls. The activity is increased in the presence of other carbohydrases. These esterases have the potential to release phenolic precursors from agricultural and food wastes such as wheat bran and sugar beet pulp.

We thank the Biotechnology and Biological Sciences Research Council and the European Union (AIR1-CT92-0026 and AIR1-CT94-7502) for funding.

2. INTRODUCTION

Ferulic acid and related phenolic acids make up a substantial portion of the cell walls of many agriculturally important products and waste materials. For example, wheat bran contains about 0.7% w/w ferulic acid[1], whereas sugarbeet pulp contains about 0.9%[2]. In wheat bran, most of the ferulic acid is (1,5)-linked to arabinosyl residues as part of the arabinoxylan component of the wall. In sugarbeet, the feruloyl groups are linked either (1,2) to the arabinosyl or (1,6) to the galactosyl side chains of the "hairy regions" of the pectin[3]. Since there are a variety of linkages and attachment positions of the feruloyl groups, a range of enzymes are required to release the ferulic acid from these molecules. The complex nature of the substrate also means that other enzymes are required to break down the cell wall in order to optimise the yield. This report summarises the forms of ferulic acid esterases we have so far purified and characterised, and indicates their specificities and potential applications.

3. MATERIALS AND METHODS

The enzymes were isolated as previously described: FAE-I and FAE-II[4]; FAE-III[5]; esterase[6]; FAE[7], CinnAE[8] (see table 1 for abbreviations for these enzymes). SAE was partially purified using hydrophobic interaction chromatography, ion exchange chromatography and gel filtration and will be published elsewhere.

Methyl esters of phenolic acids were synthesised and used to assay esterase activity using hplc[5]. Cell wall fragments from either wheat bran or sugarbeet were isolated and purified after driselase or acid hydrolysis[9].

4. RESULTS AND DISCUSSION

Ferulic acid esterase was first purified from *Streptomyces olivochromogenes*, and subsequently from *Aspergillus niger* and *Pseudomonas fluorescens*. Each enzyme was homogeneous after ammonium sulphate sulphate precipitation, hydrophobic interaction chromatography, ion exchange and gel filtration. Table 1 shows the physical properties of seven esterases from 3 different sources.

Table 1. *Properties of ferulic acid esterases*

	Source	pI	Native M_r
FAE-I	*Aspergillus niger* (commercial pectinase preparation)	3.0	132000 (2)[1]
FAE-II	*Aspergillus niger* (commercial pectinase preparation)	3.6	29000 (1)
FAE-III	*Aspergillus niger* (cultures grown on oat spelt xylan)	3.3	36000 (1)
Cinnamic acid esterase (cinnAE)	*Aspergillus niger* (cultures grown on sugar beet pulp)	4.8	145000 (2)
Sinapinic acid esterase (SAE)	Aspergillus niger (cultures grown on sugar beet pulp)	4.0	ND[2]
acetyl esterase	*Pseudomonas fluorescens*	ND	58500 (1)
FAE	*Streptomyces olivochromogenes*	7.9 8.5	29000 (1)

[1]Value in brackets is the number of subunits
[2]ND, not determined

Substrates for ferulic acid esterases have two components: the sugar moiety and the phenolic moiety. Acid- or Driselase- (a mixture of enzymes from *Trichoderma*) hydrolysis of wheat bran yields, for example, substrates in which the phenolic can be either ferulic or p-coumaric acids[10], and in which the sugar can be arabinose alone, or arabinose linked to one or more xylose residues[3]. The corresponding substrates from sugar beet are phenolic acids connected to one or more arabinosyl residues via a different linkage (1,2)[3,9]. The esterases described in table 1 have distinct and different requirements for the phenolic and carbohydrate moieties of the substrate, and also for the nature of the linkage. These are shown in simplified form in table 2.

Table 2 *Specificity of some ferulic acid esterases*

	phenolic moiety with highest activity	sugar moiety with highest activity
FAE-III	sinapinate	Ara-Xyl-Xyl
CinnAE	caffeate	Ara-Ara-Ara
acetyl esterase	ferulate	Ara-Xyl-Xyl-Xyl

Because of the differences in specificity, esterases must be carefully and appropriately chosen for any given application. For example, we wished to release as much ferulic acid from wheat bran as possible. On incubation of FAE-III with a xylanase from *Trichoderma viride*, more than 95% of the total ferulic acid from wheat bran was released into solution[11]. However, for release of ferulic acid from sugarbeet pulp, FAE-III displayed no activity whatsoever, and in this case, cinnAE was the most active[8].

In conclusion, we have isolated and characterised a range of esterases with very different specificities for release of phenolic acids from cell walls. There are a number of potential applications for these enzymes, both to release ferulic acid for subsequent bioconversion and also to modify the properties of the cell wall substrate.

References

1. M. M. Smith and R. D. Hartley, *Carbohydr. Res.*, 1983, **118,** 65-80.

2. F. M. Rombouts and J.-F. Thibault, *Carbohydr. Res.*, 1986, **154,** 189-203.

3. I. T. Colquhoun, M.-C. Ralet, J.-F. Thibault, C. B. Faulds and G. Williamson, *Carbohydr. Res.*, 1994, **263**, 243-256.

4. C. B. Faulds and G. Williamson, *Biotechnol. Appl. Biochem.*, 1993, **17**, 349-359.

5. C. B. Faulds and G. Williamson, *Microbiol.*, 1994, **140**, 779-787.

6. L. M. A. Ferreira, T. M. Wood, G. Williamson, C. B. Faulds, G. P. Hazlewood and H. J. Gilbert, *Biochem. J.*, 1993, **294**, 349-355.

7. C. B. Faulds and G. Williamson, *J. Gen. Micro.*, 1991, **137**, 2339-2345.

8. P. A. Kroon, C. B. Faulds and G. Williamson, *Biotechnol. Appl. Biochem.*, 1996, in press.

9. M.-C. Ralet, J.-F. Thibault, C. B. Faulds and G. Williamson, *Carb. Res.*, 1994, **263**, 227-241.

10. W. S. Borneman, L. G. Ljungdahl, R. D. Hartley and D. Akin, *Appl. Environ. Microbiol.*, 1992, **58**, 3762-3766.

11. C. B. Faulds and G. Williamson, *Appl. Microbiol. Biotechnol.*, 1995, in press.

FUNCTIONAL REARRANGEMENTS OF ULTRASTRUCTURE IN APPLE FRUIT CELLS ACCOMPANYING THE SENESCENCE

E. B. Maximova

Institute of Plant Physiology
Academy of Sciences, Kishinev
Republic of Moldova

I INTRODUCTION

The solution of the problems concerning soft fruit preservation and reduction of losses during long-term storage is mainly determined by the preassessment of their state.

The process of maturation-senescence of apple fruits during storage is associated with a number of complex structural reorganizations that determine metabolism pattern and direction, the general state at various stages of lifeactivity and eventually their storability. Specific peculiarities of structural rearrangements, revealed by us, are characteristic of certain stages of the fruit lifeactivity during storage.

It has been shown that even minor structural and functional transformations of organoids at early stages of fruit maturation-senescence appear as a moment of the preformation of more vast alterations which characterise fruits at the final stage of lifeactivity, i.e. overmaturation(3). These transformations are determined by the disorder in the integrity of individual organoid groups and cell population, which is reflected on the repair/destructive background correlation in fruit pulp(2).

The above peculiar features of structural rearrangements are reliable indices, that characterise the general state of fruits, particularly, the diagnostics of their storability at various stages of storage.

The parameters of submicroscopic organization of the

cell structures of fruit and the complex signs during lifeactivity stages are new in this method. It allows us to receive more generalized and effective estimation of cell state of fruit parenchyma tissue on the basis not of a single but a group of signs during long-term storage.

2 MATERIAL AND METHODS

Comparative studies on structural signs of apple fruit of winter varieties such as: Jonathan, Mantuan, Renet Simirenko have shown a row of successive changes in cell ultrastructures and their organelles at the stage of maturation and storage.

Tissues were fixed by conventional methods for electron microscopy.

Fruit samples for electron microscopy were investigated in September after harvesting and were removed from air storage at 2°C in December and in March. Tissue samples, comprising cuticle, epidermis, hypodermis were cut from the fruit equator and were fixed in glutaraldehyde 3% on 0,2M cacodylate natrium buffer follwed by osmium tetrooxide I% on the same buffer; dehydrated in series of ethanol and embedded in Epon 8I2. Ultrathin sections post-stained with uranylacetate and lead-citrate, were examined in an electron microscope.

3 RESULTS AND DISCUSSION

Electron-microscopic investigations have revealed a number of successive changes in cell ultrastructures and their organelles (fig. I, 2,a). These changes are mainly concentrated in the plastid apparatus. At the stage of fruit maturation the chloroplasts transformed into carotenoidoplasts, the first sign of that was increase in grane electron density. Accumulation of carotenoids occurs simultaneously with the changes in the granale system. There takes place a destruction of intergranale lamella and changes in the orientation of the granes. This is followed by a slight reduction in the number of granale thylakoids, an increase in the quantity of the osmiophilic globules, the dimensions of starch grains diminished. It is established, that the presence of peripheral reticulum which is most probably related to plastid dehydration and observed generally during senescence (5). Later on, these result in the gradual degradation of membranous material, osmiophilic globules filling the stroma, decrease in dimensions of the starch grains and eventually their disappearance. Thus, we have a new type of plastid - carotenoidoplast.

Following the first changes in chloroplasts changes are observed in the endoplasmic reticulum. A separation of ribosomes and membranes takes place, the cysterns of endoplasmic reticulum swell, taking a blade form. Gradually, the endoplasmic reticulum alters considerably - the cysterns are fragmented, swell or form concentric and semiconcentric profiles of the monocentrical or polycentrical type (4). In all the cases the membranes remain reticente.

Mitochondria, nucleus, plasmalemma, tonoplast are more resistant organelles. The first sign of changes in the mitochondrium structure is cristae swelling accompanied by their deformation. Further on the mitochondrium inner content becomes homogenous, elementary membranes are separated and form numerous unit membrane vesicles.

During senescence-storage of fruit in the cells was shown the concavity and protrusion of organelles with the formation of protuberances and curvature, which creates an ameboid shape. The appearance of the blade nucleus contour encountered in fruit cells during prolonged storage and possibly is also explained by its irregular dehydration.

The curvature encompasses plasmalemma with the formation of the temperate invaginations, i.e. the sinuation is a starting point in the reorganization of cell plasmalemma during long-term storage. The movement intensification of both lowmolecular and highmolecular weight substances during senescence reorganizations are related to appearance of more complicated profiles - encrease in the number of components and total dimensions of curvature and pleatedness and formation of plasmalemmasomes.

A lipophanerose of structures is the most general sign observed on a submicroscopic level during storage manifested in increase of a number and dimensions of lipoid inclusions in cytoplasm and plastids, due to the transformations of membranous and another structures. The increase in size and quantity of the osmiophilic inclusions were observed during storage. Small lipid inclusions fuse and form large globules.

A general dehydration of the protoplast is observed during the storage, which can be judged by the blade form, curvature and on the whole by a decrease in the sizes of organelles and appearance of condensates of individual inclusions, such as - protein bodies, which arise only in the cases essential changes in intracellular metabolism take place (I). It may be supposed that protein inclusions represent a storage form of protein related to dehydration which is usually characteristic of fruits during senescence.

Thus, all the cell organelles in the apple pericarp during long-term storage are subject to various phenomena, which proceed in a certain succession and may be divided into 3 stages, namely disorganization, disintegration, destruction. These changes lead to cell senescence and death.

I. Disorganization. Primary changes in organelle components and bond breaking between them occur at the stage of disorganization (fig. 2,b) : a) destruction of intergranal lamella resulting in grana reorientation in chloroplasts; b) separation of ribosomes and membranes from ergastoplasm composition, partial vesiculation of the endoplasmic reticulum cysterns; c) insignificant curvature of plasmalemma.

2. Disintegration. The above changes in organelle ultrastructure lead afterwords to cell and organoid disintegration (fig. 2,c). Disintegration is characterized by breaking the integrity of main organelles and appearance of a new, more fractioned structures in the cell, such as: a) grana impoverishment, increase in the number of osmiophilic globules in plastid stroma, decrease in starch grain dimensions; b)vesiculation of endoplasmic reticulum; c) complication of plasmalemmasomes through curvature and pleatedness enhancement - small pockets, loops the form of semicircle; d) mitochondrion cristae swelling in individual cells, accompanied by

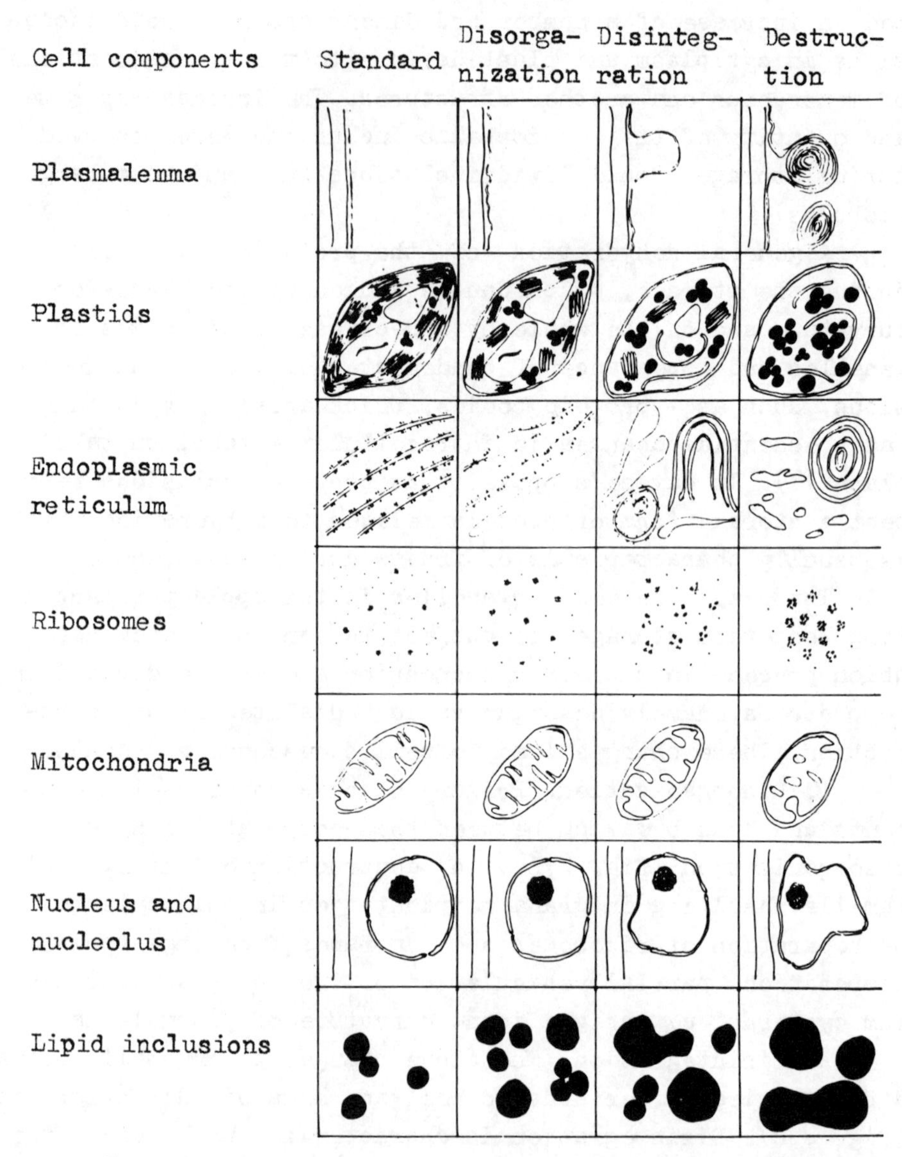

Figure I The arbitrary stages of the change directions in the organelle ultrastructure of parenchyma cells pericarp of apple fruit during senescence

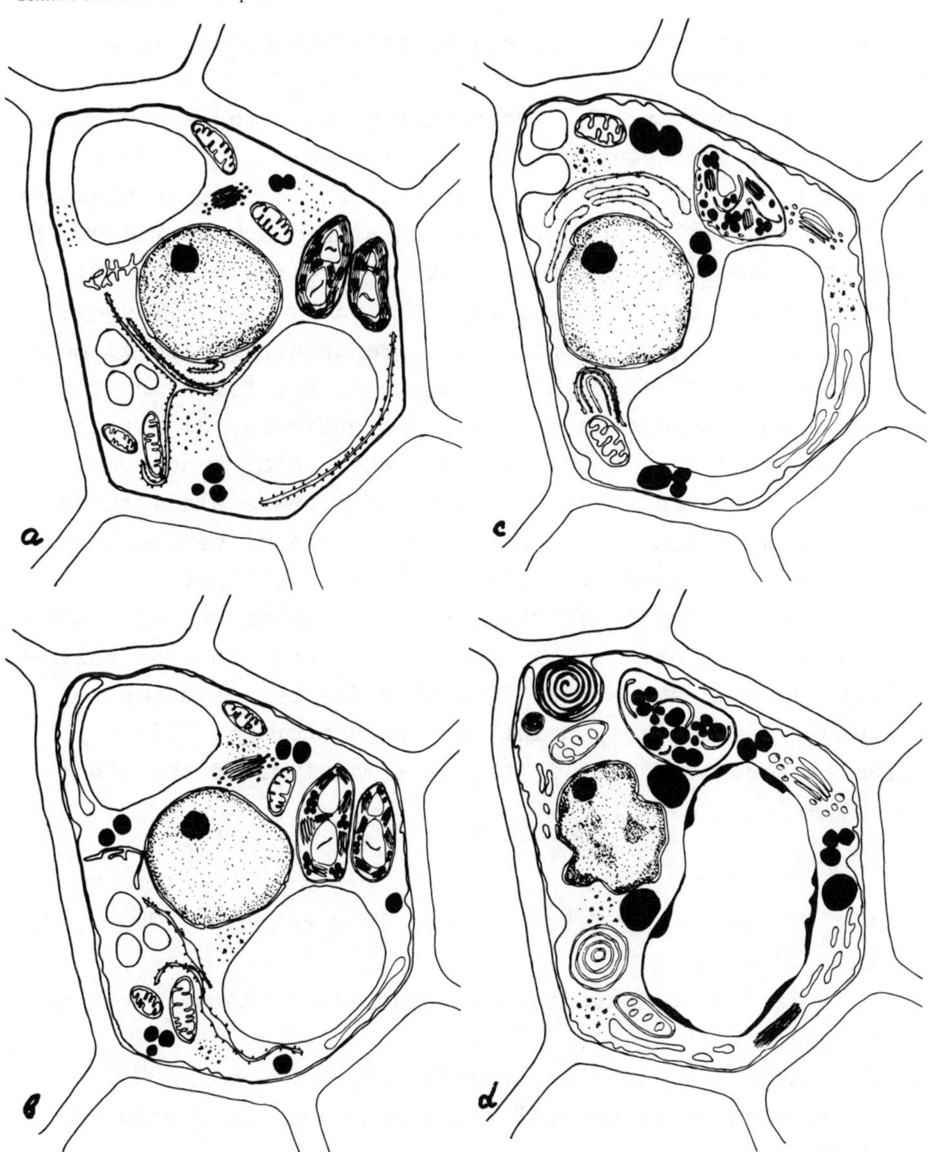

Figure 2 Scheme of apple pericarp parenchyma cell: a - standard; b - at the initial stage of disorganization; c - at the initial stage of disintegration; d - at the initial stage of destruction

their deformation; e) increase in polysome number in cytoplasm.

3. Destruction. The lipophanerose of cytoplasmic structures is the characteristic of destruction (fig. 2,d) : there are a great number of osmiophilic globules in plastids, membrane material is completely absent or represented by individual elementary membranes, starch grains are absent; b) endoplasmic reticulum is represented by of swollen cysterns and by semiconcentric and concentric figures; c) considerable curvature of plasmalemma with the formation the bladdous and concentric membranous structures; d) mitochondrion matrix is more electron-dense and contains numerous one-membranous vesicles; e) the nucleus has a blade shape and a condensation of the structural parts of heterochromatin is probably metabolically inactive on its periphery.

The registration of the structural phenomena of disorganization, disintegration, destruction increase the reliability and accuracy of the diagnosis in the assessment of the apple fruit state during long-term storage. This may serve as a control test in the determination of the plant raw material quality.

References

1. A.E. Vasiliev and I.V. Gamalei, J. Cytology, 1975, v.17, № 4, 371-389.
2. B.T. Matienco, Sravnitelnaya anatomia i ultrastructura tikvennich, Kishinev, 1969.
3. B.T. Matienco, E.M. Zagornean, E.B. Maximova, 1988, Principles of structural transformations in plants, Kishinev.
4. E.B. Matienco-Maximova, V.F. Mashanskii, B.T. Matienco, J.Botany, 1977, v.62, № 2, 267-271.
5. L.V. Metlitskii, 1970, Biochimia plodov i ovotsei, M,.

Section 5: Analytical Methods for Assessment of Food Quality

IMMUNOCHEMISTRY AND FOOD QUALITY ASSESSMENT

G. Brett, V. Bull, S. Chambers, P. Creeke, S. Garrett, S. Khokar, L. Law, H.A. Lee, E.N.C. Mills, M.R.A. Morgan, C.A. Spinks, C. M. Ward, A.P. Wilkinson, G.M. Wyatt.

Department of Food Molecular Biochemistry,
Institute of Food Research,
Norwich Research Park,
Colney,
Norwich NR4 7UA, UK.

1 INTRODUCTION

It is now generally accepted that antibodies are able to make a significant contribution to characterisation and quality assessment of agri-food materials. Such a state of affairs has come to the fore from the very low baseline of activity that existed around 1980.The unique properties of antibodies include a potential for extraordinary diversity of recognition, high affinity and specificity towards the target, and a stability and robustness well-suited to in vitro (and even 'in field') applications. Antibodies can be produced that are capable of recognising almost anything likely to be present in food, from low molecular weight compounds (with a lower limit around 80 - 90 daltons) up to cells and microorganisms. The rise of an antibody-based diagnostics industry devoted to agri-food applications has reflected the rise in research interest. Thus simple, rapid and comparatively cheap immunoassay procedures for analysis of a large number of markers of food quality are now available commercially.

Since their introduction in the clinical research area in the late 1950's[1], immunoassays have been recognised for a number of properties:
(i) performance - specificity, sensitivity, precision, accuracy and reproducibility,
(ii) high rates of sample through-put,
(iii) dealing with analytes difficult to handle by alternative procedures,
and (iv) simplicity.

Subsequently, other properties have been assimilated. Modern, non-isotopic methods such as the enzyme-linked immunosorbent assay (ELISA) have basic, comparatively cheap equipment requirements. New assay formats enable single determinations to be carried out away from the specialised laboratory; it is now possible to carry out semi-quantitative as well as quantitative determinations as required. Immunoassays can be considerably cheaper to perform than alternative methods, and can be significantly more environment friendly since solvent use for extraction and purification can be much reduced or even avoided all together. Dissemination of key antibody reagents either commercially or for research purposes is not often a problem. Finally, regulatory authorities recognise validated, collaboratively tested immunoassays for a number of different applications.

In the present article, examples of applications of antibodies to food quality assessment will be provided, ranging from safety issues to raw material functionality.

However it is first necessary to discuss the nature of antibodies and the ways in which they can be used in vitro.

1.1 Polyclonal and Monoclonal Antibodies

When a higher animal is exposed to a material it regards as 'foriegn', the potential danger is challenged in the first instance by an immune response. A number of specialised cells are stimulated to produce antibodies capable of binding to the 'invader'. Each different antibody produced will have arisen from a single clone (or line) of cells. Thus the serum of a challenged animal will contain many different antibodies capable of binding to the target; the serum is known as a polyclonal antiserum.

If antibody-secreting cells are removed from the animal, they quickly die in culture. However, if the cells are fused with tumour cells in vitro, hybrid cells can result that will secrete antibody and survive indefinitely in culture[2]. Since each hybridoma cell (as the fusion product is known) has arisen from a single antibody-secreting cell, then each hybridoma is capable of secreting only one population of chemically identical antibodies. Such an antibody is known, logically enough, as a monoclonal antibody.

Polyclonal and monoclonal antibodies each have advantages and disadvantages depending on the application. Polyclonal antibodies are cheaper and often quicker to produce. For some targets, such as certain proteins, polyclonal antibodies can be more specific than monoclonal antibodies because recognition of several sites on a target by different antibodies can be more diagnostic than recognition of a single, small site (around 10 amino acid residues) by a single antibody. Conversely, picking out a single, unique site can sometimes be easier with a monoclonal antibody, a necessary strategy when dealing with proteins that are highly heterogeneous. In many circumstances, it is the (theoretically) unlimited availability of monoclonal antibodies that is the most important property.

Materials of molecular weights below about 5000 do not stimulate the immune response directly. Antibodies to such materials (known as haptens) can be obtained by chemically linking them to a carrier molecule, such as a protein, that will themselves stimulate an antibody response. Of the many antibodies stimulated following immunisation with a hapten-protein conjugate, some will be capable of recognising free hapten.

The reader is referred to one of the many excellent texts available on antibody production for more detailed information[3]. New developments in antibody production are discussed towards the conclusion of the present article.

1.2 Immunoassay Formats

The early years of immunoassay work were dominated by the use of radioisotopically-labelled tracer compounds to determine the extent of antibody interaction with the un-labelled target. The successful use of non-isotopic methods[4,5] opened many new areas of application, particularly those related to food.

All immunoassays of the modern era are based on three basic principles (though it is not always easy to tell!), the limited reagent, the reagent excess and the homogeneous assays. Many other names are also used in a semantic minefield. The limited reagent (or competitive) immunoassay is based on the competition between the unknown target and a small, fixed amount of labelled target for a limited number of antibody binding sites. The amount of label that ends up bound to antibody will be dependent on the amount of unknown present: when lots of unknown is present it will be relatively more successful at competing for the limited number of antibody binding sites than the label - when little is present, it will be less efficient at competing. Provided antibody-bound material can be separated from antibody-free material, the concentration of unknown can then be determined by reference to standard curves. Clearly, the affinity of the antibody is a key factor in determining the sensitivity of the assay.

The reagent excess (or non-competitive, or sandwich) assay is based on the principle of an excess amount of an antibody (usually immobilised) capturing the target compound. The captured material can then be detected by exposure to an excess of a second, labelled antibody. The sensitivity obtainable with this approach, performed with excess reagents, is not limited by the affinity of the antibodies. A disadvantage is that the analyte needs to be big enough to bind two antibody molecules simultaneously; this type of assay is not suitable for haptens. An advantage is the positive signal generated by increasing amounts of analyte, in contrast to the limited reagent assay.

The homogeneous immunoassays have not found widespread application to agri-food analytes, but have the advantage that no separation step is required. The assay in its most common form is based on the inhibition of enzymic activity caused by antibody binding to an analyte-enzyme conjugate synthesised so that binding causes steric obstruction to the active site. Such inhibition can be relieved by the presence of free analyte in the sample. The direct nature and speed of the assay are countered by the comparatively poor sensitivity and difficulties of conjugate synthesis.

Immunoassays based on the principles described are widely available for application to agri-food matrices, and in a wide variety of formats - quantitative and semi-quantitative, suitable for operation in batch mode and for single samples, for use in the laboratory or in the field, for use by skilled or untrained personnel.

1.3 Examples of Applications Related to Food Safety

Not surprisingly, perhaps, many of the applications of immunoassays in the agri-food area have been related to food safety. Of the haptenic compounds, much interest has been devoted to the mycotoxins, the secondary metabolite mold products that can be found as contaminants in food and feed and which can have potent toxic bioactivities in animals and man. The group includes compounds such as the aflatoxins, the trichothecenes, and the fumonisins. Immunoassays are able to meet the need for rapid, sensitive, specific and cost-effective methods of analysis (as shown in a number of reviews,[6,7]). One example will serve as illustrative of the possibilities, that of the immunoassay of the aflatoxins [8].

Aflatoxin B1 is the most potent of the group, being generally regarded as a human carcinogen. In some regulations it is required to determine aflatoxin B1, in others determination of aflatoxins B2, G1 and G2 are required. Fortunately, it is relatively easy to generate polyclonal and monoclonal antibodies of the desired specificity [9]. A further complication, relevant to analysis of contaminants in general but seen acutely with the mycotoxins, is the need for a properly validated sampling plan. Mycotoxin contamination is extremely heterogeneous in nature, and the analysis needs to take this into account.

Methods of analysis need to be sensitive, to deal with the low levels of analyte of interest, specific, cost-effective and rapid. Immunoassay methods have been described for quantitative and semi-quantitative applications, for batch or single sample use, and for applications away from the specialised laboratory[8]. Several of these methods have received approval from regulatory bodies following collaborative trials and other assessments. Having selected the sample to be analysed, the analyte needs to be solubilised (unless, of course, a liquid sample is involved). The solvent of choice will be one that achieves maximum, efficient and reproducible extraction, but which is as safe to use and dispose of as possible. Neither the solvent nor co-extractants should interfere with the specific recognition of the antibody. For the aflatoxins, acetonitrile and methanol have usually been selected as having the desired properties. The use (and disposal) of solvents in laboratories is increasingly a factor to be considered in setting up analytical procedures, and immunoassays can help by requiring smaller volumes, by having fewer stages of manipulation, and by being more amenable to water-soluble solvents than some other procedures.

A specialised form of immunoassay is the use of immobilised antibodies to capture analyte from a complex mixture, allowing the washing away of other material, and subsequent quantification of the purified target. In the case of the aflatoxins, with their strong fluorescence, quantification can be direct and simple. As an alternative, the captured analyte can be eluted (by use of organic solvent or other chaotropic agent) and subjected to HPLC or other alternative procedure. The advantages of the latter approach is that the 'clean-up' effected by the immobilised antibodies (usually in the form of a column) is much simpler and more effective than the normally used procedures. In addition, it can be an extremely effective way of quantifying groups of analytes individually.

The aflatoxins provide an excellent example of the way in which antibody-based methods have been successfully applied to a difficult analytical problem. Many other applications exist for similar low molecular weight materials, including pesticides, phycotoxins and veterinary drugs, all well supported by the diagnostics manufacturers. Of molecules of larger size relevant to food safety, examples are provided by immunoassay of proteins characteristic of food types such as wheat gluten, peanut protein and chicken egg protein. Such examples are related to considerations of food allergy, but are also part of a wider concern related to food labelling and authenticity. For speciation of a sample of meat, immunoassay provides a rapid, simple and objective procedure that is difficult (if not impossible) to reproduce with alternative methods of analysis[10]. Dip-stick methods are now available. Whether direct use of such methods by the consumer of his/her own food, would ever be desirable or beneficial is a topic of much discussion; certainly the antibody technology can be made simple and robust enough to be used in the home, as evidenced by the wide availability and use of home pregnancy test kits.

For microbiological food safety, antibody-based procedures have played an important though not leading role despite a long history of antibody production against organisms such as Salmonella[11]. The reasons for this relate to the extreme sensitivity required (1 cell per 25g), which has been difficult to achieve without extensive amplification of cell numbers by use of time-consuming growth steps prior to detection. In more recent work with immunoassay development, assay sensitivity is being considerably reduced - down to 1 day from 4 or 5 - which bodes well for the future[12]. A reduction in the number of cases of food-borne poisoning may be more difficult to attain, but will be helped by more widespread use of HACCP techniques in the food industry. It may well be in this connection that immunoassay will play an increasing role by providing rapid, on-line methods for determination of indicators of problems at specific points in the food production plant. End-product testing is expected to play a diminishing part in this aspect of food control.

An important feature of new method development should be the possibility of applying the analysis to additional areas not previously possible. Additional information can then be provided in order to increase analytical efficiency and provide improved dietary advice. An example is provided by the analysis of food components and contaminants in human serum. The sensitivity of immunoassays make such determinations feasible, particularly because of the requirement for small volumes of blood. The information provided by such determinations relates to exposure, absorption and metabolism, and aids understanding of risk when related to food safety. Examples are provided by research on serum levels of aflatoxins[13] and the natural toxicant found in potatoes, the glycoalkaloids[14].

1.4 Examples of Applications Relating to Food Functionality

As well as the need to analyse for factors relating to food safety, much analysis is also carried out for positive attributes such as those relating to food functionality. One of the difficulties in this area is that it is often the case that the molecular basis of the functionality is not well understood. Consequently, analysis may be indirect in nature. An immunological approach can often help, not just with an improved method of analysis, but by increasing our knowledge of molecular behaviour. This can be particularly true in the study of the seed storage proteins, often complex mixtures difficult to which it is difficult to apply conventional techniques of protein chemistry. Monoclonal antibodies, by allowing single epitopes to be followed, can provide unique information as to the amount, availability and conformation of those epitopes. One example of this is provided by the characterisation and use of monoclonal antibodies against various storage proteins from soya[15,16], looking at both native and processed material, and even through denaturing processes.

Perhaps the outstanding example of such research is provided by the use of immunological techniques to study the proteins and peptides of wheat gluten. The unique technological properties of wheat gluten are utilised in bread, biscuit and pasta making. Such properties are difficult to duplicate, even by using other cereals, and are determined in any given sample by both genetic and environmental factors. Thus not all batches of a variety of wheat known to have the potential to provide 'good bread-making' flour will do so. Consequently it is necessary to determine the properties of any given batch in order to avoid mis-classification, which would have considerable economic repercussions. Monoclonal antibodies have been used to provide tests of improved performance, as well as to provide further understanding at the molecular level[17,18]. One advance is in the ability to quantify gluten itself[19], a procedure difficult to perform by any other means; others have led to the identification of new peptides involved in the bread-making process[20] and localisation studies[21]. Future research will complete the cycle of raw material and end-product analysis by involving antibody tests in the breeding programme.

Several other examples of antibody assessment of food functionality (such as barley in beer-making[22]) have been studied, but to a much lesser extent than for soya and gluten. This will be a considerable growth area in the future as the molecular basis of the functionality is more completely understood (allowing more focused antibody production and test development) and as the ability to manipulate the genetic properties of crops increases.

There is increased interest in the relationships between diet and health, and these will generate analytical opportunities. Of particular note, because they will be 'new' analytes in food quality assessment, will be the 'protective factors', dietary non-nutrients that have been linked to beneficial protection against particular diseases. Examples are the phytoestrogens and the glucosinolates, increased consumption being associated with reduced risks of certain cancers. Immunoassay development is currently underway for such compounds, not only to determine levels in food but to help determine mechanisms of action[23].

1.5 Immunoassay and Developing Countries

Many modern methods of analysis rely on expensive equipment with high running costs; such procedures are often not an option for those in developing countries. This problem is particularly relevant in food quality assessment where regulatory issues can be an effective trade barrier, and where research is often required to improve quality. Immunoassay methods for food materials have been successfully produced and applied in a number of developing countries[24]. The antibody reagents can be of a very high quality, and available in amounts allowing wide dissemination to others within the country. There is no reason why such procedures need not be of the same standard as those available in developed countries, particularly since equipment requirements and costs are comparatively low. Of course, there is a requirement for training.

1.6 Future Developments

The use of immunoassay methods for analysis of food materials has advanced dramatically over the last 10 years, and yet further expansion is likely with increasing interest in food safety, in food quality and functionality, in the links between diet and health, and in between plant breeding and end-product use. It is clear that applications across the agri-food continuum will be more apparent than is currently the case, reflecting closer research objectives. The immunoassays themselves will become simpler and faster as new formats are developed. Greater understanding of food systems will allow better definition of analytes, particularly in the area of food functionality. Similarly, mechanistic studies will better define the role of dietary components in health, which in turn will better inform analysts.

It is likely that by far the biggest impact on immunochemistry will, however, be provided by advances in antibody production (and, even further ahead, in antibody design[25]). It is now possible to produce parts of the antibody molecule as recombinant proteins in, for example, E. coli, to obtain sequence information more readily than previously possible, and to alter the sequence readily. It should, therefore, be possible to manipulate specificity and affinity. Further developments wil allow access to libraries of antibody genes in vitro, reducing the time taken for antibody production and - possibly - increasing the liklihood of identifying rare antibodies. Such developments will be certain to improve the quality of immunological reagents and the information they provide. It would be expected that food quality assessment will provide many targets for research and analytical applications.

References

1. S. A. Berson and R. S. Yalow, Ann. N. Y. Acad. Sci., 1959, **82**, 338.
2. G. Kohler and C. Milstein, *Nature*, 1975, **256**, 495.
3. J. W. Goding, 'Monoclonal Antibodies. Principle and Practice', Academic Press, London, 1986.

4. E. Engvall and P. Perlmann, *Immunochemistry*, 1971, **8**, 871.
5. B. K. van Weeman and A. H. W. M. Schuurs, *FEBS Lett.*, 1971, **15**, 232.
6. C. M. Ward, A. P. Wilkinson and M. R. A. Morgan, 'Modern Methods of Plant Analysis. Plant Toxin Analysis', Springer Verlag Berlin, Heidelberg, 1992.
7. J. J. Pestka, *Food Agric. Immunol.*, 1994, **6**, 219.
8. C. M. Ward, A. P. Wilkinson and M. R. A. Morgan, 'Chromatography of Mycotoxins. Techniques and Applications' Elsevier, Amsterdam, 1993.
9. C. M. Ward, A. P. Wilkinson, S. Bramham, H. A. Lee, H. W.-S. Chan, G. W. Butcher, A. Hutchings and M. R. A. Morgan, *Mycotoxin Res.*, 1990, **6**, 73.
10. C. J. Smith, 'Food Safety and Quality Assurance. Applications of Immunoassay Systems', Elsevier Applied Science, Barking, UK, 1992, 13.
11. G. Wyatt, H. A. Lee and M. R. A. Morgan, 'Immunoassays for Food-poisoning Bacteria and Bacterial Toxins', Chapman and Hall, London, 1992.
12. G. M. Wyatt, M. N. Langley, H. A. Lee and M. R. A. Morgan, *Apl. Env. Microbiol.*, 1993, **59**, 1383.,
13. A. P. Wilkinson, D. W. Denning and M. R. A. Morgan, *J. Toxicol.- Toxin Reviews*, 1989, **8**, 69.
14. M. H. Harvey, M. McMillan, M. R. A. Morgan and H. W.-S. Chan, *Human Toxicol.*, 1985, **4**, 187.
15. G. W. Plumb, E. N. C. Mills, M. J. Tatton, C. C. M. D'Ursel, N. Lambert and M. R. A. Morgan, *J. Agric. Food Chem.*, 1994, **42**, 834.
16. G. W. Plumb, N. Lambert, E. N. C. Mills, M. J. Tatton, C. C. M. D'Ursel, T. Bogracheva and M. R. A. Morgan, *J. Sci. Food Agric.*, 1995, **67**, 511.
17. J. H. Skerritt, O. Martinuzzi and C. W. Wrigley, *Can. J. Plant Sci.*, 1987, **67**, 121.
18. E. N. C. Mills, G. M. Brett, S. Holden, J. A. Holden, J. A. Kauffman, M. J. Tatton and M. R. A. Morgan, *Food Agric. Immunol.*, 1995, **7**, 189.
19. J. H. Skerritt and A. S. Hill, *J. Agric. Food Chem.*, 1990, **38**, 1771.
20. G. M. Brett, E. N. C. Mills, A. S. Tatham, R. J. Fido, P. R. Shewry and M. R. A. Morgan, *Theor. Appl. Genet.*, 1993, **86**, 442.
21. M. L. Parker, E. N. C. Mills and M. R. A. Morgan, *J. Sci. Food Agric.*, 1990, **52**, 35.
22. J. A. Kauffman, E. N. C. Mills, G. M. Brett, R. J. Fido, A. S. Tatham, P. R. Shewry, A. Onishi, M. O. Proudlove and M. R. A. Morgan, *J. Sci. Food Agric.*, 1994, **66**, 345.
23. M. R. A. Morgan, 'New Diagnostics in Crop Sciences', CAB International, Oxon, UK, 1995, 215.
24. A. Escobar, E. Margolles and A. Acosta, *Food Agric. Immunol.*, 1991, **3**, 57.
25. P. V. Choudary, H. A. Lee, B. D. Hammock and M.R.A. Morgan, 'New Frontiers in Agrochemical Immunoassay', AOAC International, Arlington, USA, 1995.

IMMUNOCHEMICAL TEST METHODS FOR THE DETERMINATION OF TASTE COMPONENTS IN RADISH AND CHICORY

A. van Amerongen, L.B.J.M. Berendsen, R.J. Noz, A.M. Peters and H.H. Plasman

Agrotechnological Research Institute (ATO-DLO)
P.O.Box 17
6700 AA Wageningen
The Netherlands

1 INTRODUCTION

In addition to outer quality parameters, e.g., appearance, there is an increasing interest in inner quality aspects of food and food products, such as taste. However, in contrast to most other parameters it is not possible to define an optimal taste profile. The consumer market is highly segmented and requires a supply in various taste classes.

To use taste as a possibility for additional classification of radish (*Raphanus sativus* L.) and chicory (*Cichorium intybus* L.) we have developed immunoassays to measure the components that contribute most to the specific taste of these crops. The results obtained with these assays have been correlated with sensory data.

1.1 Pungent Component of Radish

The taste of radish is due to the breakdown of 4-methylthio-3-butenyl glucosinolate (MTBGSL) by myrosinase to the pungent isothiocyanate (MTBITC).[1] In general, isothiocyanates are unstable in aqueous solution. For MTBITC in Japanese radishes it was reported that in juice 25% of the MTBITC content disappeared in 60 minutes.[2]

1.2 Bitter Sesquiterpene Lactones in Chicory

The bitter taste of chicory is caused by sesquiterpene lactones based on a guaiane skeleton.[3] Several lactones can be observed in chicory heads, such as lactucin (lac), 8-deoxy-lactucin (8dolac), lactucopicrin (lacp) and the dihydro-forms of these components (dhlac, dh8dolac and dhlacp). It has been suggested that glycosidic counterparts are also present. HPLC was used in studies in which the levels of various sesquiterpene lactones in chicory have been correlated with the bitterness perceived by a sensory panel. Correlation coefficients of bitter perception were reported for lac (0.69) and for glycosidic lac in a cellulase treated fraction (0.8).[4,5] In the last study the correlation coefficient for free lac was 0.28. However, nothing is known about the bitter perception of the glycosides. Leclercq suggested that those compounds may also be bitter.[6] It is also not known, whether glycosides are converted to their bitter aglycons by endogenous enzymes during consumption.

2 MATERIALS AND METHODS

2.1 Purification of Taste Components

MTBGSL was purified from radish roots as described.[7] To obtain MTBITC, MTBGSL was treated with a crude fraction of myrosinase.[8] Lac and lacp were isolated from the milky juice of *Lactuca virosa* L. heads. The purity of the various haptens was assessed by NMR and GC-MS.

2.2 Hapten Conjugates, Specific Antibodies and ELISA's

MTBITC was coupled to Keyhole Limpet Hemocyanin (KLH) via the isothiocyanate group. Lac and lacp were activated by coupling to p-nitrophenylchloroformate. Subsequently, the activated hapten was allowed to react with KLH. The conjugates were used to immunise rabbits and mice.
Polyclonal antisera (PAb) and monoclonal antibodies (MAb) were screened for competition with the free hapten upon binding to BSA-hapten coated ELISA plates. The best antiserum / MAb was used to develop specific immunoassays for the various taste components.

2.3 Sensory Analysis

Sensory data were obtained by expert panels. At three time points in 1994 experiments were performed to correlate sensory perception of radish puncency with the levels of MTBITC: in May six cultivars were analysed, in July eight and in October thirteen. In April 1995 sensory perception of chicory head bitterness was scored for three cultivars. The roots had been grown on two locations and with two nitrogen levels. Both, raw and cooked samples were analysed. In addition, the cooking water of the cooked samples were examined as well.

3 RESULTS AND DISCUSSION

3.1 MTBITC Specific ELISA and Correlation with Radish Pungency

A MTBITC specific ELISA was developed with MAb 20C5. The MTBITC standard curve was prepared by converting a MTBGSL stock solution with a crude fraction of myrosinase for 2 minutes just before use. A reproduceable curve was obtained in the range of 2 to 20 nmol of MTBITC per ml. The cross-reactivities of MTBGSL, sinigrin and its isothiocyanate derivative (allyl ITC) are shown in Table 1.

In the course of this study we noticed that MTBITC in Dutch radishes, as determined by HPLC, is even more labile than in Japanese radishes.[2] In juice of Dutch radishes MTBITC disintegrates with a half life of 10 to 15 minutes after an initial increase (MTBGSL conversion) during the first two minutes (results not shown). However, with the ELISA only a small decrease in MTBITC levels was detected (to 80%) for radish samples pre-incubated for two to thirty minutes after juice preparation with a household juice centrifuge and incubation for 15 minutes in ELISA. This can be explained by the binding of the MAb to breakdown products of MTBITC.[9] Compared to levels of radish samples as determined by HPLC with the same MTBITC stock

Table 1 *Cross-reactivities of MTBITC, Lactucin and Lactucopicrin Specific ELISA's*

Component	*Cross-reactivities in Percentage*		
	MAb 20C5 (MTBITC)	*PAb 455 (Lactucin)*	*MAb 4H10 (Lactucopicrin)*
MTBITC	100	-	-
MTBGSL	10	-	-
Sinigrin	<< 0.05	-	-
Allyl ITC	0.05	-	-
Lac	-	100	<< 1
Lac + Dhlac[a]	-	100	2
8Dolac + Dh8dolac[b]	-	50	2
Lacp	-	0.4	100

[a]: Mixture of lactucin and dihydrolactucin (1:2)
[b]: Equimolar mixture of 8-deoxylactucin and dihydro-8-deoxylactucin

solution as a standard, the absolute values in ELISA were a factor ± 2.3 higher. Probably, the breakdown of MTBITC in the standard solution yields other products than that in the radish samples.[9] Presumably, the affinity of the MAb varies for these different products. This may result in incorrect absolute values for the radish samples. Nevertheless, the correlation of MTBITC levels obtained by HPLC with that by ELISA was in general above 0.8.

Three experiments have been performed to assess the correlation between sensory data and the levels of MTBITC as determined by the specific ELISA. In these experiments a pre-incubation time of two minutes was used. In Table 2 the correlation coefficients are given. The average of 0.72 indicates that results obtained with the MTBITC ELISA can be used as objective values for radish pungency.

3.2 Lactucin and Lactucopicrin ELISA's and Correlation with Chicory Bitterness

Both with PAbs and MAbs the anti-lac antibodies showed a high cross-reactivity with other structurally very related sesquiterpene lactones, such as dhlac, 8dolac and dh8dolac (Table 1). Based on a correlation with HPLC data the ELISA with PAb 455 was able to detect the "lactucin-like" lactones in lac-equivalent. However, the cross-reactivity with lacp was less than 1%. Sensitivity of the lac-like ELISA was in the range of 0.5 - 50 ng per 100 µl. In the case of lacp a specific MAb (4H10) was developed that showed low cross-reactivities with the other sesquiterpene lactones. A standard curve was obtained in the range from 0.2 - 25 ng per 100 µl.

Dependent on the extraction procedure used it is possible to determine free and/or glycosidic sesquiterpene lactones with the ELISA's. Based on a recent publication[6] we have chosen to determine the total pool of free and glycosidic lac-equivalent and lacp. These values were correlated with the sensory perception. Table 2 summarises the correlations found in a first study for sensory data and ELISA results in raw and cooked samples and also in the water obtained upon cooking. On average correlation coefficients of 0.8 were obtained for raw and cooked chicory. Based on bitter threshold levels for the various lactones[10] (lac 1.7 ppm, 8dolac 1.1 ppm, lacp 0.5 ppm) a preliminary weighted ELISA score of 1 lac-equivalent and 3 lacp yielded the best correlation with sensory taste of cooked samples (0.88).

Table 2 *Correlation Coefficients (r) for Sensory Analysis of Radish Pungency and Chicory Bitterness and Specific ELISA's for Taste Components*

Radish			*Chicory*		
Exp.		*Sample*	*Lac-equivalent*	*Lacp*	*1xLac-eq+3xLacp*
	r		*r*	*r*	*r*
1	0.80	raw	0.75	0.82	0.81
2	0.59	cooked	0.81	0.85	0.88
3	0.76	cooking water	0.63	0.55	0.63

4 CONCLUSIONS

ELISA's have been developed that can be applied in all stages of the crop production chain: in plant breeding for selection on taste of various cultivars, in agronomy to optimize growing conditions and at the auction level for quality classification of bulk sales.

References

1. P. Friis and A. Kjaer, *Acta Chem. Scand.*, 1966, **20**, 698.
2. K. Okano, J. Asano and G. Ishii, *J. Japan Soc. Hort. Sci.*, 1990, **59**, 545.
3. M. Seto, T. Miyase, K. Umehara, A. Ueno, Y. Hirano and N. Otani, *Chem. Pharm. Bull.*, 1988, **36**, 2423.
4. P. Dirinck, P. van Poucke, M. van Acker and N. Schamp, 'Strategies in Food Quality Assurance: Analytical, Industrial and Legal Aspects', De sikkel, Antwerp, 1985, Vol.III, pp. 62-68.
5. K. R. Price, M. S. DuPont, R, Shepherd, H. W-S. Chan and G. R. Fenwick, *J. Sci. Food Agric.*, 1990, **53**, 185.
6. E. Leclercq, *Voedingsmiddelentechnologie*, 1992, **25**, 14.
7. M. Visentin, A. Tava, R. Iori and S. Palmieri, *J. Agric. Food Chem.*, 1992, **40**, 1687.
8. G. Ishii, R. Saijo and J. Mizutani, *J. Japan. Soc. Hort. Sci.*, 1989, **58**, 339.
9. S. Kosemura, S. Yamamura and K. Hasegawa, *Tetrahedron Letters*, 1993, **34**, 481.
10. T. A. van Beek, P. Maas, B. M. King, E. Leclercq, A. G. J. Voragen, A. de Groot, *J. Agric. Food Chem.*, 1990, **38**, 1035.

Acknowledgements

Sensory studies were initiated by the Central Bureau of the Auctions in The Netherlands. The sensory analyses of radish were performed at the Glasshouse Crops Research Station. The Research Station for Arable Farming and Field Production of Vegetables supplied the chicory cultivars used by the Product Board for Fruit and Vegetables to obtain sensory data in the framework of the project Integrated Quality Management of Chicory. Free and mixtures of sesquiterpene lactones were a gift from Prof.Dr. K.H. Gensch (Institute for Pharmacy, Freie Universität Berlin, Germany) and Dr. F.R. Visser (New Zealand Dairy Research Institute, Palmerston North, New Zealand), respectively.

APPLIED ANALYTICAL CHEMISTRY COMBINED WITH RISK ASSESSMENT - AN AID TO VALUE ADDITION

P. J. van Niekerk and A. M. Raschke

CSIR Food Science and Technology
P O Box 395
0001 Pretoria
South Africa

1 INTRODUCTION

Different criteria, which are dependent on the requirements of the consumer, are used to judge the quality of food. The overall quality of food has been described[1] as a system of partial qualities such as contents of nutrients, contents of contaminants, taste, smell, texture, availability, convenience of use etc. Value addition to food requires that one or more of these partial quality parameters be changed to increase the overall quality. Conversely, if there is a deterioration in one or more of the partial qualities, the value of the food will be reduced.

Analytical chemistry cannot increase the quality of food. However, it can be a guide during the process of value addition or the prevention of the deterioration of food quality. An analytical result, per se, has little value. It is only through a process of interpretation and reaction to the conclusions that benefit is derived. Interpretation transforms analytical data into useful information. Analytical results can be interpreted in many ways and by many techniques as illustrated by the following examples: Analysis combined with chemometric techniques[2] can prove that a food is authentic or unadulterated. Evaluation of results by a nutritionist can indicate the value of a particular food product as it relates to the total diet of consumers. Techniques for process optimization in a food factory requires accurate analytical data to ensure that a process is obtained to produce food at the lowest cost without sacrificing quality. Health risk assessment, using risk models, can be used for the interpretation of analytical results to determine the safety of food products.

2 RISK ASSESSMENT

Safety implies the absence of hazard or risk. Absolute food safety, however, is impossible because there is virtually no component of our food supply that is without some risk to some part of the population. Food hazards can be classified according to three major criteria namely severity, incidence and onset. based on these criteria, the various food hazards can be ranked from the greatest to least risk as follows[3]:

1. Food borne hazards of microbial origin.
2. Nutritional hazards.

3. Hazards from environmental contaminants.
4. Food hazards of natural origin.
5. Food and colour additive hazards.

This ranking is the direct opposite of the ranking given by consumers, who regard additives as posing the greatest risk and microbiological hazards the least.

Risk can be reduced to two generic components, i.e. an undesirable event and the likelihood of its occurrence[4]. The aim of risk assessment is to identify the undesirable event (risk identification) and to estimate its likelihood (risk analysis) as a first step in the process of risk management. Other components of risk management are risk control, risk communication and risk monitoring.

The process of risk assessment consists of four distinguishable but interacting phases, generally referred to as hazard identification, exposure assessment, dose response assessment, risk characterization and risk management. Analytical chemistry plays an important role during the phase of exposure assessment when the levels of contaminants in the food or environment have to be assessed. It is also an aid during the hazard identification phase.

3 THE APPLICATION OF RISK ASSESSMENT

The following are examples of the practical application of risk assessment, subsequent to chemical analysis, in our laboratories:

3.1 Chlorpyrifos-methyl in Maize

A person became ill after eating porridge prepared from maize which was produced by a mill in a rural area. The person claimed that he was poisoned by the maize. Samples of the maize were sent to our laboratories to establish whether any pesticide residues could be detected. Chlorpyrifos-methyl, an insecticidal, was found in four of the five samples submitted at concentrations ranging from 0.02 to 0.07 mg/kg.

During the risk assessment, the following was taken into account: The allowed daily intake (ADI) of chlorpyrifos-methyl for a person is 0.01 mg/kg body weight[5,6]. It was assumed that the average person weighs 50 kg and that he eats 1 kg of maize a day. Such a person could tolerate less than 0.5 mg of chlorpyrifos-methyl per day. According to the analytical data, eating 1 kg of maize will provide a maximum of 0.07 mg of the pesticide per day, which is c. 14 per cent of the ADI.

Based on these results, it is unlikely that the person was poisoned by the pesticide which was present in the maize.

3.2 Chlorpyrifos-methyl and Cypermethrin in Wheat Flour

This case was more complicated in that the flour samples were found to contain residues of two pesticides. The levels found were 420 μg/kg chlorpyrifos-methyl and 240 μg/kg cypermethrin. A computer software chemical hazard database, RISK*ASSISTANT was used, in conjunction with a health risk model, to assess the health risks associated with the consumption of the two pesticides in the flour.

3.2.1 Exposure Assessment. In addition to the analytical results, the following exposure parameters were assumed for this assessment and represents the exposure scenario:

Consumption rate (flour)	0.2 kg/event
Event frequency	350 events/year
Exposure period	30 years
Consumers mass	70 kg
Lifetime	70 years

3.2.2 Risk Characterization. Chlorpyrifos-methyl is an organophosphate pesticide and cypermethrin is a pyrethroid. Organophosphates are acetylcholine esterase inhibitors and can lead to neurological diseases if exposures are high and prolonged. Symptoms of acute poisoning develop within 12 hours of contact. Acute symptoms from pyrethroids result mainly due to exposure via inhalation. Nervous irritability, tremors and ataxia occur if exposed to very large amounts of pyrethrins.

At the concentrations found in the flour samples, the two pesticides show no risk of causing cancer or acute pesticide poisoning. To determine the possibility of non-cancer toxic effects, or chronic pesticide poisoning, a Hazard Quotient (HQ) is calculated for each pesticide by dividing a calculated Average Daily Dose by the Reference Dose for the particular compound. A linear combination of the HQs result in a Health Index (HI). For the levels of chlorpyrifos-methyl and cypermethrin in the flour, the calculated HI was 0.23. HIs of less than 1.0 are generally considered by the Environmental Protection Agency of the United States of America to be associated with low risks of non-cancer toxic effects.

3.2.3 Conclusion. According to the available data, the concentration levels of the two pesticides in the flour samples posed no cancer, acute or chronic health risks.

3.3 Alkaloids and Cyanic Acid in Sugar Beans

Consumers of canned sugar beans complained about a bitter taste and that they felt ill. Samples of the beans were analysed for total alkaloids and hydrocyanic acid. The alkaloid content of the samples ranged from 245 to 390 mg/kg and the hydrocyanic content ranged from 740 to 790 mg/kg.

3.3.1 Assessment. The bitter taste of the product was probably caused by the combined effect of the hydrocyanic acid and the alkaloids. High concentrations of alkaloids (500 to 800 mg/kg) have been known to cause stomach pains and cramps. However, the alkaloid content of the samples were typical of sugar beans and it is unlikely that there would be any ill effects due to their presence.

Hydrogen cyanide is notorious as a potent respiratory inhibitor and the minimum lethal dose, taken by mouth, has been estimated at between 0.5 and 3.5 mg/kg body weight. Non-fatal doses may cause headaches, a sensation of tightness in the throat and chest, palpitations and muscle weakness[7]. Cyanide in the plant kingdom is mainly found in the form of cyanogenic glucosides. Relatively high concentrations are found in certain grasses, legumes, root crops and fruit kernels. Cyanogenic glucosides, eg. linamarin in legumes, yields hydrogen cyanide, glucose and acetone upon hydrolysis or when acted upon by stomach acids or certain plant enzymes[8]. Plant products containing cyanide can be made relatively safe by adequate heat treatment and proper handling procedures.

3.3.2 Conclusion The bitter taste may have been caused by the combined effect of the alkaloids and glucosides. The illness was probably caused by incorrect preparation of the sugar beans which could have resulted in cyanide poisoning.

3.4 Metals in Wild Spinach

Species of wild spinach have been found growing on the ash dumps deposited by a chemical plant and were consumed by some of the workers. A concern existed that the high mineral content of the dump may cause the spinach to have high concentrations of metals which could be detrimental to human health.

Samples of the soil (ash), spinach and commercially cultivated spinach were analysed for their mineral content. Generally, the wild spinach contained similar or lower levels of the metals than the cultivated spinach. The soil samples showed fairly low levels for each of the elements for which it was analysed except for iron which was found at concentrations ranging from 500 to 1000 mg/100g. There was no significant difference between the iron concentrations in wild and cultivated spinach. This indicates that spinach does not absorb abnormally high levels of metals from soil which do contain high concentrations. Both wild and cultivated spinach contained chromium at levels which are above the recommended maximum limit[9].

4 CONCLUSION

The above and other examples have shown that health risk assessment is a good method of adding value to analytical results and to the foods to which they are applied. Risk assessment will assist in making decisions on the safety of food. It will alert the user when the consumption of the food could pose a danger and it will also ensure that a perfectly safe food is not discarded because of some unsubstantiated fear of contamination.

References

1. L. H. Grimme, R. Altenburger, M. Faust and K. Prietzel, in extract from *Biologi Italiani,* 1990, Vol. 3.
2. P. J. van Niekerk, PhD Thesis, University of South Africa, 1990.
3. H. R. Roberts, 'Food Safety', John Wiley &Sons, Inc., New York, 1981, p. 1.
4. N. H. Rodda and C. Moore, 'Risk assessment in the Aquatic Environment', Paper delivered at the G. J. Stander Lecture, held 12 October 1992 at the CSIR Conference Centre, Pretoria.
5. C. R. Worthing and S. B. Walker, 'The Pesticide Manual', The British Crop Protection Council, Thorton Heath, Eighth Edition, 1987, p. 179.
6. H. Kidd and D. R. James, 'Agrochemicals handbook', The Royal Society of Chemistry, Third Edition, 1991.
7. R. D. Montgomery, in 'Toxic Constituents of Plant Foodstuffs', I. E. Liener Academic Press, New York, 1969, Chapter 5, p. 143.
8. J. M. Jones, 'Food Safety', Eagon Press, St Paul, 1992, Chapter 5, p. 79.
9. M. V. Krause and L. K. Mahan, 'Food Nutrition and Diet Therapy', John Wiley and Sons, New York, Seventh Edition, 1981, Chapter 10, p. 218.

CHARACTERIZATION OF PEA SEED TRYPSIN INHIBITORS

C. Domoney and T. Welham

John Innes Centre
Colney Lane
Norwich NR4 7UH

1 INTRODUCTION

Enzyme inhibitors are considered to be important antinutritional compounds due to their negative effects on animal digestive processes and growth and, in general, these compounds limit the extent to which seeds of major legume species can be included in the diets of young animals. On the other hand, enzyme inhibitors and other antinutritional compounds are recognized as having positive attributes which means that their total elimination from crops may be unwise; such attributes include the clearly demonstrated role of some enzyme inhibitors in the protection of plants from insect attack[1, 2] and a possible preventative role in the development of animal tumorigenesis[3]. In pea seeds, the major enzyme inhibitors are the protein inhibitors of the digestive enzymes, trypsin and chymotrypsin (TI)[4]. Genetical, biochemical, molecular biological and nutritional studies of these inhibitors are aimed, firstly, at establishing the ease or otherwise with which TI levels may be manipulated in peas and, secondly, clarifying to what extent these proteins should be regarded as being antinutritional compounds.

2 GENETICAL STUDIES OF PEA SEED TI

Several distinct TI have been identified in individual pea genotypes[4] and quantitative variation in seed TI levels has been documented among pea germplasm[5]. Pea genotypes showing quantitative and/or qualitative variation in seed TI were used as parents in genetic crosses. Analysis of progeny from such crosses has demonstrated that the individual TI are encoded by a single genetic locus (*Tri*) on linkage group 5 and, furthermore, that there is a co-segregation of the structural variation and relative seed TI activity levels[6]. These data suggest that the genes encoding TI also control, or are closely linked to genes that control, seed TI levels. The inheritance of TI protein variants has been shown to reflect the segregation of DNA polymorphisms in the structural genes[6]. These genetic data have direct implications for the selection of TI variants from breeding programmes and further facilitate the generation of pea lines that are near-isogenic, except for alleles at the *Tri* locus.

3 BIOCHEMICAL AND MOLECULAR BIOLOGICAL STUDIES OF PEA SEED TI

Purification of individual pea seed TI proteins and analysis of their N-terminal sequences has shown that the different TI are very closely homologous to each other and belong to the Bowman-Birk class of inhibitors[7]. Isolation of DNA sequences corresponding to pea TI and genomic hybridization analyses have indicated that pea seed TI are encoded by two genes[8]. These genes are developmentally regulated during pea seed development, being active at the later stages of embryogenesis, and encode TI precursors that contain a long N-terminal pre-sequence (Figure 1). The derivation of the several TI isoforms in mature pea seeds has been attributed to modifications of these two primary gene products[8]. Based on electrospray mass spectrometry of TI proteins, one of these modifications has been postulated to be a removal of nine amino acids from the C-terminus of a proportion of the two primary gene products. This modification occurs at very late stages of seed development and the modified TI show a higher affinity for the target enzyme, trypsin, than the primary mature TI[8]. Variation in the extent of processing of the primary mature TI would influence relative seed TI activities and could explain much of the observed environmental variation in TI activity levels in single genotypes[9].

```
                                                             ▼
  1  MELMNKKVMM KLALMVFLLS FAANVVNARF DSTSFITQVL SNGDDVKSAC
                 ▽                          ▽
 51  CDTCLCTKSD PPTCRCVDVG ETCHSACDSC ICALSYPPQC QCFDTHKFCY

101  KACHNSEVEE VIKN
         ----- ----
```

Figure 1 *The complete amino acid sequence of a precursor to a pea trypsin/chymotrypsin inhibitor. The site of cleavage of the precursor to yield the mature protein is indicated by a closed arrowhead. The two enzyme inhibitory sites are indicated by open arrowheads. The underlined C-terminal amino acids are removed from a proportion of the mature proteins.*

4 NUTRITIONAL STUDIES OF PEA SEED TI

Several studies have indicated that animals fed diets which included peas, having a high TI content, do not perform as well as animals fed control diets[10]. However, in many animal studies, the specific effects of TI are unclear and other seed components could be responsible for the observed effects. *In vitro* digestion studies, based on target animal enzymes, offer the potential for small-scale analyses of the effects of particular plant compounds on animal digestion. The effects of purified pea TI have been compared with those of soybean TI in an *in vitro* digestion system and the results indicate that pea

TI are much less potent inhibitors of animal digestion than their soybean counterparts[11]. These data require validation in whole animal studies, precluded at the present time by the limited availability of purified pea TI. The specific nutritional consequences of pea TI in animal diets may be determined through the use of near-isogenic pea lines, differing only in their relative TI contents.

5 CONCLUSIONS

The similarity among the individual pea TI proteins and the fact that these are the products of a single genetic locus means that gene expression at this locus is amenable to modulation by a single introduced anti-sense TI gene. However, in view of the potential beneficial functions played by these proteins, in particular the protective functions that have been demonstrated for related proteins, a total abolition of TI synthesis may not be desirable. Prevention of the post-translational modification of pea TI that leads to higher affinity isoforms is likely to lower seed TI activities dramatically. However, this modification may involve enzymes crucial to the maturation of other seed proteins and specifically preventing TI modification may not be straightforward. If the modified TI also show a higher affinity for pest digestive enzymes, then expressing modified TI in non-food plant organs would be desirable. Clarification of the antinutritional status of pea and other TI is important and may allow for a wider expression of these proteins in food plants for their beneficial functions, which include a high content of the nutritionally desirable sulphur-containing amino acids (see Figure 1). The use of near-isogenic pea lines in animal feeding trials should provide this clarification.

REFERENCES

1. V.A. Hilder, A.M.R., Gatehouse and D. Boulter, 'Genetic Engineering of Crop Plants' (D. Grierson and G. Lycett, eds), Butterworths, London, 1990, p.51.

2. R. Johnson, J. Narvaez, G. An and C.A. Ryan, *Proc. Natl. Acad. Sci. USA*, 1989, **86**, 9871.

3. M.P. Le Guen and Y. Birk, 'Recent Advances of Research in Antinutritional Factors in Legume Seeds (A.F.B. van der Poel, J. Huisman and H.S. Saini, eds), Wageningen Pers, The Netherlands, 1993, p.157.

4. C. Domoney, T. Welham and C. Sidebottom, 'Recent Advances of Research in Antinutritional Factors in Legume Seeds (A.F.B. van der Poel, J. Huisman and H.S. Saini, eds), Wageningen Pers, The Netherlands, 1993, p.401.

5. C. Domoney and T. Welham, *Seed Sci. Res.*, 1992, **2**, 147.

6. C. Domoney, T. Welham, N. Ellis and R. Hellens, *Theor. Appl. Genet.*, 1994, **89**, 387.

7. C. Domoney, T. Welham and C. Sidebottom, *J. Exp. Bot.*, 1993, **44**, 701.

8. C. Domoney, T. Welham, C. Sidebottom and J.L. Firmin, *FEBS Lett.*, 1995, **360**, 15.

9. J. R. Bacon, N. Lambert, P. Matthews, A.E. Arthur and C. Duchene, *J. Sci. Food Agric.*, 1995, **67**, 101.

10. C. Jondreville, F. Grosjean, G. Buron, C. Peyronnet and J.L. Beneytout, *J. Anim. Phys. Anim. Nutr.*, 1992, **68**, 113.

11. M. Al-Wesali, N. Lambert, T. Welham and C. Domoney, *J. Sci. Food Agric.*, 1995, **68** (in press).

HIGH-PERFORMANCE LIQUID CHROMATOGRAPHY OF LUPIN AND LENTIL SAPONINS

R.G. Ruiz[1], K.R. Price[1], M.E. Rose[2], M.J.C. Rhodes[1] and G.R. Fenwick[1]

[1]*Food Molecular Biochemistry Department, Institute of Food Research, Norwich Research Park, Colney, Norwich, NR4 7UA, United Kingdom.*
[2]*Department of Chemistry, The Open University, Milton Keynes MK7 6AA, United Kingdom.*

1 SUMMARY

A quantitative method for the determination of intact saponins in *Lupinus angustifolius* and *Lens culinaris* seed by high-performance liquid chromatography (HPLC) has been developed. Soyasaponin VI, also known as soyasaponin ßg, a DDMP (2,3-dihydro-2,5-dihydroxy-6-methyl-4*H*-pyran-4-one) conjugated form of soyasaponin I, was the only saponin detected in both the lupin and lentil seed.

2 INTRODUCTION

Saponins are glycosides with terpenoid or steroid aglycones occurring primarily in plants, which are consumed by animals and man but usually processed industrially or domestically prior to consumption. A wide range of both beneficial and deleterious properties has been ascribed to saponins. However, these properties are better seen as characterizing particular types of saponin rather than being shared by all members of the chemically-complex group of saponins[1].

A number of chromatographic methods has been used for saponin analysis and special attention has focused on the use of gas chromatography[2] but it has the limitation that it can only be used for the separation and quantification of the aglycone portion of the saponin (after hydrolysis and suitable derivatization) which involves both the loss of structural information about the glycosidic portion of the molecule and potential loss of material during hydrolysis and derivatization.

Since relatively little is known about the effect of both industrial and domestic processing on the fate of the saponins in food and with recent work demonstrating a relationship between chemical structure and biological activity[3], there is a requirement for analytical methods that can measure the individual saponins as they exist in the food matrix. However, the development of techniques for the analysis of these intact saponins, such as HPLC, has been limited due firstly to the difficulties with the detection of triterpene saponins which do not contain a UV-chromophore and secondly to the lack of appropriate standards. Derivatisation procedures have increased the sensitivity of HPLC methods but suffer from selectivity and stability problems[4]. Kudou et al.[5], Yoshiki et al.[6] and Okubo et al.[7] have successfully separated, on a qualitative basis using HPLC, intact saponins from soybean (*Glycine max*), runner bean (*Phaseolus coccineus*) and American groundnut (*Apios americana*) respectively and the study reported here has further developed this work for the separation and quantification of the intact saponins present in lupin (*Lupinus angustifolius*) and lentil (*Lens culinaris*) seed.

3 MATERIAL AND METHODS

3.1 Material

The lupin and lentil seed samples were commercial Australian and Spanish varieties of *Lupinus angustifolius* and *Lens culinaris* respectively.

All solvents used during the extraction process were of AnalaR grade.

3.2 Methods

3.2.1 Extraction of saponins. The ground sample was extracted with 70% ethanol containing 0.01% EDTA at room temperature[6]. An internal standard of alpha-hederin was added to the sample prior to solvent extraction.

3.2.2 Instrumentation. HPLC analyses were performed using a Philips PU 4100 liquid chromatograph coupled to a Philips PU 4025 UV detector and a Gilson 715 data collection system. Separations were performed on a column (25 cm x 4.6 mm ID) packed with Ultratechsphere 5 μ C_{18} (HPLC Technology Ltd., Macclesfield, UK).

3.2.3 Chromatographic conditions. Chromatographic runs were carried out with an acetonitrile-water gradient elution system. Solvents were acetonitrile:acetic acid (1000:0.3 v/v) (Solvent A) and water:acetic acid: EDTA (1000:0.3:0.15 v/v) (Solvent B). The gradient was run according to the following programme: 65% A isocratically for 18 min followed by a reduction to 58% A after a further 4 min and held at 58% A for 10 min then changed to 65% A over a further 4 min and finally held at 65% A for 8 min. The flow rate was 0.9 ml/min and detection was monitored by UV absorption at 205 nm.

4 RESULTS AND DISCUSSION

HPLC analyses of soyasaponin I (SSI) and soyasaponin VI (SSVI), also known as soyasaponin ßg, which are the main saponins known to be present in *Lupinus angustifolius* and *Lens culinaris* seed[8], together with the internal standard alpha-hederin revealed well resolved peaks whose retention times were 16.8, 29.3 and 33.4 min respectively (Figure 1).

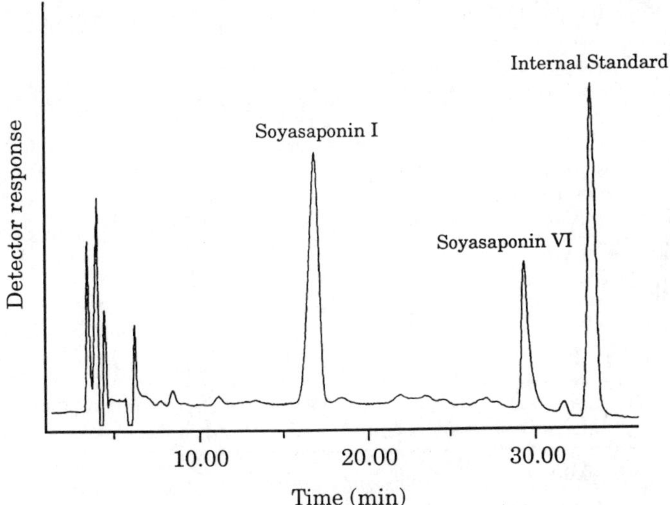

Figure 1 *Chromatogram of a mixture of soyasaponin I, soyasaponin VI, and alpha-hederin*

The chromatograms resulting from the analysis of the lupin and lentil seed extracts following the mild extraction conditions of the study reported here are shown in figures 2 and 3. Retention time comparisons relative to the internal standard alpha-hederin and confirmation by co-chromatography of the reference saponins showed the presence of only SSVI. No SSI was detected. This finding is in agreement with Kudou et al.[5] who postulated that the DDMP-conjugated saponins were, in fact, the genuine saponins in the case of intact soybeans whilst soyasaponins I-V, the saponins normally associated with soya following conventional exhaustive hot solvent extraction, were in fact artifacts derived from degradation of the DDMP saponis during extraction.

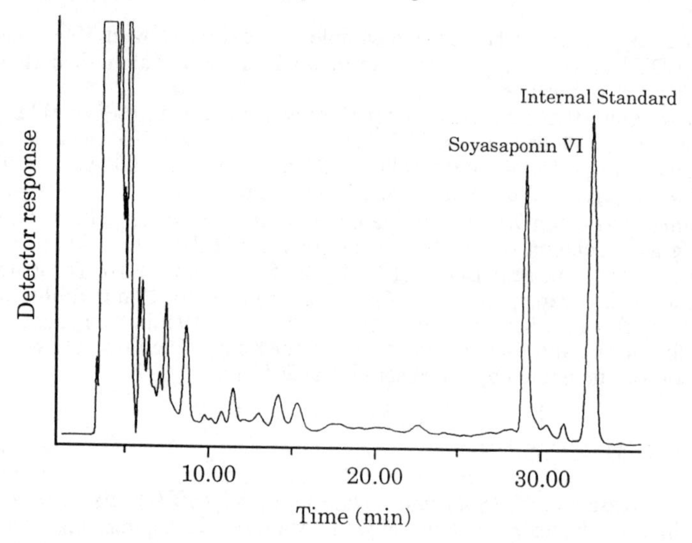

Figure 2 *Chromatogram of Lupinus angustifolius seed saponins (soyasaponin VI) with internal standard (alpha-hederin).*

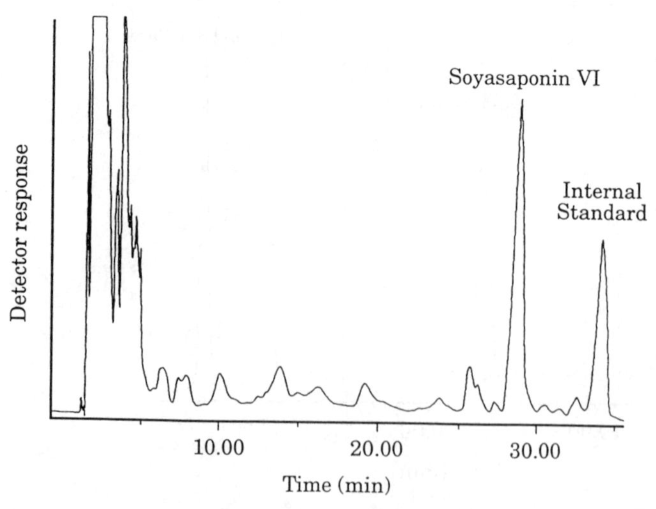

Figure 3 *Chromatogram of Lens culinaris seed saponins (soyasaponin VI) with internal standard (alpha-hederin).*

The HPLC method described here provides the quantification of intact, underivatised saponins in legumes with the use of a suitable internal standard and a proven mild quantitative extraction technique. It will therefore enable quantitative studies to be made for the first time on the effect of different kinds of processing not only on the total saponin content but also on the true saponin composition. These data will, in turn, allow the fate of the individual saponins, some of which have been shown to be relatively unstable[9] to be determined. Since some of the biological properties reported for saponins are now recognised to be dependent on their particular chemical structures[10], this type of information will allow further understanding of not only the relationship between chemical structure and bioactivity but also the impact of processing on that bioactivity in relation to both human and animal health.

5 ACKNOWLEDGEMENTS

The authors would like to thank Dr. K. Okubo and coworkers of the Faculty of Agriculture of Tohoku University for their generous gift of saponin ßg and the European Union within the Agro-Industrial Research programme, (fellowship contract number: AIR-CT92-5121) for financial support.

References

1. G. R. Fenwick, K. R. Price, C. Tsukamoto and K. Okubo, 'Toxic Substances in Crop Plants', J. P. F. D'Mello, C. M. Duffus, J. H. Duffus, Royal Society of Chemistry, London, 1992, Chapter 12, p. 285.
2. K. R. Price, C. L. Curl and G. R. Fenwick, *J. Sci. Food Agric.*, 1986, **37**, 186.
3. M. Tacheki and Y. Tanaka, *Phytochemistry*, 1990, **29**, 451.
4. W. Oleszek, M. Jurzysta, K. R. Price and G. R. Fenwick, *J. Chromatogr.*, 1990, **519**, 109.
5. S. Kudou, M. Tonomura, C. Tsukamoto, T. Uchida, T. Sakade, N. Tanamura and K. Okubo, *Biosci. Biotech. Biochem.*, 1993, **57**, 546.
6. Y. Yoshiki, J. H. Kim and K. Okubo, *Phytochemistry*, 1994, **36**, 1009.
7. K. Okubo, Y. Yoshiki, K. Okuda, T. Sugihara, C. Tsukamoto and K.Hoshikawa, *Biosci. Biotech. Biochem.*, 1994, **58**, 2248.
8. R. G. Ruiz, M. E. Rose, K. R. Price and G. R. Fenwick, 'Bioactive Substances in Food of Plant Origin', H. Kozlowska, J. Fornal, Z. Zdunczyk, Polish Academy of Sciences, Olsztyn, 1994, Volume 2, p.386.
9. S. Kudou, M. Tonomura, C. Tsukamoto, T. Uchida, M. Yoshikoshi and K. Okubo, *ACS Symposium Series Fruits and Vegetables*, 1994, **546**, 340.
10. W. Oleszek, J. Nowacka, J. M. Gee, G. M. Wortley and I. T. Johnson, *J. Sci. Food Agric.*, 1994, **65**, 35.

POLYMERISATION REACTIONS IN RED WINES AND THE EFFECT OF SULPHUR DIOXIDE

Peter Bridle, Johanna Bakker, Cristina Garcia-Viguera and Anna Picinelli.

Department of Consumer Sciences
Institute of Food Research
Reading Laboratory
Earley Gate
Whiteknights Road
Reading RG6 6BZ, UK.

1 INTRODUCTION

Young red wine pigments are essentially those of the red grapes from which the wine has been made. The red pigments are anthocyanins, but there are also many colourless phenolics present, such as phenolic acids, tannins and various classes of flavonoids. These compounds can interact during wine aging to produce changes in colour and taste characteristics. We have studied the role of some key compounds responsible for wine colour during aging, using model solutions. Of particularly interest, is the reaction between flavonoids in the presence of acetaldehyde, when rapid changes in hue and intensity occur.[1] This colour change is especially apparent during the early stages of ruby port wine maturation, when the acetaldehyde concentration is higher than in other red wines. In this work, colour changes were studied in model red wine solutions, using high performance liquid chromatography (HPLC) and colour measurements. The interaction between malvidin 3-glucoside (the main anthocyanin in red wine made from *Vitis vinifera* grapes) and (+)-catechin, was examined, alone and in the presence of acetaldehyde.

2 RESULTS AND DISCUSSION

In the models with no acetaldehyde, loss of malvidin 3–glucoside was observed, but there was only negligible loss of catechin; no new compounds were detected. Model wines with added acetaldehyde, showed a large increase in colour intensity and the mixture became more purple as the wavelength maximum shifted from 524 nm to 557 nm.[2] The progress of this reaction was monitored by HPLC and two major new compounds (labelled A and B) were observed; the reaction was accompanied by rapid losses of malvidin 3-glucoside and catechin.

The new purple-coloured compounds, having absorbance maxima at 544 nm (Figure 1), were formed by the polymerisation of malvidin 3-glucoside to catechin, via an acetaldehyde bridge, according to a mechanism reported previously.[1] A molecular ion at m/z 809 was determined by FAB MS, corresponding to a dimeric structure.

This reaction also occurred in the presence of sulphur dioxide (commonly used as an antimicrobial and antioxidant agent in wine-making), but at a much slower rate. It is possible for these small polymers to react further with other available wine phenolics to form larger polymeric products; in the presence of sulphur dioxide these condensation reactions still occurred.[3]

The rate of loss of initial reactants and the longevity and amounts of new products varied according to pH. A study of this effect,[4] revealed the conditions for optimum formation of compounds A and B. The greatest amounts and most rapid formation occurred at pH 2 (Figure 2).

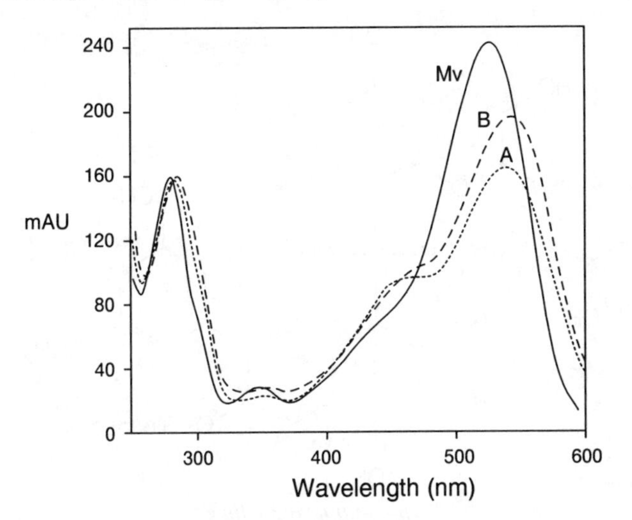

Figure 1 *Spectra of Malvidin 3-glucoside (Mv) and compounds A and B.*

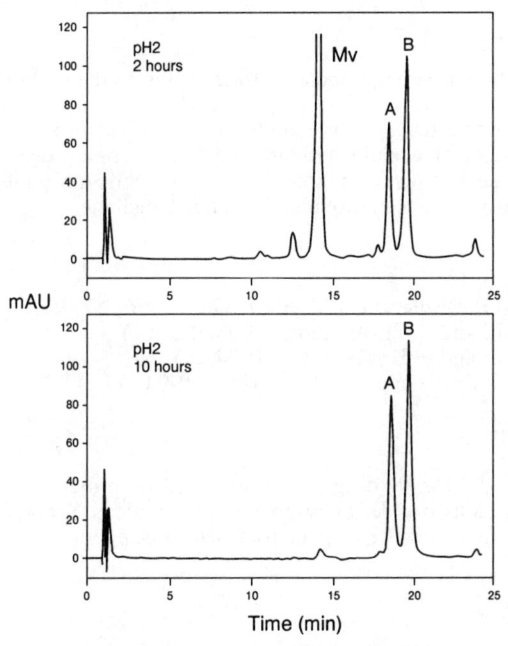

Figure 2 *HPLC chromatogram showing malvidin 3-glucoside (Mv) and the formation of compounds A and B after 2 hours and 10 hours at pH 2.*

Interestingly, the catechin concentration still decreased when there was no malvidin 3-glucoside remaining, showing that further polymerisation reactions were occurring, either between catechin molecules alone or catechin and coloured polymers.

Preparative HPLC was used to separate and purify A an B for full structural characterisation. Subsequent NMR analysis showed these compounds to be isomers based on the structure shown in Figure 3.

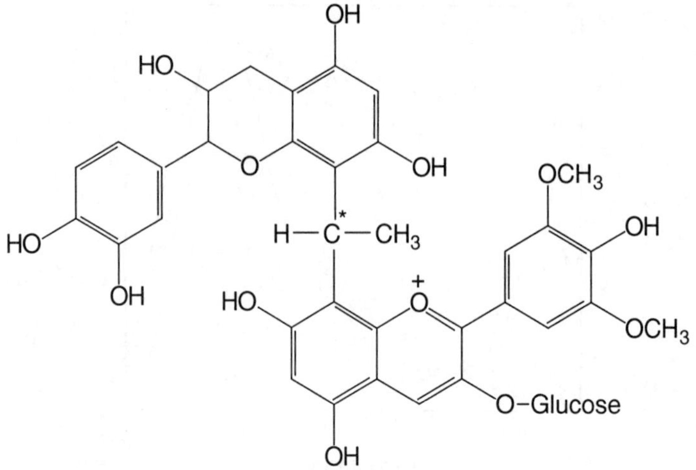

Figure 3 *Proposed structure of catechin - anthocyanin dimer.*

3 CONCLUSION

New highly-coloured dimeric compounds are formed during the early stages of red wine aging, when anthocyanins, flavan 3-ols and acetaldehyde are present. Model wine experiments showed that these new compounds were isomers, formed from anthocyanins and flavan 3-ols, linked by an acetaldehyde bridge. The reaction proceeds most rapidly at low pH and may continue with the formation of larger polymeric molecules. These reactions also occur in the presence of sulphur dioxide, but at a reduced rate.

References

1 C.F. Timberlake and P. Bridle, *Amer. J. Enol. Vitic.* 1976, **27**, 97.
2 J. Bakker, A. Picinelli and P. Bridle, *Vitis* , 1993, **32**, 111.
3 A. Picinelli, J. Bakker and P. Bridle, *Vitis* , 1994, **33**, 31.
4 C. Garcia-Viguera, P. Bridle and J. Bakker, *Vitis* , 1994, **33**, 37.

Acknowledgement

PB and JB thank the EC for funding this work (FLAIR project no. 89053), CGV is indebted to the Consejo Superior de Investigaciones Cientificas, Spain, for a grant and AP thanks NATO's Scientific Committee, Spain, for financial support.

Section 6: Post-harvest Effects and Quality Determinants

ADDED VALUE OF AGRI-FOOD PRODUCTS BY AN INTEGRATED APPROACH

Wim M.F. Jongen
Department of Food Science
Wageningen Agricultural University
The Netherlands

INTRODUCTION

The Food Science and Technology research area has developed in three phases (9). In ***the first phase*** research was product oriented and characterised by an empirical approach. On the basis of a lot of trial and error one was capable of solving existing problems. However due to a lack of insight in the underlying mechanisms it was impossible to predict the behaviour of food components during process- and product development. This has resulted in a more disciplinary approach, ***the second phase***.
Especially the developments in the areas of food chemistry, - microbiology, -physics and unit operations have significantly contributed to a further development of food science and technology. This has resulted in a situation in which food industry is capable of providing clients with safe food products of good quality for a reasonable price.

However also this approach has shown its limitations. The development of the various disciplines mentioned has gone along different lines. The integration of disciplinary knowledge, essential for adequate product development, has not received sufficient attention during this period. The continuously increasing complexity of the question of product quality has strengthened the need for a product oriented approach, ***the third phase***. Product liability for producers of food products does not end at the factory gate anymore but includes the whole production chain up to the consumer
Especially areas such as quality management, product development and packaging of food products are by definition interdisciplinary and require an integrated approach.
The role of integrated food science is therefore by definition product oriented and aims for the development of knowledge systems enabling the development of food products with optimal quality.

One essential difference from the first phase in the development of the area of food science is that this product oriented approach can now be based on disciplinary knowledge. This widens the perspectives for the development of integrated systems with a predictive potential considerably.

Integrated systems can be developed at different levels. These are summarised in Table I.

Two types of integration can be recognised. Vertical integration is integration throughout the production chain. At the level of product aspects important quality determinants can be identified such as healthiness, keepability, taste and texture.

Table 1 *Levels and types of integration in food science*

System Level	*Research Area*	*Type of Integration*
Company	*Process-Systems* *Quality Assurance* *Product Development*	**Vertical**
Product	*Packaging* *Quality Modelling*	**Vertical**
Product Aspect	*Safety* *Healthiness* *Keepability* *Freshness* *Taste* *Texture* *Nutritional value* *Appearance*	**Horizontal**

Integration at this level is horizontal integration which means integration between disciplines such as food chemistry, -physics and -microbiology.

CHANGING MARKETS

We are dealing with fast changing situations in the market for food products (3). In Table II a number of aspects are listed which will be dealt with in the following.
As a consequence of too high production levels of both primary products (raw material) as well as food product market saturation is apparent. In most western countries a saturation point is to be reached with respect to caloric intake.

We are dealing with important demographic changes such as an ageing population, changes in household composition and changes in participation in the labour market which will influence the consumption pattern considerably. Changes in portion size, the increasing demand for convenience foods are clear examples. The presence of large populations of minority groups will open up markets for exotic food products.

The role of the consumer is becoming increasingly important. Generally the consumer is better trained and informed and has higher demands concerning product quality and variety. The perception of quality has changed in that it not only comprises product quality but also production methods. Product acceptance by the consumer will result in a larger influence of environmental factors. Sustainability will be integrated in future concepts of product development. New biotechnology will play an increasingly important role in the production of new food products. Although traditionally biotechnology already has played an important role for example in fermentation processes it is important to note that consumer acceptance of new biotechnology is strongly dependent on the advantage that the consumer sees for him/herself.

In addition to the market pull aspects mentioned also technology push coming from scientific developments will result in new processing technologies and new types of food products coming on the market.

Table 2 *Some factors determining changes in the market for food products*

CHANGING MARKETS	
MARKET SATURATION	- *Production Levels*
	- *Caloric Intake*
DEMOGRAPHIC CHANGES	- *Ageing Population*
	- *Changes in Household*
	- *Minority Groups*
CONSUMER ATTITUDES	- *Constant Quality*
	- *Product Diversity*
	- *Changes in Quality Perception*
SCIENTIFIC DEVELOPMENTS	

VERTICAL INTEGRATION

Production chains

The traditional approach of the production chain was and to some extent still is characterised by a one-way trafficking from raw material production to end-product and the definition of quality. The starting point of thinking was and in a number of situations still is based on productivity and costs of production. The diverse partners in the production chain often use various definitions of product quality which sometimes are conflicting with a good quality of the end-product. With the occurrence of market saturation and the fast changes in consumer preferences the role of the consumer becomes increasingly important. There is a shift in emphasis from productivity towards quality (2). This kind of development make it necessary to come to a situation of quality control throughout the production chain. To realise this a new approach has to be followed. Starting point is detailed knowledge of the market and consumer preferences. Also the one-way trafficking should be replaced by an iterative approach.

The description of quality I would like to use is:

Quality *is to meet the expectations of the consumer.*

Two important aspects are pointed out here (1). The first is that the consumer is the starting point of thinking. The second is that the consumer doesn't work with detailed product specifications. **The** consumer doesn't exist. Neither does the average consumer exist. There is a specific consumer who, in a specific situation, has a specific need to which the producer can respond.

The consumer buys and eats a product for a number of reasons. To a large extent these reflect the view of the consumer on product quality (***intrinsic factors***). To some extent these reflect his/her view on production methods (***extrinsic factors***).

A food product as such has no quality. A food product has physical properties indicated as intrinsic factors. These factors are translated by the consumer in quality attributes. One example is the texture of a product. From a chemo-physical point of view the texture of a product can be described in terms of cell wall composition and structure. The consumer talks about things such as firmness, mealiness, toughness.

In fact there are a number of other quality attributes such as taste, nutritional value, keepability, healthiness, safety, freshness and appearance. The total of attributes determines the product quality.

Extrinsic factors are linked to aspects of production methods such as the use of pesticides, the type of packaging material used, a specific technology used during production or the use of genetically manipulated micro-organisms for the production of ingredients. They do not directly influence the physical properties of the product but can be determining factors in the choice of the consumer. The total of intrinsic and extrinsic factors determines product acceptance by the consumer. In Figure 1 a scheme is given summarising the definition of product acceptance.

Once the desired product quality is defined the next step is to study how processing conditions can be optimised, what the band-width are for specific quality determinants and how the quality determinants are interrelated. On the basis of this knowledge it will be possible to define the desired raw material properties. In this way raw material composition and properties can be directly linked to end-product quality. Also, this approach makes it possible to make a distinction between the possibilities to reach the defined end-product quality by optimisation of processing conditions and the options for improvement of raw material properties e.g. by new biotechnology. A scheme depicting this approach is given in Figure 2.

For all plant products it holds that during the growing phase there is a build up in product quality and in the post-harvest phase minimisation of quality loss is the primary aim. Adequate knowledge of the build up of product quality during the growing phase with respect to growing conditions, cultivar difference and harvest moment is of great importance for the production of food products with a constant quality. Loss of quality during the post-harvest period is caused by the mere fact that harvesting of product is a form of imposed stress. The conditions during post-harvest storage are essential for the keepability and quality also for processed products. We are dealing with living material that respires and is metabolically active. These metabolic processes determine to a large extent the quality of the product also after processing.

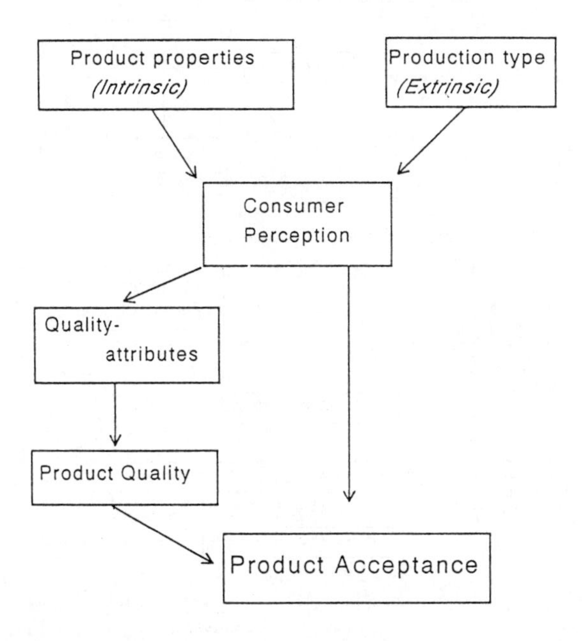

Figure 1 *Relationship between intrinsic and extrinsic factors in consumer acceptance of new food products*

HORIZONTAL INTEGRATION

Healthiness as a quality determinant of food products

The healthiness of a food product can not be seen in absolute terms of good or bad. Almost all food products consist of a complicated complex of components which each individually and all interrelated influence the wellbeing of the consumer. Some of these factors are important because of their caloric value. Others such as the vitamins and minerals contribute to vital processes in the body. Of course there are also a number of components such as the naturally occurring toxicants whose presence is not wanted because they may harm the health of the consumer if consumed in excess amounts.

The wholesomeness of a food product must therefore be considered as a balance between health promoting and health threatening factors.

The increased attention for a physiological role of the so-called non-nutrients such as e.g. some secondary plant metabolites and oligosaccharides has led to an explosion of data concerning the modulation of (human) health by these factors.

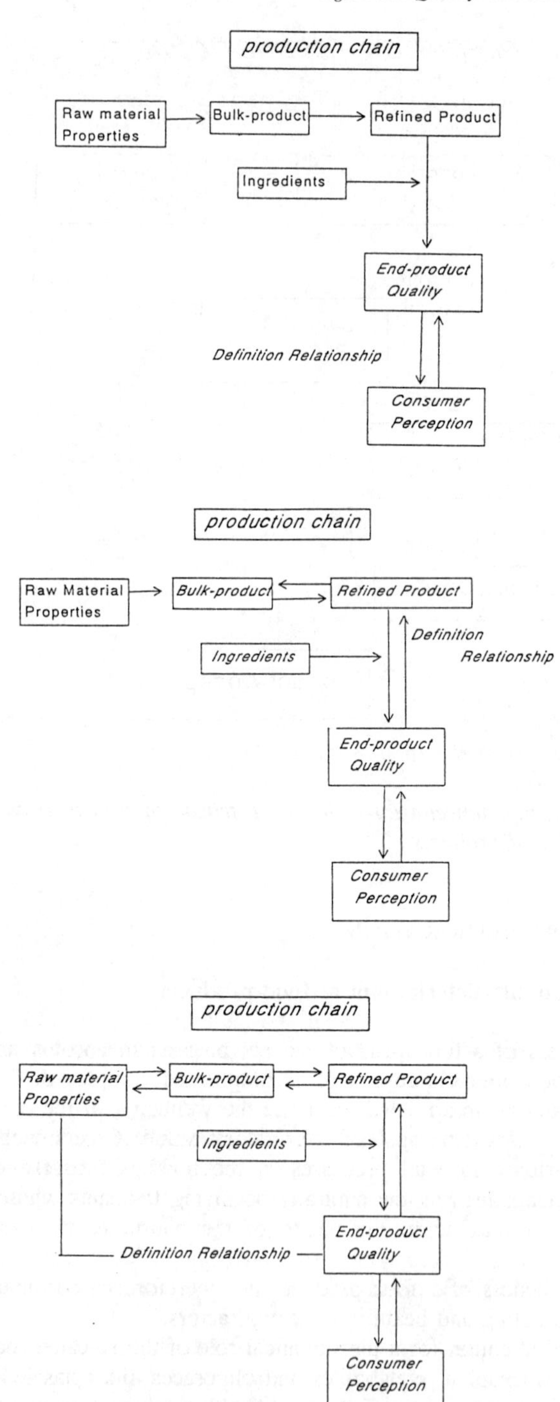

Figure 2 *Integrated approach for quality control in the production chain*

Although it is still too early to draw conclusions about any specific factor it has become clear that potentially there is a large perspective for improvement of the quality of the diet with respect to the prevention of ageing diseases (9). In most western societies we are dealing with ageing populations and a concomitant increase in the incidence of ageing diseases such as cardiovascular diseases and cancer. Improvement of the quality of the diet may result in decreases in the incidence of these types of disease and an improvement in the quality of the life of the elderly.

Another factor in this context is the increasing awareness amongst consumers about the diet and health relationship. There is an increasing demand for "healthy" food products. This has already resulted in the development for specific markets for food products with specific health claims such as nutraceuticals and functional foods (4). Both groups fall within the same category but the difference between them is that functional foods should be consumed in the form of a food product whereas for nutraceuticals the presence of the food matrix is not a prerequisite. It can be consumed in the form of a pill. In my view the latter is not advisable. Consumption in the form of a food product at least to some extent may prevent strongly motivated consumers from consuming too high doses of a specific component.

Additionally the development of this type of food product must be based on claims that meet the following criteria.

- ***They must be based on solid animal and human studies***
- ***The mode of action of the health-promoting effect in humans must be known***
- ***The effectiveness of the food product must be measurable***
- ***Consumption of (combinations of) health-promoting food products must not lead to unwanted side-effects.***

One important aspect of the development of food products with specific health claims is that the health promoting properties are and measurable and quantifiable Consequently health claims need to be linked to the presence of specific components. In that situation the mechanism of protection and the relevant biological activity is known. Systems can then be devised to measure the biological activity and quantify it. I will try to explain this with some examples.

1. Indolylglucosinolates are secondary plant metabolites occurring in cruciferous vegetables such as various types of cabbages, broccoli and brussels sprouts. They are reported to protect against the induction of cancer (8,10). Their mode of action is the potential to induce enzyme activity of the biotransformation system which is responsible for the elimination of toxic compounds. Especially induction of the conjugating enzyme systems such as the glutathione-S-transferases is important in this connection.

2. A second example comes from the flavonoids, like the glycoside routine. Flavonoids occur widespread in plant foods such as onions, grapes and apples. For this group of compounds protective effects have been reported against the incidence of cardiovascular disease (5). The mode of action is supposed to be their capacity to act as anti-oxidant.

In this way they function, in addition to the existing physiological defense systems, as oxygen radical scavengers and provide additional protection.

3. Oligosaccharides are sugar molecules occurring in many different food products such as onions, tomatoes, apples, etc. They act as preventive agents against the occurrence of cancer of the gastro-intestinal tract. The supposed mode of action lies in their capacity to modulate the composition of the micro-flora of the intestine by stimulation of growth of lactic acid bacteria and/or by prevention of attachment of toxic micro-organisms to cellular receptors (5).

4. Also some proteins and peptides are connected to health promoting properties. Lactoferrine is a clear example. It has inflammation inhibitory properties and is already in use in baby food. The milk protein casein has excellent properties in scavenging mutagenic compounds (6).

This list could be much more extensive but that was not the purpose of this exercise. The point to be made here is that as soon as health-promoting claims are linked to the presence of a specific component something can be said about the underlying mechanisms and the relevant biological activity. For all examples given systems can be developed which enable quantification of the relevant biological activity. My proposal is to call this the Health Promoting Capacity (HPC) of a given food product. In the complex matrix of the food product all kind of interaction may occur making it very difficult to predict the effect of a food product on the basis of measurements on individual components. Consequently systems to measure HPC should be based on measurement in the food matrix.

On one hand these systems can be used to optimise processing conditions in product development. On the other hand desired properties in food products can be quantified and comparison between different product will be possible. Health claims of food products can be substantiated and checked.
To be able to do so knowledge is required on the following items.

1. Occurrence	Role in raw material
	Processing behaviour
2. Biological availability	In product
	In consumer
3. Effectiveness	

Very often the non-nutrient components have a physiological role in the plant in which they occur. In order to be able to optimize levels it is important to know what this role is. For example some of the breakdown products of the indolylglucosinolates have a hormonal role in the brassica vegetables.

This means that changes in levels of these compounds may have large consequences for plant growth and development.

The effects of specific amounts of a purified component are rarely exactly comparable to the effects if the product matrix is present. Both synergistic as well as antagonistic effects have been observed. Adequate measurement of the biological availability of the physiologically active principle(s) both in the product matrix and in the consumer (food matrix) should be available.
It should be known what are the relevant doses in the human situation.

Obviously we are only at the beginning of an exciting new period in the diet and health area. Meeting the challenges lying ahead of us in an appropriate way requires close collaboration between various scientific disciplines such as food scientists, nutritionists and toxicologists.

Literature

1. M.G. van den Berg, 'Kwaliteit van Levensmiddelen', Kluwer Quality Handboeken, Deventer, Nederland, 1993.
2. W.E. Deming, 'De Crisis Overwonnen', Kluwer Quality Handboeken, Deventer, Nederland, 1994.
3. G.W. Fuller, 'New Food Product Development: From Concept to Marketplace', CRC series in Contemporary Food Science, CRC Press, Boca Raton, USA, 1994.
4. I. Goldberg, ed., 'Functional Foods: Designer Foods, Pharmafoods, Nutraceuticals', Chapman and Hall, New York, USA, 1994.
5. M.G.L. Hertog, PhD Thesis, Wageningen Agricultural University, 1994.
6. A.J. Morgan, A.J. Mul, G. Beldman and A.G.J. Voragen, *Agro-Food Industry Hi-Tech*, 1992, 35.
7. W.M.F. Jongen, M.A.J.S. van Boekel and L.W. van Broekhoven, *Fd Chem. Toxicol.*, 1987, **25**, 142.
8. W.M.F. Jongen, 'Food and Cancer Prevention: Chemical and Biological Aspects', K. Waldron, I.T. Johnson and G.R. Fenwick, eds, Royal Soc. of Chem., Cambridge, UK, 1993, p. 383.
9. W.M.F. Jongen, Inaugural Speech, Wageningen Agricultural University, 1995.
10. H.G.M. Tiedink, A.M. Hissink, S.M. Lodema, L.W. van Broekhoven and W.M.F. Jongen, *Mutation Res.*, 1990, **232**, 199.

CHANGES IN NITRATE REDUCTASE ACTIVITIES AND THE LEVELS OF NITRATES AND NITRITES IN POTATO TUBERS DURING STORAGE

E. Cieślik* and W. Praznik**

* Department of Human Nutrition, Agricultural University of Cracow, Al. 29 Listopada 46, 31-425 Cracow, Poland
** Center of Analytical Chemistry, Institute for Agrobiotechnology, Konrad Lorenzstr. 20, 3430 Tulln, Austria

1 INTRODUCTION

During the storage of potatoes there occur in them some physical, biochemical and microbiological processes affecting their chemical composition, including the levels of nitrates and nitrites[1]. The enzymatic reductase of those compounds develops with nitrate reductase taking part, which activity depends on storage conditions. The aim of the study was to observe changes in nitrate reductase activity and the nitrate and nitrite levels in the potato tubers during storage at various temperatures.

2 MATERIAL AND METHODS

In the study were used potato tubers of 2 varieties (Aster, Drop) belonging to a group of the very early ones, coming from the experiments conducted in 1993 at the Experimental Station of Varieties Evaluation at Węgrzce near Cracow, Poland. Seed-potatoes on a super elite level were planted in the third decade of April to be harvested after 120 days of growth.

Potatoes were stored for 6 months at two different temperatures: 4^0 and 10^0 C. Material was analyzed on the day of harvest and three times during storage.

In tubers the nitrate reductase activity was determined by the Jaworski method[2], the nitrite level by colorimetric method with the sulfanilic acid and the nitrate level after prior reduction to nitrites, with the metallic cadmium[3].

3 RESULTS AND DISCUSSION

Mean content of dry matter in tubers ranging from 23.2 to 25.2% changed during storage depending on temperature (Tables 1, 2). The levels of dry matter in the tubers stored at 4^0 C were increasing (Table 1) while in those stored at 10^0 C were decreasing (Table 2).

After being harvested the tubers of Aster variety were characterized by a considerably higher activity of nitrate reductase (0.61 µg NO_2^- g d.m.$^{-1}$ h^{-1}) than the tubers of Orlik variety (0.13 µg NO_2^- g d.m.$^{-1}$ h^{-1}) (Tables 1, 2). A dependence of the nitrate reductase activity on potatoe variety was also found by Davies et al.[4]. After a two-month storage period a significant decrease was found in the activity of that enzyme in the tubers of both varieties.

Table 1 *Activities of nitrate reductase (NR) and contents of nitrate and nitrite of potato tubers stored at 4°C*

storage 4° C±2°	*dry matter* %	*NR-activity* mg NO_2^- g d.m.$^{-1}$ h^{-1}	*nitrate* mg kg^{-1}	*nitrite* mg kg^{-1}
Aster				
harvest	25,2	0,6	202	0,9
2 months	25,7	0,3	188	0,0
3 months	27,4	0,5	208	0,1
6 months	32,0	0,7	366	0,0
Orlik				
harvest	23,2	0,1	293	2,6
2 months	25,6	0,1	337	0,8
3 months	24,4	0,2	216	0,1
6 months	25,6	0,4	219	0,0

Table 2 *Activities of nitrate reductase (NR) and contents of nitrate and nitrite of potato tubers stored at 10°C*

storage 10° C±2°	*dry matter* %	*NR-activity* mg NO_2^- g d.m.$^{-1}$ h^{-1}	*nitrate* mg kg^{-1}	*nitrite* mg kg^{-1}
Aster				
harvest	25,2	0,6	202	0,9
2 months	26,1	0,3	235	0,9
3 months	26,6	0,5	280	0,0
6 months	23,7	0,6	184	0,0
Orlik				
harvest	23,2	0,1	293	2,6
2 months	23,8	0,1	269	1,1
3 months	25,0	0,1	186	0,5
6 months	22,5	0,4	167	0,0

After 6 months of storage the intensity of nitrate reductase activity attained higher values than the initial ones. Statistically significant differences were found in that enzyme activity depending on variety, time and temperature. The nitrate and nitrite levels in tubers after harvest were found to be variety-dependent (Tables 1, 2). Before storage significantly fewer nitrates and nitrites were found in the tubers of Aster variety. The levels of those compounds in the tubers of Orlik variety exceeded the admissible values[5], however, they were within the limits stated by other authors[6,7].

Table 3 *Correlation coefficients between the determined components of the potato tubers*

Components	*1*	*2*	*3*
1.nitrate	1.0	0.73**	-0.63**
2.nitrite	-	1.0	-0,52**
3. activity NR	-	-	1.0

As a result of 6-month storage of potatoe tubers some changes in the nitrate level depending on variety and temperature were found. In the Orlik variety tubers a decrease was ascertained in their levels, being more noticeable in tubers stored at 10^0 C. A similar direction of changes in nitrates was noticed in the Aster variety tubers, stored at 10^0 C (Table 2). Whereas, the same tubers when stored at 4^0 C, after 6 months of storage were characterized by a higher level of nitrates than immediately after harvest. Some decreases in nitrate levels in tubers during storage were found in other studies[1].

The levels of nitrites in both varieties were systematically decreasing during storage, and after 6 months none of them was ascertained any more.

Statistical analysis has shown a significant correlation between the determined components of the tuber (Table 3).

References

1. Cieślik E., *Pol. J. Food Nutr. Sci.*,1994, **3/44**, 2, 25.
2. Jaworski E.G., *Biochem. Biophys. Research Communications*, 1971, **6**, 1274.
3. Tyszkiewicz I., *Gospodarka Mięsna*, 1986, **37**, 1.
4. Davies H.V., Ross H.A., Oparka K.J., 1987. *Annals Botany*, 1986, **59**, 301.
5. Rozporządzenie Ministra Zdrowia i Opieki Społecznej z dn. 8. X. 1993 r., Dziennik Ustaw RP, 1993, 104.
6. Grassert V., Vogel J., Neubauer W., Bartel W., *Der Kartoffelbau*,1990, **41**, 398.
7. Cieślik E., Food Chem., 1994, **49**, 233.

KINETICS OF MUSHROOM DISCOLOURATION

Harry J. Wichers and Jeroen van Leeuwen

Agrotechnological Research Institute (ATO-DLO)
Bornsesteeg 59
6708 PD WAGENINGEN, The Netherlands.

1. INTRODUCTION

The production of edible mushrooms (*Agaricus bisporus*) is a rapidly expanding industrial branch (market volume 1 400 000 tonnes in 1993, of which ≈900 000 tonnes are produced in Western Europe with an estimated value of 1.5 billion ECU). France and The Netherlands are leading producers in Europe with volumes of approximately respectively 230 000 and 200 000 tonnes per annum, whereas the UK is responsible for approximately 120 000 tonnes. A large proportion of this production is, either fresh or processed, exported, thus creating a lively international market for mushrooms and mushroom products.

Fresh mushrooms are a sensitive and rapidly deteriorating commodity. Post-harvest quality deterioration is characterized by processes such as stem elongation, cap opening, continued respiration associated with weight losses and discolouration. Senescence phenomena cause a decrease in consumer appreciation and therefore in economic value.

2. CHEMICAL AND BIOCHEMICAL BASIS OF DISCOLOURATION

2.1 Chemical Basis of Discolouration

Enzymatic browning (melanin formation) is part of the senescence process of the mushroom. It results from the oxidation of phenolic substrates by tyrosinases. The principal endogenous substrates for these discolouration reactions are the monohydroxyphenolic compounds L-tyrosine, *p*-amino phenol (pAP) and its condensation product with glutamate, γ-glutamylhydroxy benzene (GHB), all originating from the shikimic acid pathway[1,2] (Figure 1).

The enzymatic end product γ-glutamylbenzoquinone (GBQ) is non-enzymatically cleaved to yield 2-hydroxy-4-iminoquinone, which eventually polymerizes into melanin.

2.2 Biochemical Basis of Discolouration

As indicated in Figure 1, a number of enzymes may be involved in the oxidation of phenolics. Whereas the hydroxylation of compounds containing a phenolic moiety can only be catalyzed by tyrosinases (cresolase activity), the subsequent oxidation of catecholic compounds can theoretically be catalyzed by either tyrosinases (catecholase

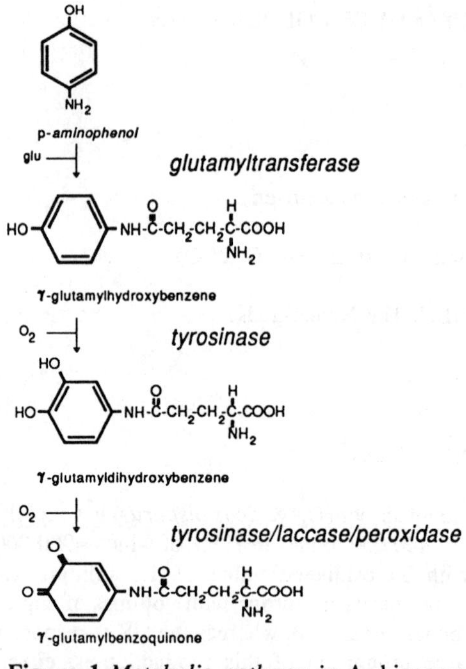

Figure 1 *Metabolic pathway in* A. bisporus *leading to enzymatic discolouration*

activity), laccases or peroxidases.

Enzymatic analysis has indicated that commercial stage fruiting bodies of *A. bisporus* U1 are characterized by very low laccase and peroxidase activities (both approx. 0.01 µkat/gDW) as opposed to their tyrosinase contents (10-15 µkat/gDW, see below). Only in the basal zone of the stipe, which is usually removed upon processing, some laccase activity may be present. It is concluded therefore, that tyrosinase (EC 1.14.18.1) is the enzyme that is responsible for enzymatic discolouration of *A. bisporus* fruiting bodies.

Tyrosinase occurs in mushroom extracts in several isoforms[4] and may occur both in a latent or in an active form[5]. This latent form may *in vitro* be activated by detergents such as SDS or by proteolytic action[5], the latter mechanism possibly having *in vivo* significance as well[6].

3. OCCURRENCE OF PHENOLIC COMPOUNDS IN *A. BISPORUS* FRUITING BODIES

Using the Folin-Ciocalteus reagent[7], the total amount of phenolic compounds in U1 stage 2-3 fruiting bodies was determined, using tyrosine as a reference for quantification.

Analysis of different tissues (cap epidermis, remaining cap tissue, gills, velum and stalk) revealed little variation in levels of phenolics when related to mushroom developmental stage. Highest levels were found in gills (41.3±11.1 µmol/gDW), which was similar to values obtained for the velum tissue (38.3±4.5 µmol/gDW) but significantly higher than values found for cap epidermis, cap tissue or stalk (28.3±4.9, 26.2±5.3 and

20.6±3.8 μmol/gDW respectively). Similar values were reported for strains D649 and U3[8], using the same method for quantification, and for specific GHB-measurements in *A. bisporus* S609[9].

4. OCCURRENCE OF TYROSINASE IN *A. BISPORUS* FRUITING BODIES

The spatial distribution of tyrosinase in U1 stage 3 fruiting bodies was measured and is presented in Figure 2. Values given were measured using L-DOPA as a substrate (catecholase activity) and represent averages of 10 determinations, standard deviations are approx. 10-20%. Values reported concern total detectable activity, as measured in the presence of 0.1% SDS; the active fraction, measured in the absence of SDS is 1-2% of total detectable activity. This active fraction increased, up to 10%, however, when homogenates were prepared with a more vigorous sonication procedure, while not increasing total detectable activity. This observation suggests that *in vivo* at least the majority of tyrosinase activity is latent, and may raise the question as to whether the small active enzyme fraction in *in vitro* preparations is not merely an artifact of the enzyme preparation procedures.

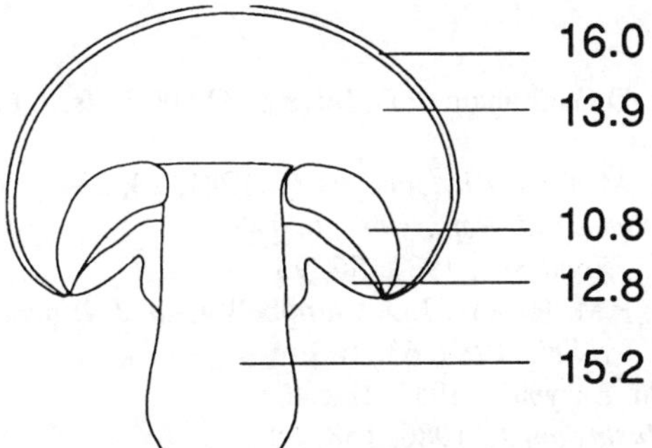

Figure 2 *Tyrosinase activity (μkat/gDW) in different tissues of* A. bisporus *U1 fruiting bodies (stage 3).*

From these measurements it was concluded that the distribution of tyrosinase activity throughout the major tissues of the fruiting bodies is not very variable, except for the gills where the activity was significantly lower than in the other tissues.

Earlier, two isoforms of tyrosinase were isolated from fruiting bodies of *A. bisporus* U1[10]. It concerned two 43 kD monomeric enzymes possessing both cresolase and catecholase activity, with respectively specific activities of 0.1 μkat/mg (tyrosine as substrate) and for the cresolase activity and 17 μkat/mg for the catecholase activity (DOPA as a substrate), indicating that the first enzymatic step that can be catalyzed by tyrosinases may take place at a two orders of magnitude lower speed than the second catalytic step.

5. MUSHROOM DISCOLOURATION

Comparison of levels of phenolics in separate fruiting body tissues with tyrosinase activity levels in the corresponding tissues suggests that complete oxidation of all phenolics present may be a matter of 1 (based on catecholase activities) to 100 seconds (based on the assumed cresolase activities, taking a two orders of magnitude lower activity into account, see paragraph 4). Furthermore, Stüssi and Rast[2] have indicated that the monohydroxyphenolic substrates (GHB, tyrosine) exceed the dihydroxyphenolic substrates (GHB) in levels by at least one order of magnitude. Taken into account these considerations, it is still evident that tyrosinase levels are unlikely to be rate-limiting in the mushroom discolouration process. The above calculations were based on the following assumptions:

- all tyrosinase occurs in the activated form
- the enzyme has free access to its substrates.

This theoretical scenario apparently does not reflect the *in vivo* situation, since mushrooms can easily be stored for at least some days without too large quality deterioration. These observations imply that the discolouration process is under strict regulatory control, by as yet unknown mechanisms, in which process the absolute amount of enzyme does not play a rate-limiting role. Possibly, compartmentalization of enzymes and substrates, association and dissociation of enzyme subunits[11] and proteolytic activation of latent forms may be crucial.

References

1. K. Boekelheide, D.G. Graham, P.D. Mize and P.W. Jeffs, *J. Biol. Chem.*, 1980, **255**, 4766
2. H. Stüssi and D.M. Rast, *Phytochemistry*, 1981, **20**, 2347
3. P.M. Dewick, *Nat. Prod. Rep.*, 1991, **8**, 149
4. W.H. Flurkey, *J. Food Sci.*, 1991, **56**, 93
5. M. Yamaguchi, P.M. Hwang, J.D. Campbell, *Can. J. Biochem.*, 1970, **48**, 198
6. K. Burton, *J. Hort. Sci.*, 1988, **63**, 103
7. J.R. Spies, *Meth. Enzymol.*, 1957, **III**, 467
8. K.S. Burton, *Mushroom J.*, 1986, **158**, 68
9. L. Soulier, V. Foret and N. Arpin, *Mycol. Res.*, 1993, **97**, 529
10. H.J. Wichers, Y.A.M. Gerritsen, G.C.F. van Duijnhoven and K. Recourt, *First International Conference on Mushroom Biology and Mushroom Products*, 1993, Hong Kong, Book of Abstracts p.82
11. K.G. Strothkamp, R.L. Jolley and H.S. Mason, *Biochem. Biophys. Res. Comm.*, 1976, **70**, 519

EXTRUSION-INDUCED MODIFICATIONS TO PHYSICAL AND ANTI-NUTRITIONAL PROPERTIES OF HARD-TO-COOK BEANS

Caroline Karanja*, Gladys Njeri-Maina*, Maria Martin-Cabrejas**, Rosa M. Esteban **; George Grant***; Arpad Pusztai***, Dominique M.R. Georget, Mary L. Parker, Andrew C. Smith and Keith W. Waldron.

Institute of Food Research, Norwich Research Park, Colney, Norwich, UK.
* Kenya Agriculture Research Institute, Nairobi, Kenya.
** Departamento Quimica Agricola, Facultad de Ciencias, University Autonoma de Madrid.
*** The Rowett Research Institute, Bucksburn, Aberdeen AB2 9SB, UK

1 INTRODUCTION

Beans are consumed widely and supply substantial amounts of protein to the diets of the populations of the developing world. Storage-induced development of the hard-to-cook (HTC) defect results in long preparation and cooking times. Grain-legumes also contain antinutritional factors which mitigate against their use. Of particular importance are the lectins which are highly toxic and increase during the development of the HTC defect[1]. Milling, extrusion cooking and incorporation into other food ingredients are options to improve their usage. However few studies of extrusion cooking of legumes have been reported. In most of these, the chief use of legumes in the extrusion cooking process is in blended foods using a mixture of a cereal and a legume[2].

The extrusion cooking of cereals or cereal mixtures is able to produce expanded products: the water in the extruder is superheated so that on leaving the die it flashes off as steam causing expansion of the starchy matrix[3]. The high temperatures involved may have the potential to inactivate antinutrients such as lectins. Furthermore, the process would be expected to alter the physical properties of the flour, such as solubility. This may facilitate the incorporation of bean flour into other products.

Extrusion also exists as a low cost technology (where there is no temperature control and mechanical energy dissipation produces a temperature rise) for use in the developing world to produce weaning foods using locally available ingredients such as maize blended with full fat soy or cottonseed flour[4,5] or black-eyed pea with maize and cassava[2].

An objective of the EC-funded STD-3 project TS3*-CT92-0085 "Improving the cooking quality of grain legumes and their products in Kenya and Cameroon" is to investigate alternative processes for exploiting hard-to-cook (HTC) *Phaseolus* beans. This paper reports a study of the extrusion cooking of milled HTC Canadian Wonder bean flour under different extrusion conditions with a bench top single screw extruder and the assessment of various properties of the extrudate. Water solubility and absorption indices, bulk density and structure of the expanded extrudates and Lectin activity were measured. The results are compared with published studies of extruded cereal flours and extrusion textured proteins.

2 EXPERIMENTAL

HTC beans (*Phaseolus vulgaris* cv Canadian Wonder) with an average cooking time of 300 minutes, were pin-milled and conditioned to water contents of 25 and 30 % (wet weight basis, w.w.b.) by slowly mixing in water and equilibrating over 24 hours. The bean flours were extruded using a Brabender bench top extruder 20DN fitted with a grooved barrel and a 1:1 compression ratio screw. It was flood fed by hand and the throughput measured for each speed setting. The throughput generally increased from 30 to 90 g min^{-1} when the screw speed increased from 49 to 133 rpm. The specific mechanical energy (SME) was measured from the mechanical power consumption and the mass throughput. The extruder barrel was heated in two zones, the first set at 60°C and the second at 100°, 120°, 140°, 160° or 180°C. A circular die of 3mm diameter was used.

The extrudate was cooled, allowed to dry at room temperature and then stored in sealed plastic bags at 1°C. The extrudate bulk density was measured from the weight and dimensions of cylindrical pieces. The water absorption and solubility indices (WAI and WSI) were measured[6] in which the ground product was suspended in water, stirred and then centrifuged. The supernatant was evaporated to give a dried solids weight which expressed relative to the original dry solids weight is the WSI. The WAI is the weight of gel obtained after removal of the supernatant per unit weight of original dry solids.

Haemagglutinating activity (lectin) was estimated in sodium phosphate extracts of the extrudate or flour[1] by a serial dilution procedure using rat blood cells[7]. The amount of material which caused haemagglutination of 50% of the erythrocytes was defined as that containing 1 haemagglutinating unit (HU). For comparison, values were expressed as HU kg^{-1} meal.

3 RESULTS AND DISCUSSION

3.1 Bulk Density

The results show that speed or throughput has little effect on density (Figure 1). Thermal energy input is dominant and the density falls with increasing barrel temperature as observed with maize[8] and soy protein[9]. The bulk density values are not as low as those of highly expanded cereal extrudates for which extrusion conditions typified by low moisture and high temperature lead to high mechanical energy dissipation, microstructural degradation and consequent high expansion[10].

The moisture content of the flour has a slight effect on the density, the higher moisture leading to the lower density. Lipids, sucrose and gluten all affect expansion in a complex way[11]. Die size, water content, extrusion temperature, screw speed and feed rate also affect expansion and bulk density as described more fully elsewhere[8,11].

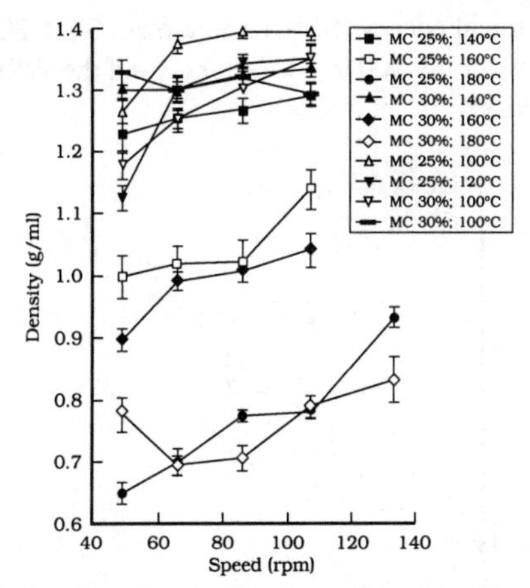

Figure 1 The bulk density of extrudates as a function of screw speed (rpm)

3.2 WSI and WAI

The WSI and WAI values are comparable with those for maize extrudates[4,12]. The WSI generally decreases with increasing barrel temperature (Figure 2). Some exceptions are observed at high temperatures and low speeds for which the WSI is high. Most published data for WSI is for high starch-containing cereals and the WSI increases with increasing temperature [10,12]. However protein denaturation is known to decrease protein solubility as measured by the Protein Dispersibility Index and the Nitrogen Solubility Index. The results here appear to exhibit characteristics of both systems, and may influence protein digestibility.

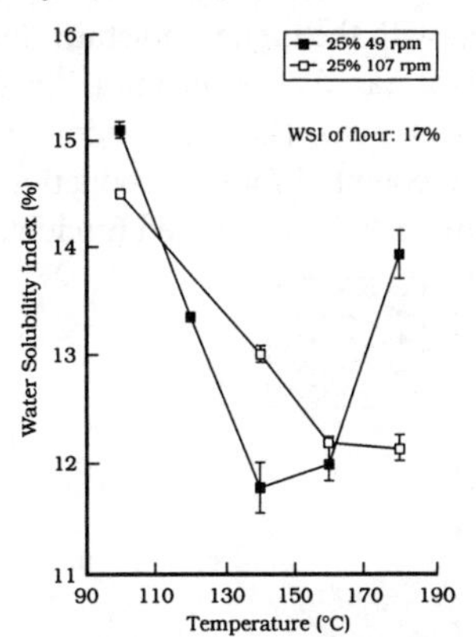

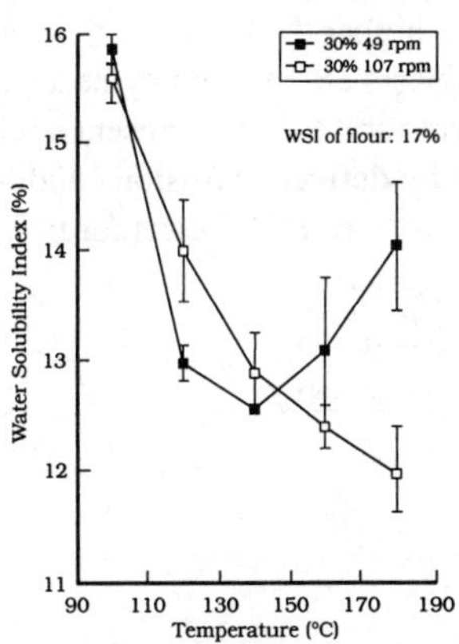

Figure 2 The water solubility index as a function of temperature for moisture contents (% w.w.b.) of (a) 25% and (b) 30% at extremes of screw speed

The WAI increases with increasing temperatures and decreasing moisture content (Figure 3). A similar response was reported for extruded soybean[9,13] and for extruded maize[6,10,12].

Microstructural changes in starchy materials are accompanied by increased solubility such that WSI increases with increasing specific mechanical energy[10]. The WAI of extruded maize showed a maximum with increasing WSI and was due to an increasing proportion of gelatinised granules reaching complete gelatinisation at the WAI peak, although this study covered a specific mechanical energy (SME) range from 50 to 200 Wh kg^{-1}. In the present study the range of SME was only 100 to 135 Wh kg^{-1} and the WSI and WAI were inversely related under most conditions.

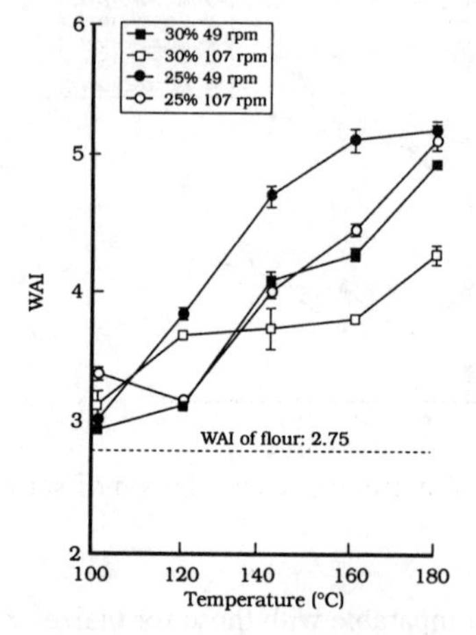

Figure 3 The water absorption index as a function of temperature for moisture contents (% w.w.b.) of 25% and 30% at extremes of screw speed

3.3 Lectin activity

The Lectin activity falls dramatically for temperatures of 120 °C and greater (Figure 4). At an extrusion set temperature of 100 °C the activity falls with decreasing screw speed; this is probably due to the increased residence time. Interestingly at temperatures of 120°C and higher the lowest activities are obtained with the higher moisture flour. Trypsin inhibitors are reduced by heat treatment and increased residence time[14] and this has been corroborated in the present study (results not shown). The inactivation of antinutrients by defined extrusion conditions gives the potential for the production of non-toxic flour from HTC beans for utilisation in cooked and non-cooked products.

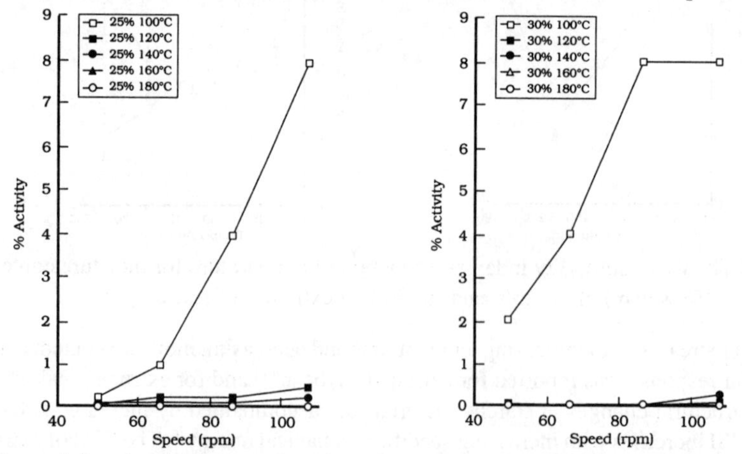

Figure 4 Lectin activity as a function of screw speed at different temperatures for moisture contents of a) 25%, b) 30%.

4 CONCLUSIONS

1) Bulk density of HTC bean flour was reduced and expansion increased by extrusion at higher temperatures; this was modulated slightly by moisture content, but throughput had little effect.
2) WSI generally decreased with increased extrusion temperature, and was probably influenced by protein denaturation. Conversely, WAI exhibited an increase with increased extrusion temperature
3) Lectin antinutrient activity was inactivated at higher temperatures (>120°C) and longer residence times. This gives the potential for producing non-toxic bean flour for use in both cooked and uncooked products.

5 REFERENCES

1. M. Martin-Cabrejas, R.M. Esteban, K.W. Waldron, G. Maina, G. Grant, S. Bardocz and A. Puztai. Hard-to-cook phenomenon in beans: changes in antinutrient factors and nitrogenous compounds during storage. J. Sci. Food Agric. (in press).
2. M.R. Molina, R. Bressani and L.G. Elias. Food Technology, 1977, **31**, 188.
3. A.C. Smith, in 'Food Extrusion Science and Technology', J.L. Kokini, C.-T. Ho and M.V. Karwe (eds.), Marcel Dekker, New York, 1992, p. 573.
4. J.M. Harper. 'Extrusion of Foods', CRC Press, Florida, 1982.
5. K. Lorenz, G.R. Jansen and J.M. Harper. Cereal Foods World 1980, **25**,161.
6. R.A. Anderson, H.F. Conway, V.F. Pfeifer and L.E.J. Griffin. Cereal Sci. Today 1969, **14**,4.
7. G. Grant in 'Toxic substances in crop plants', J.P.F. DeMello, C.M. Duffus and J.H. Duffus (eds.) Royal Society of Chemistry, Cambridge, 1991, p. 49.
8. S.I. Fletcher, P. Richmond and A.C. Smith. J. Food Eng., 1985, **4**, 291.
9. D.B. Cumming, D.W. Stanley and J.M. DeMan. Can. Inst. Food Sci Tech. J. 1972, **5**, 124.
10. A.R. Kirby, A.-L. Ollett, R. Parker and A.C. Smith. J. Food Eng. 1988, **8**, 247.
11. P. Colonna, J. Tayeb and C. Mercier, in 'Extrusion Cooking', C. Mercier, P. Linko and J.M. Harper (eds.), AACC, St Paul, Minnesota, 1989, p. 247.
12. C. Mercier and P. Feillet. Cereal Chem. 1975, **52**, 283.
13. J.M. Aguilera and F.V. Kosikowski. J. Food Sci., 1976, **41**, 647.
14. G.C. Mustakas, W.J. Albrecht, G.N. Bookwalter, J.E. McGhee, W.F. Kwolek and E.L. Griffin. Food Technol. 1970, **24**, 1290.

ACKNOWLEDGMENTS

The authors wish to thank the European Communities for their financial support. This work was also partly financed by the BBSRC, and by a SOAFD to Dr Pusztai.

INFLUENCE OF MILLING, CONCENTRATION AND TEMPERATURE DURING SOAKING ON SOLUBLE CARBOHYDRATE CONTENT AND TRYPSIN INHIBITOR ACTIVITY IN LENTIL SEEDS

J. Frias[1], C. Sotomayor[2], C. Diaz-Pollan[2], C.L. Hedley[1], R.G. Fenwick[3] and C. Vidal-Valverde[2]

[1]Dept of Applied Genetics, John Innes Centre, NR4 7UH Norwich, U.K.
[2]Instituto de Fermentaciones Industriales, C.S.I.C., Juan de la Cierva 3, E-28006, Madrid, Spain.
[3]Dept. of Food Molecular Biochemistry, Institute of Food Research, NR4 7UA Norwich, U.K.

INTRODUCTION

Lentil, as a pulse crop, is a very important component of tropical agriculture and is one of the best and cheapest sources of vegetable protein[1]. Lentil seeds provide a highly nutritious and protein-rich food, containing about 25% protein, 56% carbohydrate and 2% fat.

Although lentils are considered to be one of the most nutritious pulses, they contain several antinutritional factors which limit their consumption[2]. Included among these are trypsin inhibitors, which inhibit the proteolytic activity of trypsin and chymotrypsin, and can lead to a reduction in the availability of amino acids and to a decrease in animal growth[3]. The raffinose family of oligosaccharides, so called α-galactosides, while not being toxic are implicated in the production of flatulence, which is one of the major reasons for the limited human consumption of pulses[4,5,6].

Several procedures have been proposed to achieve a higher nutritional quality for legume meals by removing antinutritional compounds. Soaking is used as a pre-treatment to remove soluble antinutritional factors, which can be eliminated by discarding the liquid follows soaking. Vidal-Valverde et al.[7] reported that some metabolic reactions take place during soaking of lentils, which affect the content of the α-galactosides and of the monosaccharide and disaccharides. This process could be due not only to leaching but also to the metabolic mobilisation of sugars that might have been produced during the soaking procedure. Trypsin inhibitor activity (TIA), however, did reduce appreciably when whole lentil seeds were soaked[8]. The aim of this present study is to provide information about the effect of milling, temperature of soaking and concentration of the lentil flour suspension during the soaking procedure on the content of fructose, glucose, α-galactosides and trypsin inhibitor activity.

MATERIAL AND METHODS

Sample preparation.- Lentil (*Lens culinaris* var. vulgaris, cultivar Magda 20) seeds were used in all soaking experiments. Whole lentil seeds and milled lentil flour (0.25-0.05 mm) were soaked in distilled water at concentrations of either 80g/L or 250g/L, and a temperature of either 28°C or 42°C for 60 min during day light. After this time soaked samples and soaked solution were freeze-dried and kept for analysis. Each treatment was performed in duplicate.

Trypsin inhibitor analysis. Trypsin inhibitor activity (TIA) was determined using the method described by Kakade *et al.*[9] as modified by Valdebouze *et al.*[10]

Determination of monosaccharides, disaccharides and α-galactosides. The extraction and quantification of monosaccharides (glucose, fructose) disaccharides (sucrose) and oligosaccharides (raffinose, ciceritol and stachyose) were carried out using the method described by Frias *et al.*[11]

Statistical analysis: Multifactor analysis of variance was applied to the data using Statgraphics Statistical Graphics System Software ver. 5.

RESULTS AND DISCUSSION

Changes in monosaccharide, disaccharide and α-galactoside content during soaking of lentils as whole seed or milled flour, at 28°C or 42°C of temperature and at concentrations of either 80g/L or 250g/L are shown in Table 1. The presence of glucose was observed in every soaking experiment at a content ranging from 0.06 to 0.17%, although this monosaccharide was not detected in raw lentil seeds.

Sample milling played an important part in elimination of α-galactoside and reductions in raffinose, stachyose, ciceritol, and sucrose were higher when milled lentils were soaked than those in soaked lentil seeds.

Temperature had a major effect on the degradation of α-galactosides. In general, soaking experiments carried out at 42°C had greater sucrose and α-galactoside (raffinose, ciceritol and stachyose) losses than those found at 28°C. The exception was soaking experiments carried out with milled lentils at 42°C and 250g/L, which resulted in a large increase in fructose, sucrose and raffinose.

Sample concentration did not bring about changes when whole lentils were soaked, however, milled lentils soaked at a concentration of 250g/L and 42°C gave a large decrease in ciceritol and stachyose and a noticeable increase in fructose, sucrose and raffinose.

It has been shown previously that soaking modifies the soluble sugar content in lentils and that this effect was due not only to leaching but also to metabolic reactions[7]. In this paper we have shown that greater losses are observed when soaking is performed with milled lentil seeds. This may be due to an increased access to enzymes to the major sugars. α-galactosidases are involved in the degradation of the α-galactosides by splitting off terminal galactosyl residues. α-galactosides can also be degraded by β-fructofuranosidase (invertase). Some authors have reported that the optimum temperature for α-galactosidase activity is 40°C in *Cicer arietinum*[12,13], 45-50°C in *Vigna unguiculata*[14,15], and 55° in *Phaseolus vulgaris*[16] . Our results are in agreement with these reports, since maximum α-galactoside elimination was obtained when soaking was carried out at 42°C.

Table 1 *Changes in monosaccharide, disaccharide, α-galactoside content and TIA following soaking of whole or milled lentil seeds.*

Samples	Non-α-galactosides*			α-galactosides*				TIA**
	Fructose	Glucose	Sucrose	Raffinose	Ciceritol	Stachyose	Total	
Raw lentil	0.09 ± 0.01^{a}	ND	3.12 ± 0.01	0.55 ± 0.02^{a}	2.36 ± 0.01	4.01 ± 0.02	6.92 ± 0.02	5.12 ± 0.15
Soaking the whole seed at 28°C, 80g/L	0.11 ± 0.02^{ab}	0.08 ± 0.02^{a}	3.34 ± 0.04^{a}	0.52 ± 0.05^{a}	2.09 ± 0.03	3.51 ± 0.03	6.12 ± 0.04	4.01 ± 0.37^{a}
Soaking the whole seed at 42°C, 80g/L	0.15 ± 0.01^{c}	0.06 ± 0.01^{abc}	2.85 ± 0.07	0.32 ± 0.03^{b}	1.79 ± 0.06^{a}	3.09 ± 0.06^{a}	5.20 ± 0.05^{a}	3.71 ± 0.14^{b}
Soaking the whole seed at 28°C, 250g/L	0.11 ± 0.02^{ad}	0.08 ± 0.02^{ab}	3.31 ± 0.04^{a}	0.52 ± 0.02^{a}	2.03 ± 0.04	3.43 ± 0.03	5.98 ± 0.04	4.58 ± 0.22^{c}
Soaking the whole seed at 42°C, 250g/L	0.13 ± 0.02^{bcd}	0.08 ± 0.01^{ac}	2.55 ± 0.05^{b}	0.31 ± 0.02^{bc}	1.84 ± 0.03	3.09 ± 0.03^{a}	5.24 ± 0.06^{a}	3.79 ± 0.20^{b}
Soaking the lentil flour at 28°C, 80g/L	0.13 ± 0.02^{bcd}	0.06 ± 0.01^{b}	2.59 ± 0.06^{b}	0.29 ± 0.02^{c}	1.89 ± 0.06	2.43 ± 0.06	4.61 ± 0.09	3.99 ± 0.18^{a}
Soaking the lentil flour at 42°C, 80g/L	0.14 ± 0.02^{bc}	0.06 ± 0.01^{b}	2.65 ± 0.06	0.30 ± 0.03^{bc}	1.70 ± 0.05	2.21 ± 0.03	4.21 ± 0.07	3.64 ± 0.18^{b}
Soaking the lentil flour at 28°C, 250g/L	0.35 ± 0.05	0.16 ± 0.01^{d}	3.04 ± 0.02	0.40 ± 0.02	1.78 ± 0.06^{a}	3.07 ± 0.02^{a}	5.26 ± 0.08^{a}	4.52 ± 0.25^{c}
Soaking the lentil flour at 42°C, 250g/L	0.61 ± 0.03	0.17 ± 0.03^{d}	3.77 ± 0.04	0.72 ± 0.03	1.58 ± 0.03	2.76 ± 0.02	5.06 ± 0.07	3.41 ± 0.11

* % dry matter. ** Trypsin inhibitor units/mg dry matter. Values are the mean of four determinations of each replicate ± standard deviation. The same superscript in the same column means no significant difference ($P \leq 0.05$).

Variation of TIA during the soaking process is also shown in Table 1. Soaking treatments, in general, produced a significant decrease on TIA and reductions ranging between 11% and 33% were found. No significant differences ($P \leq 0.05$) were obtained between soaking experiments carried out with whole or with milled lentils, except for experiments on milled seeds at 42°C and 250g/L, where the maximum reduction in TIA was found (33%).

Temperature and sample concentration had a key effect. A reduction of 26-33% in TIA was obtained for both 80g/L and 250g/L carried out at 42°C, while soaking at 28°C and 80g/L caused a 22% decrease in TIA and at 28°C and 250g/L produced a reduction of 11-12% TIA. However, when soaking was carried out with milled lentils at 42°C a greater loss of TIA was obtained at 250g/L (33%) than at 80g/L (29%).

Some other authors have also observed TIA reductions in lentils during soaking. Vidal-Valverde *et al.*[8] reported appreciable changes on TIA after 9h soaking (11% loss), and Batra et al.[17] found that soaking lentil seeds for 24 hours resulted in a 58-66% decrease in TIA.

ACKNOWLEDGMENTS

This work has been supported by the Spanish Comision Interministerial de Ciencia y Tecnologia ALI-94-1419-E. J. Frias is grateful to EU for financial support through an individual bursary (AIR3-BM93-1118).

REFERENCES

1. G.P. Savage. *Nutr. Abs. Rev. (Serie A)*, 1988, **58**, 319.
2. D.K. Salunkhe and S.S. Kadam. *CRC Handbook of World Food Legumes: Nutritional Chemistry, Processing Technology and Utilization.* CRC Press, Inc. Boca Raton, Florida, 1989.
3. I.E. Liener and M.L.Kakade. Protease inhibitors. In: *Toxic constituents of plant Foodstuffs.* I.E. Liener Eds. Academic Press, New York, 1980, pp 7.
4. E. Cristofaro, F. Mottu and J.J. Wuhrmann. *57th Annual Meeting Am. Assoc. Cereal Chem.* 1972, **2**, 102.
5. D.H. Calloway, C.A. Hickey and E.L. Murphy. *J. Food Sci.*, 1975, **36**, 251.
6. K.R. Price, J. Lewis, G.M. Wyatt and G.R. Fenwick. *Die Nahrung*, 1988, **32**, 609.
7. C. Vidal-Valverde, J. Frias and S. Valverde. *J. Food Protection*, 1992, **55**, 301.
8. C. Vidal-Valverde, J. Frias, I. Estrella, M.J. Gorospe, R. Ruiz, J. Bacon. *J. Agric. Food Chem.* 1994, **42**, 2291.
9. M.L. Kakade, J.J. Rackis, J.E. McGhee, G. Puski, G. *Cereal Chem.* 1974, **51**, 376.
10. P. Valdebouze, E. Bergeron, T. Gaborit, J. Delort-Laval. *Can. J. Plant Sci.* 1980, **60**, 695.
11. J. Frias, C.L. Hedley, K.R. Price, R.G. Fenwick and Vidal-Valverde, C. *J. Liq. Chrom.*, 1994, **17**, 2461.
12. B. Dopico, G., Nicolas, E. and Labrador. *J. Plant Physiol.*, 1991, **137**, 477.
13 Y. Mittal and C.B. Sharma. *Plant Sci.*, 1991, **77**, 185.
14 S.R. Alani. *Dissertation Abstract International-B*, 1989, **49**, 4087.
15 M.M. King, *Dissertation Abstract International-B*, 1987, **48**, 1200.
16 V.L.S. Baldini, I.S. Draetta, Y.K. Park. *J. Food Sci.*, 1985, **50**, 1766.
17 V.I.P. Batra, R. Vasishta, K.S. Dhindsa. 1986. *J. Food Sci. Technol.* **23**, 260.

QUALITATIVE CHANGES IN SOY PROTEINS IN THE PRODUCTION OF SOYA CHEESE.

Y. Zhang, J. Kershaw and P. Ainsworth.

Department of Food and Consumer Technology
Manchester Metropolitan University
Manchester M14 6HR

1 INTRODUCTION

In the production of a wholly non-dairy cheese from soya milk, proteolysis during ripening plays an important role in development of flavour and texture, as it does in dairy cheese. The type and quantity of proteins present in the curd, and the quaternary structure of these proteins will affect proteolysis and the resulting quality of the soya cheese analogue. The effect on soy proteins of soya milk production by an industrial process was investigated by poly acrylamide gel electrophoresis (PAGE). Heat treatment is required during the industrial soya milk production process to inactivate lipoxygenase (LOX) which can produce beany and grassy flavours in the milk, and to destroy trypsin inhibitor protein (TI), an anti-nutritional factor.

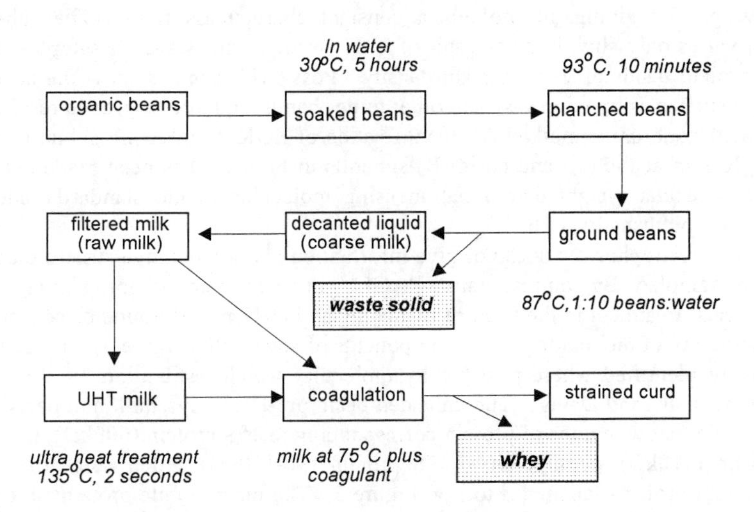

Figure 1 *Flow Diagram of Soya Milk Process and Curd Production*

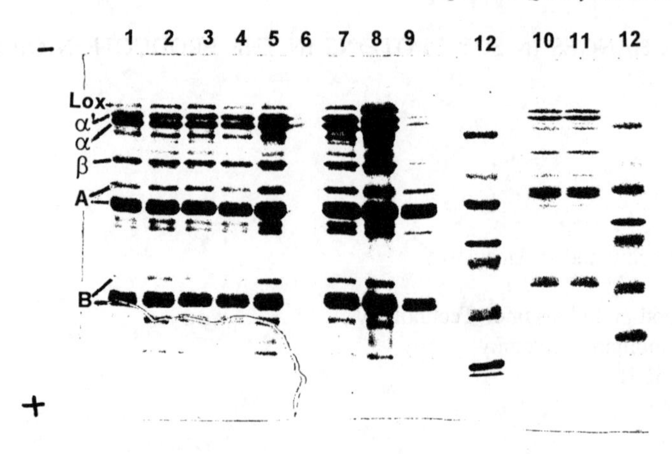

Figure 2 *SDS-PAGE of samples from soya milk production process. Lanes 1 & 2: filtered milk (19μg protein), lanes 3 & 4: decanted milk (20μg protein), lane 5: ground beans (35μg protein), lane 6: waste water, lane 7: waste solid (26μg protein), lane 8: blanched beans (110μg protein), lane 9: soaked beans (119μg protein) lanes 10 & 11: UHT milk (17μg protein), lane 12 molecular weight standards.*

2 RESULTS

2.1 Effects of soya milk production process on soya proteins

Heat treatment occurs at the blanching and ultra heat treatment (UHT) stages of soya milk production (Figure 1). Samples from all stages of the process were analyzed by discontinuous denaturing electrophoresis (SDS-PAGE) in 12.5% gels by the method of Laemmli [1] (Figure 2), which denatures proteins and separates sub-units on the basis of molecular weight by giving all proteins a constant charge:mass ratio. The sub-unit composition varies only slightly as a result of UHT, prior to this stage all samples show 22 discreet protein bands of varying stain density. Post UHT samples show the absence of the band identified as lipoxygenase and other minor bands, and also improved resolution of glycinin acidic sub-units, marked A. Identification of LOX, conglycinin subunits α, α' and β, and glycinin acidic (A) and basic (B) subunits in Figure 2 has been made here on the basis of molecular weight determination using molecular weight standards, and by comparison with published results[2,3].

Denaturing electrophoresis does not give information about the native structure of the proteins in the samples. By contrast, native PAGE[4,5] in gels of varying pore sizes (Figure 3) shows that heat treatment in the form of blanching and UHT has profound effects on the quaternery structure of the major protein components of soya milk (Figure 3). These have been tentatively identified where possible by molecular weight estimation by a method modified from Bryan[6] and Davis[5]. The unheated bean, pre-UHT soya milk and post-UHT soya milk samples show regions of protein corresponding to 15S protein (600kD), glycinin or 11S protein (290kD), conglycinin or 7S protein (90-120kD) and a range of lower molecular weight proteins labelled d to i on Figure 3. The most mobile proteins, h and i, are only seen in the unheated bean sample, while proteins labelled f and g are only seen in soya milk samples following heat treatment by blanching or UHT. Proteins labelled d and e are seen in all three samples; the molecular weight of d has been estimated at 93kD.

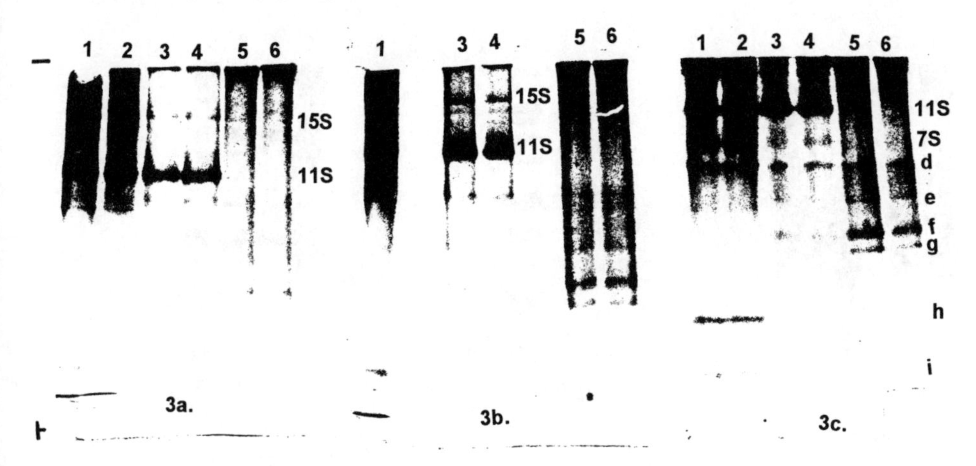

Figure 3 *Native PAGE of samples from soya milk production process.*
3a) 6% gel. Lanes 1 & 2: soaked beans (57μg protein), lanes 3 & 4: pre-UHT filtered milk (61μg protein), lanes 5 & 6: UHT milk (50μg protein).
3b) 7% gel. Lane 1: soaked beans (57μg protein), lanes 3 to 6 as 3a).
3c) 8% gel. Lanes 1 to 6: as 3a).

The most important effect of UHT is seen in the complete lack of native 15S, glycinin (11S), and conglycinin (7S) proteins in the UHT milk sample in Figure 3. In addition this sample shows a higher concentration of proteins with very low electrophoretic mobility, even in the large pore 6% gel, indicating protein complexes of very high molecular weight (HMW), in excess of 750-1,000kD. The possibility of low mobility due to positive charge on the protein was eliminated.

UHT soya milk was treated with mercaptoethanol (ME) to reduce di-sulphide bonds, and the effect can be seen in Figure 4. Di-sulphide reduction caused a decrease in the concentration of HMW protein complex at the top of the gel, and an increase in the concentration of more mobile, lower molecular weight proteins, particularly in the region of proteins d, e, f, and g. There was no protein corresponding to glycinin in this ME treated milk. These results indicate that di-sulphide bonding is involved in the structure of the HMW protein complexes formed as a result of UHT processing. Other workers have found that 11S and 7S proteins denature to subunits on heating, and subsequently reassociate to form HMW complexes[7,8]. As glycinin (11S) has 6 di-sulphide and 11 sulphydryl groups per mole it is not unlikely that di-sulphide bonding would be involved in subunit association in such complexes, as it is in native glycinin structure.

2.2 Effect of Heat Treatment and Coagulant on Soya Curd

Comparison of curds coagulated by glucono-delta lactone (GDL) or by Ca^{2+} revealed differences in protein content, texture, firmness and water retention[9]. Curds and wheys of pre- and post-UHT soya milk coagulated with GDL or Ca^{2+} were analyzed by SDS-PAGE (Figure 5). There were no differences observed between protein composition of milks and the curds formed from them. The type of coagulant did not affect sub-unit composition of the curd, however GDL precipitated more protein from both milks as shown by the concentration of sub-units in the wheys. This was confirmed by Kjeldahl protein assay.

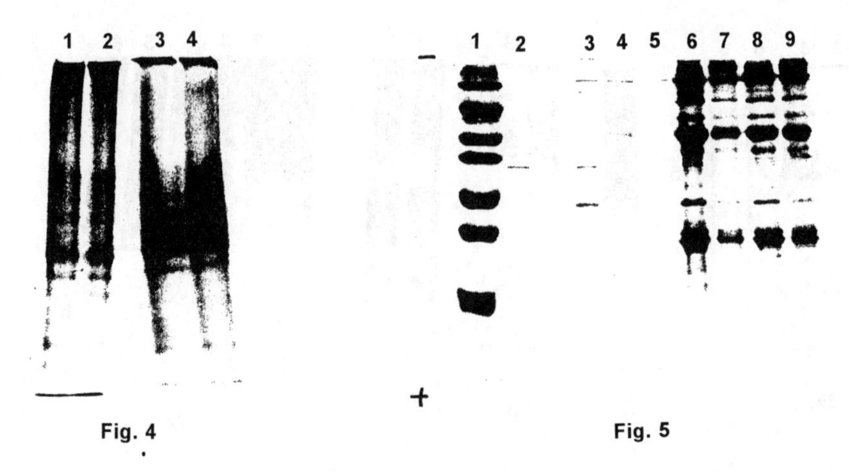

Figure 4 *Native PAGE of UHT soya milk treated with mercaptoethanol (ME). Lanes 1 & 2: UHT milk (50µg protein), lanes 3 & 4: UHT milk treated with 5mM ME.*

Figure 5 *SDS-PAGE of soya curd and whey from pre- and post-UHT milk coagulated with GDL or Ca^{2+}. Lane 1: molecular weight standards, lane 2: pre-UHT whey from GDL curd, lane 3: UHT whey from GDL curd, lane 4: pre-UHT curd from Ca^{2+} curd, lane5: UHT whey from Ca^{2+}curd, lane 6: pre-UHT GDL curd, lane 7: UHT GDL curd, lane 8: pre-UHT Ca^{2+}curd, lane 9: UHT Ca^{2+}curd.*

3 CONCLUSIONS

UHT has little effect on the sub-unit composition of proteins in soya milk, but significantly alters the arrangement of the subunits in the quaternary structure. Notably native glycinin, 30-35% of soya bean protein by weight, is not present in UHT soya milk, and there is an increase in the concentration of protein complexes of high molecular weight in excess of 750kD. Di-sulphide bonding is probably involved in the formation of these complexes.

Coagulant used in curd production affects the quality of the curd in terms of texture and amount of protein precipitated, but does not result in differences in the type of protein sub-units present in the curd.

References

1. U. K. Laemmli, 1970, *Nature (London)*, **227**, 680.
2. S. Sathe, G. Lilley, A. Mason and E. Weaver, 1987, *Cereal Chem.*, **64**, 380.
3. S. Iwabuchi and F. Yamauchi, 1987, *J. Agric. Food Chem.*, **35**, 205.
4. L. Ornstein, 1964, *Ann. NY Acad. Sci.*, **121**, 321.
5. B. J. Davis, 1964, *Ann. NY Acad. Sci.*, **121**, 404.
6. J. K. Bryan, 1977, *Anal. Biochem.*, **78**, 513.
7. K. Hashizume, M. Makamura, T. Watanabe, 1974, *Agric. Biol. Chem.*, **39**, 1339.
8. S. Utsumi, S. Damodaran, J.Kinsella, 1984, *J. Agric. Food Chem.*, **32**, 1406.
9. K. Schofield and P. Ainsworth, 1993, *Home Economics and Technology*, **38**, 15.

PLANT PROTEIN FOODS FERMENTED BY YOGHURT CULTURES

B. Tchorbanov

Department of Enzyme Engineering
Institute of Organic Chemistry with Centre of Phytochemistry
Bulgarian Academy of Sciences, Sofia 1113, Bulgaria

1 INTRODUCTION

Lactic acid fermentation is an ancient procedure for preparation and preservation of dairy foods. The mold modified proteins are traditional for East countries and the soybean fermentations are changing rapidly through modern biotechnology[1]. However, products manufactured using molds are not widely accepted.

The cultivation of lactic acid bacteria on non-dairy substrates from plant origin have been developed after the Second War starting by soybean milk as a suitable medium[2]. The investigations on the growth of lactic acid bacteria in soy milks up to 1974 are summarized shortly[3]. After the "soybean" period the studies spread to the other non-dairy substrates as follows:

a) fermentation of vegetables and juices;
b) fermentation of fruit juices;
c) fermentation of plant protein foods.

During the last 10 years the cultivation of yoghurt cultures on non-dairy substrates from plant origin was researched due to the acceptable taste as well as to the healthfulness of the products obtained.(Table 1).

The purpose of this paper is to present our point of view for more completely use of yoghurt cultures fermentation as a new tool for transformation of plant protein foods. We have more of all been interested in: a) using of whey/permeate as medium for normal fermentation; b) carrying out of enzymic hydrolysis to improved the proteins digestibility; c) the reduction of raffinose type sugars in legume seed substrates.

Table 1 Plant Protein Sources Used as Starting Materials for Lactic Acid Fermentation by Yoghurt Cultures

Source	Pretreatment and Additives	Strains	Ref.
Soybean	soybean milk	Symb	3
	glucose + 0.1 % yeast extract	Lbb	4
	pectin, carrageenan	Lbb	5
	vegetable juices	Lbb	6
	soybean milk	Symb, Lbb, Str	7
	papain-induced hydrolysis	Lbb + Str	8
Sunflower seeds	0.4 % sodium caseinate	Lbb + Str	9
Peanut	peanut milk	Lbb	7,10
	glucose	Lbb	10
Sorghum	sorghum milk	Lbb	7
Wheat	amylase and proteinase treatment	Lbb	11
Cowpea	cowpea milk	Lbb, Str	7
Oats	oats milk	Lbb, Str	12
Oats bran	extract	Lbb, Str	13
Oats	whey, subtilisin-induced hydolysis	LBL, Symb	14
Green microalgae	ethanol extraction		
Scenedesmus	subtilisin-induced hydrolysis	LBL	15
Lentil seeds	germination, grinding in whey		
	incubation for alpha-galactosidase action		
	subtilisin-induced hydrolysis		
	fermentation	Symb	16

2 STRAINS

The fermentation of *Lactobacillus delbrueckii* subsp.*bulgaricus* (Lbb) and *Stretococcus salivarius* subsp. *thermophilus* are used for yoghurt and yoghurt type foods preparation. Symbiotic cultures(Symb) are used for production of Bulgarian yoghurt. A strain *Lactobacillus* LBL 4 (LBL) isolated from yoghurt produced in a Bulgarian mountain region manifests a significant level of aminopeptidase activity[17].

3 SUBSTRATES AND FERMENTABLE CARBOHYDRATES

Practically all protein-containing substrates fermented up to now by yoghurt cultures are shown in Table1. Usually the plant materials are suspended in water, homogenized, pasteurized and fermented. Under these conditions the yoghurt cultures grow more slowly showing a slow acid development, especially on legume substrates. The carbohydrates in legume seeds are low molecular weight oligosaccharides such as sucrose and raffinose type sugars. The inability of Lbb to utilize sucrose and alpha-galactosides explains the poor growth and slow acid production.

There are several decisions of the problem raised :

a) addition of carbohydrates fermentable by yoghurt cultures [4,10] ;

b) suspension of the non-dairy substrate in fresh whey giving acceptable conditions for lactic acid fermentation ;

c) hydrolysis of raffinose type carbohydrates by alpha-galactosidase.

A combination of points a) and b) presents the addition of spray dried whey [16] which is rather easier then the application of the fresh whey. In this case the fermentation rate is very closed to the standard dairy process[14]. The application of whey/permeate from cheese manufacture has an additional advantage-the level of volatile is very closed to the data of the standard yoghurt[14]. It is shown that the aminopeptidase activity of LBL strain contributes significantly to the releasing of free amino acids. Further the threonine is transformed into acetaldehyde - the major volatile compound in yoghurt[17].

The hydrolysis of raffinose type sugars can be realized by an addition of microbial enzyme[18] , by germination of legume seeds[19] as well as by a fermentation by microbial strains isolated from the seeds surface [20]. Some lactic acid bacteria of non - yoghurt origin demonstrate a capability to utilize sucrose and alpha($1\rightarrow6$) galactosides [3].

Our experiments on the fermentation of algae protein hydrolysates showed a possibility in some cases to carry out the process without any addition of fermentable sugars[15].

Our studies on the germination of lentil seeds demonstrated a significant level of alpha-galactosidase after the first 24 hours. The hydrolysis process can continue in grounded seeds stopping the germination process [16].

4 ENZYMIC TREATMENTS

Enzymic hydrolysis of food proteins is an appropriate process for pre treatment of substrates before the lactic acid fermentation [8,15].A partial proteolysis (6 - 8 %) results in an improving of the functional properties and the released peptide fragments are without bitter taste. The enzymic treatment of proteins to a higher degree of hydrolysis (over 40-50%) reduces the bitter taste of the peptides to an acceptable level. The products obtained are suitable for preparation of health and therapeutic foods.

In some cases the aim of the enzymic pre treatment is to change the viscosity and to obtain a milk-like medium - e.g. the amylase and proteinase action on oatmeal porridge [14]. In an opposite case the addition of polysaccharides is applied to increase the viscosity [5].

Although the high protein content (about 50 %) in unicellular algae indicates a fairy good quality, the green algae *Chlorella* sp. and *Scenedesmus* sp. are practically indigestible for humans and non-ruminant animals due to their strong cell wall. Various physical and chemical means are known of pre treatment thus enhancing the protein hydrolysis. The digestion of algae, extracted with ethanol, by subtilisin afforded protein hydrolysate in yields of 60-70 % , degree of hydrolysis of 20-22 % and comparatively low bitterness[21]. The hydrolysate obtained was submitted to fermentation with inoculation of LBL strain without addition of any fermentable sugars. After 72 h of incubation the degree of hydrolysis slowly increased up to 32-34 % and the bitterness disappeared completely due to the intracellular aminopeptidase[17]. The product obtained manifested an excellent taste and can be used in high concentrations for preparation of health foods.

The further development of the plant protein foods fermented by yoghurt cultures as novel foods is connected mainly with the medium composition and the enzymic pre treatment ensuring a short fermentation as well as an acceptable taste closed to taste of natural yoghurt.

References

1. H. L. Wang and C. W. Hesseltine, in "Microbial Technology", Academic Press, New York, 1979, 96.
2. C. Gehrke and H. H. Weiser, *J. Dairy Sci.,* 1948, **31**, 213.
3. B. K. Mital, K. H. Steinkraus and H. B. Naylor, *J. Food Sci.*, 1974, **39**, 1018.
4. R. M. Pinthong, R. Macrae and J. Rothwell, *J. Food Technol.*, 1980, **15**, 647.
5. T. Wakana, T. Iwamura and K. Yotsuhashi, Jap. Patent 61 23,977, 1986.
6. A. Fukura, Jap. Patent, 62 14,249, 1987.
7. D. W. Schaffner and L. R. Beuchat, *Appl. Environ. Microbiol.*, 1986, **51**, 1072.
8. J. L. Hernandez, J. Adris, E. DeRank, R. Farias and N. Samman, *J. Am.Oil Chem. Soc.*, 1981, **58**, 510.
9. K. Fujisawa, A. Yokoyama, G. Suzukamo, US Patent, 4,563,356, 1986.
10. D. W. Schaffner, L. R. Beuchat and R. Chiou, *Food Microbiol.*, 1985, **2**, 249.
11. Japan Natural Food Co., Neth. Appl. Patent 76 06076, 1977.
12. I. Marklinder and C. Lonner, *Food Microbiol.,*1992, **9**, 197.
13. H. Salovaara, P. Kontula, A. Nieminen and S. Mantere-Alhonen, in Proc. EUROFOOD CHEM VII "Progress in Food Fermentation", Valencia, Spain, 1993, vol. **1**, 314.
14. B. Tchorbanov, L. Grozeva and B. Gyosheva , *ibid.,* 1993, vol. **1**, 235.
15. B. Tchorbanov and M. Bozhkova, *Enzyme Microb. Technol.,* 1988, **10**, 233.
16. B. Tchorbanov, A. Mincheva and N. Zlatkova (in preparation).
17. L. Grozeva, B. Tchorbanov and B. Gyosheva, *J. Dairy Res.*, 1994, **61**, 581.
18. R. Somiari and E. Balogh, *Enzyme Microb. Technol.*, 1995, **17**, 311.
19. J. Frias, C. Diaz and C. Vidal-Valverde, In Proc.EURO FOOD TOX, Olsztyn, Poland, 1994, vol. **1**, 256.
20. J. Tabera, I. Estrella, R. Villa, J. Frias and C. Vidal-Valverde, *ibid.,* 1994, vol. **1**, 268.
21. B. Tchorbanov, L. Grozeva and B. Gyosheva (in preparation).

EFFECT OF NATURAL FERMENTATION ON PHYTIC ACID AND SAPONINS OF LENTILS

C. Cuadrado, G. Ayet, L.M. Robredo, J. Tabera *, R. Villa *, C. Burbano and M. Muzquiz
Area de Tecnología de Alimentos, SGIT-INIA.
Apdo. 8111, 28080 Madrid, Spain.
* Instituto de Fermentaciones Industriales, CSIC.
Juan de la Cierva 6, 28006 Madrid, Spain.

1 INTRODUCTION

Legume-based fermented foods originated centuries ago in Southeast Asia, the Near East, and parts of Africa. The popularity of legume-based fermented foods is due to desirable changes in the flavour, aroma and appearence of these foods, but there are also improvements in the nutritive value, because partial or complete elimination of antinutitional factors may take place.[1]

Lentils are a rich source of easily available and cheap protein, both for human and animal nutrition, which can complement cereal protein in terms of several essential amino acids. Their acceptability as staple foods is limited because they contain several non-nutritive factors such as saponins and phytic acid.[2,3]

The purpose of the present work was to obtain a flour with higher nutritive quality than the raw legume by natural fermentation of lentils. With this aim we studied the effect of natural fermentation on saponins and phytates content of lentils.

2 MATERIAL AND METHODS

2.1. Fermentation

Lentil seeds (*Lens culinaris* var. vulgaris, cultivar Magda-20, Albacete, Spain, harvested in 1991) were finely ground in a ball mill and sieved, and the 0.050 - 0.250 mm fraction was collected. Suspensions of lentil flour in sterilized tap water were aseptically prepared at concentrations and temperatures set up by the experimental design and were allowed to ferment naturally for 4 days without aeration in a stirred fermentor. Time 0 of fermentation was assigned when the complete suspension was under stirring and controlled temperature. Samples were daily collected and freeze-dried before analysis.

Table 1 *Complete 2^2 factorial design for temperature (X_T) and concentration (X_C) factors.*

Batches	F_1	F_2	F_3	F_4	F_5	F_6	F_7
Temperature(°C)	28	42	28	42	35	35	35
Concentration (g/L)	79	79	221	221	150	150	150

A 2^2 complete factorial design with three replicate centerpoints[4] was used to study the effects of temperature and suspension concentration on the fermentation performance. Experimental conditions for the different fermentation batches are shown in Table 1.

2.2. Methods

2.2.1. pH, pO2, titratable acidity and lactic acid measurement. An Ingold combined electrode connected to a register allowed a continuous measurement of pH. pO_2 was determined with a polarographic electrode, connected to the same register. Titratable acidity was measured by titration with 0.1 M NaOH and expressed as percentage of lactic acid. Lactic acid was determined with a commercial enzymatic kit.

2.2.2. Saponins determination. The extraction and quantification method was as described in Cuadrado et al..[5] The saponins were extracted with methanol and the extract purificated with a column packed with SiO_2-C_8. The qualitative analysis was conducted by thin layer chromatography (TLC) and fast atom bombardment-mass spectrometry (FAB-MS).[6] The quantitative analysis was made by gas liquid chromatography of the corresponding aglycones (sapogenols), released as a result of acid hydrolysis.[5]

2.2.3. Inositol phosphates determination. The sample (0.5g) was extracted with 0.5 M HCl. The chloride extract was purified and concentrated by ion-exchange (SAX) column. The inositol tri- (IP3), tetra- (IP4), penta- (IP5) and hexaphosphate (IP6) were determined by ion-pair C18 reverse phase HPLC and detected by refractive index.[7]

2.2.4. Statistical analysis. Variance analysis was made to study the effect of fermentation process itself by comparing control (0 h) and fermented samples: F_1, F_2, F_3, F_4 and F_x (mean of F_5, F_6 and F_7) on all the compounds studied. Also the significance of time of fermentation was studied by variance analysis for inositol phosphates.

3 RESULTS AND DISCUSSION

3.1. Saponins.

TLC results showed that all the samples of lentils contained soyasaponin I (R_f=0.30 on reversed phase C_{18} and R_f=0.09 on normal phase) and the mass spectra corresponds with that of soyasaponin I standard (m/z 941, 923, 633, 457; -ve mode). These results are in agreement with those reported by Price *et al.*.[8] By using this qualitative information on the presence of individual saponins together with the quantitative data from the gas chromatographic separation and measurement of soyasapogenol B, the total saponin content for each sample was determined.The variance analysis results showed that the level of soyasaponin I and soyasapogenol B were significantly higher in the fermented samples F_2, F_3 and F_4, did not change in F_x and was slightly lower in F_1 than the raw lentil (Table 2).

Table 2 *Soyasapogenol and Soyasaponin I content (mg/g d.w. ± standard error) of raw and fermented (96 hr) lentil seeds.*

Samples	Raw	F_1	F_2	F_3	F_4	F_x
Soyasapogenol	0.34±0.01	0.27±0.01	0.85±0.01	0.97±0.07	0.74±0.04	0.54±0.19
Soyasaponin I	0.69±0.01	0.56±0.02	1.74±0.03	1.98±0.15	1.53±0.09	1.09±0.41

F_x mean value of F_5, F_6 and F_7

Scarce information has been encountered about the effects of cooking and processing on the saponins of legumes although saponins are considered to survive the rigours of cooking and food processing.[9] Our results seem to indicate an apparent increase in the saponin content in the majority of the fermentation conditions used, but these results should be treated with caution as it is unlikely that a synthesis of this compound will take place during the fermentation process. Probably concentration of saponins, by loss of other ingredients in the flour, accounts for this relative increase.

On the other hand, germination at different environmental conditions showed that the saponin level increased during germination, using the same lentil variety.[10] These results indicate that lentil saponins appear to be very resistent in general to reduction by processing.

3.2. Inositol phosphates.

The results obtained for the different inositol phosphates are presented in table 3. The phytates content present in raw seeds is lower than those reported earlier using precipitation methods,[11,12] due to a higher precission in the methodology used. Comparing the total inositol phosphate in raw lentils with that found at the end of the different fermentation processes (96 hr), a significant reduction took place in all of them. Due to the phytate reduction in this processing, the influence of fermentation on inositol phosphates was better studied following the kinetics of the different fermentations. Therefore, pH, pO_2, acidity, lactic acid and the inositol phosphate content of the slurry were analyzed each 24 hr.

Differences between the raw lentil and the control samples (0 hr) of the different fermentations were found in the total inositol phosphates (IP_t) content. IP_t content decreased in all the fermentation processes as a result of decrease of IP_6, the majority compound, and IP_5. Non differences with the control samples were observed in F_1 and F_2 conditions for IP_3 but in the other fermentation conditions the IP_3 content increased. In relation with IP_4 lower content was observed in F_1 and F_5 and higher in F_3 than the control.

The fermentation time produced a significant decrease in the IP_6 content and so in the IP_t content since the begining of the fermentation processes. For all the compounds the minimum levels except for IP_3, were obtained at 2 and 3 days.

In general, the relation meal/distilled water had a higher influence than temperature during fermentation. The highest reduction in inositol phosphate content was achieved in the lowest value of the relation meal/distilled water tested. In F_1 (28 ^{0}C, 79 g/L) a reduction of 66% was reached at 96 hr and in F_2 (42 ^{0}C, 79 g/L) the higher reduction, 63%, was reached at 72 hr. Similar degree of reduction of the total phytates content (65%) was obtained at 120 hr in the natural fermentation of maize.[13]

Acknowledgements. This study was supported by CICYT ALI 91-1092-C02-02.

4 REFERENCES

1. D.K. Salunkhe and S.S. Kadam, "Handbook of world food legumes: Nutritional chemistry, processing technology, and utilization", CRC Press, Inc., Boca Ratón, Florida, 1989, Vol. III.
2. I.E. Liener, "Chemistry and Biochesmistry of Legumes", Oxford and IBH, New Delhi, 1982.
3. M.M. Tabekhia and B.S. Luh, *J. Food Sci.*, 1980, 45, 406.
4. C.K. Bayne and I.B. Rubin, "Practical experimental designs and optimization methods for

chemists", VCH Pub. Inc., Deerfield Beach, Florida, 1986.
5. C. Cuadrado, G. Ayet, C. Burbano, M. Muzquiz, L. Camacho, E. Cavieres, M. Lovon, A. Osagie and K.R. Price, *J. Sci. Food Agric*, 1995, 67, 169.
6. M. Muzquiz, C.L. Ridout, K.R. Price and G.R. Fenwick, *J. Sci. Food Agric*, 1993, 63, 47.
7. C. Burbano, M. Muzquiz, A. Osagie, G. Ayet and C. Cuadrado, *Food Chem.*, 1995, 52, 321.
8. K.R. Price, J. Eagles and G.R. Fenwick, *J. Sci. Food Agric.*,1988, 42, 183.
9. D.E. Fenwick and D. Oakenfull, *J. Sci. Food Agric.*, 1983, 34, 186.
10. G. Ayet, C. Burbano, C. Cuadrado, C. de la Cuadra, L.M. Robredo, A. Castaño, M. Muzquiz and A. Osagie, *J. Food Sci*, In press.
11. A.R. El-Mahdy, Y.G. Moharram and O.R. Abou-Samaha, *Z. Lebensm. Unters Forsch.*, 1985, 181, 318.
12. C. Vidal-Valverde, J. Frias, I. Estrella, M.J. Gorospe, R. Ruiz and J. Bacon, *J. Agric. Food Chem.*, 1994, 42, 2291.
13. Y. Lopez, D.T. Gordon and M.L. Fields, *J. Food Sci.*, 1983, 48, 953.

Table 3 *Inositol phosphates content[a] (g/100g d.w.) of raw and fermented lentil seeds*

Samples	*IP3*	*IP4*	*IP5*	*IP6*	*Total*	*% Reduction*
Raw	-	-	-	-	0.54	
F_1						
0 h	-	0.03	0.05	0.60	0.68	
24 h	0.01	0.03	0.05	0.34	0.40	41
48 h	-	-	-	0.32	0.32	53
72 h	-	-	-	0.30	0.30	56
96 h	-	-	-	0.23	0.23	66
F_2						
0 h	-	0.03	0.05	0.50	0.57	
24 h	0.01	0.04	0.06	0.29	0.41	28
48 h	-	0.02	0.04	0.17	0.24	58
72 h	-	0.02	0.04	0.16	0.21	63
96 h	-	0.02	0.05	0.28	0.37	35
F_3						
0 h	0.01	0.03	0.04	0.40	0.46	
24 h	0.01	0.04	0.02	0.18	0.24	48
48 h	0.02	0.05	0.02	0.21	0.30	35
72 h	0.03	0.05	0.02	0.21	0.32	30
96 h	0.05	0.07	0.04	0.34	0.49	6
F_4						
0 h	0.01	0.02	0.05	0.43	0.51	
24 h	0.03	0.04	0.05	0.21	0.33	35
48 h	0.04	0.03	0.05	0.21	0.34	33
72 h	0.04	0.02	0.05	0.17	0.30	41
96 h	0.04	0.03	0.09	0.30	0.52	2
F_x[b]						
0 h	0.01	0.03	0.05	0.47	0.57	
24 h	0.02	0.01	0.03	0.30	0.37	35
48 h	0.01	-	0.03	0.29	0.33	42
72 h	0.01	0.01	0.05	0.38	0.45	21
96 h	-	-	0.03	0.31	0.34	40

[a] results are mean values of two extracts of each sample analysed in duplicate by HPLC
[b] mean values of F_5, F_6 and F_7

PHYTATE REDUCTION IN BROWN BEANS (*PHASEOLUS VULGARIS* L.)

E-L. Gustafsson and A-S. Sandberg

Department of Food Science
Chalmers University of Technology
c/o SIK, Box 5401
S-402 29 Göteborg
Sweden

1 SUMMARY

In this study[1] we investigated possible means of increasing mineral bioavailability in brown beans (*Phaseolus vulgaris* L.) by degradation of the inhibitor phytate. Brown beans (*Phaseolus vulgaris* L.) were subjected to treatments to evaluate the effects of pH, temperature, $CaCl_2$ and fermentation on degradation of phytate. Soaking was performed at 21°C, 37°C and 55°C and at pH 4.0, 6.0, 6.4, 7.0 and 8.0. Optimal conditions for phytate degradation were pH 7.0 and 55°C. After soaking 4, 8 or 17 hr at these conditions 79%, 87% and 98% of the phytate was degraded, respectively. Fermentation of pre-soaked whole beans resulted in reduction of 88% of phytate after 48 hr.

2 INTRODUCTION

A significant part of the human world population relies on legumes as a staple food for subsistence, particularly in combination with cereals. Legumes are often advocated in Western diets because of their beneficial nutritional effects and because they are a low cost source of protein. Use of legumes in the human diet may increase in less developed regions of the world and also in the Western countries. Therefore, more information is needed about the potential nutrtional implications of legume based diets.

The content of iron in brown beans (*Phaseolus vulgaris* L.) is very high (5 mg/g dried beans). However, the beans also contain high amounts of phytate which make the bioavailability of the iron low. The phytate molecule is negatively charged at physiological pH values and chelates with several mineral elements, including iron, zinc, magnesium and calcium. This forms insoluble complexes thereby making the minerals unavailable for absorption[2]. Phytases hydrolyze phytate to *myo*-inositol and inorganic phosphate via intermediate *myo*-inositol phosphates (penta- to monophosphates)[3]. In beans of *Phaseolus vulgaris* L. (e.g. Navy beans, California Small White bean, Dwarf French bean) studies have shown that phytases had pH optima at 5.2-5.3 and temperature optima at 40-60°C[4-7]. Scott[8] showed completely different pH optima. Scott extracted phytases from nine varietes of *Phaseolus vulgaris* L. (e.g. Pinto beans, Navy beans, Red kidney beans). All phytases had alkaline activity. Our objective was to determine whether the phytase in brown beans had alkaline activity and any effects of soaking, addition of 1mM $CaCl_2$ and fermentation on phytate degradation.

3 MATERIALS & METHODS

3.1 Preparation of the seeds

The beans were either ground directly (bean flour 1) with a coffee-grinder (Kenwood) or pre-soaked 17 hr in 10 parts of 37 °C de-ionized water, dried in a hot air oven (Memmert Model BKE 30) at 55 °C for 24 hr and then ground (bean flour 2). The pre-soaking water was decanted and analyzed for inositol hexa-, penta-, tetra- and triphosphates (IP_6-IP_3) before drying.

3.2 Soaking procedure

Duplicate samples (1 g) of ground beans (flour 1 or 2) were soaked in 10 mL soaking medium. The soaking medium was either 20 mM citrate-buffer pH 4.5 or pH 6.0, de-ionized water or 100 mM tris-buffer pH 7.0 or pH 8.0. $CaCl_2$ was added to one set of samples. 9.5 mL soaking medium and 0.5 mL tannase stemsolution was added to some samples. The samples were incubated in water bath at 21 °C, 37 °C or 55 °C for 4 hr, 8 hr or 17 hr.

3.3 Fermentation

The beans were fermented whole or ground (flour 2) without addition of a starter culture. The whole beans were pre-soaked for 17 hr in 37 °C de-ionized water and then dried for 24 hr at 55 °C prior to fermentation. Duplicate samples (1 g) of ground beans were soaked in 4 mL de-ionized water, covered with parafilm, placed in a 21 °C shaking waterbath and fermented for 24 hr. The temperature was then raised to 30 °C and fermentation was continued for 24 hr or 48 hr. Duplicate samples (2 g) of whole beans were fermented in 8 mL water. After fermentation of whole beans, the water was decanted and analyzed for IP_6-IP_3. Whole, fermented beans were freeze-dried, ground and then analyzed.

3.4 Determination of inositol phosphates

Duplicate samples (0.5 g) of bean flour 1, flour 2 or freeze-dried fermented beans were extracted with 0.5 M HCl (20 mL) for 3 hr. Extracts were centrifuged and the supernatant decanted, frozen overnight, thawed and centrifuged again. For soaked samples 30 mL HCl was added to stop the enzymatic activity and to extract the inositol phosphates. The strength of added HCl was adjusted depending on pH of the soaking medium to provide a pH similar to that when 0.5 M HCl was added to dry samples. The extraction was performed for 3 hr.

After the second centrifugation, an aliquot (15 mL) of the supernatant was evaporated to dryness and then redissolved in 0,025 M HCl (15 mL). The inositol phosphates were separated from the crude extract by ion exchange chromatography according to Sandberg and Ahderinne[9] and Sandberg *et al.*[10].

3.5 Determination of free *myo*-inositol

55 mL ethanol and 0.1 mg of the internal standard pentaerythritol was added to soaked

samples. They were magnetically stirred for 15 minutes and were centrifuged. The supernatants (2mL) were evaporated to dryness and then silylated according to Santa-Maria *et al.*[11]. The silylated samples were analyzed for free *myo*-inositol by capillary gas chromatography.

4 RESULTS

4.1 Effects of time, pH and temperature in soaking experiments

The content of IP_6 and IP_5 in raw material was 12.6 and 1.01 µmol/g dry sample respectively, and the content of IP_4 and IP_3 was not detectable. When bean flour 1 was soaked in de-ionized water at different temperatures (figure 1), the greatest phytate reduction occured at 55°C. After 8 hr incubation the pH dropped gradually from 6.4 to 4.8. Soaking in citrate-buffer pH 4.5 resulted in a small ($p<0.01$) reduction of the phytate that increased with increasing temperature. Soaking in tris-buffer pH 7.0 resulted in a higher degradation of phytate than soaking in pH 4.5 at 55°C ($p<0.001$) or soaking in de-ionized water at 37°C ($p<0.05$) or 55°C ($p<0.001$). Soaking in tris-buffer pH 8.0 resulted in a higher degradation of phytate than soaking in pH 4.5 ($p<0.02$) or soaking in de-ionized water ($p<0.01$) at all temperatures. After 17 hr incubation of bean flour 1 in tris-buffer pH 7.0 at 55°C, the phytate had been degraded by 98% and the sum of IP_6 and IP_5 was 0.45 µmol/g. The content of free *myo*-inostitol in samples soaked for 17 hr at 55°C and pH 6.4, pH 4.5 and pH 7.0 was analyzed. The formation of free *myo*-inositol was greatest at pH 4.5.

Phytate reduction was more extensive when beans were pre-soaked before ground to flour (flour 2) when soaking in de-ionized water at all temperatures ($p<0.01$). When bean flour 2 was soaked for 17 hr in pH 8.0 and at 55°C, the degradation of phytate was 98%, the same that was achieved after soaking flour 1 in the same manner.

4.2 Effect of calcium

Reduction of phytate when incubated at 21°C in tris-buffer pH 8.0 for 4 hr with the addition of 1mM $CaCl_2$ was not significantly different than without the addition.

4.4 Effect of fermentation

Fermentation of flour from beans that had been pre-soaked (flour 2) resulted in a reduction of phytate by 68% after 48 hr. Fermentation of whole pre-soaked beans was more favourable for phytate reduction than fermentation of ground beans ($p<0.01$). 86% phytate was degraded after 48 hr and the fermentation water had a high content of inositol IP3. Fermentation for 72 hr did not increase the phytate reduction.

5 DISCUSSION

5.1 Optimal conditions for bean phytase

Phytase of brown beans (*Phaseolus vulgaris* L.) had optimal activity at pH 7 and

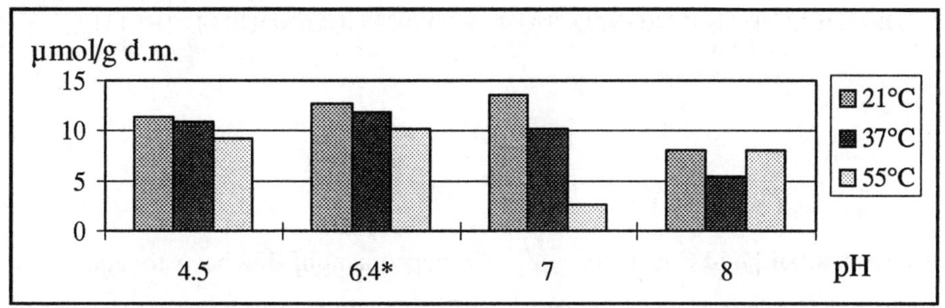

Figure 1 *Phytate content in brown beans soaked for 4h (* = de-ionized water)*

55°C. The phytase also had a high activity at pH 8 and 37°C in agreement with Scott (1991). In other studies pH optima of bean phytases were 5.2 - 5.3 [4-7, 12-13]. However, only one of those studies included neutral pH [5].

IP_4 and IP_3 did not degrade within 17 hr at pH 7 and 55°C but accumulated. Thus the bean phytase active at pH 7 may be specific for IP_6 and IP_5 as substrates or may have been inhibited by IP_4 and IP_3. The small increase in free *myo*-inositol when the bean flour was incubated at pH 7 also suggested an incomplete degradation of inositol phosphates. The content of free *myo*-inositol increased more when the flour was incubated at pH 4.5 and there was no accumulation of IP_4 and IP_3. This indicated that some phytase or phosphatase had acid activity and could use IP_4, IP_3, IP_2 and IP_1 as substrates.The sum of IP_6 and IP_5 inositol after incubation 17 hr in tris-buffer pH 7.0 at 55°C was very low (0.45 µmol/g) while the content of IP_4 and IP_3 was still high (2.33 µmol/g and 10.37 µmol/g, respectively). The contents of IP_6 and IP_5 were low enough that they would not have a strong negative effect on the bioavailability of iron[14]. The effect of IP_4 and inositol IP_3 on iron absorption is not clear.

References

1. E-L. Gustafsson and A-S. Sandberg, *J. Food Sci.*, 1995, **60**, 149.
2. K.E. Weingartner, J.W. Jr. Erdman, *Ill. Res.*, 1978, **20**, 4.
3. GCJ. Irving,Phytase, "Inositol Phosphates. Their chemistry, biochemistry and physiology", DJ. Cosgrove , Elsevier, Amsterdam, 1980, Chapter. 10, p. 85.
4. LN. Gibbins, FW. Norris, *Biochem. J.*, 1963, **86**, 67.
5. R. Becker, AC. Olson, DP. Frederick, S. Kon, MR. Gumbmann, JR. Wagner, *J. Food Sci.*, 1974, **39**, 799.
6. R. Chang, S. Schwimmer, H.K. Burr, *J. Food Sci.* 1977, **42**, 1098.
7. GM. Lolas, P. Markakis, *J. Food Sci.*, 1977, **42**, 1094.
8. J.J. Scott, *Plant Physiol.*, 1991, **95**, 1298.
9. A-S. Sandberg and R. Ahderinne, *J. Food Sci.*,1986, **51**, 547.
10. A-S. Sandberg, N-G. Carlsson, U. Svanberg, *J. Food Sci.*, 1989, **54**, 159.
11. G. Santa-Maria, A. Olano, M. Tejedor, *Chem. Mikrobiol. Technol. Lebensm.*, 1985, **9**, 123.
12. D.M. Gibson, A.H.J. Ullah, *Archives of Biochem. and Biophys.*, 1988, **260**, 503.
13. G.H. Sloane-Stanley, "Biochemists´ Handbook", C. Long, London, 1961, p. 259-262.
14. A-S. Sandberg, U. Svanberg, *J. Food Sci.*, 1991, **56**, 1330.

PRODUCTION OF MACARONI DOUGHS FROM LEGUMINOUS MATERIALS

E. Kovács* and J. Varga**

* Department of Food Chemistry
University College of Food Industry
H-6724 Szeged, Mars tér 7.
Hungary

** Department of Biochemistry and
Food Technology, Technical University
H-1111 Budapest, Műegyetem rkp. 3.
Hungary

1 INTRODUCTION

The protein needs of humans are only partly supplied by macaroni dough products with their 10-12 % protein content. Therefore the biological value of the dough can be increased by enrichment and dietetic products can be made by changing its components.

During the past years more and more attention was paid to producing macaroni doughs on dietetic and carbohydrate basis. Gluten sensetive patients must not consume wheat and rye flour products, because of their gliadin content[1]. PCR reaction can be used to detect gluten sensitivity[2]. The use of leguminous flours in producing macaroni dough is something new in the literature[3-4].

The proteins of leguminous flours are made up of albumin and globulin. So a flexible network begins to form, but its quality is not as good as that of gluten protein. The quality of dough is determined by the quality of raw material[5]. There are several possibilities for improving the quality of products: to change technology or application of additives as emulsifiers as well as to change the composition of cooking water[6-11].

The emulsifiers interact with the proteins, carbohydrates and lipids. The proteins establish hydrophobe, hydrophil and hydrogen-bridge interactions with emulsifiers. In this way they help to form a flexible network in the leguminous basis doughs. The emulsifier-carbohydrate interaction is significant which can be a complex forming with amylose or a hydrogen-bridge with the amylopectin[12]. The quantity of the complex with amylose is depending from the amount of amylose fraction. In these systems an emulsifier-protein-carbohydrate-lipid complex can be expected[13]. The rate of the individual interactions is a function of components of the given sample.

Our experiments aimed of producing good quality, not traditional basis and gliadin free macaroni doughs by a complex application of both emulsifiers and technology.

2 Materials and methods

Maize and waxy starch, two sorts of leguminous flour yellow pea and soya bean flour were used to produce the experimental samples and the emulsifiers were Amidan 250 B and Dimodan PM (Grindsted, Denmark).

2.1 Preparation of dough samples. The laboratory examination of the model systems were carried out in 1994 and the pilot scale production was continued in 1995.

When making the dough the mixture of starch, leguminous flour and water was dried to reach a 40 % moisture content. The emulsifier was added in 0-2 % concentration in relation to the mass of flour. In model systems 100 g dough was made per sample.

At pilot plant production casein was applied in colloid state too. 5 kg dough was made with a Pavan type dough press at 1 MP pressure throngh 75°C teflon matrix. The dough was dried in a drying facility (FEUTRON ILKA, Germany) at 75°C for one hour then for 30 minutes at 80 % humidity.

2.2 Examination of raw material and dough samples. Moisture, protein, carbohydrate content and amylose and amylopectin were determined from the raw material (Table 1). The cooking time, water uptake , cooking loss and the sensory assessment of dough samples are defined by test cooking[14]. Beside sensory assessment an instrumental examination of consistency was also carried out.

3 Results and evaluation

The results of laboratory examinations were evaluated with mathematical statistical method at P = 5 % level. The characteristics indicate a favourable and significant improvement when applying additives.

On the basis of laboratory experiments the most favourable results were obtained with the mixture of 60 % yellow pea, 40 % starch and 2 % emulsifier.

In 1995 during the pilot plant process further changes were made in the components: beside starch, rice flour was applied at the pea flour. To reach further improvement in the dough structure 3 % casein and 1,5 % emulsifier were applied as additives (Figure 1 and Figure 2).

Table 1 Characteristics of raw material

Raw material	Dry material, %	Carbohydrate, % Total, %	Amylose, %*	Amylopectin, %*	Protein, %
Yellow pea	88,86	37,52	75	25	24,48
Soya	93,67	22,45	87	13	38,57
Maize starch	89,25	73,15	24	76	--
Waxy starch	89,87	65,20	0	100	--

* relating to the starch amount

The laboratory and pilot plant experiments gave evidence that it is possible to produce gliadin free quality dough consists of starch and pea flour. This dough is good for patients with coeliac disease and at the same time its biological value is satisfactory.

References

1. B. Nicholl, C. F. Carthy and P. F. Fottrell, 'Perspectives in Coelic Disease' International Medical Publishers, MTP Nordgate Blackburn, Lancs,1978. p.3-15
2. M. Allmann, U. Candrian and J. Lüthy, Mitt. Gebiete Lebensm. Hyg.1992,**83**,33
3. Y. Bahnassey and K. Khan, Cereal Chem., 1986, **63**, 3, 216
4. J. S. Buck, E. C. Walker and K. S. Watson, Cereal Chem., 1987, **64**, 4, 264
5. H. Toyokawa, G. L. Rubenthaler, J. R. Powers and E. G. Schantus, Cereal Chem. 1989, **66**, 5, 387
6. J. Abecassis, Getreide, Mehl und Brot, 1989, **43**, 2, 58
7. E. Mettler, W. Seibel, K. Münzig, U. Faust and K. Pfeilsticker, Getreide, Mehl und Brot, 1991, **45**, 7, 273
8. F. Meuser, Mühle und Mischfuttertechnik, 1979, **38**, 116, 515
9. G. Schuster, Z. Untersuchung und Forschung, 1984, **179**, 190
10. W. Adams, A. Funke, H. Gölitz and G. Schuster, Getreide, Mehl und Brot, 1991, **45**, 12, 357
11. G. Schuster and W. Adams, Z. Lebensmitteltechnologie und Verfahren, 1980, **31**, 6, 265
12. B. Conde-Petit, Diss. ETH Nr 9785, Zürich, 1992
13. E. Kovács, PhD Thesis, AHS Budapest, 1992
14. Hungarian Standard, MSZ 20500/3-1986, Budapest

Figure 1 The characteristics of dough samples (heat adjustment)

Remarks
* (brackets) means: starch: peas: soy bean: rice ratios
b2, b3, b4, b11 = laboratory samples
x1 = pilot plant sample

Figure 2 The characteristics of dough samples (no heat adjustment)

Remarks
* (brackets) means: starch: peas: soy bean: rice ratios
A2, A4, A6, A7 = laboratory samples
X1 = pilot plant sample

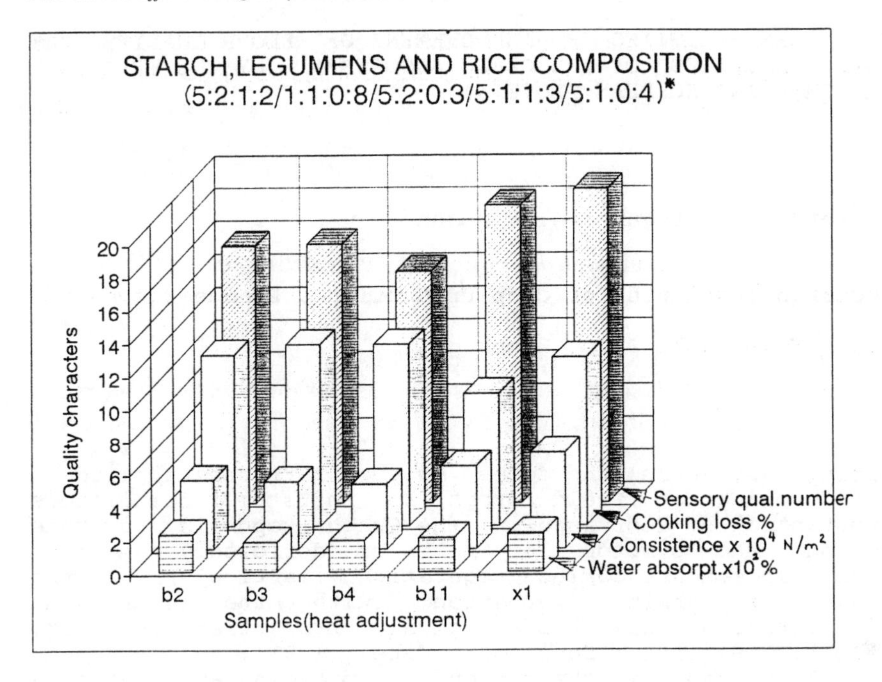

Figure 1 The characteristics of dough samples (heat adjusment)

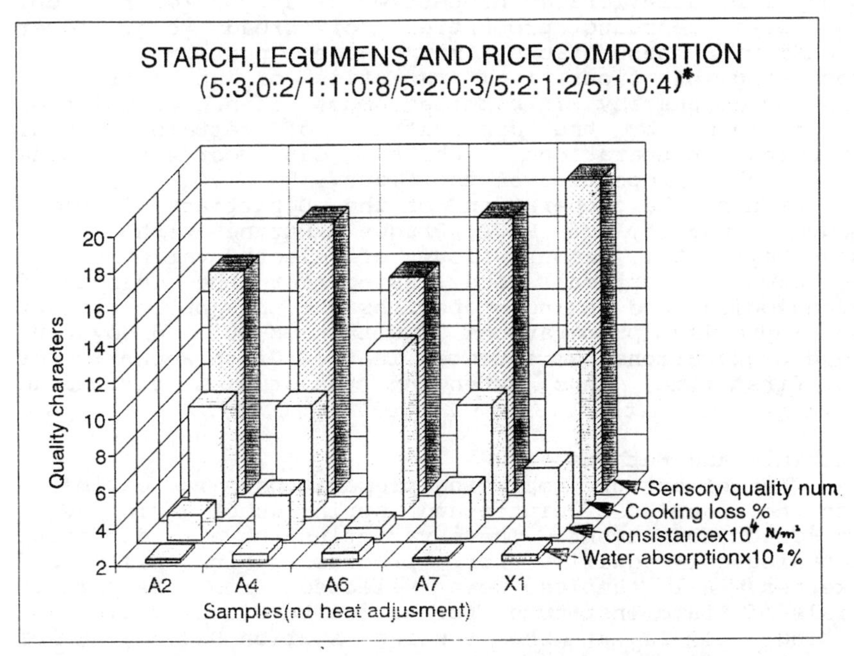

Figure 2 The characteristics of dough samples (no heat adjusment)

THE POSSIBILITIIES OF IMPROVEMENT OF FLOUR QUALITY FROM WHEAT VARIETIES OF THE CZECH ASSORTMENT BY THE EFFECT OF LECITHIN AND SOYBEAN PROTEIN.

J. Petr[1]; J. Prugar[1]; G. Klamann[2]

Research Institute for Crop Production , Prague-Ruzyně[1]

Lucas Meyer GmbH & Co, Hamburg[2]

During past years, the doses of pure nutrients applied in the growing of field crops as industrial fertilizers substantially decreased in the Czech Republic. For this reason, the symptoms of nitrogen deficiency sometimes occur in wheat during vegetation, particularly in the stage of grain formation, with the subsequent deterioration of technological properties (Brümmer, Seibel, 1991,1992 ; Schmieder, Zabel, 1991).

A decrease in the content of proteins and gluten is manifested, among others, in the lower farinographical absorption, deteriorated properties of dough, lower volume and worse sensuous properties of bread, e.g. lower porousity. These facts may cause problems in the processing of flour from varieties of an average to substandard quality in bakeries. Some chance of solution is provided by the application of certain natural improving preparations, which can correct these unfavourable properties of the flour.

Within the cooperation of the Department of Plant Products Quality of RICP, Prague - Ruzyně with Lucas Meyer Co., Hamburg flour of selected varieties of the Czech assortment were tested for the changes of technological and baking properties due the application of the improving preparations Emulthin M-501 and Soyapan. These preparations were tested in the Czech Republic for the first time, thus extending the studies carried out in Europe and in the U.S.A. (Ziegelitz, 1992).

Materials and Methods

The effect of improving preparations was estimated with the samples of winter and spring wheat varieties of the Czech assortment. These varieties have a different technological quality for processing in mills and bakeries. All samples were selected from the variety trials of State Institute for Agriculture Supervising and Testing (SIAST) at the testing station Sedlec, year of harvest 1992.

Results and Discussion

In the evaluation of baking quality of samples without the use of improving preparations applied, the quality characteristics previously determined were confirmed. A negative evaluation of the varieties with lower baking quality concerned above all a low farinographical absorption and low ability of the dough for processing in bakeries (stickiness).

An increase of the volume of bread was found in all evaluated samples where the improving preparations were used.

The highest relative increase in the bread volume (by 23 %) was observed with variety Jara, which have a low baking quality. The effect of improving preparations on the volume of bread was lowest with variety Butin. However, the sample of this variety exhibited substantionally improved properties of dough and bread.

For the sample of variety Hana, which have a very high baking quality, the application of improving agents meant only a modest improvement. However, the highly convenient technological properties in combination with the improving preparations used were marked in the test with an extended rising time where a substantial increase of the specific volume of bread was observed.

Based on the results of these tests and on the knowledge of important parameters of baking quality, it can be stated that the application of improving preparations brings about an increase in volume and quality of bread, but that it has almost no influence on farinographical absorption of flours and the yield of dough.It follows that it is still necessary to consider the properties and assortment of varieties supplied to mills and bakeries for processing.The use of improving preparations in order to increase the quality of parts of wheat composed of unsuitable varieties seems to be irrational.

Characteristic of the experimental station

Region - Middle Bohemia, Altitude - 300 m,
Sugar beet production type,Temperature - 8,2 ^{o}C
Rainfall - 501 mm , Soil type - Haplic chernozem

Description of evaluated samples

Variety		Grain quality
Hana	(winter wheat)	A-9
Livia	(winter wheat)	B-4
Sparta	(winter wheat)	B-4
Simona	(winter wheat)	B-4
Butin	(winter wheat)	B-3
Selekta	(winter wheat)	C-2
Jara	(spring wheat)	C-1

The evaluation of grain quality was carried in agreement with the international and national standards.

Characteristics of the used improving preparations

Emulthin M-501 is the natural powdery emulsifier and stabilizer for improving the quality of wheat flours. Its main component are modified phosphatides and partial glycerides developed especially for improving the flour properties.
Soyapan is the product of firm Edelsoja.It is an enzymatically active soybean protein prepared from sorted and cleaned soya beans.

References

1. BRÜMMER,J.M.-SEIBEL W.: Verarbeitungseigenschaften von Weizen aus extensiviertem Anbau,Getreide Mehl und Brot, 45,1991(11):336-441.
2. BRÜMMER,J.M.-SEIBEL W.: Extesivierter Weizenbau und seine Auswirkungen auf die Müllerei,Getreide Mehl und Brot,46,1992(4):99-102.
3. Hampl,J.: Cereální chemie a technologie.Praha-Bratislava, SNTL-ALFA 1970
4. PRUGAR,J.-HRAŠKA,Š.:Kvalita pšenice.Bratislava,Príroda 1986
5. SCHMIEDER,W.-ZABEL,S.:Einfluss einer reduzierten Stickoffdüngung auf die Qualität und den Ertrag von Backweizen.
6. ZIEGELITZ,R.:Einsatz von Lecithin in Lebensmitteln. Int.Z.Lebensm.-Techn.43,1992(1/2):22-24,26-27.
7. ČSN 46 1011. Zkoušení obilovin,luštěnin a olejnin.1988.

Table 1 **Comparation of the effect of improving prepararations used on the bread volume of evaluated flour samples**

Variety	Improving preparation used			Specific bread volume	
	Emulthin M-501 /%/	Soyapan /%/	Ascorbic acid /g per 100 kg of flour/	/cm³ per 100 of bread/	/ %/
HANA (A-9)	-	-	-	348	100
	0,15	-	1,5	385	111
	0,15	0,30	1,0	397	114
LIVIA (B-4)	-	-	-	295	100
	0,15	-	2,5	316	107
	0,25	-	2,0	336	114
SPARTA (B-4)	-	-	-	369	100
	0,25	-	1,5	404	109
SIMONA (B-4)	-	-	-	364	100
	0,15	-	2,5	412	113
	0,15	0,30	2,0	418	115
SELEKTA (C-2)	-	-	-	350	100
	0,15	-	2,0	400	114
	0,25	-	1,5	408	117
BUTIN (B-3)	-	-	-	386	100
	0,15	-	2,0	403	104
	0,15	0,30	1,5	415	107
JARA (C-1) Spring W.	-	-	-	353	100
	0,15	-	2,5	404	114
	0,15	0,30	2,0	435	123

THE INFLUENCE OF MILLING ON THE NUTRITIVE VALUE OF FLOUR FROM RYE - 1. VITAMINS -

S. Ötleş and E. Köse

Department of Food Engineering
University of Ege
Bornova Izmir Turkey

Department of Food Engineering
University of Celal Bayar
Manisa Turkey

1 INTRODUCTION

Rye is a bread grain resembling wheat in many of its characteristics.[1] There are many potential food applications where rye flour and other products can be conveniently used because of the characteristic functional properties of the rye endosperm. Rye forms an important part of the human dietary intake of vitamins such as vitamins B_1, B_2 B_6 and PP. These vitamins are mainly present in the embryo, scutellum, and aleurone layer of the seed. The milling process of rye grain to different flour extractions will, therefore, cause significant differences in vitamin composition.[2-7]

This research was undertaken to determine the vitamin composition of selected rye samples available to the consumer in five geographical locations of Turkey, and of rye flours of different extraction rates from these rye grains. Thus, information on the uniformity of vitamin composition of rye throughout the country and the degree to which the process of milling produced changes in the vitamin levels of rye flours were studied in order that data might be available for estimates of contributions of these rye products to vitamins in Turkish dietary intake.

2 EXPERIMENTAL

2.1 Row Materials

Rye samples (Secale cereale) obtained from five various locations of Turkey (1 = Denizli A, 2 = Denizli B, 3 = Erzurum, 4 = Kayseri, 5 = Polatlı) were milled into rye flours of different extractions by using of a Buhler MLU-202 laboratory mill.[8]

2.2 Analytical Methods

All analyses were made in duplicate and the results expressed on dry basis. Ash, moisture, protein, starch, thousand kernel weight and hectoliter weight were determined according to standard AOAC and ICC methods.[9-10]

The water soluble vitamins, including B_1, B_2, B_6, B_c, and PP were analyzed by high pressure liquid chromatography (H.P.L.C.). As described in previously paper,[11-12] vitamins were extracted by digestion of a milled sample with 0.05 M H_2SO_4 followed by enzymic digestion with Takadiastase and Papain, adjusting to pH 4.5. After incubation in an oven at 35±1°C overnight the solution was injected and separated on a Waters 3.9 mm ID x 30 cm μ-Bondapak C18 column, isocratically at 25°C with a water-methanol mixture (78 : 22, v/v), using a constant concentration of PIC B6 with a flow rate of 0.3 ml/min. Pantothenic acid and tocopherols were determined according to standard AVC methods.[13]

3 RESULTS AND DISCUSSION

3.1 Milling Data

Table 1 contains the levels of ash, moisture, protein, starch, thousand kernel weight and hectoliter weight of rye samples grown in five locations of Turkey. The results for rye grain and flours are in good agreement with results obtained by several workers.[1-7] The average values for the rye grain analysis were 1.69 g ash, 11.0 g protein, 60.2 g starch per 100 g dry matter, and 11.5 g moisture. As flour extraction rate decreased, the amounts of nutrients in the flour decreased with the exception of starch, which increased.

Table 1 Some Chemical and Physical Characteristics (g/100 g, dry basis) of Turkish Rye Grain and Flour of Different Extraction Rates

Area	1	2	3	4	5
Rye Grain					
Ash	1.71	1.76	1.75	1.66	1.58
Moisture	12.0	11.8	11.8	11.7	10.1
Protein	11.9	12.0	10.0	9.2	11.8
Starch	59.6	57.1	61.7	62.0	60.5
Tkw	19.4	21.3	21.9	25.5	22.4
Hw	77.2	74.7	75.9	75.3	75.2
Extraction rate (%) : 85					
Ash	1.01	1.01	1.02	0.99	1.04
Protein	10.2	10.2	8.6	7.9	10.2
Starch	72.3	69.1	74.8	73.2	68.7
Extraction rate (%) : 60					
Ash	0.65	0.65	0.62	0.65	0.61
Protein	7.4	7.5	5.8	6.3	7.5
Starch	77.4	74.0	78.3	80.1	73.5

* Tkw = Thousand kernel weight (g), Hw = Hectoliter weight (kg)

3.2 Vitamin Data

Data on the vitamin content of the rye grains and flours laboratory-milled to various extraction rates are summarized in Table 2. The vitamins are not uniformly distributed throughout in grain. Generally, vitamin B_1 is concentrated in the scutellum, vitamin PP in the aleurone layer. Vitamin B_2 and pantothenic acid are more uniformly distributed. Vitamin B_6 is located in aleurone and germ, with a little in the endosperm.[1] In our experiment, the vitamin contents of rye fell within the same range as found in other cereal grains within the exception of vitamin PP, which was found only in very small amounts of rye.

Table 2 Vitamin Composition (µg/g, dry basis) of Rye Flour of Different Extraction Rates

Area	1	2	3	4	5
Extraction rate (%) : 100					
Vitamin B_1 (thiamine)	6.9	6.8	8.1	4.7	7.3
Vitamin B_2 (riboflavin)	2.8	2.8	3.0	2.4	3.2
Vitamin B_6	3.6	3.9	3.7	3.4	3.7
Pyridoxine	3.2	3.4	3.1	3.0	3.2
Pyridoxal	0.2	0.3	0.4	0.3	0.3
Pyridoxamine	0.2	0.2	0.2	0.1	0.2
Vitamin B_c (folic acid)	0.6	0.6	0.7	0.7	0.9
Vitamin PP (niacin)	12.9	12.1	14.6	12.8	14.8
Pantothenic acid	10.3	10.6	9.8	9.5	10.1
Tocopherols (total)	32.0	34.3	58.4	47.0	51.1
Extraction rate (%) : 85					
Vitamin B_1 (thiamine)	4.8	4.6	5.9	2.5	5.0
Vitamin B_2 (riboflavin)	1.9	2.0	1.9	1.7	2.1
Vitamin B_6	1.7	1.9	1.9	1.6	1.7
Pyridoxine	1.6	1.7	1.6	1.6	1.5
Pyridoxal	0.1	0.1	0.1	trace	0.1
Pyridoxamine	trace	0.1	trace	trace	0.1
Vitamin B_c (folic acid)	0.4	0.3	0.4	0.3	0.3
Vitamin PP (niacin)	11.1	11.0	13.2	10.9	13.5
Pantothenic acid	8.7	8.9	9.1	8.5	9.2
Tocopherols (total)	7.6	7.2	8.3	7.7	8.0
Extraction rate (%) : 60					
Vitamin B_1 (thiamine)	2.6	2.5	3.5	1.4	2.6
Vitamin B_2 (riboflavin)	0.7	0.7	0.8	0.6	0.9
Vitamin B_6	0.5	0.6	0.6	0.4	0.5
Pyridoxine	0.5	0.6	0.6	0.4	0.5
Pyridoxal	trace	trace	trace	trace	trace
Pyridoxamine	trace	trace	trace	trace	trace
Vitamin B_c (folic acid)	0.1	0.2	0.2	0.1	0.2
Vitamin PP (niacin)	7.5	7.1	8.9	7.4	9.0
Pantothenic acid	4.4	4.5	4.1	4.0	4.4
Tocopherols (total)	2.5	2.5	3.0	2.7	3.0

3.1.1 Vitamin B_1. The vitamin B_1 contents of rye grains obtained from five locations were very nearly the same with the exception of area 4. Vitamin B_1 was greatly decreased by milling (Table 2), only 37 % of vitamin B_1 present in rye grain retained in the 60 % extraction flour.

3.1.2 Vitamin B_2. Vitamin B_2 was significantly affected by the various extraction rates. The lowest values were obtained in the 60 % extraction flours, ranging from 0.6 to 0.9 µg / g (d.m.).

3.1.3 Vitamin B_6. The concentrations of the B_6 vitamins were also drastically reduced in rye flours of low extractions compared to the original levels in rye grain. A reduction of 50 % or more was observed for vitamins B_6 with pyridoxal and pyridoxamine being most affected.

3.1.4 Vitamin B_c. The vitamin B_c decreased almost linearly from flour made of whole rye to the 60 % extraction flours, only 23 % of vitamin B_c present in whole rye retained in the 60 % extraction flours.

3.1.5 Vitamin PP. The vitamin PP was found in rye grain and flours in amounts expected based on literature values.[1-5] The whole-grain rye showed the highest vitamin PP value, which was significantly different from values determined for flours of different extraction rates.

3.1.6 Pantothenic acid. Pantothenic acid values determined in this study for grain and flours agree with values in the literature.[1-5] The milling of rye into flours of different extraction rates caused a decrease in the concentration of pantothenic acid. The 85 and 60 % extraction flours have a much poorer pantothenic acid values than flours made from whole rye.

3.1.7 Tocopherols. An average tocopherols content of 44.6 µg / g (d.m.) was found in the whole rye. This was significantly higher than that found in 85 and 60 extraction flours.

4 CONCLUSION

As a result, it is concluded that the vitamin composition of rye varied depending on the agronomic conditions under which the rye was grown, and the mechanical process of milling of rye grain to different flour extractions had an effect on vitamins, including vitamins B_1, B_2, B_6 and its derivates, B_c, PP, pantothenic acid, and tocopherols.

References

1. N. L. Kent, 'Technology of Cereals', Pergamon Press, Oxford, 1983.
2. K. J. Lorenz and K. Kulp, 'Handbook of Cereal Science', Marcel Dekker Inc., New York, 1991.
3. S. A. Matz, 'Ingredients for Bakers', Pan-Tech Int., Texas, 1987.
4. P. Michela and K. Lorenz, Cereal Chemistry, 1976, 53, 853.
5. A. A. Paul and D. A. T. Southgate, 'The Composition of Foods', Elsevier, New York, 1979.
6. B. Pedersen and B. O. Eggum, Qual. Plt. Fd.Hum.Nutr., 1983, 32, 185.
7. H. H. Schopmeyer, Cereal Sci. Today, 1962, 7, 138.
8. E. Köse and S. S. Ünal, E. Ü. Müh. Fak. Derg., 1992, 10, 33.
9. Anon., 'ICC Standards', I.C.C., Detmold, 1982.
10. Anon., 'Official Methods of Analysis', A.O.A.C., Washington, 1990.
11. S. Ötleş, Z. Lebensm. Unters. Forsch., 1991, 193, 347.
12. S. Ötleş and Y. Hışıl, It. J. Food Sci., 1993, 1, 69.
13. A. V. C., 'Methods of Vitamin Assay', Interscience, New York, 1966.

THE INFLUENCE OF MILLING ON THE NUTRITIVE VALUE OF FLOUR FROM RYE - 2. MINERALS -

E. Köse and S. Ötleş

Department of Food Engineering
University of Celal Bayar
Manisa Turkey

Department of Food Engineering
University of Ege
Bornova Izmir Turkey

1 INTRODUCTION

Recently studies have shown an increase in the consumption of bakery products other than bread. However, the composition of mineral elements of these products has received little prior attention. On the other hand, the increasing emphasis on food and nutrition programmes is focusing attention on the importance of expanding and updating information on the composition of mineral elements in foods and food products.[1-11]

The purpose of the present research was to investigate the mineral composition (Ca, Cr, Cu, Fe, K, Mg, Mn, Na, Ni, P and Zn) of a variety of some consumer-available rye and rye flours from five different locations of Turkey and the changes in the composition of mineral elements of rye grains and flours of different extraction rates during the milling process.

2 EXPERIMENTAL

2.1 Row Materials

Rye samples (Secale cereale) obtained from five various locations of Turkey (1 = Denizli A, 2 = Denizli B, 3 = Erzurum, 4 = Kayseri, 5 = Polatlı) were milled into rye flours of different extractions by using of a Buhler MLU-202 laboratory mill.[8]

2.2 Analytical Methods

All analyses were made in duplicate and the results expressed on dry basis. The concentration of main (Ca, K, Mg, Na,P), minor (Cu, Fe, Mn, Zn) and trace (Cr, Ni) elements were determined using a AAS / FES spectrometer (PYE Unicam SP8) with a deuterium background corrector. Lanthanum chloride was added to overcome interferences. For the determination of phosphorus in rye samples, the phosphomolybdovanadate method was used.[12]

3 RESULTS AND DISCUSSION

The distribution of minerals throughout in rye grain is non-uniform. Ash level is particularly high in the aleurone layer of rye, due primarily to the presence in these cells of aleurone bodies, which consist of phytin granules, a mixture of the magnesium and potassium salts of meso-inositol hexaphosphate surrounded by a protein-containing envelope. Other sides of the kernel contributing to the ash content are outer pericarp and aleurone cell walls.[1-7]

Data on the contents of main (Ca, K, Mg, Na and P), minor (Cu, Fe, Mn and Zn) and trace (Cr and Ni) elements of rye samples grown in five locations of Turkey is summarized in Table 1. As expected the mechanical process of milling and separation of various physical compounds of the rye had an effect on mineral content of the flours. The composition of mineral elements of rye grain and flours used in our study was in close agreement with the values published by many researchers although variety and environmental conditions affect the amounts of the mineral elements.

3.1 Calcium (Ca)

Data for mineral elements, Table 1, showed K, P, Mg, Ca and Na, in decreasing order, to be the main mineral elements of whole-rye grain. All ryes were nearly the same in content of calcium (excluded area 3). Calcium values of the whole-grain rye samples were within the range of values reported in the literature. Calcium was significantly affected by the various extraction rates. The lowest values were obtained in the 60 % extraction flours, ranging from 17 to 32 µg / g (d.m.).

3.2 Chromium (Cr) and Nickel (Ni) - Trace Elements

Chromium and nickel values of whole-grain rye samples determined in this study for rye agree with values in the literature.[1-11] There were no significant differences between grain samples. Chromium and nickel in the 60 % extraction flours was in trace amounts (not determined).

3.3 Copper (Cu)

All ryes were nearly the same in content of the copper. Copper was greatly reduced by milling (Table 1), only 30 % of copper present in whole rye were retained in the 60 % extraction flours.

3.4 Iron (Fe)

Iron content of rye flours was in close agreement with expected levels (Table 1). Analyses showed that the 100 extraction flours retained iron to a greater extend than other extraction flours (85 - 60).

Table 1 Mineral Composition (µg/g, dry basis) of Rye Flour

Area	1	2	3	4	5
Extraction rate (%) : 100					
Main					
Ca	43	41	67	36	39
K	480	454	415	524	468
Mg	118	120	135	126	79
Na	18	15	15	24	29
P	395	386	356	368	375
Minor					
Cu	0.8	0.7	0.9	0.8	0.8
Fe	3.9	3.6	5.7	3.1	2.8
Mn	5.6	6.7	5.9	6.1	5.6
Zn	2.6	2.9	3.1	2.9	2.1
Trace					
Cr	0.3	0.3	0.1	0.2	0.1
Ni	0.1	0.1	0.2	0.1	0.1
Extraction rate (%) : 85					
Main					
Ca	36	30	49	29	30
K	251	216	198	268	239
Mg	62	56	68	53	43
Na	15	13	14	19	20
P	243	231	199	213	224
Minor					
Cu	0.6	0.5	0.6	0.5	0.6
Fe	3.1	3.0	4.3	2.5	2.0
Mn	3.2	3.9	3.4	3.5	3.2
Zn	2.1	2.4	2.4	2.3	1.9
Trace					
Cr	0.1	0.1	trace	0.1	trace
Ni	trace	trace	0.1	trace	trace
Extraction rate (%) : 60					
Main					
Ca	20	18	32	17	18
K	166	134	112	185	145
Mg	19	22	26	16	12
Na	6	5	5	8	12
P	127	111	84	91	94
Minor					
Cu	0.3	0.2	0.4	0.3	0.2
Fe	2.1	1.9	2.6	1.5	1.3
Mn	1.7	2.6	1.8	1.8	1.6
Zn	1.2	1.3	1.3	1.2	1.0
Trace					
Cr	trace	trace	trace	trace	trace
Ni	trace	trace	trace	trace	trace

3.5 K (Potassium), P (Phosphorus) and Mg (Magnesium)

Data for potassium, phosphorus and magnesium indicated that the levels of these mineral elements were drastically decreased in rye flours of low extraction rates (60-85 %) compared to the original levels in rye grain (100 %). A decrease of 81 % or more was observed for magnesium being most affected.

3.6 Na (Sodium)

The ryes from five different location of Turkey were very nearly the same in content of sodium. Sodium was significantly affected by milling process with various extraction rates.

3.7 Mn (Manganese) and Zn (Zinc)

The degree of flour extraction had effect on manganese and zinc levels in rye flours. Manganese and zinc decreased almost linearily from flour made of whole rye to the 60 % extraction flours. The lowest values of manganese were obtained in the 60 % extraction flours, ranging from 1.6 to 2.6 μg/g, similarly of zinc in the 60 % extraction flours, ranging from 1.0 to 1.3 μg/g.

4 CONCLUSION

As a result, it is concluded that the mineral composition of rye was comparable to that found in other cereal grains by several workers.[1-11] As in other cereal grains, the milling the rye grain to various flour extractions caused significant differences in composition of mineral elements. As rye flour extraction decreased, the portions of mineral elements retained in the flour decreased.

References

1. N. L. Kent, 'Technology of Cereals', Pergamon Press, Oxford, 1983.
2. K. J. Lorenz and K. Kulp, 'Handbook of Cereal Science', Marcel Dekker Inc., New York, 1991.
3. S. A. Matz, 'Ingredients for Bakers', Pan-Tech Int., Texas, 1987.
4. P. Michela and K. Lorenz, Cereal Chemistry, 1976, **53**, 853.
5. A. A. Paul and D. A. T. Southgate, 'The Composition of Foods', Elsevier, New York, 1979.
6. B. Pedersen and B. O. Eggum, Qual. Plt. Fd.Hum.Nutr., 1983, **32**, 185.
7. H. H. Schopmeyer, Cereal Sci. Today, 1962, **7**, 138.
8. E. Köse and S. S. ünal, E. ü. Müh. Fak. Derg., 1992, **10**, 33.
9. J. R. Turnlund, Cereal Foods World, **27**, 152.
10. E. L. Wheeler and R. E. Ferrel, Cereal Chem., **48**, 312.
11. V. R. Young and M. Janghorbani, Cereal Chem., **58**, 12.
12. Anon., 'Official Methods of Analysis', A.O.A.C., Washington, 1990.

A MODEL ON THE RESPIRATION OF VEGETABLE PRODUCE DURING POSTHARVEST TREATMENTS.

L.M.M. Tijskens

Agrotechnological Research Institute (ATO-DLO),
Wageningen, The Netherlands.

1 INTRODUCTION

All efforts in storage research and all technical storage facilities are in one way or another aimed at prolonging the natural life of perishable produce with as little loss of quality as possible. Prolonging natural life means diminishing the overall activity of those products. That activity is commonly expressed as respiratory activity. The respiration of vegetable produce during storage (air and CA) has therefore been a research topic for a long time[6,5]. Recently, considerable progress has been made the area of modified air packaging (MAP) which is based on respiratory activity.

A description of the interrelations in aerobic conditions between product behaviour and these gases has only recently been provided by Lee et al.[7]. Peppelenbos et al.[8, 9] have already highlighted the interaction between aerobic respiration and anaerobic fermentation and its consequences on product behaviour. Based on the same premises, Andrich et al.[1, 2] developed a respiration model and applied it to MAP conditions. Rao et al.[10] presented an empirical logistic relation between the heat of respiration and temperature, valid for more than 70 different commodities.

In none of the models presented so far, the effects of neither various and varying temperatures, product maturity nor climacteric phase on the respiratory behaviour have been formulated. In this study a consistent mathematical system will be presented that is based on a simplified representation of the respiratory processes describing:

- the mutual influence between **oxygen** and **carbon dioxide** levels,
- the response of product respiration in **aerobic** conditions,
- the response of product fermentation in **anaerobic** conditions,
- influence of **temperature** on all processes involved,
- the link with product **stage of maturity** or state of activity.

2 THE RESPIRATORY PROCESSES

2.1 Why does a Vegetable Product Respire?

It is commonly accepted that the main reason for the plant or plant parts to respire, is to provide the energy (ATP) necessary to drive all kinds of biochemical reactions. Part of the reactions in the product are maintenance reactions necessary to stay alive. Like all living species, the urge to stay alive is essential to vegetable produce. So, the product will try to generate sufficient energy (ATP) to keep maintenance going on at all cost. Another part of energy consumption are the reactions that inevitably cause eventually decay. So, as far as the driving forces are concerned, the key point in the model is the equilibrium between production and consumption of energy (ATP) in combination with a minimal need for maintenance.

The very complex biochemical scheme of the glycolysis, the Krebs cycle and the oxidative phosphorylation can not be modelled in full, partly by lack of knowledge, partly by lack of parameter calibration. Fortunately, only those processes really important in explaining the phenomena should be considered. Respiration is assumed to consist of two distinct parts: aerobic respiration and anaerobic fermentation. As discrete changes in mechanisms do normally not occur in nature, both processes are assumed to occur simultaneously. The total oxygen consumption and carbon dioxide and ATP production encountered in living produce are than the sum of both aerobic and anaerobic processes (see eq. 1).

$$\frac{dCO_{2.total}}{dt} = \frac{dCO_{2.aer}}{dt} + \frac{dCO_{2.ana}}{dt}$$
$$\frac{dO_{2.total}}{dt} = \frac{dO_{2.aer}}{dt} \qquad 1$$
$$\frac{dATP_{total}}{dt} = \frac{dATP_{aer}}{dt} + \frac{dATP_{ana}}{dt}$$

2.2 Aerobic Respiration

The many processes occurring during aerobic respiration are simplified into one enzymatic process: the combustion of carbohydrate substrates by gaseous or dissolved oxygen in an enzymatic reaction. The notional intermediate active complex is inhibited by CO_2[7]. ATP is also known to inhibit respiration[12]. The reaction scheme can then be represented by Scheme 1.

$$E + S + O_2 \underset{k_{-1}}{\overset{k_1}{\rightleftharpoons}} AC_1 \xrightarrow{k_{1p}} CO_2 + x\,ATP + E$$
$$AC_1 + CO_2 \underset{k_{-2}}{\overset{k_2}{\rightleftharpoons}} AC_2$$
$$E + ATP \underset{k_{-3}}{\overset{k_3}{\rightleftharpoons}} AC_3$$

Scheme 1

If we assume a steady state for all of the intermediate complexes ($dAC_i/dt=0$), a constant supply of necessary substrate S (present in abundance), and simplify the expression, using the same reasoning as for the derivation of the Michaelis-Menten equation, one comes up with equation 2.

$$\frac{dCO_{2.aer}}{dt} = \frac{k_s \cdot O_2}{1 + K_{mo} \cdot O_2 + K_{moc} \cdot O_2 \cdot CO_2 + K_{ma} \cdot ATP} \qquad 2$$

Except an extra term for ATP in the denominator these equations are quite the same as those proposed by Lee et al.[7] and by Andrich et al.[1, 2]. The extra term describes the slowing down of the respiration rate whenever large amounts of ATP are available, and its speeding up in case of low amounts of ATP.

2.3 Anaerobic Conditions

A model for anaerobic conditions has been developed along comparable lines of reasoning and based on the assumption that no abrupt switch over between aerobic respiration and anaerobic fermentation exists. As we will concentrate on aerobic applications, the model is not deduced here but the result is merely shown (eq. 3).

$$\frac{dCO_{2.ana}}{dt} = \frac{k_{s.ana}}{1 + K_{mo.ana} \cdot O_2 + K_{ma.ana} \cdot ATP} \qquad 3$$

The occurrence, effects and behaviour of anaerobic processes have already been described[8, 9].

2.4 Effects of External Factors

2.4.1 Temperature

The most important external factor is temperature. As the rates of all chemical and biochemical reactions are susceptible to temperature, temperature will affect the rates of the respiration processes as normal: all reaction rates are assumed to depend on temperature according Arrhenius' law. In equations 2 and 3, k_i represent true reaction rates. K_i, however, represent a (complex) ratio of the constituting reaction rates. A ratio between reaction rates (equilibrium constant) also depends on temperature according Arrhenius' law, but the energy of activation is the difference between the energies of both constituting reaction rates. This difference is very likely to be smaller than that of the true reaction rates in the same system. Hence, the major effect of temperature on the apparent rate of respiration will be apparent in the numerator of equations 2 and 3.

This implies that the relative respiration rate, defined as the ratio between the actual respiration at any aerobic gas condition and the respiration in air (21% O_2 and 0% CO_2) at the same temperature and the same ATP level, is quite independent of temperature: the major temperature depending parameter (k_s) disappears by elimination (see eq. 4).

$$RelResp = \frac{O_2}{21} \cdot \frac{1 + K_{mo} \cdot 21}{1 + K_{mo} \cdot O_2 + K_{moc} \cdot O_2 \cdot CO_2} \qquad 4$$

2.4.2 State of Activity and Respiration Quotient

The activity of a product is commonly measured as respiratory activity. Hence, the model can not be validated with respect to ATP. In statistical analyses the term for the ATP inhibition has therefore to be removed from the equation. Data sets should be grouped at comparable state of activity and analyzed separately. Any discussion about this aspect is rather theoretical and philosophical. It can be taken from the model formulation, that the state of activity, as determined by factors others than storage conditions (CA, MA) like for example wound healing and climacteric phase, has an effect on respiration through the ATP inhibition factor in both the denominators of equations 2 and 3. As the level of ATP seems to be quite stable in horticultural products (it is the steady state between production and consumption), the coefficient K_{ma} has to be rather large to enhance the action of ATP on the respiration.

From the combined equations for aerobic respiration and anaerobic fermentation, an expression for the respiration quotient can be derived as the ratio between total CO_2 production (eq. 2 and 3) and the fermentative CO_2 production (eq. 3):

$$RQ = 1 + \frac{k_{s.ana}}{k_s O_2} \cdot \frac{1 + K_{mo} \cdot O_2 + K_{moc} \cdot O_2 \cdot CO_2 + K_{ma} \cdot ATP}{1 + K_{mo.ana} \cdot O_2 + K_{ma.ana} \cdot ATP} \qquad 5$$

3 STATISTICAL ANALYSIS

Statistical analysis is conducted using nonlinear regression (eq. 2) on a number of experiments. Without going too deep into details of experiments and analysis, results of the production of CO_2 and consumption of O_2 in strawberries (cv. Elsanta), general relative respiration in apples and heat production of Iris bulbs are show in Table 1.

Table 1 Results of statistical analysis.

	Strawberries[15]		Apples[14]		Iris bulbs[11]	
	Estimate	s.e.	Estimate	s.e.	Estimate	s.e
k_{sref}	20.[a]	-	-	-	61.70	7.94
E_s/R	10916.	1102.	-	-	5622.	277.
K_{mo}	0.1400	0.0174	0.2109	0.0213	0.3144	0.0469
K_{moc}	0.00647	0.00129	0.05320	0.00609	0.01528	0.00759
Nobs	37		40		21	
R^2_{adj}	90.1%		97.0%		97.9%	

[a] fixed value

The percentage variance accounted for (R^2_{adj}) is high for all experiments. The standard error of estimates is 10% or less of the estimate value. These results, however small in number of experiments, provide a fair validation of the model and the underlying assumptions. Also the deduction of the temperature independent relative respiration is validated (apples).

Table 2 Results Apples Cox Orange Pippin.[3]

Time range (days)	N_{obs}	k_{sref}	E_s	K_{mo}	K_{mco}	R^2_{adj}
15-45	108	17.5	6613	0.169	0.036	90.2
50-95	135	19.8	5703	0.229	0.042	90.5
100-200	262	24.8	2426	0.359	0.067	75.8

Nonlinear regression analysis has been carried out on the data of Fidler and North[4], kindly provided by the Horticulture Research International (East Malling), on several apple cultivars. There were roughly three regions of activity during long term storage: an active initial region (climacteric phase), a subdued activity during prolonged storage (dormant phase) followed again by a period of increasing activity (terminal phase). Some results of the analysis of the three phases for Cox O.P. are shown in Table 2.

4 CONSEQUENCES AND APPLICATIONS

4.1 Effect of Temperature

The concept of relative respiration has the advantage that the temperature dependency of maintenance and decay reactions remains in effect as normal without interference of the respiration activity. The apparent reaction rate of for example quality decreasing reactions then can be decomposed into a part specific for that particular reaction, and a part attributed to the physiological activity, expressed as relative respiration:

$$k_{app} = k_{spec} \cdot RelResp \qquad 6$$

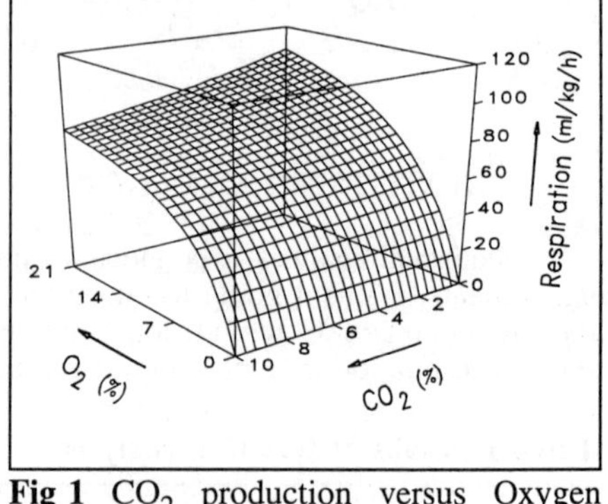

Fig 1 CO_2 production versus Oxygen and Carbondioxide level.

This concept, although only marginally validated, has successfully been applied in several models describing quality behaviour in respiring products (apples, flowers, chicory, unpublished) in CA and MA conditions. In Fig 1 a 3-D example is given for the simulated behaviour of the relative respiration as a function of the level of O_2 and CO_2. The parameters are taken from the strawberries in Table 1.

4.2 State of Activity and Coefficient of Respiration

From the model formulation, the importance of ATP level as an indication for state of activity (e.g., climacteric state, wound healing, senescence etc.) seems quite clear. Logical reasoning reveals, however, that separate information about ATP production and ATP consumption is what is really needed, to include completely this aspect of product behaviour. In Fig 2 a 3-D example is given for the CO_2 production in function of O_2 and ATP levels. The parameters are taken from the strawberries in Table 1 with 0.5 for K_{ma}.

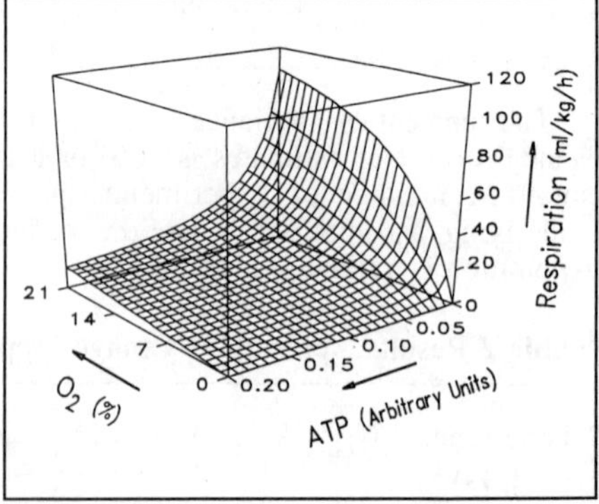

Fig 2 CO_2 production versus Oxygen and ATP level.

As a consequence of the gradual increase in anaerobic fermentation with decreasing O_2 levels, the stage of maturity or development of the produce as well as the actual temperature have an influence on the occurrence and the rate of anaerobic fermentation. The gradual increase in relative importance of anaerobic fermentation with decreasing oxygen levels is reflected in the gradually increasing respiration quotient.

4.3 Respiration as a quality index

The fact that respiration can be considered a measure for the general activity of the product at any moment of its life time, can advantageously be used for generating a general index for quality. In the equations 2 and 3 respiration is presented as a differential equation. Integration of this differential equations in time, or the cumulative respiration, is the total amount of CO_2 produced. This total amount is a measure for the total amount of sugar combusted during the life time of the product. It is therefore equivalent to the loss in quality: the higher the activity at any moment, the faster the total combustion increases and the more quality is lost. Brash et al.[3] reported the keeping quality of asparagus spears to be linear related to cumulative respiration. We should, however, bear in mind that the conversion factor from total respiration to quality loss, is based on energy demands and is defined by the mass to be converted *(how many moles)* in the product, combined with the specific amount of energy required for that conversion *(how much ATP per mole)*. This type of general index will rely heavily on mass conversion like pectin and cell wall biopolymer decay (e.g. softening) and less on inactivation of enzymes or loss of colour, vitamins and flavour compounds (e.g. taste and appearance). So for products with firmness as (one of) the major quality attribute, the cumulative respiration should work fine as a quality index, for products with taste or appearance as major attribute, the usefulness as a quality index is rather doubtful.

REFERENCES

1. Andrich G., Fiorentini R., A. Tuci, Zinnai A., Sommovigo G. 1991. A tentative model to describe the respiration of stored apples. *J. Amer. Soc. Hort. Sci.* **116**,478-481.
2. Andrich G., Zinnai A., Balzinik S., Fiorentini R. 1993. Kinetics of respiration and fermentation in Golden Delicious apples during storage in controlled atmosphere. *COST94 Workshop, September 1993, Leuven, Belgium.*
3. Brash D.W., Charles C.M., Wright S., Bycroft B.L. 1995: Shelf-life of stored asparagus is strongly related to postharvest respiratory activity. *Postharvest Biology & Technology* **5**, 77-81
4. Fidler , North. *J. Hort. Sci* (1967) **42**,189-206;207-221; (1968) **43**,421-428;429-439; (1971) **46**,229-235;237-243;245-250
5. Kader A.A., Zagory D., Kerbel E.L. 1989. Modified atmosphere packaging of fruits and vegetables. *Crit. Rev. in Food Sci. and Nutr.* **28**, pp 1-30
6. Kidd F. 1917. The retarding effect of carbon dioxide on respiration. *Proc. Royal. Soc. London, B,* **87**, pp 136-156
7. Lee, D.S., Haggar, P.E., Lee, J., Yam, K.L. 1991. Model for fresh produce in modified atmospheres on principles of enzyme kinetics. *J. Food Sci.* **56**,6,1580-1585
8. Peppelenbos H.W., Van 't Leven J., Van Zwol B.H., Tijskens L.M.M. 1993. The influence of O_2 and CO_2 on the quality of fresh mushrooms. *Sixth International Controlled Atmosphere Research Conference, Ithaca, USA.*
9. Peppelenbos H.W., Tijskens L.M.M., van 't Leven J., Wilkinson, E.C. 1995. Modelling carbondioxide production of fruits and vegetables based on enzyme kinetics and ATP production. Submitted *J. Exp. Bot.*
10. Rao N., Flores R.A., Gast K.I.B. 1993. Mathematical relationships for the heat of respiration as a function of temperature. *Posth. Biol. Techn.* **3**,173-180.
11. Rudolphij J.W., Verbeek W., Fockens F.H. 1977. Measuring heat production of respiring produce under normal an CA-storage conditions with an adiabatic calorimeter. *Lebensm. Wiss. und Technol.* **10**,153-158.
12. Stryer L. 1981. Biochemistry. p. 301, Sec. Ed. W.H. Freeman & Co, San Francisco.
13. Tijskens L.M.M., Peppelenbos H.W. 1995. Respiration of vegetable produce. A fundamental and mathematical approach. Submitted *Ann. Appl. Biol.*
14. Van der Bliek W., van Overbeeke J., Schmitz W., Verhoeven D. 1989. Van koelcel naar ULO-bewaring. P. 3-5. CAD Fruitteelt in de volle grond, Wilhelminadorp, The Netherlands.
15. Van Schaik A., Boerrigter H.A.M. (1991) unpublished data ATO, Wageningen, The Netherlands.

THE EFFECT OF COOKING ON CABBAGE GLUCOSINOLATES

B. W. Kassahun, J. Velíšek and J. Davídek

Department of Food Chemistry and Analysis
Institute of Chemical Technolgy
166 28 Prague, Czech Republic

1. INTRODUCTION

The family *Brassicaceae* contains a class of structurally uniform compounds known as glucosinolates. Upon hydrolysis by a co-occurring thioglucosidase enzyme, called myrosinase (thioglucoside glucohydrolase EC 3.2.3.1), they give rise to a wide range of biologically active breakdown products.[1] The character of the side chain (the aglucone) and other factors influence the quality of glucosinolates breakdown products.[2] Adverse effects on animal health fed on rapeseed meal,[3] exhibition of both acute and chronic toxicity of glucosinolates and their breakdown products in mammalian systems including their chemoprotective activity have been well documented.[4] Glucosinolate degradation products also participate in the formation of the flavour of meals prepared from Brassica vegetables. These facts have made glucosinolates and their breakdown products one of the most extensively studied secondary plant metabolites in the Brassica family.

Glucosinolates have been investigated in raw and processed cabbage. Different processing procedures resulted in changes of the initial glucosinolate levels.[5, 6]

Variation in glucosinolate composition and content in cooked, fermented and stored cabbage was also reported.[5 - 8]

In the present experiment we wanted to measure not only the glucosinolate level remaining in the cooked cabbage and cooking liquor, but also to identify the glucosinolate breakdown products arising during the cooking process.

2. MATERIALS AND METHODS

Fresh heads of cabbage (early summer variety) were bought from local market. Extraction, purification, GLC and HPLC analysis of glucosinolates were carried out according to Spinks et al.[9] Volatile degradation products of cooked cabbage were identified and determined following vapour phase extraction of these compounds. Freshly sliced cabbage (150 g) was mixed with 150 ml of distilled water or with 150 ml acetic acid solution and put in a 500 ml round bottom flask. A Likens-Nickerson apparatus was fitted to this flask on one side, and to a flask containing 150 ml diethyl ether on the other side. Both flasks were put in a heating mantle and boiled for 15, 30 and 60 min. The organic phase was concentrated to a volume of 20 ml using a Snyder column at 40°C, dried over sodium sulfate, and concentrated under a stream of nitrogen to a volume of 0.5 ml. One µl of this concentrate was injected into the gas chromatographic column.

2.1 Chromatographic Analysis

A Hewlett Packard 5890A gas chromatograph was equipped with a fused silica capillary column with a SE-54 stationary phase (25 m x 0,25 mm i.d., film thickness 0,25 μm) and flame ionization detector. For the resolution of the breakdown products the column temperature was held at 40°C for 10 min and programmed to 250°C (5°C.min $^{-1}$). GC/MS analysis of the glucosinolate breakdown products was performed on the above gas chromatograph equipped with a Hewlett Packard quadrupole mass spectrometer model 5972 and a fused silica capillary column with HP-5 silicone (25 m x 0,25 mm i.d., film thickness 0,25 μm). The column temperature was held at 45°C for 5 min and increased to 250°C at a rate of 5°C.min^{-1}. Mass spectra were obtained by electron impact ionization at 70 eV. The ion source temperature was 180°C.

For the HPLC analysis, Thermo Separation Constametric 3200 gradient pump system (Riviera Beach, FA, USA) with a Constametric control panel, a reversed phase C_{18} column (4.6 x 250 mm, 4 μm, Waters-Nova-Pack, Milford, MA, USA) and Thermo Separation auto injection system (Spectra Series AS100) was used. Detection was made by a SpectroMonitor 3200 spectrophotometric detector (Thermo Separation Products), at a wavelength of 230 nm.

3. RESULTS AND DISCUSSION

The total content of glucosinolates determined in the fresh and cooked samples is given in Table 1 and the content of their degradation products, which was calculated from the calibration curve of allyl isothiocyanate, in Table 2. The major glucosinolates found were allyl (sinigrin), 3-methylsulfinylpropyl (glucoiberin) and 3-indolylmethyl (glucobrassicin). They represent 44, 23 and 13 %, respectively, of the total cabbage glucosinolate composition. Pent-4-enyl (glucobrasicanapin), 2-hydroxypent-4-enyl (gluconapoleiferin), 4-methylsulfinylbutyl (glucoraphanin) and 4-hydroxy-3-indolylmethyl glucosinolates were the least abundant or only found in traces.

The decrease in total glucosinolate content in cabbage cooked in water is in agreement with our previous studies.[8] Leaching of individual glucosinolates into the cooking water or acetic acid solution took place in different proportions. This does not agree with the findings of Rosa and Heaney[7] who found losses or leaching of individual glucosinolates into the cooking liquor in similar proportion. Loss reported in these studies[7, 8] in total glucosinolate content of cabbage cooked for 5 and 10 min are consistent with each other, even though different cooking procedures were used.

Loss in the content of individual glucosinolates in cabbage samples cooked in 3% and 6% acetic acid solution is greater than the loss found in the samples cooked in 1.5% acetic acid solution within the same length of time. The leaching effect of these acid solutions is also higher than the leaching effect of the water (Table 1). Total glucosinolate content found in the sample cooked for 15 min in 1.5, 3.0 and 6.0 % of acetic acid solution represent 32, 33 and 13 %, respectively, of the total glucosinolate content of the sample cooked in water (for 15 min). Higher losses in total glucosinolate content occurred in the material cooked for 60 min in water and in 1.5 and 6.0 % acetic acid solution.

The total amount of glucosinolate found in the cabbage cooked for 30 min was higher than that found in the cabbage cooked for 15 min. The reason might be attributed the combined action of heat and the endogenous myrosinase in the beginning of cooking whereas the enzyme is destroyed with prolonged cooking.

Table 1 *Total content (mg.kg $^{-1}$ of dry weight) of glucosinolates(GLs) in fresh and cooked cabbage*

Acid conc. (%, w/v)		*Cooking time (min)* 0	*15*	*30*	*60*
0	in cooked cabbage	8872	2267	2466	118
0	in the cooking water	-	1951	2251	2286
15	in cooked cabbage	-	732	1005	50
15	in the cooking acidic solution	-	635	1349	386
30	in cooked cabbage	-	277	886	374
30	in the cooking acidic solution	-	1381	1918	723
60	in cooked cabbage	-	289	474	120
60	in the cooking acidic solution	-	437	1817	739

Table 2 *Content (mg.kg $^{-1}$ of dry weight) of breakdown products of GLs determined in the volatile portion of cooked cabbage in water*

Breakdown product of GLs	*Cooking time (min)* 15	30	60
1. Allyl cyanide	263	415	245
2. Allyl thiocyanate	54	22	22
3. Allyl isothiocyanate	269	568	628
4. 3-Butenylisothiocanate	8	8	-
5. 3-Methylthiopropyl cyanide	292	593	233
6. Benzeneacetonitrile	7	23	22
7. Benzenepropionitrile	6	20	tr
8. 4-methylthiobutyl cyanide	tr	16	tr
9. 3-methylthiopropyl isothiocyanate	97	233	155
10. Benzyl isothiocyanate	tr	tr	tr
11. 4-Methylthiobutyl isothiocyanate	28	5	70
12. 2-Phenethyl isothiocyanate	1	12	-
13. 5-Vinyloxazolidine-2-thione (goitrin)	-	12	18
14. 3-Methylsulfinylpropyl cyanide	5	tr	15
15. 3-Methylsulfinylpropyl isothiocyanate	4	-	58
16. 3-Indolylacetonitrile	97	367	450

Cooking of cabbage either in aqueous or in acidic medium causes a decrease in glucosinolate content not only by leaching of these compounds into the cooking liquor but also by degradation. The majority of these degradation products are constituents of the volatile portion of the cabbage.[10] Alkyl or alkenyl isothiocyanates are volatile, except those bearing a sulfoxy terminal side chain and indole moiety such as 3-methylsulfinylpropyl isothiocyanate and 3-indolylacetonitrile. Many of the breakdown products identified both in volatile portion of the sample cooked in the vinegar and water are identical but they differ in their quantity. The identity of the breakdown products formed in the cabbage cooked in the acetic acid solution (vinegar) resembles the enzymatic hydrolysis products of glucosinolates in acidic medium - nitriles predominate (data not shown). The level of nitriles formed in the sample cooked in water was also high, but lower than the level of isothiocyanates (Table 2).

The major breakdown products, found in the volatile portion, in the sample which was cooked in water, were allyl isothiocyanate and allyl cyanide (from sinigrin), 3-methylthiopropyl isothiocyanate and 3-methylthiopropyl cyanide (from glucoibervirin). 3-Indolylacetonitrile (from glucobrassicin) was the main non-volatile compound in the cooking liquor. Traces (< 1 $mg.kg^{-1}$ of dry weight) of N-methoxy-3-indolylmethylthiomethane and 4-methoxyindole were also identified in the vapour phase extract of the sample which was cooked in water. In addition, benzeneacetonitrile and benzyl isothiocyanate were detected. This indicates the presence of the parent glucosinolate, benzyl glucosinolate (glucotropaeolin), in some cabbage varieties as it was proved by VanEtten et al.[10] Other constituents of the volatile portion in both cooking liquors were alkyl sulphides, dimethyl sulphide, dimethyl disulphide, dimethyl trisulphide and dimethyl tetrasulphide which have been known as volatiles of cooked cabbage.[6]

As determined in the fresh cabbage, 3-methylsulfinylpropyl glucosinolate represents the second major glucosinolate. Its breakdown products, 3-methylsulfinylpropyl isothiocyanate and 3-methylsulfinylpropyl cyanide, were present in unexpectedly lower quantities, the latter compound was found even in traces either in the sample cooked in water or in vinegar. Further investigation is necessary to prove whether these breakdown products undergo subsequent degradation or transformation to other compounds.

Cooking cabbage in vinegar and water for more than 30 min resulted not only in reduction of glucosinolate content but also in the content of the breakdown products.

References

1. D. I. McGregor, W. J. Mullin and G. R. Fenwick, *J. Assoc. Off. Anal. Chem.* 1983, **66**, 825.
2. G. R. Fenwick, R. K. Heaney and R. Mawson, In: Toxicants of Plant Origin, Vol. II, Glucosinolates, P. R. Cheeke (ed.), 1989.
3. G. R. Fenwick and R. F. Curtis, *Anim. Feed Sci. and Technol.*, 1980, **5**, 255.
4. J. M. Betz and W. D. Fox, In: Food Phytochemicals I: Fruits and Vegetables, 1994.
5. K. Sones, R. K. Heaney, G. R. Fenwick, *J. Sci. Food Agric.*, 1984, **35**, 712.
6. R. H. De Vos and W. G. H. Blejleven, *Z. Lebensm. Unters. Forsch.*, 1988, **187**, 525.
7. E. A. S. Rosa and R. K. Heaney, *J. Sci. Food Agric.*,1993, **62**, 2598.
8. B. W. Kassahun, J. Velíšek and J. Davídek, In: Proceedings of the Int. Euro Food Tox. IV Conference-Bioactive Substances in Food of Plant Origin, Vol. 1, 1994.
9. E. A. Spinks, K. Sones and G. R. Fenwick, *Fette, Seifen, Anstrichm.*, 1984, **6**, 228.
10. C. H. VanEtten, M. E. Daxenbichler, P. H. Williams and W. F. Kwolek, *J. Agric. Food Chem.*, 1976, **24**, 452.

EFFECTS OF PREPARATION PROCEDURES AND PACKAGING ON NUTRIENT RETENTION IN SHREDDED CHINESE CABBAGE

Margareta Hägg[1], Ulla Häkkinen[1], Jorma Kumpulainen[1], Eero Hurme[2] & Raija Ahvenainen[2]

[1]Agricultural Research Centre of Finland, Laboratory of Food Chemistry, FIN-31600 Jokioinen, Finland.
[2]VTT, Biotechnology and Food Research, P.O. Box 1500, FIN-02044 VTT, Finland.

1. INTRODUCTION

The use of prepeeled and presliced vegetables has been advocated by the catering industry in order to save on time. Chinese cabbage is one of the most common vegetables used in Finland as a compound of various salads. It has significance in the supply of vitamins in the diet. In order to centralize the manufacture of shredded cabbage, new methods for retaining its sensory and microbiological quality are needed[1]. This study dealt with the effects of preparing procedures such as shredding, washing, and packaging, packaging materials, and storage time on the nutrient quality of Chinese cabbage. Vitamin C, ß-carotene, dietary fiber, sugars, ash and moisture contents were determined in fresh and packaged shredded cabbage after 1, 3 (or 4) and 7 d storage.

2. MATERIALS AND METHODS

The trials were performed on the summer cultivar Kasumi and the winter cultivar Kingdom. The Kingdom cultivar was determined both in autumn and in early spring after winter storage in 1991-1992 and 1992-1993. The Kasumi cultivar was determined in early autumn 1991 only. The cabbages were stored before the trials at 0-1 °C, RH 85 %. All external parts were first removed from the Chinese cabbages after which it was mechanically shredded into 6 mm strips. In the 1991 trials cabbage was washed in water (+6°C, 3l/k), spin dried and packaged in oriented polypropylene (OPP)-film (40 μm). The O_2 permeability of the OPP-film was 1100 cm^3/m^2 24 h 101.3 kPa, 23 °C, RH 0 % and the CO_2 permeability was 3900 cm^3/m^2 24 h 101.3 kPa, 23 °C, RH 0 %. In the 1992 and 1993 trials the O_2 permeability of the OPP-film was 1200 cm^3/m^2 24 h 101.3 kPa, 23 °C,

Table 1. *Washing methods used for shredded Chinese cabbage*

sample no.	*washing method and temperature*	*washing time*
1.	water +6 °C	1 min
2.	water +6 °C twice	1 min + 1 min
3.	water +30 °C and water +6 °C	0.5 min + 1 min
4.	chlorinated water (0.01 % free chlorine) +6 °C	1 min
5.	chlorinated water +6 °C and water +6 °C	1 min + 1 min
6.	chlorinated water +30 °C and water +6 °C	0.5 min + 1 min
7.	water with citric acid (1 g/l) +6 °C and water +6 °C	1 min + 1 min

RH 0 % and the CO_2 permeability was 3900 cm^3/m^2 24 h 101.3 kPa, 23 °C, RH 0 %. One kilo of shredded cabbage was packed into bags in normal air. Various washing methods for shredded cabbage were employed in the 1992-1993 trials. The different washing methods are presented in Table 1. The amount of water was 3 l/k or 3 + 3 l/k. The storage conditions for the packaged shredded Chinese cabbage were +5°C at 75 % relative humidity.

All nutrient were determined on the same day and all determinations were performed in duplicate. Vitamin C was determined according to Speek *et al* [2] with minor modification according to Hägg *et al* [3]. ß-carotene were determined according to Ollilainen *et al* [4] with some slight modifications according to Hägg *et al* [3]. Dietary fiber content was determined according to the method of Prosky *et al* [5]. Sugars were determined according to Li and Schuhman[6] with some modifications according to Haila *et al* [7].

3. RESULTS AND DISCUSSION

The summer cultivar Kasumi was studied in August 1991. The vitamin C content of fresh cabbage was 25 mg/100 g fresh weight (FW) and after 7 d the vitamin C content was 16 mg/100 g FW. Vitamin C content of packaged shredded Chinese cabbage was 16 mg/100 g FW after 1 d and 13 mg after 7 d. Trials were also performed after keeping the cabbage in serving dishes at room temperature or at 10 °C for 3 h. After 3 h, the vitamin C content of fresh shredded cabbage was 17 mg/100 g while that of packaged cabbage did not change after 3 h. The ß-carotene content in packaged cabbage after 1 d storage was the same as after 7 d; 30 µg/100 g FW. The water content of cabbage was 96 % and the ash content was 0.6 g/100 g FW. The dietary fiber content was 1 % of the fresh weight.

The winter cultivar Kingdom had a vitamin C content of 17 mg/100 g FW in December 1991 and that of packaged shredded cabbage was 15 mg/100 g FW after 1 d. Thus, the decrease was much lower than that found in the summer cultivar. Vitamin C content in packaged shredded cabbage was about the same after the 7 d storage. The vitamin C content of fresh Chinese cabbage was 14 mg/100 g FW in December 1992 and in packaged shredded samples it was about 12-13 mg/100 g after 1 d, depending on the washing method, and about 9-12 mg/100 g after 7 d. The highest decrease in vitamin C content occurred in samples washed with water at 6 °C whereas washing at higher temperatures preserved the vitamin C content best. Trials were performed after 6 wk winter

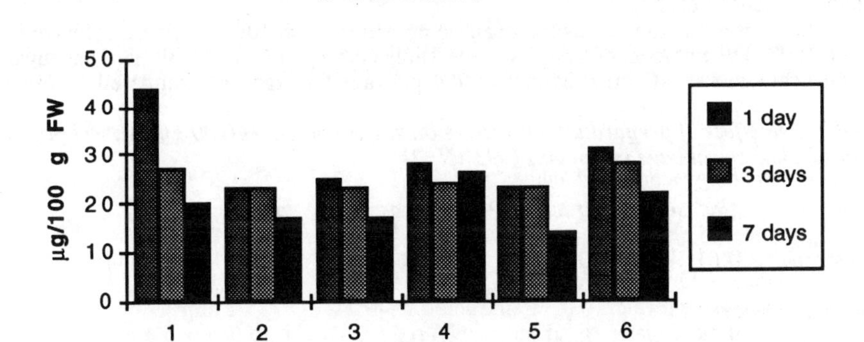

Figure 1. *The effect of preparation procedure on β-carotene content of packaged, shredded Chinese cabbage (The numbers of samples correspond with Table 1)(N=2)*

Table 2. *The effect of preparation procedure on sugar content (g/100 g fresh weight) of shredded Chinese cabbage in autumn 1992 (N=2)*

sample	*glucose*	*fructose*	*sucrose*	*total sugar*
fresh cabbage	0.59	0.50	0.06	1.15
packaged and stored for 7 d				
1.	0.53	0.38	0.04	0.95
2.	0.51	0.38	0.02	0.91
3.	0.54	0.41	0.03	0.98
4.	0.50	0.38	0.04	0.92
5.	0.51	0.38	0.04	0.93
6.	0.49	0.36	0.03	0.88

storage also. The vitamin C content of fresh cabbage was 13 mg/100 g FW and only 1 mg had been destroyed during the winter storage. The packaged samples had a vitamin C content of 10 mg/100 g FW after 1 d storage and about 8-9 mg after 7 d storage, depending on washing method. Thus, vitamin C content was not strongly affected by the washing method in January 1993. Vitamin C content was also determined after cabbage had been kept in dishes for 3 h at room temperature. It was found that 3 h at room temperature did not notably affect the vitamin C content of shredded cabbage.

The ß-carotene content of fresh cabbage was only 19 µg/100 g FW, being still lower in packaged cabbage; 15 µg/100 g FW in December 1991. The results for ß-carotene in 1992 are presented in Figure 1. In December 1992 the ß-carotene content of packaged shredded cabbage varied between 20 and 30 µg/100 g FW, mainly after 1 d storage. The content slightly decreased after 7 d storage, ranging between 15-25 µg ß-carotene/100 g FW. The results for ß-carotene are given in Figure 1. The ß-carotene content of cabbage decreased considerably during 6 wk winter storage. About 5 µg/100 g FW was retained and this amount decreased further during the 7 d packaged storage. The various washing methods used did not appear to affect the ß-carotene content in cabbage. Neither was ß-carotene content notably changed by keeping the shredded cabbage for 3 h in dishes at room temperature.

The water content of fresh Chinese cabbage was 95.7 % in December 1991 and 96.0 % in December 1992. After the 6 wk winter storage the water content was 95.7 %.

The dietary fiber content of fresh Chinese cabbage was 1.2 % in December 1991, 1.1 % in December 1992 and 1.2 % of the fresh weight in January 1993. Neither the 7 d packaged storage nor the various washing methods appeared to affect the dietary fiber content of shredded cabbage.

Table 2 presents the results for sugar composition in shredded Chinese cabbage in autumn 1992. The numbers of samples in the Table correspond with Table 1. Total sugar content decreased in all samples during 7 d packaged storage as compared to fresh

Table 3. *The effect of preparation procedure on sugar content (g/100 g fresh weight) in shredded Chinese cabbage in January 1993 (N=2)*

sample	*glucose*	*fructose*	*sucrose*	*total sugar*
fresh sample	0.61	0.56	0.07	1.24
packaged and stored 7 d				
1.	0.58	0.50	0.04	1.12
5.	0.50	0.43	0.02	0.95
7.	0.45	0.38	0.01	0.84

cabbage. Washing with water at 30 °C and 6 °C preserved best sugar content during the storage period and washing with chlorinated water decreased sugar content slightly more than other washing methods. Fructose had the highest decrease. Sugar content was slightly higher after the 6 wk winter storage in January 1993. Table 3 presents the results for sugar composition in January 1993. At that time all of the trials were not performed, but a new washing method involving the addition of citric acid into the water (1 g/l water) was tried. Washing in plain water only was found to preserve the sugar content best and the washing method employing the citric acid addition lowered sugar content the most.

4. CONCLUSIONS

The retention of nutrients in packaged, shredded Chinese cabbage was good. Vitamin C content was practically unchanged during the 7 d storage. Vitamin C content was affected by the washing methods employed. The ß-carotene content of packaged samples decreased during storage. Dietary fiber content was not affected by storage or by the different washing methods. Sugar content decreased during the 7 d packaged storage. It is therefore possible to retain good nutrient quality in packaged, shredded Chinese cabbage during 7 d storage at +5 °C.

Acknowledgement

Many thanks to Mrs. Tuula Kurtelius and Mrs. Marja-Terttu Wiisak for their skillful technical assistance. This study was financed by the Ministry of Agriculture and Forestry, the Foundation for the Promotion of Food Production and by the Finnish Packaging and Food industry.

References

1. E. Hurme, R. Ahvenainen, A. Kinnunen, E. Skyttä, Proceedings of the Sixth International Symposium of the European Concerted Action Program COST 94 'Post-harvest treatment of fruit and vegetables'. In press.
2. A. Speek, J. Schrijver, W. Schreurs, J. Agric. Food Chem. 1984, **32**, 352.
3. M. Hägg, S. Ylikoski, J. Kumpulainen, J. Food Comp. Anal. 1994, **7**, 252.
4. V. Ollilainen et al., J. Food Comp. Anal. 1988, **1**, 178.
5. L. Prosky et al., J. AOAC 1988, **71**, 1017.
6. B. Li and P. Schuhmann, J. Food Sci. 1980, **45**, 138.
7. K. Haila et al., J. Food Comp. Anal. 1992, **5**, 100.

Modification of mineral element content in green asparagus (*Asparagus officinalis*, L.) during storage by freezing.

M.A. Amaro López, R. Moreno-Rojas, G. Zurera-Cosano, C. Cañal-Ruíz and R.M. García-Gimemo

Departamento de Bromatología y Tecnología de los Alimentos.
Facultad de Veterinaria y Ciencia y Tecnología de los Alimentos.
Avda. Medina Azahara 9
14005-Córdoba

1 INTRODUCTION

The freezing process of vegetables and their subsequent storage may cause variations in their mineral composition[1-3]. This fact, together with an increase in the commercialization of frozen asparagus, has led to the formulation of a study on the possible modifications of the mineral content in asparagus throughout their storage by freezing.

The aim of this research was to determine the changes in the concentrations of copper, iron, zinc, manganese, nickel, chromium, calcium, magnesium, sodium, potassium and phosphorus in green asparagus as a result of their storage by freezing at -18°C during 45 and 90 days.

2 MATERIAL AND METHODS

2.1 Samples

This study was carried out in samples of green asparagus of the UC-150 variety which were classified in two groups according to their diameter: <11 mm and >14 mm. The asparagus in each sample group were sliced 21 cm long and divided into two portions for their subsequent analysis: the apical portion or bud (7cm from the tip of the asparagus) and the basal portion or remainder (remaining 14 cm of stem). The samples were taken at the following time intervals: 0, 45 and 90 days of freezing storage (-18°C).

2.2 Analytical Procedure

The turions were first washed with deionized water to remove soil and dust from the surface of the asparagus. The portions of the turions (apical and basal) were weighed separately and dried at 100°C to constant weight. For analysis of all the mineral elements, the dry ashing method[4] was applied. Two grams of the homogenized and dried sample were weighed in porcelain crucibles and mineralized overnight in muffle furnace (460°C). Subsequently, the ash was extracted with a mixture of $HCl-HNO_3-H_2O$ deionized (2.5 ml) at a ratio of 1:1:2 (v/v). The solution obtained was dried on thermostatic plate and placed in muffle furnace (460°C) for 1 h. The resulting white ash was suspended with 5 ml of mixture $HCl-HNO_3-H_2O$ deionized (1:1:2) and the solution was placed in a 15 ml-volumetric flask made up to volume with deionized water. The analytical determinations of copper, iron, zinc, manganese, nickel, chromium, calcium and magnesium were performed with a Perkin-Elmer Model 2380 Atomic Absorption Spectrophotometer, using air-acetylene flame and single element hollow cathode lamps. Sodium and potassium were determined by flame atomic emission spectrophotometry. The analysis of the phosphorus was carried out by visible-ultraviolet spectrophotometry according to the ammonium vanadate-molybdate colorimetric method[5] and a wavelength of 400 nm was used.

The sensitivity obtained (mg/l) for copper, iron, zinc, manganese, nickel, chromium, calcium, magnesium and phosphorus was 0.094, 0.743, 0.363, 0.061, 0.142, 0.078, 0.077, 0.088 and 0.051 respectively. The precision of the method was obtained by calculation of variation coefficients[6,7]; the results were Cu=1.36%, Fe=0.98%, Zn=1.16%, Mn=1.35%, Ni=3.32%, Cr=6.40%, Ca=1.22%, Mg=1.20%, Na=3.90%, K=1.35% and P=2.12%. The accuracy was evaluated by standard additions of known concentrations of trace elements and by an analysis of the standard reference material *"Citrus Leaves"*, Standards Measurement and Testing, Brussels, (BCR 1572). The recovery percentages of the standard additions were Cu=94.3%, Fe=99.9%, Zn=97.6%, Mn=99.8%, Ni=99.1% and Cr=99.4%. The mean value recoveries of mineral elements by Standards Measurement and Testing, Brussels (BCR 1572) were Cu=103.8%, Fe=102.6%, .Zn=100.3%, Mn=99.3%, Ni=105.0%, Cr=108.0%, Ca=102.4%, Mg=99.1%, Na=108.1%, K=97.5% and P=99.0%. For the calculation of the detection limit. the definition and criteria established by I.U.P.A.C. were followed[8,9]. The concentration limits obtained (minimum concentrations detectable in dry weight) were 0.053, 0.843, 0.775, 0.218, 0.549, 0.040, 11.82, 17.96 and 1.04 (mg/kg dry weight) for copper, iron, zinc, manganese, nickel, chromium, calcium, magnesium and phosphorus respectively.

2.3 Statistical Study

The concentrations obtained in the chemical analysis for the samples were evaluated statistically using analyses of variance studies and the Scheffé multiple range test ($p<0.05$), which permitted the formation of homogeneous groups by an association of classes of statistically similar concentrations[10,11]

3 RESULTS AND DISCUSSION

The results obtained are shown in Table 1. Lower levels of moisture were observed in the apical portions of the asparagus with respect to the rest of stem. The analysis of variance for moisture established statistically significant differences ($p<0.001$) between portions, diameters and throughout the frozen storage period. Three different Scheffé homogeneous groups were formed ($p<0.05$) of which the fresh asparagus displayed the highest percentage of moisture whilst the frozen asparagus stored for 45 days showed the lowest moisture content.

The mean concentrations found for each element investigated were generally within the ranges indicated by different authors[12-15]. Three-factor (diameter, time and portion) analyses of variance for each element were performed and the results obtained showed statistically significant differences for most of the elements ($p<0.001$, in calcium $p<0.01$) throughout frozen storage except for iron, zinc and chromium with non-significant differences ($p>0.05$). The homogeneous groups formed in the Scheffé multiple range analysis ($p<0.05$) made it possible to observe how, after 45 days of freezing, levels decreased in most of the elements (except for copper which fell at 90 days) and a common homogeneous group was formed in the controls at 45 and 90 days of freezing. The green turion considered as being fresh was included in another independent group with a higher concentration. The behaviour of phosphorus stood out as it formed three different homogeneous groups for each control interval established. Not all the elements manifested this general behaviour since the manganese content underwent a notable increase at 90 days` freezing which was similar to that of sodium whose concentration increased during the storage time.

To sum up, during the frozen storage period of green asparagus, the mineral elements investigated evolved in different ways and a diminution was noted in some of them. In general, the different authors report few changes in the mineral concentration in fresh or frozen products. However, a decrease in calcium content in fruit and vegetables was observed after freezing[1] similar to that described for green asparagus. Other authors[3] proposed that the mineral losses due to the freezing process were only significant in the case of copper in green beans and peas. On the other hand, minerals such as manganese and sodium displayed an increase in their levels throughout the freezing period which coincided in this sense with the increase in the concentrations of some minerals, such as zinc and calcium, observed by other authors[2] during the freezing process of green beans.

The diameter factor had a varying influence according to the mineral considered and statistically significant differences ($p<0.001$) were established for copper, iron, zinc, calcium and potassium, whilst the remainder of the elements analyzed did not show any significant differences ($p>0.05$). Statistically significant differences were found between portions for all the elements ($p<0.001$ and sodium $p<0.01$), with the exception of manganese, chromium and potassium ($p>0.05$).

References

1. C.J. Wyatt and K. Ronan, *J. Agric. Food Chem.*, 1983, **31 (2)**, 415.
2. A. Lopez and H.L. Williams, *J. Food Sci.*, 1985, **50**, 1152.
3. M.V. Polo, M.J. Lagarda and R. Farré, *J. Food Composition and Analysis*, 1992, **5**, 77.
4. G. Zurera-Cosano and R. Moreno-Rojas, *Food Chemistry*, 1990, **38**, 133.
5. AOAC (Association of Official Analytical Chemist), "Official methods of analysis" (1991). 15th de; 2nd. supplement, 991.25, 101.
6. A. Alegría, R. Barberá and R. Farré, *Journal of Micronutrient Analysis*, 1988, **4**, 229.
7. R. Barberá, R. Farré and M.J. Roig, Anales de Bromatología, 1990, **42(2)**, 345.
8. G.L. Long, and J.D. Winefordner, *Anal. Chemestry*, 1983, **55**, 712A.
9. Analytical Methods Committee, *Analyst*, 1987, **112**, 199.
10. J.R. Piggott, "Statistical Procedures in food reserch", London and New York: Elservier Applied Science, London, 1986.
11. A. Molina-Alcalá, J.V. Delgado-Bermejo, J.M. Rodero-Fraganillo and R. Moreno-Rojas, "Introducción a la estadística descriptiva e inferencial para investigadores. Procedimientos S.A.S.", Centro de Cálculo-Instituto de Zootecnia, Universidad de Córdoba, 1992.
12. L. Brown and R. Carolus, *Proceeding American Soc. Hortic. Sci*, 1965, **86**, 332.
13. E. Lubet and C. Juste, *CR Academia de Agriculture Francaise*, 1974, **60.**
14. J.A. Espejo-Calvo "Influencia de la fertilización fosfórica y longitud de corte en la productividad y calidad del espárrago verde de Huetor-Tejar", Tesina de Licenciatura, Facultad de Ciencias, Universidad de Córdoba, 1991.
15. K.K: Nielson, A.W. Mahoney, L.S. Willians and V.C. Rogers, *J. Food Composition and Analysis*, 1991, **4**, 39.

Table 1 *Percentage content and Cu, Fe, Zn, Mn, Ni, Cr, Ca, Mg, Na, K and P concentrations (mg/kg dry weight) according to freezing storage, diameter and portion green asparagus (mean ± SD).*

	Moisture	Cu	Fe	Zn	Mn
Freezing					
0 days	93.1 ± 1.63^{A}	64.7 ± 24.8^{A}	68.9 ± 11.4^{A}	109.8 ± 31.8^{A}	34.9 ± 3.52^{B}
45 days	92.2 ± 1.59^{C}	60.2 ± 25.4^{A}	64.9 ± 12.5^{B}	95.6 ± 30.1^{B}	33.1 ± 3.48^{B}
90 days	92.5 ± 1.43^{B}	53.6 ± 18.7^{B}	65.0 ± 10.7^{B}	99.3 ± 30.2^{B}	37.2 ± 2.88^{A}
Diameter					
<11 mm	92.2 ± 1.38	64.7 ± 24.5	69.6 ± 11.7	103.3 ± 33.3	35.2 ± 3.94
>14 mm	93.0 ± 1.66	54.2 ± 20.5	63.0 ± 10.0	99.9 ± 27.7	34.9 ± 3.39
Portion					
Apical	91.2 ± 0.44	79.2 ± 13.8	76.4 ± 5.08	130.0 ± 8.82	37.9 ± 2.07
Basal	94.0 ± 0.77	39.7 ± 4.59	56.2 ± 3.72	73.2 ± 6.74	32.2 ± 2.18
TOTAL	92.6 ± 1.54	59.5 ± 22.5	66.3 ± 11.2	101.6 ± 30.0	35.0 ± 3.59

Table 1 *(continue)*

	Ni	Cr
Freezing		
0 days	5.42 ± 1.12^{A}	0.68 ± 0.08^{A}
45 days	4.11 ± 0.67^{B}	0.64 ± 0.05^{A}
90 days	3.80 ± 0.84^{B}	0.62 ± 0.00^{A}
Diameter		
<11 mm	4.28 ± 1.43	0.66 ± 0.07
>14 mm	4.61 ± 0.71	0.63 ± 0.04
Portion		
Apical	5.07 ± 1.13	0.67 ± 0.08
Basal	3.81 ± 0.69	0.62 ± 0.00
TOTAL	4.44 ± 1.12	0.64 ± 0.06

Table 1 *(continue)*

	Ca	Mg	Na	K	P
Freezing					
0 days	5996 ± 529^{A}	2208 ± 294^{A}	290 ± 28^{C}	37862 ± 3253^{A}	8812 ± 2455^{A}
45 days	5693 ± 314^{B}	1857 ± 234^{B}	312 ± 36^{B}	28286 ± 4330^{B}	7973 ± 2532^{B}
90 days	5703 ± 147^{B}	1855 ± 162^{B}	342 ± 27^{A}	30244 ± 5538^{B}	7243 ± 2022^{C}
Diameter					
<11 mm	5948 ± 474	1972 ± 349	307 ± 38	29670 ± 6406	7830 ± 2387
>14 mm	5647 ± 154	1975 ± 212	322 ± 41	34592 ± 4600	8189 ± 2376
Portion					
Apical	6032 ± 391	2161 ± 250	344 ± 29	29344 ± 5133	10159 ± 929
Basal	5563 ± 157	1786 ± 166	285 ± 23	34918 ± 5666	5860 ± 689
TOTAL					

A, B, C, **Scheffè homogeneous groups ($p<0.05$) between ripening state.**

VEGETABLES AS POTENTIAL SOURCES FOR MEAT PRODUCTS

Z. Simkevicienė and D. Kazemėkaitytė

Department of Food Technology
Kaunas University of Technology
Kaunas, Lithuania

INTRODUCTION

Coronary heart disease, cancers, strokes, diabetes mellitus and atherosclerosis have been associated with dietary excesses and nutrient imbalance /1/. Meat products made traditionally are high in fat and in cholesterol and need significant improvement according the increased demand of low-fat items. Dietary recommendations noted necessary for increased consumption of foods high in fibers, vitamins and minerals, also. Gums, starch, cellulose - based derivatives have been tested in formulated meat products. Unfortunately, products containing the gum produced a significant off - flavor and those with cellulose or gum had a distinct graininess/flouriness attribute /2,3/. The comparable study of consumption of naturally high fiber foods and foods with fiber isolates showed that benefits that result from inclusion of vegetables in the diet have largely been proven. However in many cases of manufactured high-fiber foods they have not the same protective benefits as their parent products /4/. One can find reports on the effect of added vegetables for the nutritive value of foods /4 -5/, but there is a lack of information about their functionality in meat systems.

In the present study the commonly used vegetables as potential ingredients for low-fat meat products were analyzed.

MATERIALS AND METHODS

Parsley *(Petroselinum crispum Nym.)*, parsnip *(Pastinaca sativa L.)*, leek *(Allium porum L.)*, celery *(Apium graveoleus L.)* and topinambur *(Helianthus tuberosium L.)* were collected from Lithuanian Institute of Horticulture. The roots were cleaned, washed, ground and dried at ambient temperature with an activated ventilation. Dried material was milled till the particle size 1 mm and stored in bags until use.

Water and fat adsorption were conducted according to the procedures described by /5/. Content of adsorbed liquid expressed as a percentage of the original weight of dried vegetables.

Emulsifying property (emulsion capacity) was analyzed after preparing the emulsion from 3,5 g of dried vegetables and 50 ml of water and 50 ml of fat (mixing - 5 min., speed -3,600 rev/min, temperature - 20°C). Emulsions were filled to centrifuge tubes and volume of emulsion was measured after centrifugation (speed 2000 rev/min). Results expressed as a percentage of the volume of initial emulsion.

Proximate composition of dried vegetables was determined according to AOAC methods /6/ while mineral composition - by spectra emission photographic method (spectrograph ISP - 28 (Russia), Ca, Mg, Fe, Zn, Cu, Mn - with atomic absorption spectrophotometer (model Perkin - Elmer), and Na, K - after mineralization with spectrophotometer PFM (Hungary).

RESULTS AND DISCUSSION

Water and fat adsorption, emulsifying capacities are the key properties for ingredients of meat systems. Several factors, such as type and chemical composition affect the functional properties of products. Evaluation of water adsorption (Figure 1) shows that dried vegetables absorbs 4.5 to 6.5 times its own weight of water. The best water adsorption contributes to leek (647 %). The effect probably is attributable to highest content of proteins (till 20.4%).

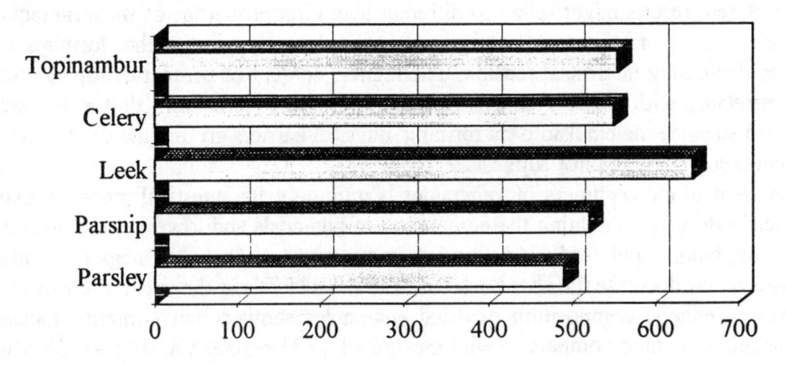

Figure 1 *Water adsorption capacity of dried vegetables, %*

The results indicated in Figure 2 showed the similarities of fat adsorption compared with the all analyzed vegetables and they did not exceed 83%. Differences may be related to different hydrophobity of proteins while carbohydrates constituents of vegetables have no marked influence on oil binding.

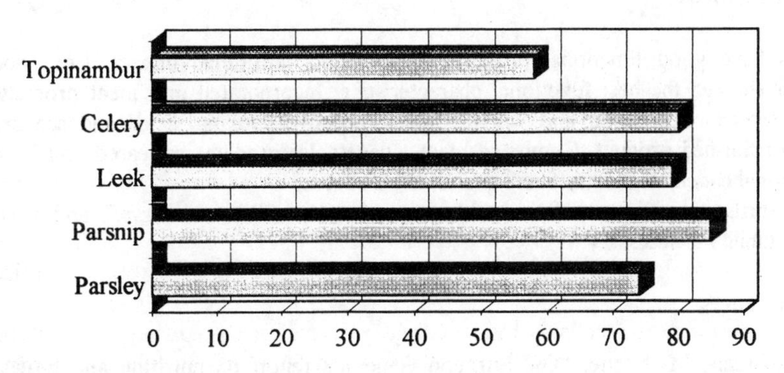

Figure 2 *Fat adsorption capacity of dried vegetables, %*

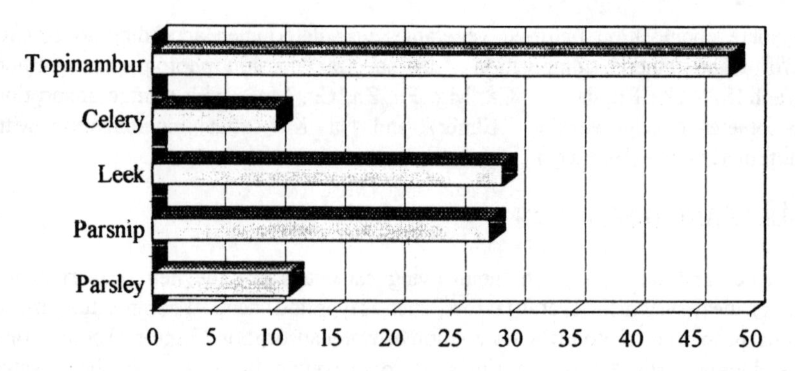

Figure 3. *Emulsifying capacity of dried vegetables, %*

These different results might relate to different hydration properties of the vegetables and their chemical composition. Proteins and phospholipids aid in the forming of emulsions by decreasing interfacial tension. The highest content of protein of topinambur and leek correlates with highest emulsion stability. It was noted /3/, that plant cell fragments also stabilize the emulsions by forming physical barriers around the oil droplets, but our experimental data did not suggest it.

High content of dietary fibers of vegetables is important for intestinal mobility, may have beneficial effect by facilitating the elimination of bile acids and lowering the level of cholesterol in blood and reducing insulin requirements. On the negative side, polysaccharides may decrease the absorption of essential minerals in the small intestine /1/. The analyses of mineral composition of dried vegetables showed that content of main elements are relatively high compared with meat: 0.11 -0.31 g/100g Ca, 0.11 - 0.21 Mg, 5.5 - 7.0 mg/kg Cu, 27.5 - 35.0 mg/kg B, 2.5 - 13.0 mg/kg Mn. Potassium lessens risk of cardiovascular diseases, and the highest content of this element was found in parsnip and topinambur - 3.1 and 2.9 g/100g dried vegetables respectively. The high content of minerals of analyzed vegetables appeared will not reduce the mineral level of vegetable - meat product.

CONCLUSION

Vegetables have good functional properties as well as nutritional impact. Leek and topinambur showed the best functional characteristics. Incorporated into meat products they may reduce released fat and water content, improve cooking yields and sensory attributes of finished product. Combined meat - vegetables products appeared will have high nutritional benefit as dietary fibers and minerals sources

The further experiment of vegetables' constituents effects on "raw" and meat emulsions stability is necessary.

References

1. J.M. McGinnis, M. Nestle, 'The Surgeon General's report on nutrition and health: policy implications and implementation strategies'. Am.J.Clin.Nutr., 1989, **49**, 23.

2. J.I.Harland. The addition of dietary fiber to foods and beverages. Food Technology international Europe, 1992, 219.

3. Keeton, J.T. 1991. Fat Substitutes and Fat Modification in processing. Proceedings 44 - th Annual reciprocal meat conference. June 9 - 12, 1991, Kansas State University Manhattan, KS.AMSA, p.79.

4. D.L.Huffman, W.Russell Egbert, Chiao - min Chen, D.P.Dylewski. Technology for Low - Fat Ground Beef. Proceedings 44 -th Annual reciprocal meat conference. June 9 - 12, 1991, Kansas State University Manhattan, KS.AMSA, p.73.

5. Д.Жилинскайте.Химический состав пищевых волокн в литовских пищевых продуктов и их применение в питании. Автореферат диссертации. Москва, 1992, с. 22

6. AOAC. Official Methods of Analysis of the Assoc. of Analytical Chemists, 14th edn., Arlington, 1984.

THE ROLE OF EXPECTATIONS ON PASSIONFRUIT JUICE PERCEPTION

R. Deliza, H.J.H.MacFie, D. Hedderley and C. Howard

BBSRC - Institute of Food Research
Whiteknights Rd., Earley Gate,
Reading RG6 6BZ
UK

1 INTRODUCTION

The label plays an important role on food selection because it is the major source of information for consumers[1], permitting them to make better choice decisions in the marketplace. Many studies dealing with different label aspects have been carried out, including food label legislation[2], brand name and its role on quality and sensory perception[3], and the impact of nutritional information towards the product[4].

Although liking for a food product is determined by its sensory attributes[5], external factors may affect the expected food product quality, with direct effects on liking[6]. Until now, no study has been published focusing on the sensory attributes and the features on the packaging which could affect consumers' perception. This experiment investigated how manipulation of several aspects of fruit juice packaging affected consumer expectations of the passionfruit juice sensory attributes.

2 MATERIAL AND METHODS

2.1 Subjects

The subject population of 94 consumers was recruited locally, and consisted of: 50% male, 50% female; 50% with children, 50% with no children at home. All consumers were British and had never tried passionfruit juice. They were aged between 19 and 63 years. The subjects received £5.00 for their participation.

2.2 Packaging manipulation and image display

The images were created based on the salient features on the fruit juice packaging which were achieved from focus group sessions and the repertory grid method[7]. The features and their levels are in table 1.The actual features used were obtained from commercial fruit juice packagings which were photographed with colour slide film and then scanned into a photographic image retouching program. They were editing by using the Adobe Photoshop™ Macintosh® software and transformed into slides, which were presented to the consumers in the sensory booths for a standardized time. An experimental design using SAS software was used in order to test a reasonable number of packages with

Table 1 *Packaging variables*

Feature	*Level*
Background colour	Orange and White
Picture	Photograph and Drawing
Level of Information	None, Medium and Lots
Brand	None, Minor and Major
Language	Name of the product in English and Foreign
Shape	Normal and Unusual

a maximum efficiency. The final design had twenty-four packages which were tested by all the consumers.

2.3 Sensory evaluation

Consumers evaluated the expected sweetness, sharpness, refreshingness, and liking only by looking at the fruit juice packages, without tasting the juices. They used a non-structured horizontal nine-centimetre line scale labelled with "not at all" on the left and "extremely" on the right side. The liking scale was labelled with "dislike extremely" and "like extremely" at the left and right hand poles, respectively.

2.4 Statistical analysis

The data were analyzed by using Conjoint and Cluster Analysis[8]. Internal preference mapping (MDPREF)[9] analysis was also used with the liking data.

3 RESULTS

Six factor effects from each consumer were used to form five clusters of similar patterns. A five-segment solution was chosen based on the dendogram and the sixth one was formed grouping all the individual scores (n=10) which did not fit into these five segments (these results are not presented here).

For the liking scores, a four-segment solution was chosen and a fifth one was formed for those whose data did not fit into the previous segments (n=14).The relative ranges of the part-worth functions were compared across features within segments to estimate the relative importance of each feature.

The results of the part-worth analysis for the sensory and liking and the preference mapping for the liking data are presented below.

3.1 Sweetness

Segment 1, which includes 7% of the consumers, considers brand to be the most important packaging characteristic on the packaging. Packaging with a major brand name was expected to be sweeter than with no name. The second most important significant characteristic was shape and consumers expect the unusual shape to be sweeter than the normal one. Picture had a significant effect for this segment and the drawing was expected to be sweeter than the passionfruit photograph. Language had a smaller relative importance and background colour and information had no effect on the evaluation of the expected sweetness.

Segment 2 (size 29%) regards brand as the most important feature on the packaging when evaluating the expected sweetness. Major brand name was expected to be sweeter than either no name or minor one. Language was significant for that segment and the packaging written in English was expected to be sweeter than the foreign label. The other features had no effect on the expected sweetness.

Segment 3, which comprises 14% of the consumers, uses picture as the most important characteristic for the evaluation of the passionfruit juice sweetness. The packaging with a drawing was expected to be sweeter than with a photograph. Shape had a significant effect and normal shape was expected to be sweeter than the unusual one. No brand and major brand were expected to be less sweet than minor brand name.

Segment 4, which includes 33% of the consumers, uses background colour to evaluate the expected sweetness and rated orange background colour sweeter than the white one. Picture also had an effect for this segment and photograph increased expected sweetness relative to the drawing.

The smallest segment which includes 6% of the consumers, cued primarily on information and background colour to evaluate the sweetness, and they expected the packaging with lots of information to be sweeter than either no or medium information. Orange background colour was expected to be sweeter than a white one.

3.2 Sharpness

Segment 1 uses mainly background colour to evaluate the expected passionfruit juice sharpness and consumers in this segment expected the orange background colour to be sharper than the white one. Language was significant for this segment and foreign label was expected to be sharper than English label.

Although information had the greatest relative importance in segment 2, it was not significant. The only significant packaging characteristics on the evaluation of the expected sharpness was language and, consumers in this segment expected the foreign language to be sharper than the English one.

Segment 3 uses picture as the most important characteristics on the sharpness evaluation. Packaging with a photograph was expected to be sharper than with a drawing.

Segment 4 considers background colour, information and picture as the most important features on the sharpness assessment. Orange background colour, photograph, and none and medium information were expected to deliver a less sharp juice than the white background, drawing and lots of information.

The features had no effect on the evaluation of the expected sharpness for the consumers in segment 5.

3.3 Refreshingness

For segment 1 brand was the most important feature and consumers expected the major brand name to be more refreshing than both none and minor brand name. Picture had a significant effect for this segment and packaging with a drawing was expected to be more refreshing than with a photograph.

Segment 2 uses information to evaluate the expected refreshingness. As the amount of information increases, the expected refreshingness increases as well.

Consumers in segment 3 regard information as the most important feature on the packaging when evaluate refreshingness. As the amount of information increases, the expected refreshingness also increases. Picture, brand, and shape are relevant and consumers expected the packaging with a drawing, major brand name and normal shape to be more refreshing than the alternative features.

Segment 4 uses information and picture for evaluating the expected refreshingness. Consumers expect a more refreshing juice when the amount of information increases and when a drawing is on the packaging.

Segment 5 uses information and brand as the most important characteristics on the expected passionfruit juice refreshingness.This attribute is expected to increase as the amount of information increases and a brand name (either minor or major) is presented.

3.4 Liking

The Conjoint analysis on the liking data revealed that segment 1, which includes 47% of the consumers, considers information the most important package characteristic for the evaluation of the expected passionfruit liking. Package with no information was expected to be less liked than either with medium or lots of information. The second most important feature on the package for this segment was background colour: white was preferred over the orange one. Although having a smaller relative importance, other package characteristics were statistically significant for the consumers in this segment, such as brand, language and picture.

Segment 2 (size 13%) regards information as the most important feature on the package. As in segment 1, consumers consider the more the amount of information the more the expected liking. For the consumers in this segment, the shape of the package also had a considerable relative importance, and the normal shape was expected to be more liked than the unusual one.

Segment 3, which is the smallest one (size 8.5%), uses information and brand as the most important characteristics on the evaluation of the passionfruit juice liking. As the amount of information increases, the expected liking increases as well, and a major brand name is expected to be more liked than a minor or no name.

Segment 4 (size 17%) considers picture as the most important package attribute on their expected liking. Consumers expect the package with the drawing to be more liked than with the photo. The second most relevant factor for these consumers is the information. They expect the package with lots of information to be more liked than ones with no or medium amount of information.

According to the results of the study, it is possible to infer that the features on the packaging had a very important role in the evaluation of the expected sensory attributes for most of the segments. For the expected liking, information was significant for all the segments.

Figure 1 shows the subject and product plots of the first two orthogonal axes, from de MDPREF. The majority of the consumers for whom the regression was significant, have the expected liking in the positive direction of dimension 1, and in both positive and negative directions in dimension 2. Most of the consumers expected to like more the juice with lots of information (packagings 9-12 and 21-24) and either white (packagings 1-12) or orange (13-24) background colour .

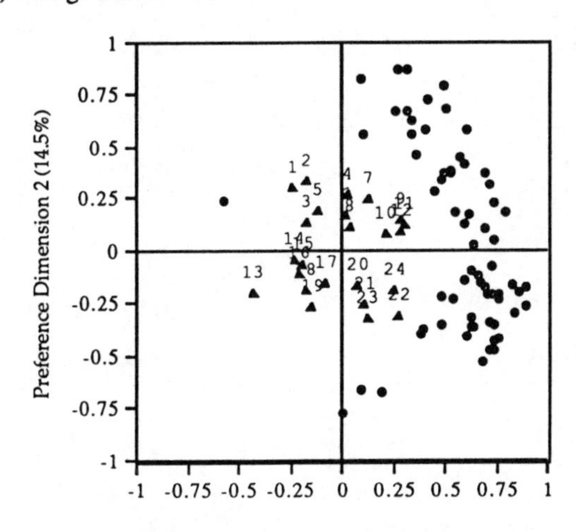

Figure 1 *Preference map showing the expected preferences for passionfruit juices from the packagings (1-24 denote packagings and the dots denote subjects)*

4 CONCLUSIONS

Packaging had a significant effect on the expected passionfruit juice sensory attributes and liking. Consumers were differently affected by the features on the packaging, as revealed by the segments after clustering. Further research is necessary to investigate the effect of the packaging on sensory perception when consumers taste the juice.

References

1. ANONYMOUS, Food Australia, 1989, **41**, 930.
2. N.H. MERMESLSTEIN, Food Tech., 1993, **47**, 81.
3. D. MARTIN, Food Sci. and Tech.Today, 1990, **4**, 44.
4. S. FULLMER; C.J. GEIGER and C.R.M. Parent, J. Am. Diet Assoc., 1992, **91**, 66.
5. A. LIGHT, H. HEYMANN and D.L.HOLT, Food Tech., 1992, **47**, 54.
6. A.V. CARDELLO and F.M. SAWYER, J. Sensory Stu., 1992, **7**, 253.
7.W.A.K.FROST and R.L.BRAINE, Commentary, 1967, **9**, 161.
8. P.E. GREEN and V.R. RAO, J. Marketing Research, 1971, **8**, 355.
9. K. GREENHOFF and H.J.H MacFIE, Measurement of food preferences', HJH MacFie and DMH Thomson (ed),Blackie Academic & Professional, London, 1994, Chapter 6, p.137.

This work was funded by EMBRAPA - Brazilian Enterprize of Agricultural Research

Section 7: Anti-oxidants in Food Plants

HEALTH-ENHANCING INGREDIENTS OF SOYBEAN IN FOODS: OXYGEN RADICAL SCAVENGING CHARACTERISTICS OF DDMP SAPONINS AND THE RELATED SUBSTANCES

Kazuyoshi Okubo

Faculty of Agriculture
Tohoku University, Japan

1 INTRODUCTION

The quality of food is often evaluated by its function such as primary, secondary and third functions. Since any food is basically a nutrient supplier, its nutritional function is understood to be of primary importance(primary function).This understanding may be accepted almost in any segment of the world in any period of history. It is also understood that any food should have a palatable taste(secondary function). Actually, the eating quality of most foods depends on their function to our sensory systems. Such being the case, the function of food has long been studied mostly from the aspects of both nutrition and preference. A great many studies conducted in recent years have shown that there are substances of food-origin with the "third" function involved in preventing diseases by modulation of immune, endocrine, neuron, circulation, and digestion systems. Conceptually, the cellular system may be included. Actually, a number of substances possibly contributing to modulation of these physiological systems have been elucidated for chemical structure and physiological function.

In principle, physiologically functional foods are designed and constructed based on substances with the third function. They lie in a position between conventional foods and medicines, targetting the intermediate, semi-healthy state which is understood as a premonitory. To be brief, any physiologically functional food should be used for prevention of a particular disease at the stage of premonitory, not for remedy of the disease at the stage of development. This concept was proposed and followed up in three national projects[1-3] sponsored by the Japan Ministry of Education, Science and Culture. Meanwhile, the Japan Ministry of Health and Walfare has initiated an investigation of the physiologically functional food from an administrative standpoint and has been officially approving several items defined in terms of "food for specified health use".

Soybean seeds have been consumed by humans for thousands of years because they are rich in good nutritional protein and oil. Most East Asians consume soybean seeds regularly from childhood via a variety of soybean products. The incidence of breast and colon cancer in Oriental people is considerably lower than in those living in Western countries[4], who seldom eat soybean products. Additionally, vegetarians, who are also at decreased risk of breast and colon cancer, frequently consume soybean-based meat substitutes.[5].

Recent works in Japan on hypolipidemic, hypocholesterolemic and hypotentative ingredients of soybean seeds and the products are introduced, and oxygen radical scavenging characteristics of DDMP(2,3-dihydro-2,5-di hydroxy-6-methyl-4*H*-pyran-4-one) conjugated saponins and the related substances in soybean seeds and the products will be the focus of this presentation. Active oxygen species can cause damage to biomolecules including protein and DNA.[6-10] Chemiluminescence in the presence of aldehyde and superoxide anion, hydroxy radical, hydrogen peroxide or lipid hydroperoxide is a new useful method for detection and characteristic of oxygen radical scavengers.

2 CHEMICAL STRUCTURE OF SAPONINS

Soybean saponins are oleanone-type triterpene saponins. Two nomenclature systems(from Kitagawa and Okubo, Table 1) have been used to identify various soyasaponins[11]. For the sake of consistency, only the Okubo system will be used here.

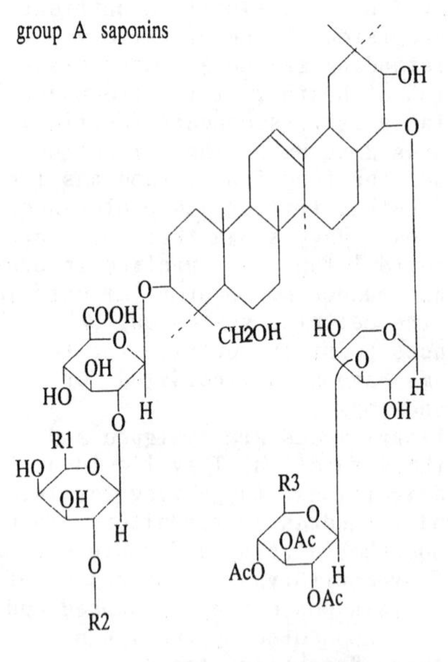

	R1	R2	R3
soyasaponin Aa	CH2OH	β-D-Glc	H
soyasaponin Ab	CH2OH	β-D-Glc	CH2OAc
soyasaponin Ac	CH2OH	α-L-Rha	CH2OAc
soyasaponin Ad	H	β-D-Glc	CH2OAc
soyasaponin Ae	CH2OH	H	H
soyasaponin Af	CH2OH	H	CH2OH
soyasaponin Ag	H	H	H
soyasaponin Ah	H	H	CH2OH

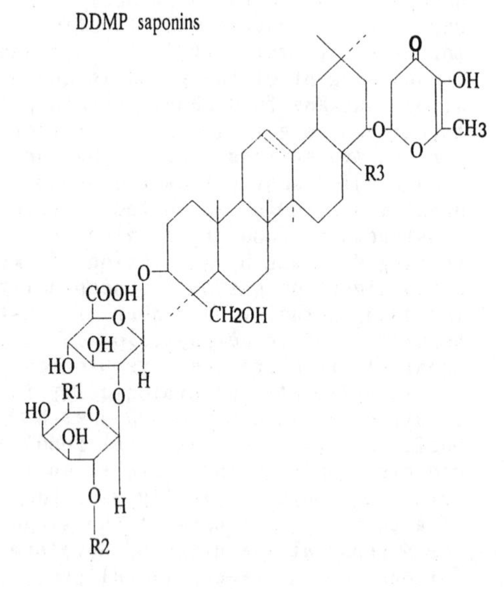

	R1	R2	R3
soyasaponin αg	CH2OH	β-D-Glc	CH3
soyasaponin αa	H	β-D-Glc	CH3
soyasaponin βg	CH2OH	α-L-Rha	CH3
soyasaponin βa	H	α-L-Rha	CH3
soyasaponin γg	CH2OH	H	CH3
soyasaponin γa	H	H	CH3
lablab saponin I	CH2OH	α-L-Rha	CHO

Figure 1 *Chemical structure of Soyasaponins and Lablab Saponin I*

Table 1 *Soyasaponin Nomenclature Assigned by the Research Group of Kitagawa and Okubo*

	Kitagawa	Okubo
A group — acetylated[a]		
glc-gal-glcUA-A-ara-xyl (2,3,4-triAc)	acetylA4	Aa
glc-gal-glcUA-A-ara-glc (2,3,4,6-tetraAc)	acetylA1	Ab
rha-gal-glcUA-A-ara-glc (2,3,4,6-tetraAc)	—	Ac
glc-ara-glcUA-A-ara-glc (2,3,4,6-tetraAc)	—	Ad
gal-glcUA-A-ara-xyl (2,3,4-triAc)	acetylA5	Ae
gal-glcUA-A-ara-glc (2,3,4,6-tetraAc)	acetylA2	Af
ara-glcUA-A-ara-xyl (2,3,4-triAc)	acetylA6	Ag
ara-glcUA-A-ara-glc (2,3,4,6-tetraAc)	acetylA3	Ah
A group — deacetylated[a]		
glc-gal-glcUA-A-ara-xyl (2,3,4-triAc)	A4	deacetyl Aa
glc-gal-glcUA-A-ara-glc (2,3,4,6-tetraAc)	A1	deacetyl Ab
gal-glcUA-A-ara-xyl (2,3,4-triAc)	A5	deacetyl Ae
gal-glcUA-A-ara-glc (2,3,4,6-tetraAc)	A2	deacetyl Af
rha-gal-glcUA-A[b]	—	—
ara-glcUA-A-ara-xyl (2,3,4-triAc)	A6	deacetyl Ag
ara-glcUA-A-ara-glc (2,3,4,6-tetraAc)	A3	deacetyl Ah
B group[c]		
glc-gal-glcUA-B	V	Ba
rha-gal-glcUA-B	I	Bb
rha-ara-glcUA-B	II	Bc
gal-glcUA-B	III	Bb′
ara-glcUA-B	IV	Bc′
E group[d]		
glc-gal-glcUA-E		Bd
rha-gal-glcUA-E		Be

[a] Sugar chains on the left of A are linked to 3-*O* and those on the right to 22-*O*.
[b] Isolated by Curl *et al.* and given the name soyasaponin A3.
[c] Soyasapogenol B contains a hydroxyl moiety at C-22.
[d] Soyasapogenol E contains a ketone moiety at C-22.

Many soybean saponins have been isolated and charaterized.[11-14] They are divided into three groups such as A, B and E(Fig. 1) according to their respective aglycone, soyasapogenol A, B or E. Group A saponins are bis desmoside saponins that contain two sugar chains attached at the C-3and C-22 positions of soyasapogenol A. Group B and E saponins had been thought to be mono-desmoside saponins that contain a sugar chain attached to the C-3 position alone, but recently it has been found that they contain a 2,3-dihydro-2,5-dihydroxy-6-methyl-4*H*-pyran-4-one(DDMP) moiety at the C-22 position of soyasapogenol B or E(Fig. 1)[15]. Saponin Bb belong to the group B saponins, and saponin Be is included in the group E saponins. Soyasaponin βg (saponin BeE) contains DDMP at the C-22 position of saponin Bb and Be, must be the genuine saponin of "soybean saponins Bb and Be", because the purified soyasaponin βg produces saponin Bb and Be after heating [16]. Furthermore, other DDMP saponins, αg, βa, γg and γa, which correspond to each of the four group B saponins, Ba, Bc, Bb´ and Bc´, respectively, have now been identified [17] This means that group B and E saponins must have been derived from the saponin which contain DDMP at the C-22 position of soyasapogenol B.

2.1 Group A Saponins

It has been observed that the saponin composition in the soybean seeds is not affected by a difference in cultivation conditions but is peculiar to the variety. [18] Soybean saponin Aa and Ab are the major constituents of group A saponins. Most soybean varieties contain either soybean saponin Aa or Ab. [18] Saponin Aa contains a xylose residue at the C-22 position of soyasapogenol A, whereas saponin Ab contains a glucose residue at this position. [19] The existence of soybean saponin Aa and Abis controlled by codominant alleles at a single locus, [20] and most soybean varieties can be divided into the xylose(X) and glucose(G) types

2.2 DDMP saponins

Recently we have isolated genuine saponins, soyasaponin αg, βg, βa, γg and γa from soybean, [17] and soyasaponin αa from scarlet runner bean. [21] Also soyasaponin βg was isolated from rootstock of the American groundnut(*Apios americana*). [22] The structure was identified by ^{1}H-NMR and ^{13}C-NMR and by chemical techniques. The distribution of this DDMP saponin in the rootstock was detected as the brown color produced by their action with $FeCl_3$. A high concentration of DDMP saponin was observed around the cell in fibrovascular bundle connecting the stem to plumule. These date suggest that group B saponins might be widely distributed in legumes as DDMP conjugated forms.

3 CHEMILUMINESCENCE OF PHENOLIC COMPOUNDS IN THE PRESENCE OF ACTIVE OXYGEN SPECIES AND ACETALDEHYDE.

The occurrence of chemiluminescence (CL) in the presence of active oxygen species and acetaldehyde (MeCHO) was closely related to the radical scavenging activity. Phenolic compounds such as flavonoids, anthocyanins and catechins are known to act as superoxide (O_2^-) and hydroxyl radical (HO•) scavengers. [23,24] The CL of phenolic compounds was measured in the presence of active oxygen species and MeCHO. We investigated the relationship between the chemical structures of phenolic compounds and their CL intensities and/or radical scavenging activities.

3.1 Chemiluminescence of Flavonoids [25]

The CL of flavonoids was strong in the presence of HO• which formed by the Fenton reaction, and MeCHO, however in the presence of hydrogen peroxide (H_2O_2) and MeCHO, anthocyanins and catechins exhibited stronger CL than that in the presence of HO•. CL decreased following methylation of the hydroxyl groups in the B-ring. These results indicated that a vicinal hydroxyl group in the B-ring was important for CL. The presence of a double bond between C-2 and C-3 contributed only slightly to the stronger CL. The CL intensity increased with the C-3 sugar chain: hydroxyl(quercetin) < glucose(isoquercitrin) < rhamnosylglucose (rutin).

3.2 Chemiluminescence of Anthocyanins [26]

The higher CL of anthocyanins in the presence of H_2O_2 and MeCHO, may suggest that anthocyanins having a pyrylium nucleus than the flavonoids having a γ-pyrone ring. The CL of anthocyanins and their related compounds in the presence of *tert*-butyl hydroperoxide(t-BuOOH) and MeCHO were measured. To investigated the effects of aglycone form, pH 2.6, 7.0 and 9.0 buffers were used in reaction mixture, respectively, because it was well known that anthocyanins might be converted into pseudobase and chalcone forms depending on the pH of the aqueous medium. The CL intensities of these anthocyanins at basic pH increased about 2 ~ 7 fold, when compared with those measured at neutral and acidic conditions. These results suggested that the chalcone of anthocyanins was more important for strong CL, rather than was formation of the flavylium cation.

3.3 Chemiluminescence of Catechins

The CL of catechins were measured in the presence of active oxygen species and 2% MeCHO. (-)-Epigallocatechin(EGC) exhibited strongest CL among the 8 kinds of catechins. When the reaction mixture composed of, EGC, MeCHO and H_2O_2 was subjected to HPLC and LC-MS, it was comfirmed that EGC was almost not lost during the reaction. EGC itself did not show CL. Therefore, this reaction may proceed via a route similar to that described Gotoh, N. and Niki, E. [27].

active oxygen + phenolic compound → phenolic compounds • 1)

phenolic compound • + acetaldehyde →

light + phenolic compound + products 2)

CL intensity of the reaction mixture was dependent on the concentrations of H_2O_2 (X ; active oxygen species), phenolic compound(Y ;catalytic species) and MeCHO (Z ;receptive species) (Fig. 2). We proposed that the CL intensity, P, in the presence of H_2O_2 and MeCHO is given by $P = k\ [X]\ [Y]\ f(z)$. The calculated photon constants(log k) are summarized in Table 2. Log k in the presence of HO • and t-BuO • was calculated using H_2O_2 and t-BuOOH concentrations. In the presence of H_2O_2 as X, k of catechins was closely related to the characteristics of their chemical structures showing good agreement with the flavonoids and anthocyanins. CL of phenolic compounds was increased by the presence of following partial structures: (a) the pyrogallol structure rather than catechol structure in the B-ring, (b) free hydroxyl group at C-3, (c) the stereoscopic structure between C-3 hydroxyl group or glycoside and B-ring.

Radical scavenging activity of antioxidants is generally expressed from a point of view how rapidly they react with active oxygen to scaveng them, however, it may be necessary to investigate precisely for the mechanism to eliminate high potential energy, which may be produced in the presence of active oxygens and antioxidants in the biological systems.

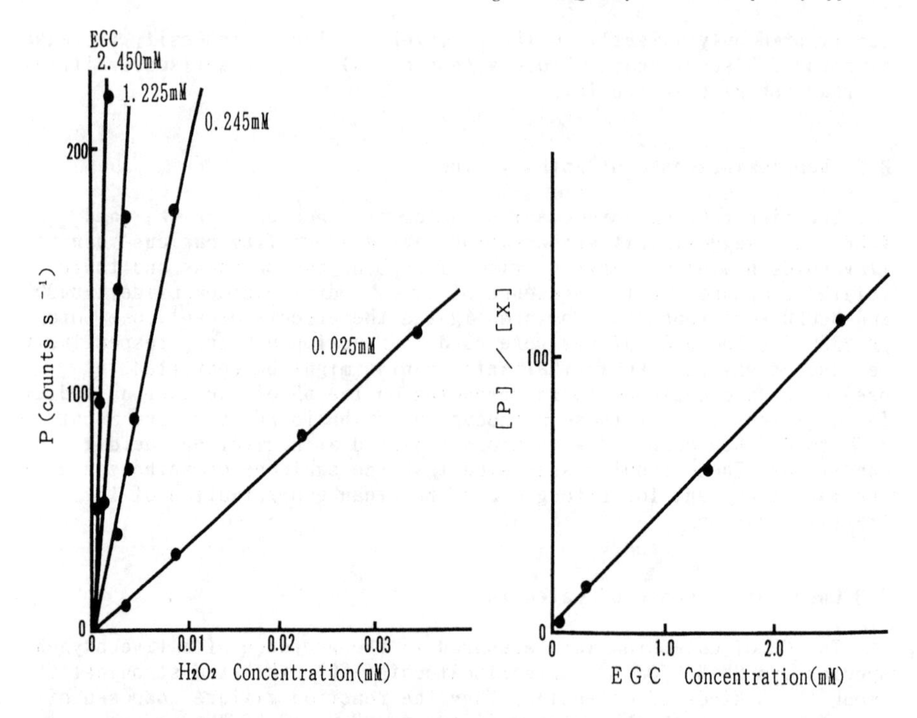

Figure 2 *Effect of EGC(Y;catalytic species) and H_2O_2 (X;active oxygen species) Concentration on Photon Intensity(P) in the Presence of Acetaldehyde (Z;receptive species)*

Table 2 *Photon Constants of Catechins or Gallic acid(Y;catalytic species) in the Presence of Acetaldehyde (Z;receptive species) and Active Oxygen Species(X), in M^{2} s^{1} counts at pH 7.0 and 23℃.*

Name	Substituent at		log K (K ; photon constant)			
	C-3	C-5´	H_2O_2	HO・	t-BuOOH	t-BuO・
(-)-Catechin	OH	H	6.52	6.38	5.98	5.72
(-)-Epicatechin	OH	H	7.47	6.50	6.25	6.20
(-)-Gallocatechin	OH	OH	7.99	6.39	6.93	7.46
(-)-Epigallocatechin	OH	OH	10.64	7.12	6.90	7.73
(-)-Catechin gallate	O-Ga	H	6.46	5.85	5.80	6.23
(-)-Epicatechin gallate	O-Ga	H	6.85	5.69	5.79	6.26
(-)-Gallocatechin gallate	O-Ga	OH	7.39	5.98	6.72	6.27
(-)-Epigallocatechin gallate	O-Ga	OH	9.41	5.51	6.42	6.77
gallic acid			6.38	6.84	5.46	6.71

Ga:galloyl group
C-3 and C-5´ are substituted as follow

or

4 CHEMILUMINESCENCE OF SOYBEAN SAPONINS

DDMP saponins showed oxygen radical scavenging activity when assayed using the xanthin oxidase-NH_2OH method, electron spin resonance(ESR) and chemiluminescence.[28] One mg of DDMP saponin/mg exhibited superoxide scavenging activity of a degree equivalent to 17.1 units of superoxide dismutase/ml with ESR spin trapping method(Fig.3). This scavenging activity of DDMP saponin is caused by DDMP moiety attached to the tri-terpene aglycone since soybean saponin Bb which is derived from soya-saponin βg, but which lacks this group, did not show the scavenging activity.

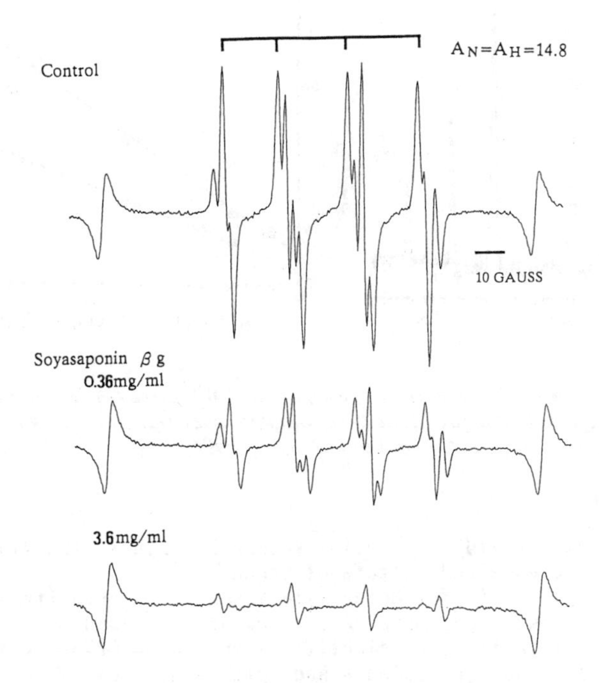

Figure 3 *Superoxide Scavenging Activity of Soyasaponin βg by ESR spin Trapping Method*

The CL was good agreement with the radical scavenging results showed by enzymatic and ESR spin trapping methods. The CL exhibited in the presence of active oxygen species(X), catalytic species(Y) and receptive species(Z) is most convenient to investigate the scavenging function of active oxygen species. We have found that group A and DDMP saponins are Y equivalent to EGC and Z equvalent to MeCHO, respectively. The photonintensity(P) in the presence of 1.12mM soyasaponin βg increased linearly depending on the concentration of H_2O_2 and soyasaponin Ab(Fig.4). It is important that soybean saponin emitted CL in the same system; $[P]=K[X][Y]f(Z)$. Especially, in the presence of soyasaponin βg exhibited stronger CL about 100 fold than that of MeCHO as Z. Thisresult suggested that the combination of group A and DDMP saponins was important in the scavenging function of active oxygen species, because, in the presence EGC as Y, soyasaponin βg exhibited about 10 fold comparison with that of MeCHO.

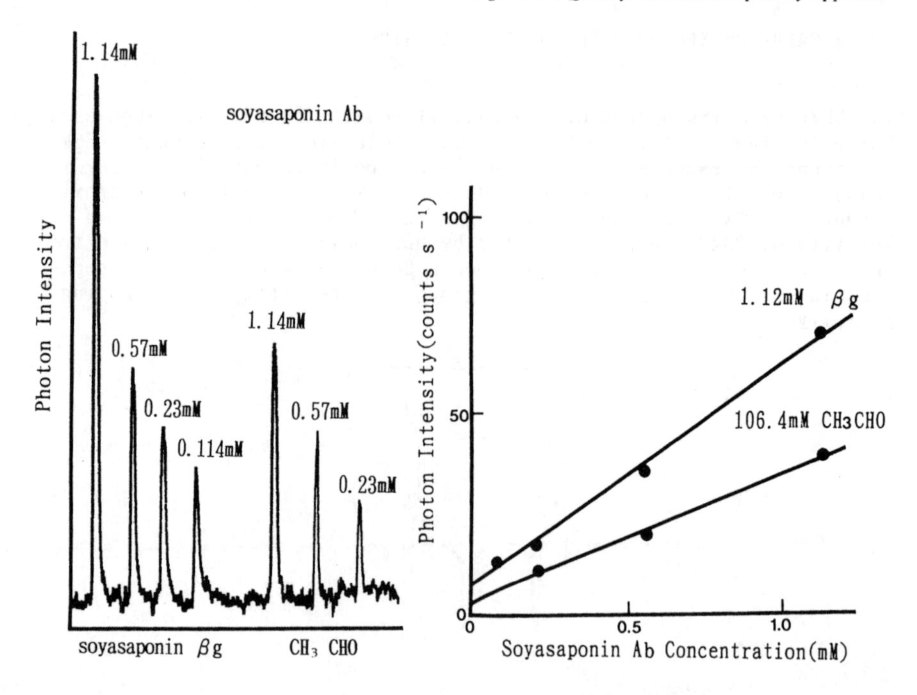

Figure 4 *Chemiluminescence of Soyasaponin Ab(Y;catalytic species) in the Hydrogen Peroxide(X;active oxygen species) and Soyasaponin βg or acetaldehyde(Z;receptive species)*

References

1 1984-86 Grant-in-Aid for Special Research Project, The Ministry of Education, Science and Culture of Japan.

2 1988-91 Grant-in-Aid for Scientific Research on priority Area, The Ministry of Education, Science and Culture of Japan.

3 1992-94 Grant-in-Aid for Scientific Research on priority Area, The Ministry of Education, Science and Culture of Japan.

4 M.Kurihara, K.Aoki and F.Hisamichi, UICC Publication, Nagoya University Press, Nagoya, Japan, 1989

5 P.N.Nair, N.Turjuman, G.Kessie, B.Calkins, G.T.Goodman, H.Davitvitz and G.Nimmagadda, *Am. J. Clin. Nutr.*, 1984, 40, 927.

6 H.S.Basage, *Biochem. Cell Biol.*, 1990, 68, 989.

7 A.Audic and P.U.Giacomoni, *Photochem. Photobiol.*, 1993, 57, 508.

8 O.I.Aruoma, B.Halliwell, E.Gajewski and M.Dizdaroglu, *J. Biol. Chem.*, 1989, 264, 20509.

9 G.Witz, *Proc. Spc. Exp. Biol. Med.*, 1991, 198, 675.

10 T.F.Slater, *Biochem. J.*, 1984, 222, 1.

11 G.R.Fenwick, R.Price, C.Tsukamoto and K.Okubo, Toxic Substances in Crop Plants, J.P.F.D'Mello, C.M.Duffus and J.H.Duffus, Royal Society of Chemistry, Cambridge, 1991, pp.285.

12 C.Tsukamoto, A.Kikuchi, S.Kudou, M.Tonomura, K.Harada, T.Iwasaki and K.Okubo, ACS Symposium Series 546, Food Phytochemicals for Cancer Prevention I, 1993, pp372.

13 K.Okubo, K.Kudou, T.Uchida, Y.Yoshiki, M.Yoshikoshi and M.Tonomura, ACS Symposium Series 546, Food Phytochemicals for Cancer Prevention I, 1993, pp330.
14 S.Kudou, M.Tonomura, C.Tsukamoto, T.Uchida, M.Yoshikoshi and K.Okubo, ACS Symposium Series 546, Food Phytochemicals for Cancer Prevention I, 1993, pp340.
15 S.Kudou, M.Tonomura, C.Tsukamoto, M.Shimoyamada, T.Uchida and K. Okubo, *Biosci. Biotech. Biochem.*, 1992, 56, 142.
16 Y.Yoshiki, C.Tsukamoto, K.Harada and K.Okubo, Abstract of World Soybean Research Conference V, Chiang Mai, Thailand(21-27 Feb. 1994) pp.113.
17 S.Kudou, M.Tonomura, C.Tsukamoto, T.Uchida and K.Okubo, *Biosci. Biotech. Biochem.*,1993, 57, 546.
18 M.shiraiwa, K.Harada and K.Okubo, *Agric. Biol. Chem.*,1991, 55,323
19 M.shiraiwa, S.Kudou, M.Shomoyamada, K.Harada and K.Okubo, *Agric. Biol. Chem.*, 1991, 55, 315.
20 M.shiraiwa, F.Yamauchi, K.Harada and K.Okubo, *Agric. Biol. Chem.*, 1990, 54, 1347.
21 Y.Yoshiki, J.H.Kim, K.Okubo, I.Nagoya, T.Sakabe and N.Tamura, *Phytochem.*, 1995, 38, 229.
22 K.Okubo, Y.Yoshiki, K.Okuda, T.Sugihara, C.Tsukamoto and K.Hoshikawa, *Biosci. Biotech. biochem.*, 1994, 58, 2248.
23 J.Baumann, G.Wurm, and V.F.Bruchhausen, *Arch. Pharm.*, 1980, 313, pp340.
24 A.Puppo, *Phytochem.*, 1991, 31, 85.
25 Y.Yoshiki, K.Okubo, and K.Igarashi, *Phytochem.*, 1995, in press.
26 Y.Yoshiki, K.Okubo, and K.Igarashi, *J. Biolumi. Chemilumi.*, 1995, in press.
27. N.Gotoh, and E.Niki, *Biochim. Biophy. Acta*, 1992, 1115, 201. 330.

HYDROPHILIC OXYGEN RADICAL SCAVENGERS IN THE LEGUMINOUS SEEDS AND DERIVED FOODS

Yumiko Yoshiki, Tomoaki Sirakura, Keiko Okuda, Kazuyoshi Okubo, Terumi Sakabe*, Ichiro Nagoya * and Nobuhiro Tamura*

Faculty of Agriculture,
Tohoku University, Japan
* Analytical Research Center,
Asahi Chem. Ind. Co., LTD., Japan.

1 INTRODUCTION

The photon emission(chemiluminescence;CL) of radical scavenging compounds such as phenolic compounds in the presence of active oxygen species and acetaldehyde(MeCHO) was corroborated to occur non-enzymatically at room temp. and at neutral condition. By using phenolic compounds such as flavonoids, anthocyanins and catechins, the efficacy was related to the number of hydroxyl groups on the B-ring, and to the substituent at C-3 of phenolic compounds, showing good agreement with the radical scavenging results[1,2] reported by enzymatic and ESR spin trapping methods. From CL result, we have found that the CL observed in the presence of active oxygen species (X), catalytic species (Y) and receptive species (Z), XYZ system, is most convenient to investigate the scavenging function of active oxygen species and that most effective Y or Z was (-)-epigallocatechin (EGC) or MeCHO, respectively. Using the XYZ system, Hydrophilic oxygen radical scavengers in the leguminous seeds and derived foods were investigated.

2 MATERIALS AND METHODS

2.1 Materials

Hydrogen peroxide and acetaldehyde were purchased from Santoku Chem. (Tokyo, Japan) and Sigma Chem., respectively. (-)-Epigallocatechin from *Camellia sinensis* was purchased from Kurita Co. Ltd. (Tokyo, Japan).

2.2 Measurement of Chemiluminescence

The CL was measured by filter-equipped photon counting-type spectrophotometer(CLD-110, Tohoku Electronic Ind.). The photons counted in the wavelength between 300 and 650 nm were computed as total spectral intensities.

3 RESULT AND DISCUSSION

3.1 Chemiluminescence of Foods

When the CL of vegetables and the foods was measured in the presence of H_2O_2 or $HO\cdot$ and MeCHO. These results indicated that hot-water extract exhibited stronger intensity of CL than ethanol-extract, and futher indicated that leguminous foods, generally showed strong intensity of CL. It is well known that hot-water extract of vegetables and plants suppress tumor promotion, as well as to scavenge lipid peroxide radicals *in vitro*.[3] CL result well agreed with lipid peroxide scavenging activity.

3.2 Chemiluminescence of Soybean

When the CL of soybean seed were measured in the presence of H_2O_2 and MeCHO. The hot-water extract exhibited strong CL than that of ethanol extract. Futhermore, CL was stronger accompanied with germination untill 4 days. To search radical scavenger in soybean seedling, soybean (*Glycine max* var. Miyagisirome) seeds were grown in complete darkness for 4 days at 25℃. The extracts with 50 mM phosphate buffer (pH 7.0) were obtained from segments which removed cotyledons. The CL in the presence of H_2O_2 and MeCHO was measured in the fractions fractionated using Sephadex G-25 column. Three fraction were obtained dependent on CL intensity of every tube. From result of TLC analysis using by $CHCl_3$: MeOH : water/ 65 : 35 : 10 as development solvent, one was suggested to be polysaccharide such as constituents of cell wall which was more than 5,000 MW because this fruction eluted at void sephadex G-50 column. The CL of this fraction was strong as Y in the presence of H_2O_2 and MeCHO. Two other fractions exhibited strong

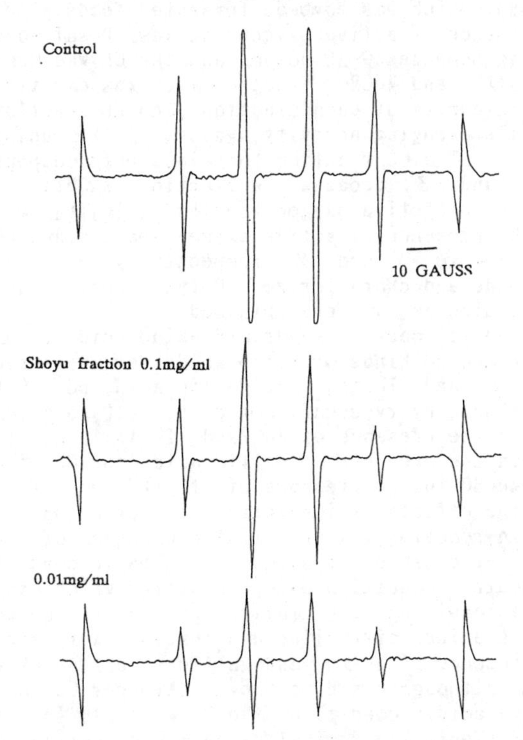

Figure 1 *Hydroxyl radical scavenging activity of shoyu fractionated with sephadex G-25 column by ESR spin trapping method.*

Table 1 *Chemiluminescence(photon counts s^{-1}) in X(active oxygens), Y (catalytic species) and Z(receptive species). pH 7.0, 23℃.*

	X(HO•) Y(GA)	X(H_2O_2) Y(EGC*)		X(HO•) Y(GA)	X(H_2O_2) Y(EGC*)
acetaldehyde	100	236	D-arabinose	63	68
L-arabinose	49	71	D-lyxose	40	68
L-lyxose	48	65	D-mannose	41	72
L-mannose	57	65	D-lactose	43	70
D-galactose	45	74	D-ribose	38	64
L-ribose	38	73	2-deoxy-D-ribose	47	87
D-xylose	34	63	L-xylose	45	64
2-deoxy-D-glucose	50	63	D-glucose	53	66
mannitol	52	60	sucrose	51	67
pectin	74	53	glucuronic acid	33	43

GA; gallic acid, EGC; (-)-epigallocatechin

CL in the presence of H_2O_2 and EGC in comparison with that in the presence of H_2O_2 and MeCHO. This result suggested to act as Z. These fractions are suggested to contain amino group related with function as Z, because these fraction reacted with ninhydrin.

3.3 Chemiluminescence of Fermented Foods

Soy-source (shoyu) which was soybean fermented foods also exhibited strong CL in the presence of active oxygen species. Desalted shoyu was fractionated by using Sephadex G-25 column and the CL was measured in the presence of H_2O_2 and MeCHO. Fractionation was carried out in compliance with CL intensity of each fraction. The CL fraction from shoyu showed radical scavenging activity against HO• using ESR spin trapping method (Fig. 1). The CL fraction contained both components corresponding to Y and Z, because the fraction exhibited strong CL either in the presence of active oxygen species (H_2O_2 or HO•) and MeCHO, and in the presence of active oxygen species and EGC. Compounds corresponding to Y and Z, respectively, were isolated from soybean fermented foods and characterized. Polysaccharides as Y, and phenylalanine and uridine as Z were obtained.

To investigate the efficacy of sugar and amino acid as Z, the CL of 19 kinds of sugar and 20 kinds of amino acids were measured in the presence of EGC as Y and H_2O_2, or gallic acid and HO• by the Fenton reaction. Although every sugar showed the 1/10 fold comparison with that of MeCHO in the presence of EGC and H_2O_2, CL of sugar exhibited 1/2 fold in the presence of gallic acid. The CL of amino acids was also weak than MeCHO in the presence of H_2O_2 or HO• (Table 1).

To investigate the effects of ionization, buffer of pH 2.6, pH 7.0 or pH 9.0 was used in reaction mixture. In the presence of H_2O_2 and EGC, all amino acids were exhibited strong CL in basic condition than that in acidic or neutral condition except cysteine which exhibited strong CL in neutral condition. Especially, homoserine, isoleucine, proline, threonine, alanine, citrulline and leucine, were exhibited strong CL. In the presence of HO• and gallic acid, the CL was strong in acidic condition. Although cysteine showed stronger CL than that of MeCHO about 2 fold in acidic condition, the CL was 1/10 fold of MeCHO CL in neutral condition (Table 2). Radical oxidation of certain plant polysaccharides or proteins is important,[4,5] since it can alter the solubility, solution viscosity, and gel-forming properties of polymers

Table 2 *Chemiluminescence(photon counts s^{-1}) in X(active oxygens), Y (catalytic species) and Z(receptive species) at 23 ℃.*

	X(HO•) Y(GA)			X(H_2O_2) Y(EGC)		
	pH2.6	pH7.0	pH9.0	pH2.6	pH7.0	pH9.0
acetaldehyde	92	105	163	75	293	256
arginine•HCl	40	29	35	45	46	
cysteine	204	9	188	8	69	5
phenylalanine	151	39	53	47	48	93
glycine	60	27	35	32	33	87
histidine	87	31	82	29	53	
homoserine	67	33	43	39	41	121
isoleucine	68	35	32	43	45	139
proline	76	33	35	43	52	127
threonine	87	35	45	44	49	145
valine	79	37	52	45	47	141
ornithine	63	25	31	36	40	70
alanine	67	36	42	43	32	129
aminobutyric acid	43	23	29	31	10	109
citrulline	83	36	50	46	45	123
leucine	72	36	55	46	46	135
methionine	59	17	32	39	40	92
uridine		54			45	
cytosine		182			88	
inosine		113			90	
thymidine		51			56	

GA; gallic acid, EGC; (-)-epigallocatechin

employed in the food industry. The CL results of amino acid and sugar which have function as Z in the presence of active oxygen and catalytic compounds, suggest that free amino acid and sugar may prevent serious consequences of polysaccharides and proteins. This also suggests that active oxygen species may attack readily such amino and sugar group in polysaccharides and proteins.

REFERENCES

1. Y.Yoshiki, K.Okubo, and K.Igarashi, *Phytochemistry*, 1995, in press.
2. Y.Yoshiki, K.Okubo, and K.Igarashi, *J.Biolumi.Chemilumi.*, 1995 in press.
3. H.Maeda, T.Katsuki, T.Akaike, and R.Yasutake, *Jpn.J.Cancer Res.*, 1992, **83**, 923.
4. B.C.Gilbert, D.M.King, and B.Thomas, *Carbohydr.Res.*, 1984, **125**, 217
5. H.W.Gardner, *J.Agric.Food Chem.*, 1979, **27**, 220.

CHEMILUMINESCENCE OF PHENOLIC COMPOUNDS IN THE PRESENCE OF ACTIVE OXYGEN SPECIES AND ACETALDEHYDE

Yumiko Yoshiki, Kiharu Igarashi * and Kazuyoshi Okubo

Faculty of Agriculture,
Tohoku University, Japan
* Faculty of Agriculture,
Yamagata University, Japan

1 INTRODUCTION

The occurrence of chemiluminescence (CL) in the presence of active oxygen species and acetaldehyde (MeCHO) was closely related to the radical scavenging activity. Phenolic compounds such as flavonoids, anthocyanins and catechins are known to act as superoxide (O_2^-) and hydroxyl radical (HO•) scavengers.[1,2] The CL of phenolic compounds was measured in the presence of active oxygen species and MeCHO. We investigated the relationship between the chemical structures of phenolic compounds and their CL intensities and/or radical scavenging activities.

2 MATERIALS AND METHODS

2.1 Materials

Flavonoids, anthocyanins and other reagents were prepared described by Y.Yoshiki *et al*.[3,4] Catechins of *Camellia sinensis* were purchased from Kurita Co.Ltd.(Tokyo,Japan).

2.2 Measurement of Chemiluminescence

The CL of phenolic compounds was measured by filter-equipped photon counting-type spectrophotometer (CLD-110,Tohoku Electronic Ind.). The photons counted in the wavelength between 300 and 650 nm were computed as total spectral intensities. The phosphate buffer (pH 7.0) was used as reaction mixture.

3 RESULT AND DISCUSSION

3.1 Chemiluminescence of Flavonoids

The CL of flavonoids was strong in the presence of HO• formed by the Fenton reaction, and MeCHO, however in the presence of hydrogen peroxide (H_2O_2) and MeCHO, anthocyanins and catechins exhibited stronger CL than that in the presence of HO• (Table 1 and 3). The CL decreased following methylation of the hydroxyl groups in the B-ring. This indicated that a vicinal hydroxyl group in the B-ring was important

Table 1 *Chemiluminescence Intensity of Flavonoids in the Presence of Hydrogen Peroxide and Hydroxyl Radical*

Name	Substituent at						CL in the presence of	
	C-3	C-5	C-7	C-3'	C-4'	C-5'	HOOH	HO·
Flavonol or flavanonol:								
Kaempferol	OH	OH	OH	H	OH	H	2	243
Quercetin	OH	OH	OH	OH	OH	H	3	612
Myricetin	OH	OH	OH	OH	OH	OH	75	1197
Dihydroquercetin	OH	OH	OH	OH	OH	H	2	529
Astilbin	ORha	OH	OH	OH	OH	H	0	181
Isoquercitrin	OGlc	OH	OH	OH	OH	H	5	1155
Rutin	ORha-Glc	OH	OH	OH	OH	H	16	7101
Isorhamnetin	OH	OH	OH	OMe	OH	H	0	125
Isorhamnetin 3-*O*-glucoside	OGlc	OH	OH	OMe	OH	H	0	584
Rhamnetin	OH	OH	OMe	OH	OH	H	0	194
Flavone:								
Apigenin	H	OH	OH	H	OH	H	90	273
Flavone	H	H	H	H	H	H	23	52
Luteolin 4'-*O*-glucoside	H	OH	OH	OH	OGlc	H	0	97
Diosmetin 7-*O*-glucoside	H	OH	OGlc	OH	OMe	H	0	368
Flavanol:								
(+)-Catechin	OH	OH	OH	OH	OH	H	0	140
Anthocyanin:								
Nasunin	ORha Glc-*p*-coumaric acid	OGlc	OH	OH	OH	OH	2347	891
Delphinidin	OH	OH	OH	OH	OH	OH	794	313
Rubrobrassicin	OGlc-Glc	OGlc	OH	OH	OH	H	33	424
Cyanidin	OH	OH	OH	OH	OH	H	588	111
Malvin	OH	OGlc	OH	OMe	OH	OMe	110	111
Malvidin	OGlc	OH	OH	OMe	OH	OMe	77	83

Flavonol, Flavanonol, 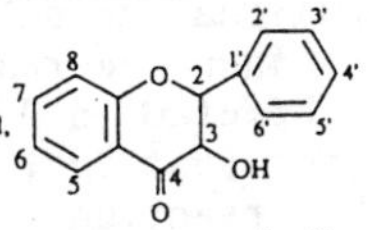 Flavone, 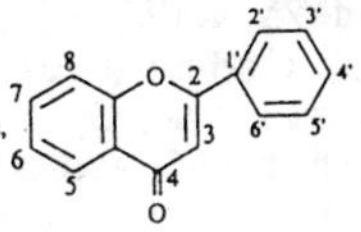 Flavanol, 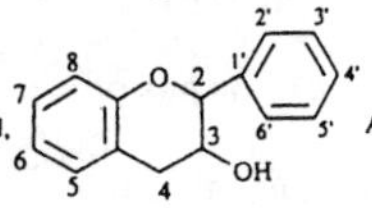 Anthocyanin

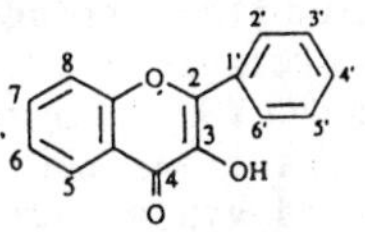

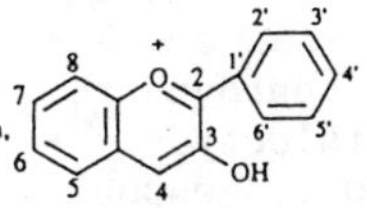

C-3, C-5, C-7, C-3', C-4' and C-5' are substituted as above.

for CL. The presence of a double bond between C-2 and C-3 contributed only slightly to the stronger CL. The CL intensity increased with the C-3 sugar chain : hydroxyl(quercetin) < glucose(isoquercitrin) < rhamnosylglucose (rutin)(Table 1).

3.2 Chemiluminescence of Anthocyanins

The higher CL of anthocyanins in the presence of H_2O_2 and MeCHO, may suggest that anthocyanins having a pyrylium nucleus than the flavonoids having a γ-pyrone ring. The CL of anthocyanins and their related compounds in the presence of *tert*-butyl hydroperoxide(t-BuOOH) and MeCHO were measured. To investigate the effects of aglycone form, pH 2.6, 7.0 and 9.0 buffers were used in reaction mixture, respectively, because it was well known that anthocyanins might be converted into pseudobase and chalcone forms depending on the pH of the aqueous medium. The CL intensities of these anthocyanins at basic pH increased about 2 ~ 7 fold, when compared with those measured at neutral and acidic conditions(Table 2). These results suggested that the chalcone of anthocyanins was more important for strong CL, rather than was formation of the flavylium cation.

Table 2 *Chemiluminescence Intensities of Anthocyanins in the Presence of Acetaldehyde and tert-Butyl Hydroperoxide*

Name	Substituent at C-3	C-5	C-7	C-3´	C-4´	C-5´	pH 2.6	pH 7.0	pH 9.0
Nasunin	O-Rham-Glc -*p*-coumaric acid	O-Glc	OH	OH	OH	OH	107	*1000	2285
Delphinidin	OH	OH	OH	OH	OH	OH	118	197	752
Rubrobrassicin	O-Glc-Glc	OH	OH	OH	OH	H	58	255	420
Cyanidin	OH	OH	OH	OH	OH	H	77	118	346
Malvin	O-Glc	O-Glc	OH	OMe	OH	OMe	51	192	258
Malvidin	OH	OH	OH	OMe	OH	OMe	28	95	654

*arbitrarily set as standard.

C-3, C-5, C-7, C-3´, C-4´ and C-5´ are substituted as above. Glc : glucose.

3.3 Chemiluminescence of Catechins

The CL of catechins was measured in the presence of active oxygen species and 2% MeCHO. (-)-Epigallocatechin(EGC) exhibited strongest CL among the 8 kinds of catechins. When the reaction mixture composed of EGC, MeCHO and H_2O_2 was subjected to HPLC and LC-MS, it was comfirmed that EGC was almost not lost during the reaction. EGC itself did not show CL. Therefore, this reaction may proceed via a route similar to that described Gotoh, N. and Niki, E.[5]

active oxygen + phenolic compound → phenolic compounds • 1)

phenolic compound • + acetaldehyde →
light + phenolic compound + products 2)

CL intensity of the reaction mixture was dependent on the concentrations of H_2O_2 (X ; active oxygen species), phenolic compound(Y ;catalytic species) and MeCHO (Z ;receptive species) (Fig. 1). We proposed that the CL intensity, P, in the presence of H_2O_2 and MeCHO was given by $P = k\,[X]\,[Y]\,f(z)$. The calculated photon constants(log k) are summarized in Table 3. Log k in the presence of HO • and t-BuO • was calculated using H_2O_2 and t-BuOOH concentrations. In the presence of H_2O_2 as X, k of catechins was closely related to the characteristics of their chemical structures showing good agreement with the flavonoids and anthocyanins. The CL of phenolic compounds was increased by the presence of following partial structures: (a) the pyrogallol structure rather than catechol structure in the B-ring, (b) free hydroxyl group at C-3, and (c) the stereoscopic structure between C-3 hydroxyl group or glycoside and B-ring.

Radical scavenging activity of antioxidants is generally expressed from a point of view how rapidly they react with active oxygen to scaveng them, however, it may be necessary to investigate precisely for the mechanism to eliminate high potential energy, which may be produced in the presence of active oxygens and antioxidants in the biological systems.

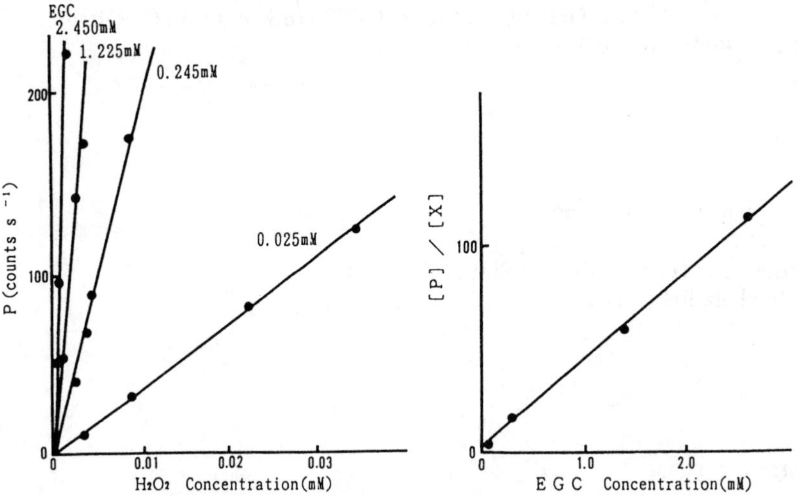

Figure 1 *Effect of EGC(Y;catalytic species) and H_2O_2 (X;active oxygen species) concentration on photon intensity(P) in the presence of acetaldehyde (Z;receptive species)*

Table 3 *Photon Constants of Catechins or Gallic acid(Y;catalytic species) in the Presence of Acetaldehyde (Z;receptive species) and Active Oxygen Species(X), in M^{-2} s^{-1} counts at pH 7.0 and 23℃.*

Name	Substituent at C-3	Substituent at C-5´	log K (K; photon constant) H_2O_2	HO・	t-BuOOH	t-BuO・
(-)-Catechin	OH	H	6.52	6.38	5.98	5.72
(-)-Epicatechin	OH	H	7.47	6.50	6.25	6.20
(-)-Gallocatechin	OH	OH	7.99	6.39	6.93	7.46
(-)-Epigallocatechin	OH	OH	10.64	7.12	6.90	7.73
(-)-Catechin gallate	O-Ga	H	6.46	5.85	5.80	6.23
(-)-Epicatechin gallate	O-Ga	H	6.85	5.69	5.79	6.26
(-)-Gallocatechin gallate	O-Ga	OH	7.39	5.98	6.72	6.27
(-)-Epigallocatechin gallate	O-Ga	OH	9.41	5.51	6.42	6.77
gallic acid			6.38	6.84	5.46	6.71

Ga:galloyl group
C-3 and C-5´ are substituted as follow

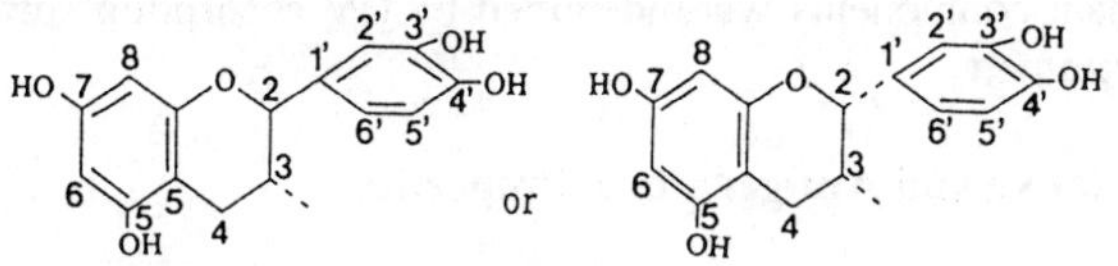

REFERENCES

1. J. Baumann, G. Wurm, and V. F. Bruchhausen, *Arch. Pharm.*, 1980, **313**, 330.
2. A. Puppo, *Phytochemistry*, 1991, **31**, 85.
3. Y. Yoshiki, K. Okubo, and K. Igarashi, *Phytochemistry*, 1995, in press.
4. Y. Yoshiki, K. Okubo, and K. Igarashi, *J. Biolumi. Chemilumi.*, 1995, in press.
5. N. Gotoh, and E. Niki, *Biochim. Biophy. Acta*, 1992, **1115**, 201.

ANTIOXIDATIVE AND ANTIGENOTOXIC PROPERTIES OF FLAVONOIDS PREVAILING IN VEGETABLE

B. Raab, J. Hempel, and H. Böhm

German Institute of Human Nutrition Potsdam-Rehbrücke
D-14558 Bergholz-Rehbrücke
Germany

1 INTRODUCTION

Flavonoids represent a large group of plant secondary metabolites, which are regularly consumed by humans as components of food. They exert numerous effects on mammalian cell functions. Owing to their polyphenolic structure they are able to act as antioxidants and ought to decrease DNA damage resulting from oxidative stress, that is to have antigenotoxic properties. To assess the possible role of flavonoids in the prevention of degenerative diseases caused by reactive oxygen species (ROS), especially in the initiation of cancer, it is necessary to investigate their natural composition in plant food, to elucidate their structure and to evaluate their antioxidative and antigenotoxic properties.

2 MATERIALS AND METHODS

2.1 Quantitative and Qualitative Determination of Flavonoids

Fruits and bulbs, respectively, of 6 varieties in each case of yellow and green beans (*Phaseolus vulgaris L.*) and 14 of onions (*Allium cepa L.*) were freeze-dried and extracted with methanol/water (70:30, v/v). After selective adsorption on polyamide and fractionizing desorption the flavonoid glycosides were separated and characterized by RP-HPLC with diode array detection. The main components were identified by UV absorption spectrum, retention time, and mass spectrometry[1].

2.2 Investigation of Antioxidative and Antigenotoxic Properties

Flavonoids used were quercetin (Q), kaempferol (K), quercetin-3-O-rutinoside (Q-3-O-R), and kaempferol-3-O-rutinoside (K-3-O-R) (all commercially available) as well as quercetin-3-O-glucuronide (Q-3-O-G) and kaempferol-3-O-glucuronide (K-3-O-G) (isolated from beans by HPLC).

The antioxidative properties were evaluated by an enhanced chemiluminescence (ECL) technique using horseradish peroxidase, perborate, luminol, and para-iodophenol for the generation of chemiluminescene and its quenching by addition of antioxidants[2].

The antigenotoxic/genotoxic properties of the relevant flavonoids were evaluated by determination of single strand breaks (SSB) in the DNA by means of the alkaline version of the comet assay[3,4], using the L929 connective tissue (mouse) cell line as *in vitro* test system.

Cells suspended in RPMI 1640 medium without serum and phenol red were treated with different concentrations of hydrogen peroxide (H_2O_2) as ROS and/or flavonoids for 30 min at 37 °C, spun down and embedded in agarose gels on microscopic slides. After lysis, alkaline unwinding, microgelelectrophoresis, and staining with ethidium bromide DNA damage was quantified by interactive image analysis (Perceptive Instruments, Halstead, UK) on a fluorescence microscope (Leica, Wetzlar, FRG) equipped with video camera.

For each data point 200 randomly selected cells from two different slides were scored. Quantification of the flourescence intensity in the tail (TI) as percentage of the whole comet intensity provided the most accurate evaluation of the DNA damage. Data were analysed by two-way analysis of variance with replication (ANOVA).

3 RESULTS AND DISCUSSION

The total flavonoid content, estimated from peak areas, strongly depends on varieties (Table 1). The main derivatives of beans were identified as 3-O-glucuronides and 3-O-rutinosides of Q and K.

Table 1 *Content of Flavonoids in Bean and Onion Varieties [µg/g fresh weight]*

Flavonoid calculated as	*Yellow Beans (n = 6)*	*Green Beans (n = 6)*	*Onions (n = 14)*
Q-3-O-R	23.4 - 99.4	32.0 - 224.9	–
K-3-O-R	5.8 - 15.3	6.3 - 14.8	–
Isoquercitrin*	–	–	154.8 - 1197.6

* Commercially available flavonoid representative detected in onions

According to our findings from quantitative and structural analysis we selected the glycosides mentioned above to study their effects. Additionally we included the corresponding aglycones Q and K in our investigations, because they are widespread in vegetable and released from flavonoid conjugates in the gut by glycosidases and ß-glucuronidases from intestinal bacteria[5].

The antioxidative activities of 0.5 µM flavonoid solutions expressed as Trolox (water soluble tocopherol analogue) equivalents are 2.4 (Q), 1.6 (Q-3-O-R), 0.6 (K), 0.1 (Q-3-O-G), and 0.0 (K-3-O-R and K-3-O-G). This sequence shows that the hydroxyl group in position 3 and the neighbouring ones in 3' and 4' position are of great importance for antioxidative (radical scavenging) properties of the flavonols concerned. The kind of O-glycosidic substituent at C3 also seems to be influential. That's why isoquercitrin from onions as monosaccharide conjugate should be characterized next.

Q, the flavonoid with the strongest antioxidative activity produced protective effect against ROS and significantly reduced genotoxicity of H_2O_2 (Figure 1). However there is a residual level of DNA damage with both concentrations of Q. This could be caused by the action of oxygen radicals produced by autoxidation , adduct formation with DNA or ROS resulting from metabolisation of the compound.

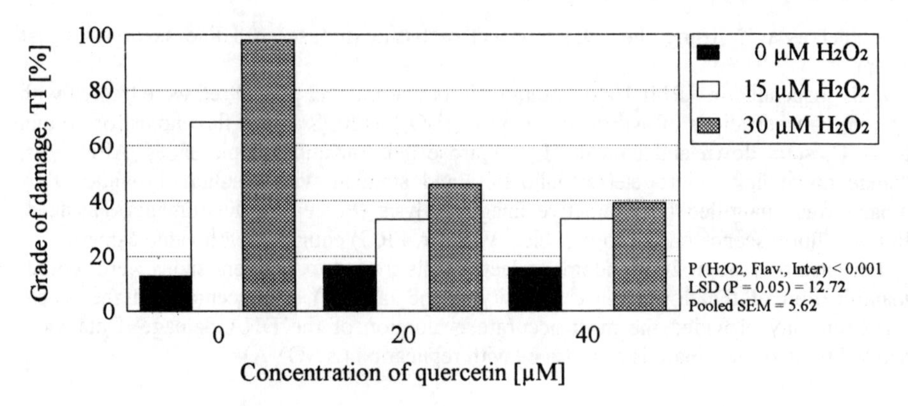

Figure 1 *Antigenotoxicity of Quercetin*

In case of the aglycone K TI values show the tendency to increase with concentration in absence of H_2O_2. In combination with the ROS the damage is significantly increased (Table 2). Similar effects can be observed with the glucuronides of Q and K.

Table 2 *Adverse Effects of Flavonoids*

Treatment H_2O_2/Flav. [µM]	*Grade of Damage TI [%] (mean of 200 cells)*				
	*K**	*Q-3-O-R*	*K-3-O-R*	*Q-3-O-G*	*K-3-O-G*
0/ 0	5.26	13.18	13.18	6.64	10.93
0/20	8.71	–	–	12.00	14.09
0/40	13.96	83.36	89.30	12.22	14.48
15/ 0	42.75	58.95	58.95	61.27	34.01
15/20	58.52	–	–	74.65	39.84
15/40	62.33	94.29	93.34	79,32	45.02
30/ 0	75.38	91.76	91.76	93.74	91.98
30/20	85.28	–	–	92.36	82.16
30/40	92.12	92.85	93.32	95.24	94.84
Significance of variance ratio [P]					
H_2O_2	<0.001	<0.001	<0.001	<0.001	<0.001
Flavonoid	<0.001	<0.001	<0.001	<0.001	0.066
H_2O_2*Flav.	<0.075	<0.001	<0.001	0.012	0.142
LSD (P=0.95)	6.34	10.59	10.59	6.15	9.85
Pooled SEM	2.80	4.68	4.68	2.72	4.35

* In case of K concentrations of H_2O_2 and flavonoid for the treatment were 0, 10, and 15 µM and 0, 10, and 30 µM, respectively.

Q-3-O-R with antioxidative activity higher than Trolox and K-3-O-R without measurable potency in the ECL test are genotoxic. They cause SSB by themselves and increase the effect of H_2O_2 significantly. That means, they exert pro-oxidative effects.

The results of our experiments do not confirm the supposed correlation between antioxidative and antigenotoxic properties of the flavonols investigated. Pro-oxidative activities seem to be responsible for the genotoxic effects of both the 3-O-rutinosides and the exacerbating effects of K and the above 3-O-glucuronides in combination with H_2O_2 shown by means of the comet assay.

Nuclear DNA damage and lipid peroxidation induced by flavonols were reported by Sahu et al.[7-9], too. Polyphenolic flavonoids obviously play a dual role depending on the redox state in the *in vivo* test system and complex biological environment, respectively.

This would explain the contradictory findings with respect to (anti)mutagenic or (anti)carcinogenic properties of flavonoids[10]. The predominant lack of adverse effects of flavonoids *in vivo* may be based on interactions with other compounds present in biological systems and the development of efficient chemical defense mechanisms of the human organism during the evolution.

It is necessary to clarify the mechanisms of actions of flavonoids and other bioactive secondary metabolites for better risk/benefit assessment, especially with respect to recommendations for higher intake of such substances in the diet.

Aknowledgement

We are grateful to Dr. Hoppe and Dr. Erhardt for statistical advice.

4 REFERENCES

1. J. Hempel and H. Böhm, *Lebensmittelchemie,* 1995, **49,** 39.
2. T. P. Whitehead, G. H. G. Thorpe, S.R.J. Maxwell, *Anal. Chim. Acta,* **266,** 265.
3. H. P. Singh, M. T. McCoy, R. R. Tice, E. L. Schneider, *Exp. Cell Res.*,1988, **175**, 184.
4. B. L. Pool-Zobel, R. Lambertz, M. Knoll, P. Schmezer, 1992, 'Biotechnologie im Ernährungsbereich. Berichte aus der Bundesforschungsanstalt für Ernährung', K. D. Jany and B. Tauscher (Eds.), BfE Karlsruhe,1995, p. 86.
5. A. M. Hacket, 'Plant Flavonoids in Biology and Medicine: Biochemical, Pharmocological,and Structure-Activity Relationships', V. Cody, E. Middleton, and J. B. Harborne (Eds.), Alan R. Liss, Inc., New York, 1986, p. 177.
6. B. J. Brandwein, *J. Food Sci.*, 1965, **30**, 680.
7. S. C. Sahu and M. C. Washigton, *Cancer Lett.*, 1991. **58,** 75.
8. S. C. Sahu and G. C. Gray, *Cancer Lett.*, 1993, **70, 73.**
9. S. C. Sahu and G. C. Gray, *Cancer Lett.*, 1994, 85, 159.
10. A. Das, J. H. Wang , and E. J. Lien, *Prog. Drug Res.*, 1994, **42,** 134.

ANTIOXIDANT PROPERTY OF POLYPHENOLIC COMPOUNDS OF CULINARY HERBS AND MEDICINAL PLANTS

A. LUGASI, E. DWORSCHAK and J. HOVARI

National Institute of Food Hygiene and Nutrition, Budapest, Hungary

1 INTRODUCTION

The recent consumer's interest in "natural" products requires natural antioxidative substances to replace conventional antioxidants such as BHT and BHA. Antioxidants are classified into four types according to the mechanism of action: chain breaker (or free radical inhibitor), peroxide decomposer, metal inactivator, or oxygen scavenger [1,2,3]. The antioxidant effect of an antioxidant on lipid peroxidation might be attributed to its properties of scavenging free radicals and active oxygen species [3,4]. Numerous spices and medicinal plants are well-known for their antioxidative properties, especially rosemary , sage and others species from the Labiatae family have been mentioned to have strong characteristics[5-8]. Plant phenolics and flavonoids have been reported to have multiple biological effect such as antioxidant activity, anti-inflammatory and anti-mutagenic effects[9-11].

The objective of our work was to investigate the free radical scavenging activity and the reducing power of the methanolic extracts of culinary herbs and medicinal plants and the connection between the polyphenolic content and the properties mentioned above.

2 MATERIALS AND METHODS

Spices (clove, cinnamon, marjoram, oregano and cumin) and medicinal plants (bearberry, rosemary, lemon-balm, peppermint and hyssop) were purchased at the local drugstores. All chemicals were reagents of analytical grade.

The concentration of the total phenolic compoundsa present in dry, ground samples was determined spectrophotometrically using Folin-Denis method[12]. Methanolic extracts of the samples were prepared after extracting the fat with petroleum ether.

The reducing power of the plant extracts was measured according to the method described by Yen and Chen[4]. The 1,1-diphenyl-2-picrylhydrazyl (DPPH) radical scavenging effect of the methanolic extract of the plants was determined spectrophotometrically at 517 nm[13]. The DPPH scavenging activity was calculated as the reciprocal of the amount of the dry sample that is requierd for 50% of inhibition of DPPH in the reaction mixture. All test and analyses were run in replicates and averaged. Statistical analyses were performed by using Student's t-test, and the linear correlation coefficient was determined.

3 RESULTS AND DISCUSSION

Spices and medicinal plants have high content of total phenolic compounds as it can be seen in Table 1. Clove and bearberry leaves have the highest level of the phenolics.

Table 1 Concentration of total phenolic compounds (g/100 g) in spices and medicinal plants

Spices	%	Medicinal plants	%
Clove	17.7	Bearberry	18.1
Oregano	11.2	Rosemary	9.2
Marjoram	5.8	Lemon-balm	7.3
Cinnamon	5.2	Peppermint	6.3
Cumin	1.3	Hyssop	4.4

Phenolic substances are supposed to be responsible for inhibition of lipid peroxidation[14,15]. Since the lipid peroxidation is induced and developed by free radicals and active oxygen species, free radical scavengers and oxygen quenchers may inhibit lipid peroxidation. Phenolic compounds that exhibit scavenging efficiency on free radicals are numerous and widely distributed within the plant kingdom[1,2,9].Due to the high content of phenolic compounds found in spices and herbs, the investigation of the reactivity with free radicals and the measurement of reducing power will be helpful in understanding the mechanism of antioxidant behavior of these plants.

As shown in Fig 1 and 2, the reducing powers (absorbance at 700 nm) of the plants depended on the concentration of the sample in the reaction mixture. The greatest reducing power was observed in the case of clove and bearberry in connected with the high content of phenolic compounds. Reducing power of the plants is smaller than that of ascorbic acid. The results show that the methanolic extracts of these plants are electron donors and can react with free radicals to convert them to more stable products and terminate radical chain reactions. A high linear correlation coefficient was calculated between phenolic content and reducing power of the plants ($r= 0.9302$). The results emphasize the importance of phenolic compounds in the antioxidant behavior of spices and medicinal plants.

The scavenging activity of methanolic extract of spices and herbs on the 1,1-diphenyl-2-picrylhydrazyl (DPPH) radical is shown in Table 2. Radical scavenging activity of all plants was very high but lower than that of well-known synthetic antioxidants BHA and BHT. With respect to results, the methanolic extracts of spices and herbs are free radical inhibitors, primary antioxidants that react particularly with the hydroperoxide radicals, which are the major propagator of the chain autoxidaton of fats, so breaking the chain[16]. The scavenging activity on DPPH radicals of the plants depends on their phenolic content, but significant correlation could not be determined.

In conclusion, the results demonstrated that methanolic extracts of spices and medicinal plants originally having high content of phenolic compounds bear great ability to be oxidized, strong hydrogen-donating activity and they are effective scavengers of free radicals. These properties seem to be important in explaining how the antioxidant activity of spices and medicinal plants arises.

Fig 1 and 2 Reducing power of methanolic extract of various spices and medicinal plants

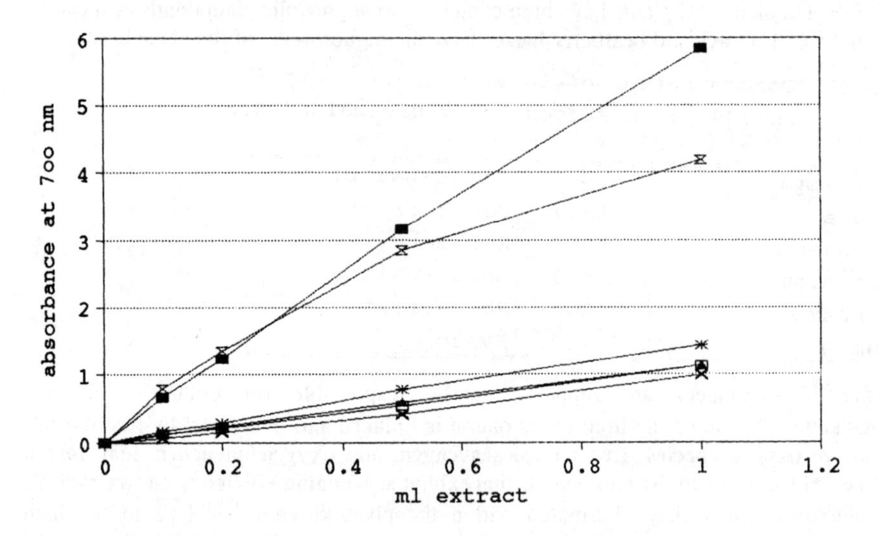

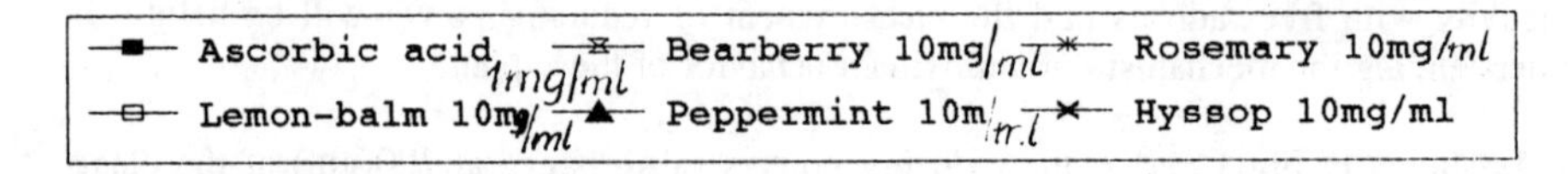

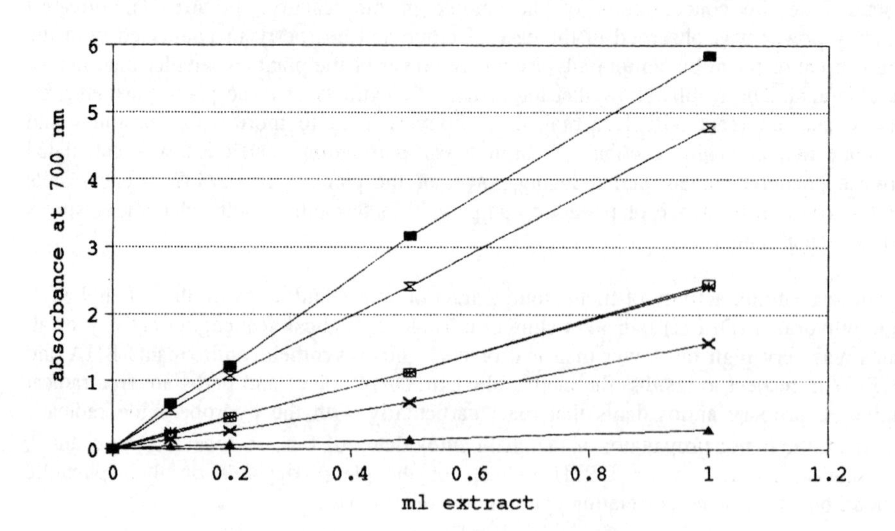

Table 2 Scavenging effect of various plant extracts on DPPH radical

Plant	The amount of the dry sample /µg/ that is requiered to 50 % of inhibition of DPPH radical in the reaction mixture (A)	Scavenging activity on DPPH radical 1000/A
Clove	570	1.7×10^{0}
Oregano	687	1.4×10^{0}
Cinnamon	821	1.2×10^{0}
Marjoram	1015	9.8×10^{-1}
Cumin	31008	3.2×10^{-2}
Bearberry	60	1.7×10^{1}
Rosemary	832	1.2×10^{0}
Lemon-balm	1910	5.2×10^{-1}
Peppermint	1280	7.8×10^{-1}
Hyssop	3426	2.9×10^{-1}
BHA	10	1.0×10^{2}
BHT	19	5.2×10^{1}
Catechin	21	4.8×10^{1}

References

1. R. A. Larson, *Phytochemistry*, 1988, 27, 969.
2. H. Macheix and A. Fleuriet, *Proceedings of the EURO FOOD TOX IV Conference*, Olsztyn, Poland, 1994, p. 97.
3. G. Yen and P. Duh, *J. Agric. Food Chem.*, 1994, 42, 629.
4. G. Yen and H. Chen, *J. Agric. Food Chem.*, 1995, 43, 27.
5. O. I. Aruoma, B. Halliwell, R. Aeschbach and J. Loeliger, *Xenobiotica*, 1992, 22, 93.
6. M. Minnuni, O. Muller, U. Wolleb, A. Pfeifer and H. U. Aeschbach, *Mut. Res.*, 1992, 269, 193.
7. M. E. Cuvelier, C. Berset and H. Richard, *J. Agric. Food Chem.*, 1994, 42, 665.
8. T. Tsuda, K. Ohshima, S. Kawakishi and T. Osawa, *J. Agric. Food Chem.*, 1994, 42, 248.
9. M. Huang and T. Ferrero, *ACS Sympsyum Series 507*, 1992, p. 8.
10. C. T. Ho, Q. Chen, H. Shi, K. Q. Zhang and R. T. Rosen, *Prev. Med.*, 1992, 21, 520.
11. J. Kanner, E. Frankel, R. Granit, B. German and J. Kinsella, *J. Agric. Food Chem.*, 1994, 42, 64.
12. AOAC: *Official Methods of Analysis*, 14th ed., 1984, p. 187.
13. M. S. Blois, *Nature*, 1958, 181, 1199.
14. E. A. Decker and A. D. Crum, *J. Food Sci.*, 1991, 56, 1179.
15. L. Ramanathan and N. P. Das, *J. Agric. Food Chem.*, 1992, 40, 17.
16. C. H. Lea, *J. Sci. Food Agric.*, 1958, 9, 621.

ANTIOXIDATIVE PROPERTIES OF LEGUME SEED EXTRACTS

R. Amarowicz, A. Troszyńska,
M.Karamać and H. Kozłowska

Division of Food Science,
Polish Academy of Science,
Centre for Agrotechnology and
Veterinary Sciences,
10-718 Olsztyn, Poland

1 INTRODUCTION

Phenolic compounds encompass a wide variety of compounds characterized by the presence of an aromatic ring with one or more hydroxyl groups and a variety of substituents. Many of phenolic compounds are primary antioxidants that act as free radical receptors and chain breakers.

Our knowlege on phenolic compounds of legume seeds concerns mainly tannins which are ascribed with antinutritional properties[1,2,3] Only a few publications deal also with the presence of flavanols and phenolic acids in legume seeds [4,5]. As these compounds are strong antioxidants it seemed worth undertaking studies on the antioxidative properties of extracts from legume seeds.

2 MATERIALS AND METHODS

Seeds of Polish cultivars of the following legumes were studied: white bean, pea, everlasting pea, lentil, broad bean and faba bean. Extraction of phenolic compounds from whole seeds and from seed hulls was conducted with 80% acetone according to an earlier established method[5] Following distillation of acetone in a rotary evaporator the remaining water solutions were lophylized. In thus obtained extracts the content of total phenolic compounds was determined using catechin as standard [6]. Antioxidative properties of the extracts were studied in a β-carotene-linoleate system[7]. Approximately 4 mg of ß-carotene was dissolved in 10 ml of chloroform. One ml of this solution was pipetted into a round-bottom flask. After removal of chloroform with rotary evaporator 20 mg of purified linoleic acid, 200 mg of Tween 40 and 50 ml of oxygenated distilled water were added to the flask with vigorous stirring. Aliquot (5 ml) of prepared emulsion was transferred to a series of tubes containing 4 mg of each extract or 0.2 mg of BHA. As emulsion was added to each tube, the zero time absorbance was read at 470 nm. Subsequently, absorbance readings were recorded at 15 min intervals by keeping the samples in a water bath at 50°C during 120 min.

3 RESULTS

The content of total phenolic compounds in the extracts from whole seeds was diversified (Table 1). The lowest content was found in the extract from lentil seeds. In the extracts from the seed coats the level of total phenolic compounds was similar. The exception was the extract from the seed coats of everlasting pea seeds which was very poor in these compounds.

Table 1 *The content of Total Phenolic Compounds in the Extracts from Whole Seeds and from Seed Coats of Legume Seeds (mg / g dry weight)*

Legume	*Whole seeds*	*Seed coat*
White bean	10.8	627
Pea	34.8	469
Everlasting pea	9.7	26
Lentil	114	700
Broad bean	60.1	531
Faba bean	80.9	648

The results of the antioxidation test (Figures 1 and 4) indicate that the extracts from pea, faba bean, lentil, everlasting pea and broad bean seeds have similar antioxidative activity whereas the extract from white bean seeds is clearly less active. After 60 and 120 min ncubation in samples with addition of more active extracts about 75-80% and 61-66% of β-carotene, respectively, remained unoxidated. For white bean the respective values were as low as 58.8 and 32.4%.

The extracts made from the seed coats of faba bean, lentil an pea seeds had similar antioxidative properties (Figures 2 and 4) less active. Very low activity was found for the extract from the seed coats of everlasting pea and white bean seeds. After 60 and 120 min incubation with addotion of the extract from the seed coats of faba bean, lentil and pea seeds about 76-81% of β-carotene remained unoxidated. For everlasting pea the respective values were 50.0 and 27.7% and for white bean as low as 37.7 and 25.4%.

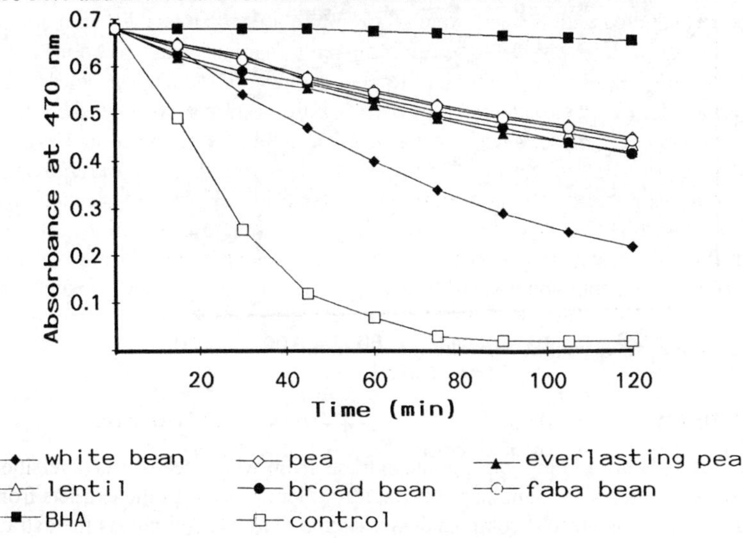

Figure 1 *Antioxidative activity of legume seed extracts*

From the results comprised in Table 1 and on Figures 2 and 3 it follows that the antioxidative properties of the extracts from whole seeds did not depend directly on the content of phenolic compounds in extracts. For instance, the extracts from white bean and everlasting pea seeds contain the same amounts of phenolic compounds but antioxidative activity of the latter was much higher.

As in the extract obtained from the seed coats the content of total phenolic compouds was about 2.5-13 times higher than in the extracts from whole seeds and with respect to the results on Figures 1-4 it can be concluded that most of them do not have antioxidative properties determined with the kind of test used.

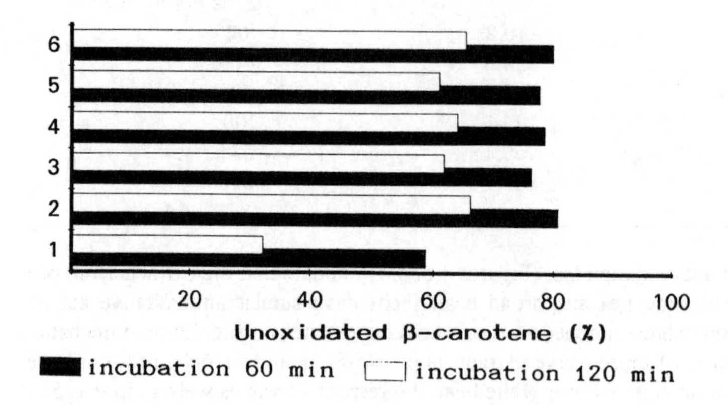

Figure 2 *Unoxidated β-carotene after incubation samples with extracts from white bean (1), pea (2), everlasting pea (3), lentil (4), broad bean (5) and faba bean (6) seeds*

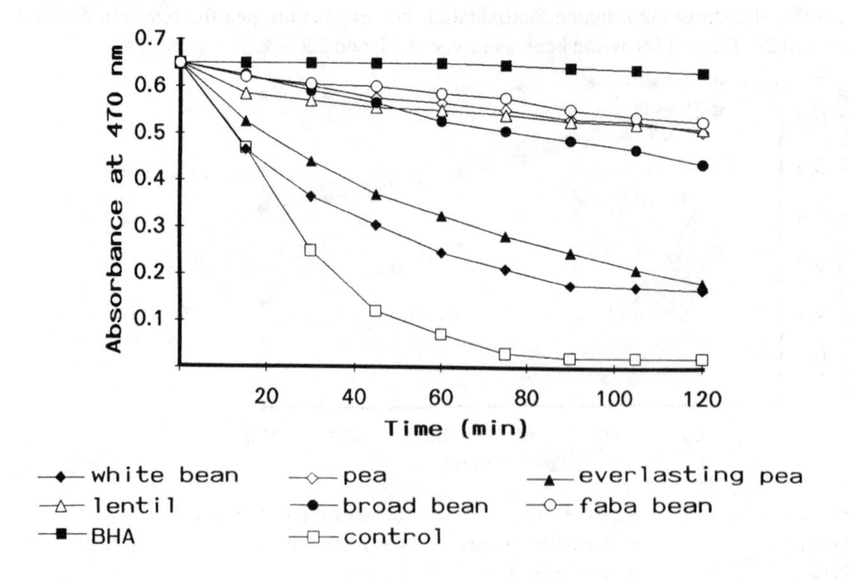

Figure 3 *Antioxidative activity of legume seed coats extracts*

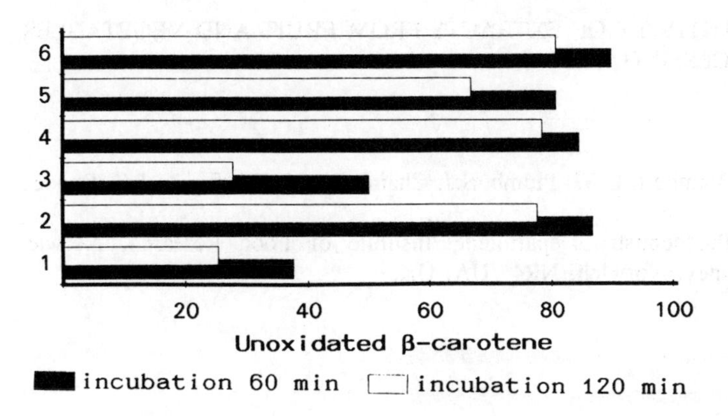

Figure 4 *Unoxidated β-carotene after incubation samples with extract from white bean (1), pea (2), everlasting pea (3), lentil (4), broad bean (5) and faba bean (6) seed coats*

The results of this study show that though the extracts from whole legume seeds contain small amounts of phenolic compounds compared with the seed coats, they have high antioxidative properties, comparable with the antioxidative activity of canola[8] , rapeseed[9] , white mustard[10] and flax[11] seeds.

References

1. D.A. Bond, *J. Agric. Sci.*, 1976, **86**, 561.
2. J.P.F.G. Helsper, J.M. Hoogendijk, A. Van Novel and K. Burger-Meyer, *J. Agric. Food Chem.*, 1993, **41**, 1058.
3. A.J.M. Jansman, M.W.A. Verstgen and J. Huisman, Anim. Feed Sci., 1993, **43**, 239.
4. J. Wilska-Jeszka and A. Stasiak, in: 'Bioactive Substances in Food of Plant Origin',Polish Academy of Sciences , Olsztyn, 1994.
5. R. Amarowicz, M. Piskula, J. Honke, B. Rudnicka, A. Troszynska and H. Kozlowska, *Pol. J. Food. Nutr. Sci.*, 1995, in press.
6. M. Naczk and F. Shahidi, *Food Chem.*, 1989, **31**, 159.
7. H. E. Miller , *J. Amer. Oil Chem. Soc.*, 1971, **45**, 91.
8. U.N. Wanasundara, R. Amarowicz and F. Shahidi, *J. Agric. Food Chem.*, 1994, **42**, 1285.
9. H. Nowak., K. Kujawa, R. Zadernowski, B.Roczniak and H. Kozłowska , *Fett Wiss. Technol.*, 1992, **94**, 149.
10. F. Shahidi, U.N. Wanasundara and R. Amarowicz, *Food Res. Intern.*, 1994, **27**, 489.
11. R. Amarowicz, U.N. Wanasundara, J. Wanasundara and F. Shahidi, *J.Food. Lipids*, 1993, **1**, 111.

ANTIOXIDANT ACTIVITY OF EXTRACTS FROM FRUIT AND VEGETABLES: EFFECT OF PROCESSING

G. Williamson, N. Lambert, G.W. Plumb, S.J. Chambers, N. Tawfiq, M.J.C. Rhodes.

Food Molecular Biochemistry Department, Institute of Food Research, Norwich Research Park, Colney, Norwich, NR4 7UA, UK.

1 ABSTRACT

We have examined the antioxidant properties of a range of extracts from fruit, vegetables and herbs. The methods used were induction of a phase II marker enzyme (quinone reductase), inhibition of enzymic and non-enzymic peroxidation of human liver microsomes, iron reduction, iron chelation, and hydroxyl radical scavenging. The results demonstrate that processing of *Brassica* vegetables can have a dramatic effect on these parameters.

We thank the Biotechnology and Biological Sciences Research Council, UK, and the Ministry of Agriculture, Fisheries and Food, UK, for funding.

2 INTRODUCTION

There is increasing evidence that increased consumption of fruit and vegetables leads to a decreased risk of cancer and other chronic diseases[1,2]. Many of the properties of the foods which are associated with this protective effect are linked to the antioxidant properties of components of these foods[3]. The antioxidant effect arises from (a) antioxidant vitamins (b) non-vitamin antioxidants (eg flavonoids, phenolics) and (c) components that stimulate antioxidant defences and detoxifying enzymes (eg suphur-containing compounds such as glucosinolates from Brassicas, alliins from oninons and garlic, and some flavonoids). In this paper, we have examined the antioxidant properties of some extracts from *Brassica* vegetables, from apple and from rosemary, and compared these to the ability of the foods to induce defensive mechanisms (quinone reductase) in cultured cells.

3 MATERIALS AND METHODS

3.1 Preparation of extracts

Extracts from fruit and vegetables were made as previously described[4], involving extraction of freeze dried powders with 70% methanol. All of the methanol was removed before analysis in any assay, and the extract made up to a fixed volume with water. Autolysis was allowed to proceed in some cases for 1 h, by addition of water to the freeze dried powder, before addition of methanol. This procedure allows enzymes such as myrosinase, which are present in the plant tissue, to act on the compounds and produce a range of degradation products. Samples which were "cooked" were boiled for 15 min, and then freeze dried, and extracted directly into 70% methanol. "Raw" samples were extracted directly after freeze drying into 70% methanol, and this treatment allows extraction of components with minimal enzymic modification or breakdown.

3.2 Induction of phase II enzymes

Induction of quinone reductase (NQO) activity was measured in hepa1c1c7 cells[4,5]. The results are shown as a "-fold" induction, which is the ratio of the treated cells to the untreated control cells. For the induction of NQO, the concentration of food extract was 2 mg dry weight/ml culture medium, and under these conditions, induction of more than 10% is significant[4], and each value presented is the mean of at least 8 determinations.

3.3 Inhibition of peroxidation

The ability of the extracts to inhibit NADPH/iron- or ascorbate/iron-induced lipid peroxidation in human liver microsomal membranes was measured using the TBARS test[6]. Human liver microsomes were washed in 150 mM Tris/Cl pH 8.0 to remove cytosolic proteins. Peroxidation was initiated by the addition (in the order stated) of either 0.4 mM NADPH/0.2 mM $FeCl_3$/0.8 mM ADP/150 mM KCl or 0.05 mM ascorbate/0.2 mM $FeCl_3$/150 mM KCl. In both cases, peroxidation was allowed to proceed for 40 min, and the reaction stopped with 2 volumes of 20% trichloroacetic acid/0.04% thiobarbituric acid/0.25 M HCl and 0.025 volumes of 5% w/v butylated hydroxytoluene. The production of TBARS was measured after 15 min at 80°C at 535 nm.

3.4 The deoxyribose assay

The presence of components able to chelate iron or to scavenge hydroxyl radicals was measured using the deoxyribose assay[7]. The ability of the test substances to affect the reduction of Fe(III) by ascorbate was assessed as described by Lambert et al[8].

4 RESULTS

The ability of extracts from Brussels sprouts and from Broccoli to induce phase II enzymes is shown in table 1. Both of these *Brassica* vegetables contain compounds able to induce NQO in cultured cells. It is clear that changes on autolysis and, to a lesser extent, cooking, increase the number or concentration of compounds which are able to induce these enzymes. Glucosinolates are found in high concentrations in many Brassica vegetables[9]. On autolysis, these compounds breakdown to form a wide range of products, many of which are able to induce phase II enzymes[10], and glucosinolates comprise the majority of the activity of the *Brassica* extracts to induce NQO. Extract from rosemary was also an inducer of phase II enzyme, but extract from most fruits (apple is shown as an example) are not.

The direct free radical scavenging activity of the extracts were also measured using the deoxyribose assay. Addition of ferric-EDTA to hydrogen peroxide and ascorbate at pH 7.4 leads to the production of hydroxyl radicals which can degrade deoxyribose into thiobarbituric acid reactive substances (TBARS). Compounds added to the reaction mixture may compete with the deoxyribose for hydroxyl radicals and inhibit sugar degradation. When iron is added to the mixture as ferric chloride instead of iron-EDTA, some of the Fe(III) ions bind to deoxyribose. Thus the damage to the sugar becomes site specific and the hydroxyl radicals formed by the bound iron immediately attack the deoxyribose. The ability of a compound to inhibit deoxyribose degradation under these conditions reflects its iron chelating ability and the ability of its iron complex to participate in Fenton chemistry[11].

The effect of extracts on these assays is shown in table 2. Rosemary is the most potent antioxidant in these systems. Further, Brussels sprouts which have undergone autolysis contain more components which are able to act as antioxidants in this system, which shows that modification by endogenous enzymes in the Brassica vegetables alters the antioxidant properties in addition to the efficacy at induction of NQO.

The ability of the same extracts to inhibit lipid peroxidation was also examined (table 3). Again rosemary is most effective, whereas the extracts from Brussels sprouts is slightly pro-oxidant in this system. The activity of rosemary is most probably due to carnosol and carnosic acid[13], but the components responsible for the pro-oxidant activities are not known. The ability of the extracts to reduce iron was also measured. Only rosemary showed a significant difference to the control value, which shows an efficient reduction of iron by some component(s) of rosemary extract.

5 DISCUSSION

The assessment of the potential health benefits of foods in humans is dependent on the measurement of biochemical effects both *in vitro* and *in vivo*. The assays described here have been developed to give an indication of the antioxidant and anticarcinogenic properties of whole extracts and of single or mixed components of foods[3,5]. It is, of course, a large jump from these assays to long term health benefits. However, these assays are intended as an initial study to determine which are the most active foods for further study. Bearing these limitations in mind, we can conclude that processing has a marked effect on Brassica vegetables to score positive in anticarcinogenesis assays (phase II enzyme induction). Processing also affects the antioxidant properties of these foods. Extract from apple shows some antioxidant properties, which are at least partly due the high content of chlorogenic acid in these samples[14]. Apple estract did not induce phase II enzymes. Rosemary was an effective antioxidant and also induced phase II enzymes, which indicates rosemary as a food with potential long term benefits. It should be borne in mind, however, that rosemary only forms a minor part of the diet, whereas apple and Brassicas are usually consumed in much larger quantities. For this reason, in a normal Western style diet, Brassicas probably contribute most to the induction of phase II enzymes, whereas fruits contribute most to the direct antioxiant action.

Table 1 *Ability of extracts to induce phase II defences*

Material tested[1]	*Induction of NQO (%)*[2]
Brussels sprouts (autolysed)	50
Brussels sprouts (cooked)	30
Brussels sprouts (raw)	NI
Broccoli (autolysed)	40
Broccoli (raw)	40
Rosemary (water extract)	20[3]
Apple (raw)	NI

[1]Processing method in brackets
[2]From Tawfiq et al[4] (concentration of extract was 2.1 mg dry weight/ml culture medium).
[3]Extract at 1 mg dry weight/ml culture medium.
NI, no induction
Values are ±10% for NQO induction (see Materials and Methods).

Table 2. *Free radical scavenging properties of extracts.*

	A_{535} *(+EDTA)*	A_{535} (-EDTA)
Brussels sprouts (autolysed)	0.61±0.05	0.18±0.01
Brussels sprouts (cooked)	0.90±0.08	0.33±0.02
Broccoli (raw)	0.83±0.06	0.28±0.02
Rosemary	0.43±0.02	0.11±0.01
Apple (raw)	0.63±0.05	0.25±0.01
Control	1.80±0.12	0.34±0.03
Hypotaurine	0.35±0.03	0.12±0.01

Deoxyribose assay performed as described in Materials and Methods section. Control experiment is water, and positive control is hypotaurine (50 mM). Data from refs. 8 and 12.

Table 3. *Effect of extracts on lipid peroxidation and iron reduction*

	NADPH/Fe-induced lipid peroxidation	*Ascorbate/iron-induced lipid peroxidation*	*Iron reduction*
Brussels sprouts (autolysed)	500[1] (pro)	>700	0.92[2]
Brussels sprouts (cooked)	170 (pro)	>700 (pro)	0.81
Broccoli (raw)	260	>700	1.06
Rosemary	1	50	1.69[3]
Apple (raw)	330	495	0.87

[1]Amount of extract (μg dry weight) required to give 5% inhibition (or promotion = pro) of peroxidation at a microsome concentration of 0.3 mg protein/ml. Thus, the most potent inhibitor or promoters are those with the lowest numbers.
[2]Control value with water alone = 0.90.
[3]Value using 25 μg dry weight of extract in the iron reduction assay instead of 700 μg as for other samples. Data from refs. 8 and 12.

References

1. G. Block, B. Patterson, and A. Subar, *Nutr. Cancer*, 1992, **18**, 1-29.

2. K. A. Steinmetz and J. D. Potter, *Cancer Causes and Control*, 1991, **2**, 325-357.

3. B. Halliwell, *Lancet*, 1994, **344**, 721-724.

4. N. Tawfiq, S. Wanigatunga, R. K. Heaney, G. Williamson and G. R. Fenwick, *Eur. J. Cancer Prev.*, 1994, **3**, 285-292.

5. H. J. Prochaska, A. B. Santamaria and P. Talalay, *Proc. Natl. Acad. Sci. USA*, 1992, **89**, 2394-2398.

6. J. A. Buege and S. D. Aust, *Methods Enzymol.*, 1978, **30**, 302-310.

7. O. I. Aruoma, *Methods Enzymol.*, 1994, **233**, 57-66.

8. G. W. Plumb, S. J. Chambers, N. Lambert, S. Wanigatunga, G. R. Fenwick, O. I. Aruoma, B. Halliwell and G. Williamson, *Free Radical Res.*, 1995, in press.

9. G. R. Fenwick, R. K. Heaney, and W. J. Mullin, *CRC crit rev Food Sci Nutr.*, 1983, **18**, 123-201.

10. N. Tawfiq, R. K. Heaney, J. A. Plumb, G. R. Fenwick, S. R. R. Musk, and G. Williamson, *Carcinogen.*, 1995, **16**, 1191-1194.

11. B. Halliwell, J. M. C. Gutteridge and O. I. Aruoma, *Analyt. Biochem.*, 1987, **165**, 215-219.

Section 8: Enhancing Nutrients and Nutritional Quality

NUTRITION CHALLENGES IN FUTURE: A NEED FOR NEWTRITION

Prof.Dr. J.G.A.J. Hautvast

Department of Human Nutrition
Wageningen Agricultural University
6700 EV Wageningen
The Netherlands

Introduction

There is a growing awareness that the outcome of nutrition sciences in the present decade is contributing significantly to health and development of mankind. Some people even argue that nutrition sciences is entering in a second renaissance (1). The first renaissance was observed in the first decades of this century when the discovery of vitamins led to an enormous excitement both in science and with the public. In this paper some of the present highlights in nutrition sciences are given. Special emphasis will be given to the question that nutrition has everything to do with food as such and with food composition in its broadest sense. This topic will be followed by some remarks on the state of nutrition training and the need for nutrition leadership in Europe. To start with a few remarks will be made on opportunities and challenges in nutrition sciences.

Opportunities and challenges

Recently I had to review a book on "Opportunities in the Nutrition and Food Sciences - Research Challenges and the Next Generation of Investigators" (2). I enjoyed very much reading this book. It confirms my observation that nutrition should come closer to food sciences and of course the reverse is also true. In this book one can read about the need to learn more about the physical and chemical complexity of food components, their inter-relationships and bioavailability in the human body and their effects and functions on the short and long term. There is no doubt that this whole system is of intriguing complexity. The outcome of this all contributes significantly to health or disease, adequate or retarded growth, normal or premature aging, good vision or blindness, adequate or retarded psychomotoric development. Because of this outcome mankind is showing or should show great interest in this aspect of life.

What nutrition research keeps me going?

In the next paragraph I will report about a number of topics in nutrition sciences which we studied in our department and which kept me "going" in the past two decades with increasing enthusiasm.

Children on macrobiotic diet

My colleague Professor Wija van Staveren was for a long time already interested in alternative dietary lifestyles. These lifestyles are characterized by some degree of avoidance of foods of animal origin. Concern was regularly expressed about the risks of nutritional deficiencies, especially in children on diets containing little animal food. We became after some time especially interested in children on macrobiotic diets. Such diets are relatively very monotonous. The daily regime consists of a diet of cereals (rice), cooked vegetables, pulses, sea vegetables, nuts and seeds. At a weekly basis one may add (cooked) fruits and sometimes a bit of fish. The diet is based on a philosophy aimed to balance the complementary forces yin and yang. Such diet must have consequences for child development and health as was preliminary shown by some groups in the U.S.A. We were very pleased that we could appoint the PhD-student Pieter C. Dagnelie to study in much depth the nutritional-health consequences in children on a macrobiotic diet (3). He started with an inventarization of families with young children who actively were using macrobiotic diets. A few hundred families, living in the whole country could be identified after a time consuming search. The first step then became to perform a cross-sectional study in all children age 0-8 years (n= 243) who were reared on and using the macrobiotic diet. The families on this diet belong in majority to the highest education levels (college or university training). We found that the reported birth weights of these children were 150 gr lower than the Dutch reference. We further observed that from 6-8 months onwards, growth stagnation occurred in both sexes, which was most marked in girls. In the period from 0-6 months the growth was found close to the median of the Dutch reference.
The second step in this research project was to start a mixed-longitudinal study in which we did compare weaning pattern, nutrient intake and health status of infants on macrobiotic diets and matched omnivorous control infants. The age period covered was from 4-18 months and three age-cohorts of the children were formed (Figure 1).

We found that:

- 96 per cent of the macrobiotic infants and 74 per cent of the control infants had been breast fed, but breast-feeding continued longer in the macrobiotic group (13.6 versus 6.6 months)
- the macrobiotic weaning diet tended to be bulky and had a low energy density (2.4 kJ/g, controls: 3.4 Kj/g).

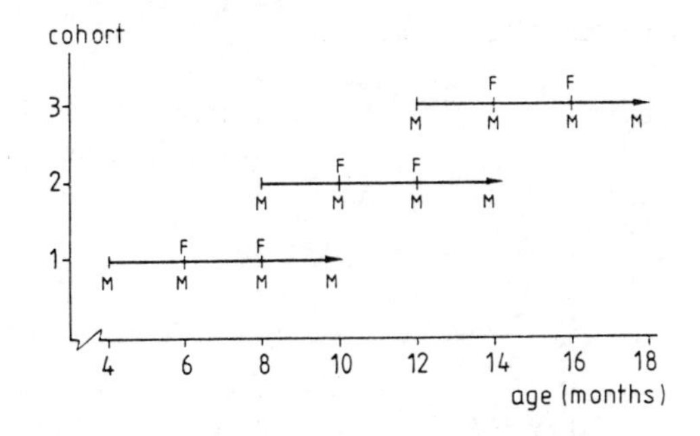

Figure 1. *Study design.*

- the intake of energy, protein, fat, calcium, riboflavin and vitamin B12 was significantly lower in the macrobiotic infants, whereas their intake of polysaccharides, fibre, iron and thiamine was higher than that of the control infants.
- growth retardation was again observed (Figure 2).
- with regard to the iron status no differences were observed between both groups although a low iron status was observed more in macrobiotic infants than in the control group (respectively 11 per cent versus 2 per cent)
- folate blood values were higher in the macrobiotic group than in the control group (respectively 32 versus 21 µmol/l), while the opposite was found for vitamin B12 concentrations (respectively 170 versus 431 µmol/l)
- in late summer physical symptoms of rickets were present in 28 per cent of the macrobiotic group; in early spring 55 per cent of these macrobiotic infants showed these symptoms. This was confirmed by low plasma levels of 25-OH-cholecalciferol (Figure 3).

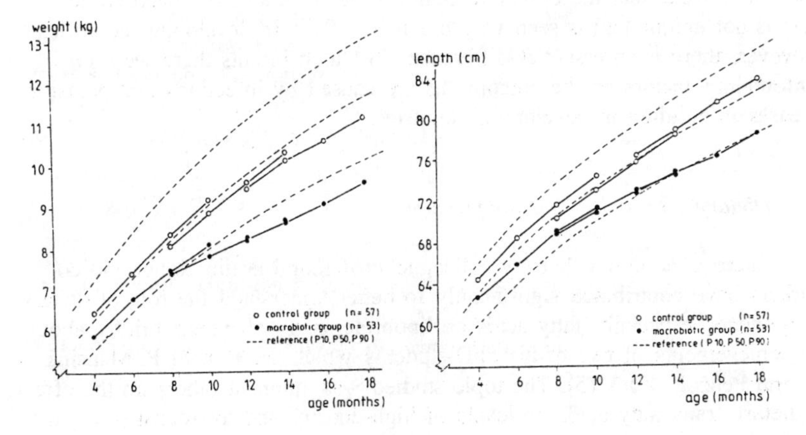

Figure 2. *Weight for age and length for age.*

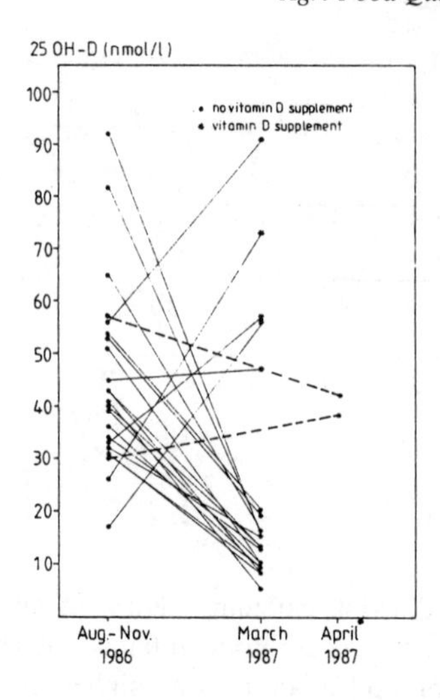

Figure 3. *Plasma 25-OH-vitamin D concentrations of 25 macrobiotic infants in August - November 1986 (age 10 - 20 months) and at follow-up in March - April 1987.*

After this study we only could conclude that nutritional deficiencies are present in a high proportion of macrobiotic infants. Then we started arguments with the macrobiotic "leaders" and strongly recommended them to adapt the macrobiotic diet and to include fatty fish (100-150 g/week), oil (20-25 g/day) and dairy products (one serving per day). Whether we were very successful we do not know. At present we are revisiting these children which is about 10 years after our study. We are studying these infants and are especially interested in the later consequences of a macrobiotic diet. Questions studied are whether catch-up growth took place and also bone density is measured, using the DEXA method.
One could argue that the observed malnutrition in infants on macrobiotic diets is not unique and is seen very much in infants in developing countries. However, there is an essential difference. In Dutch infants there were no confounding factors on the outcome as are caused by infections and parasitic diseases in children in developing countries.

Dietary fatty acids and lipoproteins

There is no doubt that my colleague Professor Martijn Katan and co-workers have contributed significantly to better understand the role of dietary components, especially fatty acids on lipoproteins. I will report briefly about the achievements of two of his PhD-students which are Ronald P. Mensink (4) and Peter L. Zock (5). The topic studied was amongst others on the effect of dietary trans fatty acids on levels of high-density and low-density cholesterol in healthy men and women. Many foods contain fatty acids in both the cis configuration but also in the trans configuration. Trans fatty acids are unsaturated fatty acids in which the carbon moieties are positioned on the two sides of the double bond in opposite directions which in cis double bonds these are positioned at the same side ((Figure 4).

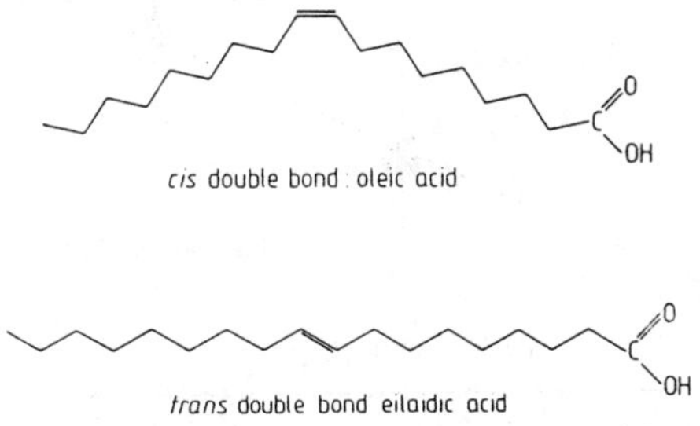

O
C
OH
cis double bond : oleic acid

O
C
OH
trans double bond eilaidic acid

Figure 4. *Configuration of cis and trans fatty acids.*

Trans fatty acids are found in small amounts in ruminant fats. Much larger amounts are found in certain types of margarines. In these margarines the trans fatty acids are formed when vegetable and marine oils, rich in polyunsaturated fatty acids, are hardened by a process called hydrogenation to produce fats that have the firmness and plasticity desired by food manufacturers and consumers. The discussion when we started the studies was whether these trans fatty acids elevate serum cholesterol levels relative to the cis-isomer with special attention to LDL and HDL cholesterol levels. In the first study about 11 energy per cent was provided as one of the treatments as trans $C_{18:1}$ monounsaturated fatty acid.
The effect we found indicated that trans fatty acids are unfavourable for the serum lipoprotein profile as HDL-cholesterol decreased and the LDL-cholesterol increased. As such it was concluded that the effects are at least as unfavourable as that of the cholesterol-raising saturated fatty acids.
The findings caused much alarm and health concerns. In addition we were told that the levels of trans fatty acids in the experimental diets were unnatural high compared to daily dietary patterns in the western world.
The decision was taken to repeat the study and to use 7,5 energy per cent in the experimental diet as trans fatty acids.
This amount is usually eaten in our western dietary pattern. The findings in this study confirmed earlier findings in our institute: trans fatty acids in the diet significantly lowered HDL-cholesterol and raised LDL-cholesterol. In Figure 5 information of both studies are given and results suggest a linear dose-response relation.
In the same period Professor Katan became involved in the discussion why

people consuming unfiltered coffee showed higher serum cholesterol levels than when the coffee was filtered during preparation.

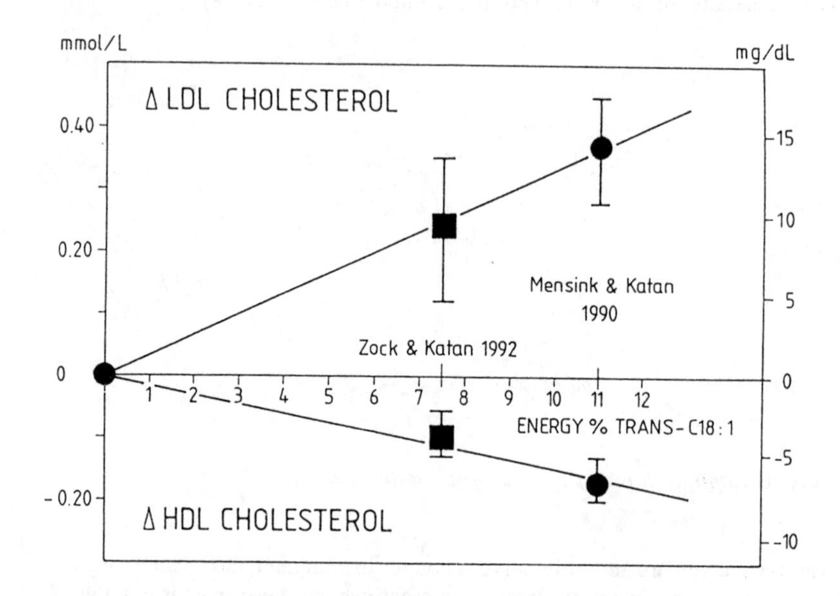

Figure 5. *Effects of trans fatty acids on LDL and HDL cholesterol in the present trial compared with those in a previous trial. Circles represent data from a comparison of trans C18:1 with oleic acid (10). Squares are based on the present comparison between trans C18:1 and linoleic acid. Bars denote 95% confidence intervals. In addition to the two experiments, the origin provides a third point, because a zero change in intake will produce zero change in lipoprotein levels.*

In that period there was a lot of speculation on this topic and Peter Zock and Martijn Katan were the first ones to show that the very small lipid-rich fraction from boiled coffee must probably cause the increase in total serum and LDL cholesterol (Figure 6). In the past years research on coffee oil has been intensified and we are gradually learning more about the main lipid increasing substances which are kawheol and cafestol.

Vitamin a and carotenoids

Vitamin A deficiency causes increased ill-health and mortality. This situation is especially observed in developing countries. A recent review has estimated that a 23 per cent reduction in mortality can be expected when children in vitamin A deficient areas are supplemented with vitamin A. My colleague Professor Clive West has been fascinated for years in the vitamin A problems in these developing countries and did research in several countries such as Nigeria, Ethiopia, Vietnam and Indonesia. I want to highlight two recent observations.

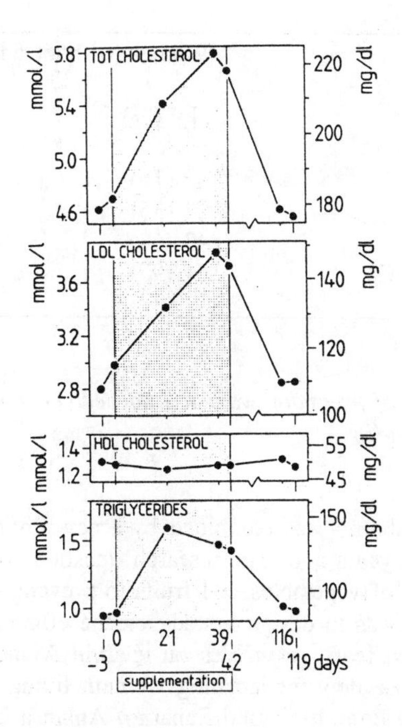

Figure 6. *Mean serum total, LDL, and HDL cholesterol, and triglyceride levels before, during and after supplementation of lipid-rich coffee fraction.*

His PhD-student Djoko Suharno studied in Indonesia the role of vitamin A in nutritional anaemia in pregnant women in West Java. In a cross-sectional study we found that 49 per cent of normal pregnant women were anaemic. On the basis of serum retinol values we observed that 2.5 per cent of the women were vitamin A deficient and 31 per cent had a marginal vitamin A status. Further analysis showed that there was a significant positive association between serum retinol concentrations and haemoglobin, haematocrit and serum iron concentrations. We then decided to test whether this anaemia would respond to improvement in vitamin A supply. A large intervention study was designed in anaemic pregnant women and the main results are shown in Table 1.

These findings clearly indicate that in areas with sub-optimal vitamin A status in people nutritional anaemia should involve improvement both in the

iron status and also in the vitamin A status.
The next question in this research program was how to improve the vitamin A status. A general recommendation is that the consumption of dark green leafy vegetables should be increased. Such foods are often much available and are cheap compared to vitamin A rich foods of animal origin.

	Women without anaemia	
Intervention	No. (%)	95% Cl
Placebo (n = 62)	10 (16)	7 to 29
Vitamin A (n = 63)	22 (35)	22 to 48
Iron (n = 63)	43 (68)	54 to 79
Vitamin A and iron (n = 63)	61 (97)	88 to 99

Table 1. *Proportion of pregnant women who become non-anaemic (Hg $\geq$ 110 g/L)*

The research of Djoko Suharno was continued by a new PhD-student Saskia de Pee who started a few years ago. The research question formulated now was on the effectiveness of vegetables and fruits to prevent vitamin A deficiency. The first step was to design a study on the effect of an additional daily portion of dark-green leafy vegetables on vitamin A and iron status again in women, we choose now for lactating women, living in the same area as where the studies were done by Djoko Suharno. Again a large group of lactating women became enroled in this study. The following treatments were given: one group received stir-fried vegetables; a second group received a wafer enriched with β-carotene, iron, vitamin C and folic acid; and a third group received a non-enriched wafer to control for additional energy intake. The surprising finding was that in the enriched-wafer group there were increases in serum retinol, breast milk retinol, and serum β-carotene. These changes were significantly different from those in the other two groups in which only small significant increases were observed in breast milk retinol in the control-wafer group and in serum β-carotene in the vegetable group. Changes in iron status were similar in all three groups (7). We came therefore to the conclusion that an additional portion of dark-green leafy vegetables did not improve the vitamin A status and that there is much need for a re-examination of recommendations regarding dark-green leafy vegetables. Is there a low bioavailability of β-carotene in dark-green leafy vegetables? There are several hypotheses with regard to this bioavailability. We think that the physical inaccessibility of carotenoids in plant tissues may be involved. In green leaves, β-carotene molecules are organized in pigment-protein complexes located in cell chloroplasts, and in fruits, β-carotene is found in lipid droplets and chromoplasts. In the mean time Saskia de Pee has started a next study in which fruits (papaya and mango) are being tested instead of dark-green leafy vegetables on β-carotene bioavailability. I want to acknowledge the excellent cooperation with Dr Muhilal (Bogor) in the above-mentioned studies.

Other research programmes?

The above-mentioned topics discussed indicate that food and nutrition sciences should interact and that this provides a lot of opportunities and new challenges. There is much more research carried out or in operation in our department in which food and nutrition sciences come in some ways more or less close together. I always find it very fascinating to be as fully as possible involved in the research plans and achievements in our group.

Nutrition sciences and scientists

Although we do not have a good overview of (post)graduate training in human nutrition in Europe I do not think that the present situation is favourable. In 1959 FAO/WHO organized a review and a meeting on the status of nutrition training in Europe. It was strongly recommended that each country should develop graduate training programmes in human nutrition. If we review the achievements since 1959 then progress on training in Europe is very modest and could have been much better developed. In addition to this we do observe quite a large range in the way graduate programmes are outlined.

I think that some of the available training programmes might need a redefinition and become directed towards what we consider as modern nutrition sciences. We also must keep in mind that nutrition sciences obviously has its own identity and objectives. According to me nutrition sciences is concerned with all aspects what food does in the human body, including e.g. food choices and health consequences. Further nutrition sciences need to attract excellent students and must try to offer to at least some of them interesting career perspectives.

In our University a Department of Human Nutrition and a MSc-degree programme in Human Nutrition started in 1969. In 1974 the first students graduated with a MSc-degree. Since 1974 about 1000 students received a MSc-degree which is a considerable number. You may think that this is probably too much. However, our University offers unique training programmes with many electives in order that students can select courses according to their interest and in line with job expectations in future.

Around 1980-'85 the PhD-programme in our Department became gradually established. In 1978 the first MSc-Nutrition graduates obtained the PhD-degree which was followed by more than 50 of such students. The PhD-students have established themselves as a strong group with an own scientific bulletin which they call "NEWTRITION".

Cooperation and interactions with European colleagues led to the european initiative to start "The European Nutrition Leadership Programme". These colleagues are especially: Prof. W.P.T. James, Prof. K. Pietrzik, Prof. N.G. Asp, Prof. A. Ferro-Luzzi and Dr. O. Korver. It was found that e.g. nutrition scientists should understand the responsibilities when being involved in dialogues on nutrition sciences between industry and consumer or other interested groups. In fact what nutrition sciences also need are good communicators. It was decided to start a programme through advanced training seminars on topics such as communication of nutrition science; future strategies and new frontiers in nutrition research; nutrition science and nutrition health in Europe and nutrition within European health policies. The model we choosed was the organization of a 7-days seminar in which excellent final year PhD-students and postdoctoral fellows from all over Europe were being trained in the awareness and skills required for future nutrition leaders and their roles in advocacy and development planning. The European Nutrition Leadership Programme started in 1994 with a one-week seminar held in Luxembourg and which was financially supported by the European Union and by a consortium of European Food Industries. The programme was repeated in 1995 and money has been secured for the years 1996 and 1997. In each programme 30 excellent young people involved in nutritional sciences were and are going to be invited after being selected using a critical external referee system. The evaluations after the 1994 and 1995 programmes showed that we have been very successful and we think that we fulfil the objectives as formulated at the start of the programme. We hope that a programme like The European Nutrition Leadership Programme will support nutrition sciences in Europe, and will become an important basis for new nutrition leadership in Europe in future.

REFERENCES

1. J.G.A.J. Hautvast. "Renaissance in de Voedingswetenschap" ("Renaissance in Nutrition Sciences"). Wageningen Agricultural University. Annual Lecture March 9, 1994, Wageningen, The Netherlands.
2. P.R. Thomas and R. Zarl. "Opportunities in the Nutrition and Food Sciences". National Academy Press, Washington D.C., 1994.
3. P.C. Dagnelie. "Nutritional Status and Growth of Children on Macrobiotic Diets: a Population-Based Study". PhD-thesis Wageningen Agricultural University, 1988, Wageningen, The Netherlands.
4. R.P. Mensink. "Effect of Monounsaturated Fatty Acids on High-Density and Low-Density Lipoprotein Cholesterol Levels and Blood Pressure in Healthy Men and Women". PhD-thesis Wageningen Agricultural University, 1990, Wageningen, The Netherlands.
5. P.L. Zock. "Dietary Fatty Acids and Risk Factors for Coronary Heart Disease - Controlled Studies in Human Volunteers". PhD-thesis Wageningen Agricultural University, 1995, Wageningen, The Netherlands.
6. Djoko Suharno. "The Role of Vitamin A in Nutritional Anaemia: a

Study in Pregnant Women in West Java, Indonesia". PhD-thesis Wageningen Agricultural University, 1994, Wageningen, The Netherlands.

7. S. de Pee, C.E. West, H. Muhilal, D. Karyadi, J.G.A.J. Hautvast. "Lack of Improvement in Vitamin A Status with Increased Consumption of Dark-Green Leafy Vegetables". Lancet 1995, **346**, 75-81.

POSSIBLE ROLE OF A PEA PROTEIN CONCENTRATE AS A NUTRICEUTICAL FOR HEALTH PURPOSES

A. Fernández-Quintela[1], M.T. Macarulla[1], A.S. del Barrio[1], M. Jiménez[2], T. Ganzábal[2] y J.A. Martínez[3]
[1]Department of Nutrition and Food Science. University of País Vasco. Marqués de Urquijo s/n. 01006 Vitoria, Spain.
[2]Biochemistry Laboratory. Txagorritxu Hospital. Jose Achotegui s/n. 01009 Vitoria, Spain.
[3]Department of Physiology and Nutrition. University of Navarra. Irunlarrea s/n. 31008 Pamplona, Spain.

1 INTRODUCTION

Legumes are widely used either in animal and human nutrition. However, the nutritive value of these seeds is limited by its low content on some amino acids and by the presence of antinutritional factors. Thus, considerable efforts have been made to improve its nutritional performance including genetic selection, fermentation, germination, physico-chemical treatments or protein extraction.[1-3] Moreover, glycemic response and lipemic levels following ingestion of dried legumes are generally reduced.[4] These characteristics have promoted the use of these legumes as nutriceuticals for the dietary management of diabetic and hypercholesterolemic patients.[5] On the other hand, as sulphur amino acids are limiting in legumes, protein metabolism may be altered in animals fed with this nutrient source.[6] In this context, the preparation of protein concentrates could support a new research avenue, exploiting the beneficial effects on metabolism and reducing the occurrence of some undesiderable alterations.[7]

Thus, in the present study we have attempted to determine whether the preparation of a pea protein concentrate could improve the nutritional and metabolical performance of the initial seed by analysing different plasma nutritional indicators.

2 MATERIALS AND METHODS

Protein concentrate from peas was elaborated by a protein solubilitation and isoelectric point precipitation method followed by liophilization as previously described.[8-10]

Thirty days old male Wistar rats (Granjas Jordi. Barcelona, Spain) weighing about 90-100 g were housed in metabolic cages. Room temperature was maintained between 20 and 22 °C, with controlled humidity and a 12 h day-night cicle. After three days of acclimatation, the rats were fed on different nutritionally balanced diets containing casein supplemented with methionine (Control; n=4), pea (Seed; n=6) and the pea protein concentrate (Concentrate; n=6) as protein sources during 12 days. At the end of this period, animals were decapitated and blood samples collected on EDTA, centrifugated at 2500 *g* for 15 min to separate the plasma and stored at -80 °C until analysis. All plasma nutritional indicators were analized through conventional enzymatic assays in a Hitachi/BM™ 717 Autoanalyzer using Boehringer Mannheim™ kits, and free plasma amino acids were measured by reversed phase chromatography and fluorimetric detection using Norleucine as internal standard.[11]

Data are expressed as mean values and the pooled standard error. The significance of differences between groups was first assessed with the Kruskal-Wallis rank sum test, and, if significant, the Mann-Whitney rank sum test was used on each pair of data,[12] with a Statview™ program on a MacIntosh™ computer. $P \leq 0.05$ was judged as statistically significant.

3 RESULTS AND DISCUSSION

Biochemical analysis showed marked differences in several nutritional indicators (Table 1). The reduction in antinutritional factors previously described for this pea protein concentrate,[10] showed no influence on the growth response to the protein concentrate feeding. Triacylglycerols levels were significantly reduced in animals fed on diets containing seeds and the protein concentrate. Furthermore, the pea concentrate intake caused the lowest glycemic levels. Concerning cholesterol and uric acid levels, there were not differences between the control and experimental groups although reduced cholesterol levels in hipercholesterolemic animals fed on legumes has been described. Nevertheless, a noticeable decrease in plasma uric acid level has been detected in rats fed on the concentrate. It is now well established that animals fed on vegetable protein have lower plasma uric acid levels than those fed with an animal protein source, probably due to the urine alcalinity detected in the former group. These results are in accordance with those presented by other authors with several different legumes.[13,14]

The pea amino acid composition could be altered by processing, which may have some impact on the metabolic outcome. Thus, protein metabolism in rats fed on protein concentrate was altered since the amino acid distribution is modified. Plasma albumin levels, an usual index of nutritional status in relation to protein quality or intake, was lower in this experimental group than in the seed fed rats. Moreover, higher plasma urea level has been observed in the concentrate group. Several authors have reported immunological disturbances in other animal models giving the same legume.[15]

Plasma amino acid differences observed between control and experimental diets support the idea of a deficiency or unavailability of sulphur amino acids (Table 2). Figure 1 represents plasma amino acid levels which were statistically different among experimental groups. The total amino acid levels in animals fed on the concentrate group was lower than in control rats. This was mainly due to a decrease in essentials (Arg, His, Ile, Leu, Lys, Met, Phe, Thr, Trp and Val) rather than to non-essentials (NE) amino acids (Ala, Asn, Asp, Glu, Gln, Gly, Ser and Tyr) in pea fed rats. In relation to essential amino acids Met, Lys and Tau plama levels were reduced in both legume fed rats while Arg plasma level was enhanced. On the other hand, the NE plasma amino acids showed different response: Asn, Gly and Ser plasma levels were enhanced, while Glu and Gln were decreased.[16]

Table 1 *Plasma biochemical profile of rats given pea (Seed or Concentrate) or control diets*

	Control	Seed	Concentrate	Pooled SE
Glucose (mg/dL)	152[a]	147[a]	126[b]	5
Triacylglycerols (mg/dL)	359[a]	145[b]	130[b]	26
Cholesterol (mg/dL)	80	76	83	3
Uric acid (mg/dL)	0.90	0.81	0.68	0.07
Protein (g/dL)	5.38[a]	5.39[a]	4.78[b]	0.08
Albumin (g/dL)	2.97[a]	3.09[a]	2.72[b]	0.04
Urea (mg/dL)	31[a]	31[a]	39[b]	2

[a,b]Data within a row with different superscript show statistical differences ($P<0.05$).

Table 2 *Free plasma amino acids (μmol/L) grouped by different criteria of rats given pea (Seed or Concentrate) or control diets*

	Control	Seed	Concentrate	Pooled SE
Totals	4401.4[a]	3916.2[a,b]	3732.7[b]	98.1
Essentials	1931.1[a]	1557.6[b]	1515.2[b]	71.2
Non-essentials	1870.6[a]	2027.2[b]	1919.3[a,b]	56.2

[a,b]Data within a row with different superscript show statistical differences ($P<0.05$).

Total amino acids remained unchanged in seed group; however, a fall in essentials and an increase in NE amino acid levels were observed in this group. Furthermore, plasma NE amino acids are generally more stable than most essential amino acids because NE amino acids are less sensitive to changes on dietary protein quality.[17]

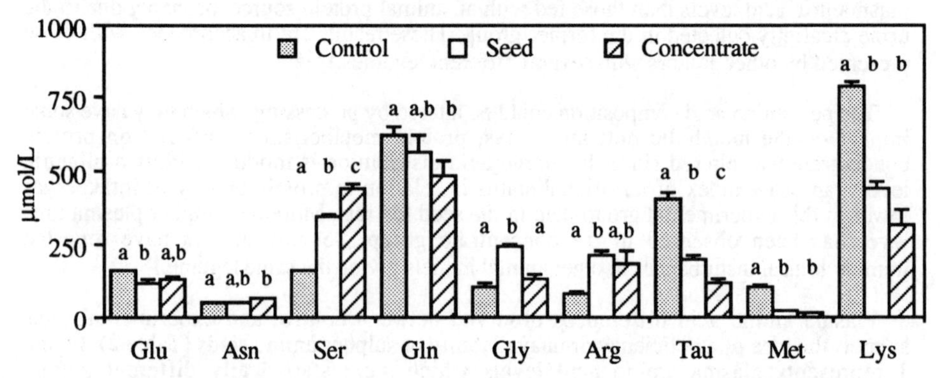

Figure 1 *Free plasma amino acids (μmol/L) from rats given experimental (Seed or Concentrate) or control diets*

4 CONCLUSION

It has been described a possible hypoglycemic effect of the pea protein concentrate, as well as hypotriglyceridemic effects from both seeds and their protein concentrate. Moreover, uric acid levels suffer a slight decrease adscribed to the pea concentrate feeding. Protein metabolism indices are also affected by the pea concentrate intake, probably associated to the pea concentrate feeding. These findings could be of interest and they provide an alternative avenue of therapeutic intervention as nutriceutical products in metabolical processes such as gout, inborn errors of metabolism, food allergies, etc.

5 ACKOWLEDGMENTS

This research was financially supported by the University of País Vasco (UPV 101.123-EA140/94) and Government of País Vasco (Marco Cooperaci6n Euskadi-Aquitania). A. F.-Q. was sponsored by a predoctoral grant from Ministerio de Educaci6n y Ciencia of Spain.

6 REFERENCES

1. J. C. Anderson, A. O. Idowu, U. Singh and B. Singh, *Plant Foods Hum. Nutr.*, 1994, **45**, 371.
2. C. Vidal-Valverde, J. Frias, I. Estrella, M. J. Gorospe, R. Ruiz and J. Bacon, *J. Agric. Food Chem.*, 1994, **42**, 2291.

3. A. Uzzan, '*Developments in food proteins-6*', Editor BJF Hudson, Elsevier Applied Science, New York, 1988, pp. 73.
4. R. M. Bakhit, B. P. Klein, D. Essex-Sorley, J. O. Ham, J. W. Erdman, Jr. and S. M. Potter, *J. Nutr.*, 1994, **124**, 213.
5. D. J. A. Jenkings, T. M. S. Wolever, A. L. Jenkings, R. L. Thorne, J. Kalmuski, R. Reichert and G. S. Wong, *Diabetologia*, 1983, **34**, 257.
6. L. U. Thompson, *J. Food Sci.*, 1977, **42**, 202.
7. J. Mouécoucou, C. Villaume, H.-M. Bau, J.-P. Nicolas and L. Méjean, *J. Sci. Food Agric.*, 1992, **60**, 361.
8. J.-P. Melcion and A. F. B. van der Poel, '*Recent advances of research in antinutritional factors in legume seeds*', Wageningen Pers, Wageningen, The Netherlands, 1993, pp. 419.
9. S. M. McCurdy and J. E. Knipfel, *J. Food Sci.*, 1990, **55**, 1093.
10. A. Fernández-Quintela, M. T. Macarulla and J. A. Martínez, *Rev. Esp. Cienc. Tecnol. Alim.*, 1993, **33**, 185.
11. G. Georgi, C. Pietsch and G. Sawatzki, *J. Chromatogr.*, 1993, **613**, 35.
12. J. V. Bradley, 'Distribution-free Statistical Tests', Prentice Hall, Englewood Cliffs, NJ, 1968, pp. 111.
13. S. Shinjo, L. Asato, S. Arakari, T. Kina, T. Kohrin, T. Mori and S. Yamamoto, *J. Nutr. Sci. Vitaminol.*, 1992, **38**, 247.
14. R. W. Peace, G. Sarwar and S. P. Touchburn, *Food Res Int.*, 1992, **25**, 137.
15. J. A. Martínez, M. L. Esparza and J. Larralde, *Br. J. Nutr.*, 1995, **73**, 87.
16. K. Nielsen, J. Kondrup, P. Elsner, A. Juul and E. S. Jensen, *Br. J. Nutr.*, 1994, **72**, 69.
17. Y. Cai, D. R. Zimmerman and R.C. Ewan, *J. Nutr.*, 1994, **124**, 1088.

FERMENTABLE NON-ABSORBED CARBOHYDRATES MODIFY THE GASTROINTESTINAL ENDOCRINE RESPONSE TO FOOD IN RATS AND IN HUMAN SUBJECTS

J. M. Gee, W. Lee-Finglas, G. W. Wortley and I.T. Johnson

Institute of Food Research
Norwich Research Park
Colney
NORWICH, NR4 7UA

1 INTRODUCTION

The alimentary tract contains a complex variety of endocrine tissues, but many of their functions, and the details of their response to food intake, remain poorly understood. In this study we describe some effects of soluble dietary fibre on enteroglucagon (EG), which is a collective term for a group of peptides secreted into the blood by the L-cells of the distal small intestine and colon. EG is thought to play a role in the control of gastric secretion and motility [1] and it may also function as a trophic hormone for the intestinal mucosa[2].

In previous studies we have observed that guar gum, a galactomannan polysaccharide which is used both as a food thickener and as a pharmaceutical material, causes a sustained rise in plasma enteroglucagon in the rat[3]. At the outset we envisaged two possible mechanisms whereby consumption of guar gum might cause a rise in plasma EG. Nutrients introduced into the distal small intestine can stimulate the release of EG from L cells in the ileal mucosa[4]. Guar is highly viscous and might amplify this post-prandial mechanism by delaying absorption of glucose and lipids, and displacing them into the distal jejunum and ileum[5]. However it has recently been shown that the short chain fatty acid butyrate causes the release another product of the L cell, peptide PYY, from the perfused rabbit colon[6]. This raises the possibility that fermentation of carbohydrate might release EG from colonic L cells. To distinguish between these two mechanisms we compared the effects of guar gum with those of two other carbohydrates used as food additives, hydroxypropylmethyl cellulose, (HPMC) which is very viscous but completely resistant to fermentation, and lactitol, which is highly fermentable but has no effect on intraluminal viscosity. We also conducted a preliminary experiment to assess the effect of fermentable carbohydrate on EG release in humans.

2. MATERIALS AND METHODS

Animals and diets

For each experiment, 60 male Wistar rats were kept singly in polypropylene cages with wire bottoms and tops in an environmentally controlled animal house, Animals were randomly assigned to two groups of thirty. Both groups were fed a semisynthetic powdered diet containing sucrose (30%), casein (20%), maize oil (8% or 12%), starch (26% or 22%). The control diet contained Solkafloc®; (cellulose; 10%); in the treatment diets this was replaced by guar gum, HPMC or lactitol. After 10 d, food was withdrawn and 5 animals from each group were deeply anaesthetized at 0, 1.5, 3, 4.5, 6 and 7.5 h. A blood sample (5 ml) was withdrawn and the animals were killed by cervical dislocation.

Human subjects

Informed, consenting volunteers (4 males, 25y-45y; 4 females 27y-49y) fasted overnight and attended a metabolic suite at 08.30 h on two occasions separated by at least 7 d. An indwelling cannula was inserted into a brachial vein, and an initial blood sample (5 ml) was obtained. After 15 min each subject consumed either a control drink (250ml) consisting of a low-calorie fruit cordial, or a similar treatment drink containing 10 g of lactitol. Subjects were unaware of the order of administration of the drinks. Further blood samples were taken at 1.5, 3.0, 4.5, 6.0 and 7.5 h after consumption of the drink. Breath hydrogen samples were taken to confirm the time course of fermentation of the lactitol. .

Peptide analysis

Plasma samples were prepared and stored in liquid nitrogen after the addition of aprotinin and heparin. Plasma pancreatic glucagon and total plasma glucagon like immunoreactivity were measured using commercial assay kits containing specifically reactive antisera (Novo Laboratories, Basingstoke) and enteroglucagon was determined by difference and expressed as pg per ml of plasma. For gastrin analysis each blood sample was placed in a tube without anticoagulant and centrifuged at 12,000g for 4 min, yielding approximately 0.5 ml of serum, which was then stored at - 20°C. Gastrin levels were determined in 100 μl samples using a commercial radio-immunoassay kit (Beckton Dickinson, Oxford).

Fermentation of non-starch polysaccharides

The susceptibility of the non-starch polysaccharide gums to fermentation by caecal microorganisms was determined using an anaerobic incubation procedure *in vitro*. Briefly, samples (60 mg) of insoluble cellulose, HPMC, guar gum and lactitol were added to incubation flasks containing a basal medium. The flasks were sampled before the addition of a rat caecal inoculum, and at intervals during incubation under strictly anaerobic conditions for 168h. The samples were centrifuged and stored at -20° C prior to carbohydrate analysis by the anthrone method.

Viscosity

The viscosities of guar gum and HPMC were determined by rotary viscometry using aqueous solutions (20g/l) at a shear-rate of 50/s and a temperature of 37° C (Rheomat 15; Contraves AG, Zurich).

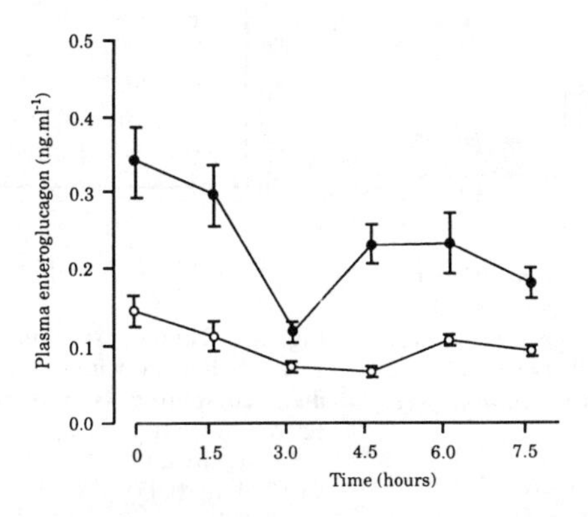

Fig. 1. Time-course for the change in plasma EG concentration following withdrawal of feed from rats fed diets containing guar gum (●) or cellulose (o). Each point is the mean of 5 animals, differences were significant at each time-point ($p<0.05$).

3. RESULTS AND DISCUSSION

All the animals accepted the feed readily and gained weight during the experiments. The groups given guar gum or lactitol tended to consume less food than their respective controls but there was no effect of HPMC on food intake. The viscosities of aqueous media containing (20g/l) guar gum and HPMC were 1730 cP and 1763 cP respectively. The viscosity of the solution containing lactitol at this concentration was not measurably different from water. The fermentability of the guar and lactitol, and the resistance to fermentation of Solkafloc® and HPMC were confirmed by the *in vitro* study.

The time-course of an experiment in which the treatment group received guar gum, is shown in Fig 1. The animals drawn from the treatment group had a significantly higher level of EG in the plasma than the controls at every time point. For subsequent experiments the areas under the curves were calculated using individual animals selected randomly at each time-point. In Fig 2 the areas under the curves obtained for the three test materials (shaded columns) were compared with their respective controls (open columns). Evidently both guar gum and lactitol caused an increase in plasma EG compared to the rats fed cellulose, but there was no effect of HPMC. These data demonstrate that guar gum causes a sustained rise in plasma EG levels and show that fermentation, rather than viscosity, is the principle stimulus for the release of the peptide under these conditions. Blood gastrin levels were determined under the same conditions, and areas under the time-course were calculated as for enteroglucagon. The gastrin response was significantly lower in animals fed lactitol, compared to their control group fed Solkafloc®, there was a similar but statistically insignificant trend in animals fed guar and no effect of HPMC. The time course for the gastrin response in rats fed lactitol or Solkafloc® is shown in Fig 3.

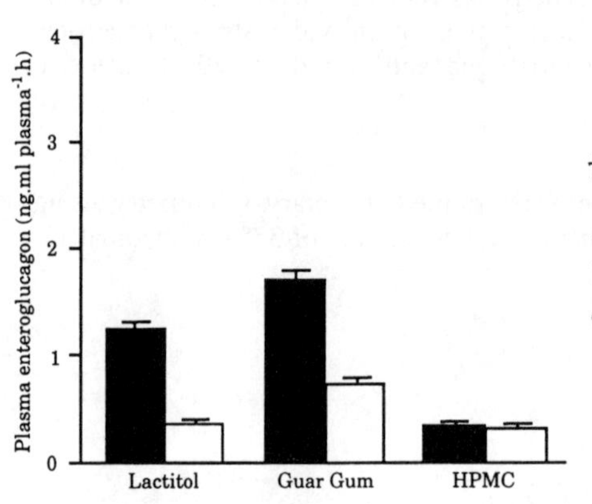

Fig. 2. Areas under the post-prandial EG curves for animals fed lactitol, guar gum or HPMC (shaded columns), and their respective controls (open columns). Differences between lactitol, guar gum and their respective controls were statistically significant ($p<0.05$).

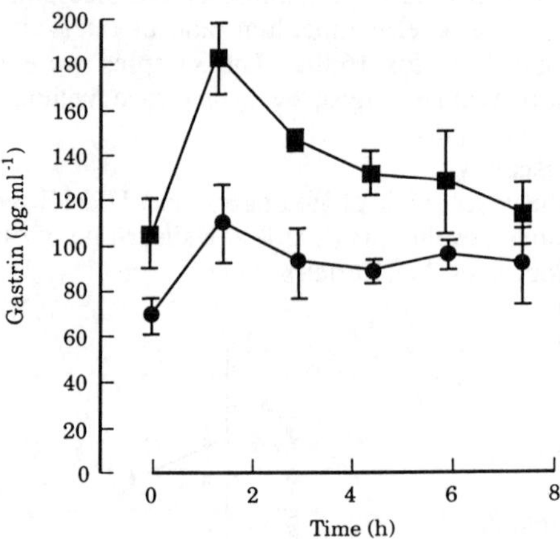

Fig. 3. Time-course for the change in serum gastrin following withdrawal of feed from rats fed diets containing lactitol (●) or cellulose (o). Each point is the mean of 5 animals, differences were significant at all time-points other than 6 and 7.5h ($p<0.05$).

In the study with human subjects, Plasma EG values were higher after lactitol, but the difference was statistically significant only at 4.5h after the test drink (p=0.02). This result suggests a trend toward increased postprandial EG in subjects fed lactitol, but further studies on the dose-response relationship between lactitol intake and plasma EG are needed. There was no effect of lactitol on the gastrin response in humans.

There can be little doubt that the breakdown products of unabsorbed dietary carbohydrate, presumably short chain fatty acids or their metabolites, are the main cause of the rise in plasma EG seen in rats fed guar gum and other types of readily fermentable dietary fibre. The colon should therefore be regarded as a putative endocrine organ which releases regulatory peptides in response to consumption of fermentable carbohydrate. What though is the function of EG in this context? There is evidence from studies in humans that EG down-regulates gastrointestinal motility and suppresses gastric acid secretion[1]. Inoue et al[7] observed a rise in the post-prandial gastrin response in dogs after colonic resection, and proposed that the colon may produce an endocrine factor which suppresses gastrin secretion. Conversely Sasaki et al[8] observed a reduction in gastrin levels following ileo-jejunal transposition in colectomised dogs, and suggested that this was due to a concommitant rise in rise in serum EG. The present results are consistent with that hypothesis. This study raises the possibility that functional foods containing fermentable carbohydrate can be used to manipulate some aspects of gastrointestinal function in human beings, but further research is needed to discover whether this approach is feasible at acceptable levels of intake.

Acknowledgements

This work was supported by the Ministry of Agriculture Fisheries and Food for England, Wales and Northern Ireland, and by the Biotechnology and Biological Sciences Research Council.

References

1. B.Schjoldager, P.E. Mortensen, J. Myhre, J. Christiansen and J.J. Holst, *Dig.Dis.Sci.*, 1989, **34,** 1411.
2. S.R. Bloom and J.M. Polak . *Scand. J. Gastroenterology,* 1982, (suppl) **74**, 93.
3. I.T. Johnson, J.M. Gee and J.C. Brown, *Am.J.Clin. Nutr,* 1988,**47**, 1004.
4. J.J. Holst, J. Christiansen and C. Kuhl, *Scand. J. Gastroenterol,* 1976, **11**, 297.
5. N.A. Blackburn, I.T. Johnson, *Brit. J. Nutr.,* 1981, **46**, 239.
6. W.E.Longo, G.H. Ballantyne, P.E. Savoca, T.E.Adrian, A.J. Bilchik, and I.M. Modlin, *Scand.J.Gastroenterology,* 1991, **26,** 442.
7. K. Inoue, I. Wiener, G.M. Fried, P. Lilja, L.C. Watson and J.C. Thompson. *Annals of Surgery* , 1982, **196**, 691.
8. I. Sasaki, T. Tuchiya, H. Naito, Y. Funayama, M. Toda, Y Suzuki, T. Sato, and A. Ohneda,*Tohoku Journal of Experimental Medicine,* 1987, **151**, 419.

EFFECT OF THE INTERACTION OF BEAN (*PHASEOLUS VULGARIS*) PROTEIN FRACTIONS ON DIGESTIBILITY AND METHIONINE BIOAVAILABILITY

M.I. Genovese, V.M.H. Del Pino and F.M. Lajolo

Departamento de Alimentos e Nutrição Experimental
Universidade de São Paulo
Cx.Postal 66083 - 05389-970 São Paulo, SP - Brazil

1 INTRODUCTION

Bean (*Phaseolus vulgaris*) proteins have a low nutritional value due to its reduced digestibility and low methionine content and bioavailability[1]. Improvement of bean for increased methionine content will not solve the problem if the biological utilization remains low. Information on the mechanisms involved are important for selection of new varieties and development of processing technologies.

The digestibility of bean proteins in humans is low, about 40-60%, and seems to depend on several factors involving both protein and non protein fractions, besides storage conditions[2]. Digestibility associated to other factors seems also to determine the low availability of methionine (average 50%) since exogenous added methionine is well utilized[3].

Among the non protein interacting factors tannins have a role and its influence on the overall digestibility of bean nutrients has been already reviewed[4].

We previously reported[5] that heating bean albumins, contrarily to what was expected, caused a decrease of the digestibility, just the opposite of what happened with the globulin fraction. Feeding studies with rats using isolated protein labelled with N^{15} and S^{35} indicated the presence of fractions with low digestibility[6].

In this paper we investigated further the reactions occuring during heating the albumin fractions and also the effect of interaction of tannins with the globulin (phaseolin).

2 RESULTS AND DISCUSSION

In vitro digestibility of bean proteins. Extracted total proteins (albumin + globulins), isolated albumins and globulins were digested in vitro by a pepsin-pancreatin system and the digestibility compared to that of heated casein. Table 1 shows that, while globulin digestion is improved by heating to the extent of casein, the digestibility of albumins is reduced to 20% or lower, even after autoclaving for 15-120 min at 121°C. Heating the albumin solution in the presence of the globulin or heating the total protein did not alter the observed effect.

Also, heating the albumins at different pHs (2.0 to 10.0) did not alter significantly what was observed indicating that ionic interactions should not be involved.

Trypsin inhibitor activity. The native albumin fraction showed high inhibitory activity which was partially eliminated at 99°C and totally destroyed only at 121°C. The residual inhibitory activity may explain part of the reduced digestibility but not the decrease caused by heating (Table 1).

Table 1 *Pepsin-Pancreatin Digestibility and Trypsin Inhibitory Activity (TIA) of Native and Heated (99°/30') or Autoclaved Bean Proteins.*

	% Hydrolysis	*TIA/mg*
Native Phaseolin	23	Traces
Heated Phaseolin	87	0
Native Albumins	32	170.4±2.6
Heated Albumins		
(99°/30')	20	117.9±2.8
3h	16	106.1±1.3
Autoclaved Albumins		
15'	16	94.3±3.9
30'	14	44.1±2.0
1h	14	30.2±1.4
2h	16	0
Total Protein Extract (native)	25	52.0±6.8
Total Protein Extract (heated)	35	50.0±7.0
Casein	89	-

Effect of the carbohydrate moiety . Bean albumins contain highly glycosylated proteins such as lectins and amylase inhibitors. To investigate the effect of the carbohydrate chain, chemically deglycosylated albumins were heated and submitted to pepsin-pancreatin. The digestibility of both native and denaturated proteins increased respectively by 15% and 10%, probably because of the elimination of steric hindrance. Isolated lectins and amylase inhibitor were poorly digested (less than 30% hydrolysis) even after heating. We also reversibly blocked lysine residues by citraconic anhydride and observed that, after heating the modified protein and removing the citraconic group, digestibility did not drop, indicating that Mailllard type reactions or other cross linking mechanisms involving lysine may partially explain the observed reduction of digestibility with heating, but are not the main cause of albumins low digestibility.

Sulfhydryl and dissulfide contents and modification. Contents of sulfhydryl/dissulfide groups for native and heated albumins are shown in Table 2.

Table 2 *Sulfhydryl and Disulfide Contents of Native and Heated (99°/30') Albumins.*

	SH (µmol/g)	SS (µmol/g)
Native	13.9±0.2	22.1±0.1
Heated	5.6±0.1	26.0±0.3

Heating caused the formation of new disulfide bonds which can be related to the decrease of digestibility, as a more compact structure formed may impair protease access to labile peptide bonds. To check further SH/SS participation, albumins were heated in the presence of β-mercaptoethanol which caused a decrease of total SS bonds from 22.1 to 9.2 µmol/g and a parallel increase of digestibility to more than 33%.

Effect of chemical modifications. Chemical dissociation of native and deglycosylated albumins by urea, mercaptoethanol and N-ethyl maleimide increased digestibility from 30% to almost 70% and from 44% to 80%, respectively. Albumins have a compact structure maintained by disulfide bonds and hydrophobic interactions, which along with attached polyssacharide chains, impairs digestibility. Heating caused physico-chemical interactions that did not allow opening the structure as it was possible by chemical denaturation.

Methionine bioavailability. Unoxidized methionine released by pepsin-pancreatin digestion was taken as a measure of its availability. While all the methionine was released from heated phaseolin and from casein, only 50% of the total (120 g/100 g obtained by ion exchange chromatography) was released from heated albumins (Figure 1).

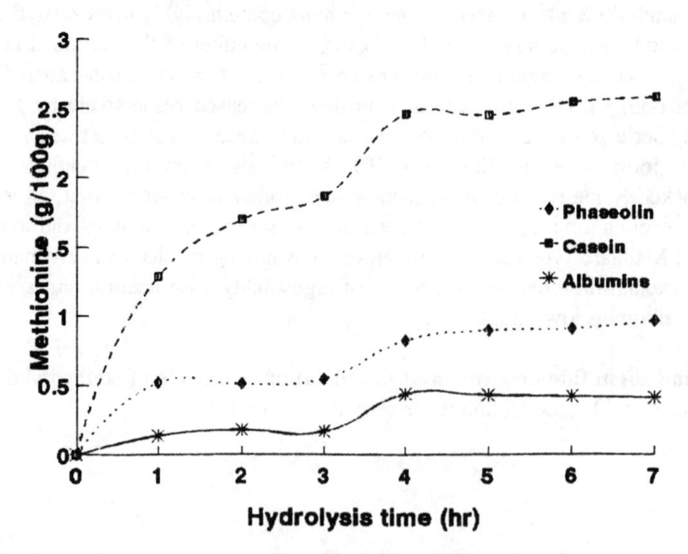

Figure 1. Methionine release during in vitro enzymatic hydrolysis

Interaction with tannins. Isolated bean globulin and isolated tannins bind strongly at different pHs (from 2 to 10) forming complexes having a 21:1 and 12:1 tannin:protein stoichiometry, respectively for the native and denaturated protein. The complex is stabilized by hydrophobic interactions as shown by SDS dissociation.

Differential scanning calorimetry analysis of the native globulin-tannin complex indicated a reduced heating stability with a 5°C decrease of the denaturation temperature.

The isolated globulin-tannin complexes are very resistant to trypsin, chymotrypsin and pepsin-pancreatin hydrolysis. A resistant structure seems to be formed even at high enzyme to protein ratio and prolonged digestion (21h), and it has a 20 KDa MW as shown by electrophoresis in dissociating conditions. Only when the enzyme concentration is higher than the globulin concentration digestibility can be lightly improved.

3 CONCLUSION

The reduced digestibility of bean albumins seems to be multicausal. Glycosylation and presence of trypsin inhibitors are important for native proteins. Hydrophobic interactions, formation of dissulfide bonds and interaction with tannins are involved in the reduction of digestibility during heating, associated to the presence of heat resistant trypsin inhibitors.

4 REFERENCES

1. Bressani, R. Food Rewiews International, 1993, 9, 237.
2. Bressani, R. Qual. Plant. Foods Hum. Nutr., 1983, 32, 101.
3. Evans, R.J.and Bauer, D.H. J. Agric. Food Chem., 1978, 26, 779.
4. Reddy, N.R. and Pierson, M.D. J. Am. Oil Chem. Soc., 1985, 62, 541.
5. Marquez, U.M.L. and Lajolo, F.M. J. Agric. Food Chem., 1981, 29, 1068.
6. Marquez, U.M.L. and Lajolo, F.M. J. Food Sci. Agric., 1991, 39, 1211.

NUTRITIONAL EVALUATION OF NEW CULTIVARS OF HIBISCUS ESCULENTUS (OKRA) RECENTLY INTRODUCED IN EUROPE.

E. Quattrucci, P. Fantauzzi, R. Francisci, J. Krizanovic and C. De Luca*

National Institute of Nutrition
Via Ardeatina, 546, 00178 Rome, Italy

* S. Gallicano Institute (IFO)
Via S. Gallicano, 25/A, 00153 Rome, Italy

1 INTRODUCTION

Hibiscus Esculentus (H.E.), also known as Okra, Bamia, Gumbo and Lady's finger, is an annual plant belonging to the Malvaceae which grows in tropical and subtropical areas. Fruits (pods) and seeds of Okra are widely used as food commodities in Western Africa, India and China. In recent years, they have been introduced in Europe, especially in Greece. Both pods and seeds contain some components which have interesting nutritional and/or technological properties. However, the size, shape and colour of the pods and seeds vary a great deal and, consequently, so do their chemical composition, as the data found in literature suggest [1,2]. Okra can be considered a good source of certain vitamins and minerals. The pods are harvested when they are unripe and tender. Once harvested, they are consumed within 24 hours or after a maximum of 10 days when stored at 7-10°C. Lastly, they can undergo various treatments such as canning, drying, etc. In addition, the pods contain a mucilage consisting of an aqueous mixture of pectin and carbohydrates which is locally used as a thickener.

Mature Okra seeds can either be eaten roasted or used for oil extraction. In fact they contain about 20% of edible oil, consisting mainly of oleic (45%) and linoleic (20%) acid. The seed powder remaining after the oil extraction has a high protein content and can be used in animal feeds. In India, Okra seeds are used as coffee substitute whereas in Egypt, the seed flour is sometimes added to maize flour. Additionally, it has to be mentioned that the cellulose of the stem and ripe pods is of a very high quality for paper making so that pratically the whole plant can be utilized.

Therefore, it would seem worthwhile to evaluate some H.E. varieties in order to see whether either pods and seeds or their single components could be used to produce new commodities of good nutritional density. Consequently, proximate chemical composition, vitamin content, dietary fibre and composition of lipids were determined on H.E. varieties grown in Italy.

2 MATERIALS AND METHODS

Five Greek varieties of Okra were considered in this study. Okra pods and seeds were obtained from the Department of Agronomy of the University of Padua, Italy. The varieties are: Bogiatiou, Veludo, Pyleas, Levadia, Kilkis. Three different lots for each variety were conventionally cultivated, whereas the Pyleas variety was also cultivated without fertilization.

Moisture, protein, ash and total lipid content were determined according to the AOAC methods [3]. Vitamin contents of Okra fruits were determined by reverse-phase HPLC [4]. Soluble and total dietary fibre were determined according to the enzymatic-gravimetric method of Prosky et al. [5]. Fatty acid composition of seed lipids was determined by GC-MS (total ion SCAN mode) and fatty acids, as methyl esters, were quantified on the basis of the total area of peaks of the standards obtained from the Sigma Chem. Co (USA). Squalene and tocopherols (TMS derivatives) were quantified from the saponified extract by GC-MS, with SIM (Selected Ions Monitoring) technique [6].

3 RESULTS AND DISCUSSION

In Table 1 the average composition of raw pods (% f. m.) is reported; only little differences among the chemical composition of samples coming from three different lots were found. Okra pods have a high water and fibre contents, and a low protein content, around 3-4%. In Pyleas samples without fertilization protein content decreased, as expected; on the contrary, dietary fibre significantly increased in Pyleas without fertilization.

Table 1 *Average Composition of Raw Okra Pods (% on f.m.), 1994 Harvest ([a] =Pyleas without Fertilization)*

Sample	*Moisture*	*Protein* (Nx6.25)	*Ash*	*Fibre*			*Carbohydrates* (by diff.)
				soluble	*insol.*	*total*	
Veludo	87.3	2.9	1.09	1.7	3.1	4.8	3.8
Bogiatiou	86.6	3.1	1.22	1.6	3.6	4.6	3.6
Levadia	87.2	2.8	1.30	1.9	3.6	5.5	3.1
Pyleas	86.6	3.2	1.44	1.2	2.3	3.6	5.2
Pyleas[a]	87.4	2.9	1.32	1.8	3.4	5.2	3.0
Kilkis	89.1	2.4	1.12	1.4	3.0	4.4	2.9

In Table 2 the average composition of Okra pods, after cooking (% of fresh product) is reported. The only relevant difference is related to the apparent decrease of the insoluble fraction of fibre in cooked samples and consequently a lower TDF value. In fact, the starch fraction sterically unaccessible to the termamyl and mesured as insoluble fibre in the raw samples, is digested in the cooked samples which are more accessible to the enzymic action.

Table 2 *Average Composition of Cooked Okra Pods (% on f.m.), 1994 Harvest ([a] =Pyleas without Fertilization)*

Sample	*Protein* (Nx6.25)	*Ash*	*Fibre*			*Carbohydrates* (by diff.)
			soluble	*insol.*	*total*	
Veludo	2.8	1.06	1.5	2.0	3.5	5.5
Bogiatiou	3.1	1.09	1.4	2.2	3.6	5.1
Levadia	2.7	1.17	1.5	2.2	3.7	5.6
Pyleas	3.3	1.36	1.4	1.9	3.4	5.3
Pyleas[a]	3.0	1.25	1.4	2.1	3.6	4.7
Kilkis	2.6	1.08	1.4	1.9	3.2	4.7

Vitamin levels (Table 3) were determined on fresh products. The Pyleas variety showed the highest content of vit. C and thiamin. In Figure 1 the percentage of the average vitamin requirements for adults provided by 100g of fresh Okra pods is reported (SCF, EC, 1993). One hundred grams of Okra pods provides 43-60% of the average requirements of vit. C, 17-19% of thiamin and 12-14% of riboflavin for adult men.

Table 3 *Average Vitamin Content of Pods (mg/100 g) ([a] =Pyleas without Fertilization)*

Sample	*Veludo*	*Bogiatiou*	*Levadia*	*Pyleas*	*Pyleas*[a]	*Kilkis*
Vitamin C	13	13	15	18	14	13
Thiamin	0.13	0.13	0.15	0.15	0.13	0.15
Riboflavine	0.18	0.18	0.17	0.17	0.16	0.15

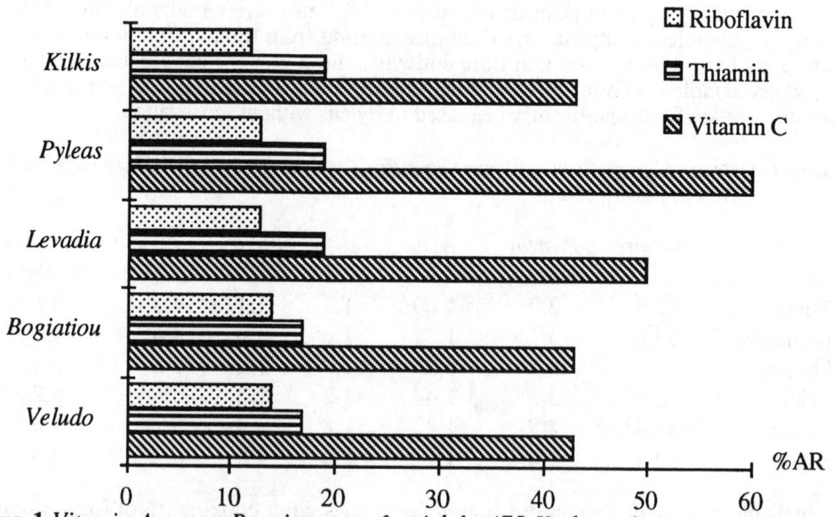

Figure 1 *Vitamin Average Requirements for Adults (75 Kg b.w.; light activity lifestyle) Provided by 100g of Pods*

Table 4 shows the average chemical composition of the five varieties of Okra seeds. The protein content ranged from 19 to 26% (d.m.). Lipid content, also quite high (between 16 and 26%) and fibre content (around 50%) showed some variability.

Table 4 *Average Composition of Okra Seeds (% on d.m.), 1994 Harvest ([a] =Pyleas without Fertilization)*

Sample	*Moisture*	*Protein* (Nx6.25)	*Total Lipids*	*Ash*	*Fibre*		
					soluble	*insol*	*total*
Veludo	8.1	22.3	18.8	4.30	2.5	49.5	52.0
Bogiatiou	6.3	20.1	20.3	4.47	3.4	49.1	52.5
Levadia	7.8	25.3	26.4	4.92	3.4	42.0	45.5
Pyleas	6.5	18.8	16.7	4.13	3.0	53.5	56.5
Pyleas[a]	4.5	22.5	16.0	4.74	4.2	42.5	46.7
Kilkis	8.3	25.9	21.2	4.96	2.4	37.8	40.3

In table 5 the fatty acids composition of Okra seeds is reported. Table 6 shows a comparison of relative fatty acid compositions of Okra and the most commonly used oils, i.e. soya bean, peanut, olive and maize. Okra oil has a high content of palmitic acid (C16:0) (about 30%), which is greater than that found in the other oils (olive oil 13.7; peanut oil 9.9; soya bean oil 9.8). The linoleic acid (C 18:2) content, around 48% is also very high in comparison with olive, peanuts and other vegetable oils, but it is lower than the level found

in soya bean oil (51%). Linoleic and linolenic acids are important components for their nutritional significance.
Tocopherols and squalene content of Okra seeds are interesting for their antioxidant properties (Table 7). A lower content of these compounds was found in not fertilized Pyleas.

Table 5 *Average Fatty Acid Composition of Okra Seed Lipids (% on d.m.), 1994 Harvest ([a] =Pyleas without Fertilization)*

Sample	*C14:0*	*C16:0*	*C16:1*	*C 18:0*	*C18:1*	*C18:2*	*C 18:3*
Veludo	0.17	30.36	0.26	2.97	15.14	48.82	0.14
Bogiatiou	0.16	29.72	0.29	2.90	15.51	49.26	0.16
Levadia	0.17	30.00	0.28	3.01	15.00	48.09	0.15
Pyleas	0.20	30.47	0.36	3.28	15.93	47.52	0.14
Pyleas[a]	0.18	29.45	0.37	3.10	16.10	47.78	0.17
Kilkis	0.17	29.39	0.37	3.80	16.65	46.86	0.17

Table 6 *Comparison of Relative Fatty Acid Composition of Okra and Other Vegetable Oils*

Oil	*C14:0*	*C16:0*	*C16:1*	*C 18:0*	*C18:1*	*C18:2*	*C 18:3*
Okra	0.17	29.90	0.32	3.18	15.72	48.05	0.15
soy	0.15	9.79	0.30	3.68	22.26	51.36	7.60
olive	-	13.67	0.82	2.23	73.63	7.85	0.99
peanut	0.27	9.91	-	2.53	51.30	27.87	-
maize	0.57	11.69	0.40	2.30	29.88	49.83	0.60

Table 7 *Squalene and Tocopherols Levels of Okra Seeds , 1994 Harvest ([a] =Pyleas without Fertilization)*

Sample	*Squalene (mg/100g)*	*s.d.*	*Tocopherols (mg/100g)*	*s.d.*
Okra seeds	17.26	3.7	210.05	56.3
Pyleas	15.37		229.44	
Pyleas[a]	12.89		53.85	

Acceptability tests were also performed on pods, casserole-style cooked. Colour, texture, taste and flavour have been judged quite satisfactory by the testing people.

The pods of the Pyleas variety seem to present the better nutritional quality. However, the results of the agricultural and technological evaluation that is still under study will allow a more complete picture of the tested Okra cultivars.

Acknowledgement

This research was supported by the EC-E.E.I.G. Project AIR 3-CT-1236

References

1. P.A. Savello, F.W. Martin and J.M. Hill, *J. Agric. Food Chem.*, 1980, **28**, 1163.
2. N. Nahar, M. Mosihuzzaman and S.K. Dey, *Food Chem.*, 1993, **46**, 397.

3. AOAC, "Official Methods of Analysis", Washington, DC, 1970.
4. C. Hasselmann, D. Franck, P. Grimm, P.A. Diop, C. Soules, *J. Micronutr. Anal.*, 1989, **5**, 269.
5. L. Prosky, N.G. Asp, I. Furda, J. De Vries, T. Schweizer and B. Harland, *J. Assoc. Off. Anal. Chem.*, 1985, **68**, 677.
6. S. Passi, A. Morrone, C. De Luca, M. Picardo and F. Ippolito, *J. Dermatol. Sci.*, 1991, **2**, 171.

IMPROVING THE FOOD QUALITY OF KHESARI DHAL (*Lathyrus sativus*) BY PROCESSING

Santosh Khokhar[1], Shiwani Srivastava[1] and J Fornal[2]

[1] Department of Foods and Nutrition
CCS Haryana Agricultural University
Hisar, India

[2] Centre for Agrotechnology and Veterinary Sciences
Olsztyn, Poland

1 ABSTRACT

Khesari dhal (*Lathyrus sativus*) was processed by following the tradition methods of cooking, soaking, cooking of pre-soaked seeds, pressure-cooking, germination and fermentation. The processed material contained higher digestibilities of protein and starch and availability of essential minerals (Ca, Fe and Zn), ß-ODAP levels were significantly reduced which emphasises the possibility of detoxification of these seeds.

2 INTRODUCTION

Khesari dhal is an important source of vegetable protein, energy and minerals, and is consumed by over 100 million people in the Indian subcontinent and Ethiopia. A major factor limiting the more extensive utilisation of this grain is the presence of a neurotoxic amino acid, ß-ODAP [ß-N-oxalyldiaminopropionic acid]. Attempts by the Indian Government to ban the crop have, however, been unsuccessful. Breeding programmes are underway to produce lines with much reduced levels of ß-ODAP.

Grain legumes in India are processed in a variety of forms-most commonly, curry cooking. Other traditional methods include germination, fermentation, and pressure cooking. In this study we have processed khesari dhal using such methods, and have measured the effect on quality indicators such as digestibility of protein and starch, availability of minerals and ß-ODAP levels.

In the study, four lines of varying ß-ODAP content (obtained from a breeding programme at the Indian Agricultural Research Institute, New Delhi) were used. However, since the overall trends were qualitatively similar, data on only one line will be presented here. Further details are available from the authors.

3 METHODS

Seeds were soaked for 12 hours in either drinking water (process A) in Figure 1 and Tables 1 and 2, in tamarind solution (process C), or for 2 hours in freshly boiled water (B). Seeds were dehusked (D), allowed to germinate for 36 h (E), allowed to ferment naturally (36 h/30°C) after soaking (F), cooked normally after soaking (G) or subjected to pressure cooking (15 psi for 15 minutes; H). All processing was carried out to local practice and all products were of acceptable taste and texture.

Total Ca, Fe and Zn contents were measured after digestion by atomic absorption spectrophotometer. The same technique was used to determine availability after suitable incubation with pepsin acid (Rao and Prabhavathi, 1978). ß-ODAP was measured according to Rao *et al* (1978) and Briggs *et al* (1985), the digestion of protein and starch were determined using standardised procedures.

4 RESULTS AND DISCUSSION

The effect of processing on ß-ODAP levels (g per kg) is shown in Figure 1. All processes significantly reduced levels of the neurotoxin, with pressure cooking being most effective. Processing also reduced other anti-nutrients such as oligosaccharides, phytate, polyphenolics (tannins), and trypsin- and amylase inhibitors; ordinary cooking and pressure cooking of pre-soaked seeds were again found to be the most effective techniques. The data on these compounds has been published in detail elsewhere (Srivastava and Khokhar, 1996).

The *Lathyrus* line reported on here is a low ß-ODAP line which contained 28.1 ± 0.5 g protein and 50.0 ± 0.9 g available carbohydrate per 100g dry weight, respectively. The non-starch polysaccharide content comprised less than 0.1 g soluble NSP and 7.9 g insoluble NSP per 100 g dry weight.

Figure 1: *Effect of processing on the levels of β-ODAP*

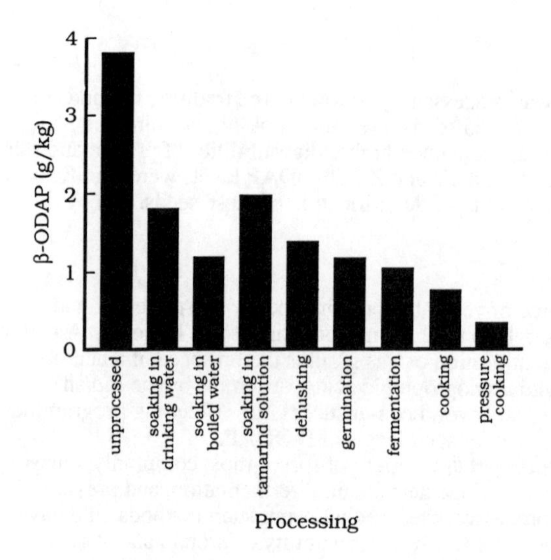

5 PROTEIN AND STARCH DIGESTIBILITY

Digestibility of both protein (48-53%) and starch, expressed as mg of maltose released per g of seed, (40-50%) in unprocessed seeds were similar to other legumes such as mung bean, lentils and lower than others like soyabean. Variations in the starch digestibility of different legumes could be due to different levels of resistant starch which would resist digestion much more strongly than others (Fuwa *et al*, 1980) and may also be due to differential levels of anti-nutrients.

Processing of seeds improved the digestibility of both protein and starch significantly (Table 10) and also showed some changes in protein and starch structures (Figure 2), as shown by SEM. Soaking of seeds in tap water was less effective than soaking in boiled water or in tamarind solution. Maximum increase in both cases (protein up to 90% and starch up to 92%) were obtained after ordinary cooking or pressure-cooking which might be due to denaturation and opening up the structure of the proteins (Kamalakanan *et al*, 1981). Fermentation also increased the digestibilities, possibly as a result of increase in proteolytic activity (Odunga, 1983). Germinated seeds also had higher protein as well as starch digestibilities.

Increases in starch digestibility of cooked seeds could be as a result of swelling and rupturing of starch granules, thus facilitating the action of alpha-amylase; alternatively,

Figure 2: *Changes in Structures of Protein and Starch After Processing*

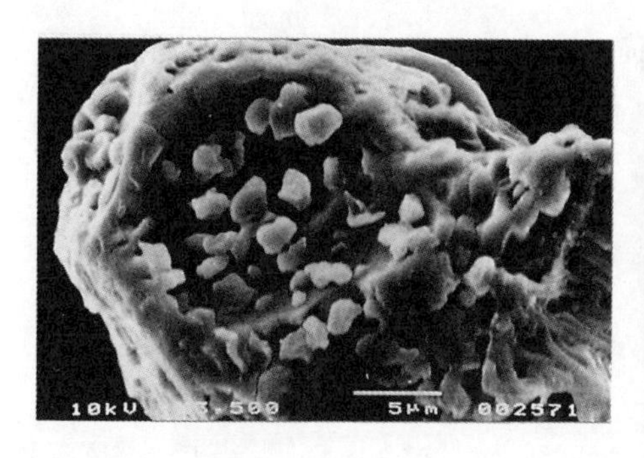

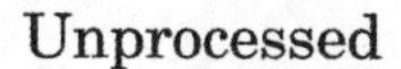
Unprocessed

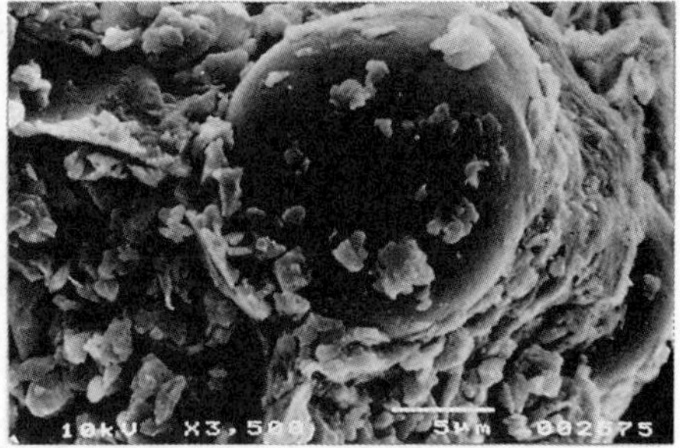

Soaking in drinking water [12 h]

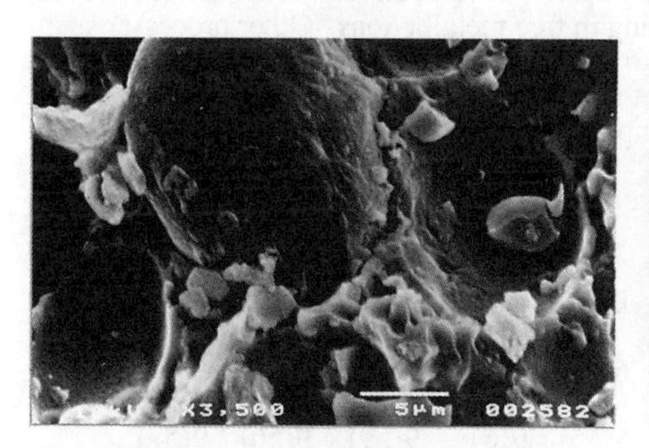

Soaking in freshly boiled water [2 h]

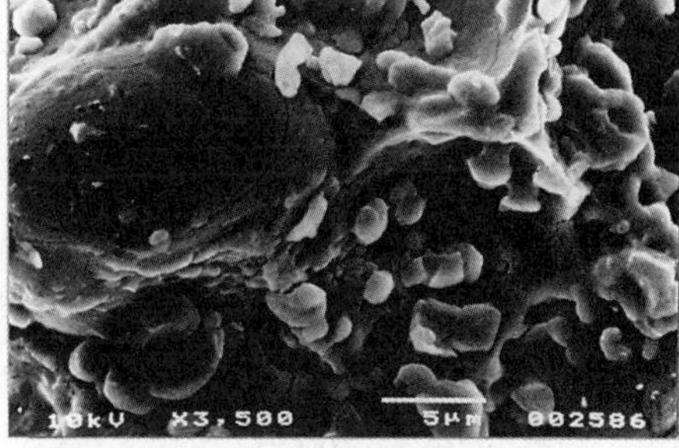

Germination [36 h/30°C]

during fermentation, increases could result from partial breakdown of starch by microflora or enzymes inherent in legumes (Gronk *et al*, 1977).

Table 1 *Effect of Processing on the Digestibilties of Protein (%) and Starch (mg maltose/g seed)*

	Unprocessed	A	B	C	D	E	F	G	H
Total Protein (%)	28.1	27.8	25.8	25.5	25.0	26.7	26.4	25.4	26.6
Protein Digestibility (%)	48.3	58.6	71.0	68.3	67.8	75.8	74.3	85.7	87.9
Starch Digestibility	42.4	66.8	63.4	67.3	70.3	83.1	90.9	74.6	79.2

6 MINERAL AVAILABILITY

Lathyrus seeds also contained some of the essential minerals like Ca, Fe and Zn (Table 2). The processing of seeds affected the availability of minerals without affecting the total mineral content of the seed. The availability of Ca (42%), Fe (39%) and Zn (33%) in unprocessed seeds was higher than some other legumes such as chickpea and cowpea (Kumar *et al*, 1978). Maximum increase in the availability of these minerals was obtained after germination; 62%, 52% and 56% for Ca, Fe and Zn, respectively. This increase could be due to increase in activity of phytase resulting in free metallic ions. Other processes such as fermentation and pressure-cooking were also very effective in improving the availability of these minerals and this may be as a result of reducing the levels of phytate and, thus, reducing the formation of phytate-mineral complexes.

Table 2 *Effects of Processing on Mineral Levels (mg/100g) and Availabilities (%, in parenthesis)*

	Unprocessed	A	B	C	D	E	F	G	H
Ca	68(42)	67(51)	64(57)	67(59)	60(52)	68(76)	68(65)	67(62)	65(66)
Fe	9(39)	9(62)	9(56)	9(53)	8(55)	9(61)	9(59)	9(56)	9(57)
Zn	4(33)	4(40)	4(43)	4(40)	4(40)	4(45)	4(59)	4(46)	4(55)

7 CONCLUSIONS

In the present study, all of the components examined were affected to a varying extent by processing. The thermal instability of ß-ODAP (which results from formation of the much less toxic alpha-isomer) and water solubility indicated that it could be reduced through processing. Given the data reported here it might be envisaged that traditional processing would improve the nutritional value of *Lathyrus* seeds.

8 ACKNOWLEDGEMENTS

We are grateful to the staff at Institute of Food Research, UK for their help during the preparation of this manuscript.

References

C. J. Briggs, N. Parreno, and C. J. Campbell, *Planta Medica*, 1985, **47**, 188-190.

T. C. Gronk, K. H. Steinkraus, L. R. Hackler, and L. P. Mattick, *Appl. Environ. Microbiol.,* 1977, **33**, 1067-1073.

H. Fuwa, T. Takaya, and R. Sugimoto, 1980. Degradation of various starch granules by amylase, in "Mechanism of saccharide polymerization and depolymerization", ed. I.S. Marshall, Academic Press, New York pp 73-100.

K. G. Kumar, L. V. Venkataraman, T. V. Jaya and K. S. Krishanamurthy, *J. Food Sci.,* 1978, **43**, 85-88.

V. Kamalakanan, A. V. Safhyamonshy and D. B. Moslong, *J. Sci. Food Agric.,* 1981, **39**, 961-967.

S. A. Odunga, *Qualitas Plantarum: Plant Foods Hum. Nutr,* 1983, **32**, 3-10.

S. L. N. Rao, *Anal. Biochem.*, 1986, **86,** 386-395.

B. S. N. Rao and T. Prabhavathi, *Am. J. Clin. Nutr.*, 1978, **31**, 169-175.

U. Singh and R. Jambunathan, *J. Food Sci.*, 1981, **46**, 1364-1367.

S. Srivastava and S. Khokhar, *J. Sci. Food Agric.*, 1996, **70**, in press.

EFFECT OF GERMINATION ON LYSINE, HISTIDINE, TYROSINE, TOTAL AND NON-PROTEIN NITROGEN CONTENTS AND PROTEIN DIGESTIBILITY OF LENTILS

M.A. Sanz[1], T. Hernández[1], G. Castillo[1], C. Vidal-Valverde[2], and A. Hernández.[1]

[1] Departamento de Nutrición y Bromatología, Universidad de Alcalá de Henares, 28871-Alcalá de Henares (Madrid), Spain.
[2] Instituto de Fermentaciones Industriales (C.S.I.C.), Juan de la Cierva, 3, 28006-Madrid, Spain.

1 INTRODUCTION

Plant products are of great importance in the human diet. Although world production of legumes is small relative to cereals, the contribution of legumes to human nutrition is noteworthy, because of the higher protein content of legumes as compared to cereals. Furthermore, legumes are good sources of carbohydrate, several water-soluble vitamins, and minerals. On the other hand, the potential nutritive value of legumes is limited by certain antinutritional factors that can affect nutrient availability[1].

Germination has often been proposed as a means of reducing such antinutritional factors in legumes. It is an inexpensive method that requires neither sunlight nor soil, and production yield is high[2].

Lentils are considered one of the major food crops in tropical and subtropical regions. Changes in antinutritional factors during germination of lentils have been reported by many investigators[3-5]. However, little research has examined the influence of germination on the content of nitrogen compounds in lentils.

Thus, the object of the present study was to determine the effect of germination under different conditions of time, light and hidratation grade in seed on the contents of total and non-protein nitrogen and three essential amino acids (lysine, histidine, and tyrosine) and on protein digestibility in lentils.

2 MATERIALS AND METHODS

2.1 Samples

Lens culinaris var. *vulgaris* cv. Magda-20 were used in the germination experiments. The seeds were washed with 0.7 % sodium hypochlorite and soaked in distilled water at room temperature for 6 h, during which time they were shaken every 30 min. The water was then drained off, and the seeds were transferred to a separating funnel, where they were germinated under the conditions described in table 1. Two replicates of germination were performed at 20 °C. A total of 90 to 100 % of the seeds germinated, and the sprouts grew to 5 to 8 cm in length. Sprouts and seeds were ground and freeze-dried for analysis.

2.2 Methods

Total nitrogen (TN) was analysed by the Kjeldahl method with endpoint potentiometric at pH 4.6. A factor of 6.25 was used for conversion to protein.

The **non-protein nitrogen (NPN)** determination was carried out by the copper sulfate method. Sample proteins were precipitated using 10 % aluminum-potassium sulfate and 3 % copper sulfate solutions. The solution was then filtered through Albet paper and the residue washed in distilled water.

Table 1. *Experimental germination conditions for the lentil seeds*

Batch	Germination time (d)	Light exposure (h of light/d)	Number of rinses (4mL water/g lentil)
Control	0	0	0
TDD	3	0	3 (daily)
TLD	3	6	3 (daily)
TDA	3	0	1 (alternate days)
TLA	3	6	1 (alternate days)
SDD	6	0	6 (daily)
SLD	6	6	6 (daily)

All supernatant liquids were filtered into a Kjeldahl flask and the non-protein nitrogen determined by the Kjeldahl method. The **protein nitrogen (PN)** was calculated as the difference between the total and non-protein nitrogen.

Lysine, histidine and **tyrosine** were determined using an isocratic HPLC procedure described elsewhere[6]. Appropriate amounts of lentil samples were hydrolyzed in 6M HCl in a nitrogen atmosphere at 110 ºC for 24 h.

Precolumn derivatization of amino acids was performed using dansyl chloride, and the separations run on a column packed with Spherisorb ODS thermostatted to 40 ºC. The mobile phase was 39:61 acetonitrile:0.01M phosphate buffer (pH 7.0) at a flow rate of 1.5 ml/min. Detection was at 254 nm.

Quantitation was accomplished using L-lysine HCl, L-histidine HCl, and L-tyrosine HCl as external standards. The standars were derivatized in the same manner as in the case of the sample hydrolysates, except that the concentration of the dansyl chloride solution used was weaker.

True protein digestibility (TD) was estimated using the *in vitro* Pedersen-Eggum pH-stat procedure[7]. A solution containing 23,100 units of trypsin, 186 units of chymotrypsin, and 0.052 units of peptidase was made up and the pH adjusted to 8.0 at 37 ºC. The activity of this solution was checked daily using an aqueous suspension of sodium caseinate (1 mg N/ml water).

Digestibility for the duplicate samples and the sodium caseinate was assessed according to the Pedersen and Eggum method. The uncorrected protein digestibility value (UTD) was calculated as follows:

UTD = 79.28 + 40.74 B (B = ml 0.1N NaOH consumed in 10 min)

UTD values for each sample were corrected on the basis of sodium caseinate digestibility to yield the TD.

3 RESULTS AND DISCUSSION

Table 2 sets out the results for TN, NPN, PN and crude protein in the control batch of ungerminated lentils and the sample batches of germinated lentils.

As already reported by other workers, the germinated seeds had higher TN and crude protein contents than the ungerminated seeds. Dagnia *et al.*[8] attributed the increase in protein to the use of fats and carbohydrates as energy sources by the developing sprouts. Comparing these results to the different germination conditions tested showed that daily rinsing of samples, exposure to light, and longer germination time increased the total nitrogen and crude protein contents in the germinated seeds.

Table 2 shows that germination brought about an increase of NPN and a concomitant decrease in PN and that this effect was more pronounced as germination time increased. Thus, the NPN/TN ratio was 14-20 % in seeds germinated for 3 d and 19-25% in seeds germinated for 6 d. Various other researchers have also reported higher NPN values as a result of germination, possibly caused by hydrolysis of the reserve proteins in the seeds into their constituent amino acids prior to use in the formation of proteins in the newly developing parts[9].

As regards the influence of the number of rinses, rinse frequency was inversely proportional to NPN content, hence the sample batches that were rinsed daily had lower NPN contents than the batches rinsed on alternate days. On the other hand, exposure to light during germination appeared to produce only a slight increase in NPN content. Hurts and Sudia[10] reported that light can affect the mobilization and distribution of several storage components in germinated soybeans. No reports have been found in the literature concerning the influence of the number of rinses or exposure to light during germination on the NPN content of germinated lentil seeds.

Table 3 presents the lysine, histidine, and tyrosine contents of the lentils before and after germination.

The effect of germination on the lysine and histidine contents was similar. Rinse frequency and germination time had little influence, whereas exposure to light during germination resulted in substantial decreases in the content of these two amino acids in the germinated seeds. The decrease in the lysine content represents an appreciable loss in the protein quality of lentil seeds, in view of the importance of lentils in supplementing cereals as a source of lysine. Conversely, exposure to light during germination did not appear to influence the tyrosine content, which decreased slightly as a result of germination.

King and Puwastein[11] reported that growing embryos use carbohydrate degradative systems during the early stage of germination and later switch to protein degradative systems as germination proceeds. A change in the amino acid profile is therefore expected as a result of protein turnover during germination.

Table 4 gives the TD values for ungerminated and germinated lentils. Germination slightly decreased the TD value for the lentils, irrespective of germination conditions. The results obtained were similar to the values recorded by El-Mahdy *et al.*[3] in germinated lentils, although those authors reported that germination increased the TD value. In contrast, Venkataraman *et al.*[12], Nnanna and Phillips[13], and Dagnia *et al.*[8] found that TD values for cowpeas and chickpeas did not improve with germination. However, it should be pointed out that these studies all involved different seed varieties, germination conditions, and methods of analysis. The increase in insoluble fiber in germinated seeds[5] can be responsible for the decrease in the TD value on germination. Insoluble fiber is known to have a deleterious effect on TD[14].

Table 2. *Total (TN), non-protein (NPN), and protein-nitrogen, (PN) and crude protein contents in lentils (% dry matter)*

Batch	%TN	%Protein	%NPN	%PN
Control	4.42	27.63	0.57	3.85
TDD	4.56	28.50	0.64	3.92
TLD	4.75	29.69	0.65	4.10
TDA	4.46	27.88	0.71	3.75
TLA	4.59	28.69	0.91	3.68
SDD	4.77	29.81	0.92	3.85
SLD	5.00	31.25	1.11	3.89

Table 3. *Lysine, histidine, and tyrosine contents (g/16g N) in lentils ($\bar{x}\pm\sigma_{n-1}$, n=2)*

Batch	Lysine	Histidine	Tyrosine
Control	6.53±0.02	2.75±0.03	2.93±0.14
TDD	6.46±0.20	2.70±0.00	2.67±0.09
TLD	5.56±0.02	2.56±0.01	2.25±0.02
TDA	6.35±0.11	2.87±0.08	2.47±0.14
TLA	5.79±0.11	2.68±0.08	2.44±0.13
SDD	6.20±0.06	2.88±0.10	2.45±0.01
SLD	5.73±0.05	2.62±0.01	2.40±0.04

Table 4. *In vitro true digestibility (%TD) for lentil ($\bar{x}\pm\sigma_{n-1}$, n=2)*

Sample	%TD
Control	92.39±0.17
TDD	92.48±0.13
TLD	91.45±0.30
TDA	92.20±0.34
TLA	92.05±0.31
SDD	92.10±0.03
SLD	92.10±0.24

References

1. S. S. Deshpande, *Crit. Rev. Food Sci. Nutr.*, 1992, **32**, 333.
2. L. H. Chen, C. E. Wells and J.R. Fordham, *J. Food Sci.*, 1975, **40**, 1290.
3. A. R. El-Mahdy, Y. G. Moharram and O. R. Abou-Samaha, *Z. Lebensm. Unters. Forsch.*, 1985, **181**, 318.
4. V. I. P. Batra, R. Vasishta and K. S. Dhindsa, *J. Food Sci. Technol.*, 1986, **23**, 260.
5. C. Vidal-Valverde and J. Frías, *Z. Lebensm. Unters. Forsch.*, 1992, **194**, 461.
6. M. A. Sanz, G. Castillo and A. Hernández, *J. Chromatogr.*, 1995, in press
7 B. Pedersen and B. O. Eggum, *Z. Tierphysiol., Tierernährg. u. Futtermittelkde.*, 1983, **49**, 265.
8. S. G. Dagnia, D. S. Petterson, R. R. Bell and F. V. Flanagan, *J. Sci. Food Agric.*, 1992, **60**, 419.
9. L. H. Chen and R. Thacker, *J. Food Sci.*, 1978, **43**, 1884.
10. C. J. Hurst and T. W. Sudia, *Am. J. Bot.*, 1973, **60**, 1034.
11. R. D. King and P. Puwastien, *J. Food Sci.*, 1987, **52**, 106.
12. L. V. Venkataraman, R. V. Jaya and K. S. Krishanmurthy, *Nutr. Rep. Int.*, 1976, **13**, 197.
13. I. A. Nnanna and R. D. Phillips, *Plant Foods Hum. Nutr.*, 1989, **39**, 187.
14. T. Hernández, C. Martínez and A. Hernández, *J. Sci. Food Agric.*, 1995, **68**.

Acknowledgments. This work was supported by CICYT ALI-91-1092-C02-01.

NUTRITIONAL VALUE OF THE STARCH CONTENT OF SOME SPANISH RICES.

J. Ortuño, M.J. Periago, G. Ros, C. Martínez and G. López.

U.D. Bromatología e Inspección de Alimentos.
Facultad de Veterinaria.
Campus de Espinardo. 30071- Murcia (SPAIN).

1 INTRODUCCION

It is generally accepted that cooking quality of rice (*Oryza sativa*) is strongly influenced by the composition and physical properties of the starch component[1]. Cereal starches are very digestible in the small intestine, only a small fraction escapes into the large bowel[2]. The reason for this is that cooked starches will retrograde to produce α-amylase resistant structures both *in vitro*[3] and *in vivo*[4]. Factors affecting the rate and extent of retrogradation include the source of the starch, amylose: amylopectin ratio, etc[5].

2 MATERIALS AND METHODS

In this work several Spanish rices with different amylose contents were studied[6]: two brown short grain rices (16 % amylose) (B) and three white rices, one long grain with a high-amylose content (27 %) (WH) and two short grain with a low-amylose content (<20 %) (WL). Resistant Starch (RS), Rapidly Digestible Starch (RDS), Slowly Digestible Starch (SDS) and Total Starch (TS) fractions, the Starch Digestibility Index (SDI)[7] and the Water Uptake (WU)[8] after 15 and 20 min, were determined also.

3 RESULTS AND DISCUSSION

Figure 1 presents the mean water uptake at 15 (WU 15) and 20 (WU 20) min. of cooking of the rices assayed. Because of the absence of the bran layer, which deters water absorption, the highest values for water uptake are observed in white rices[9]. Water uptake during cooking is considered to be inversely related to amylose content[10]. This fact can be observed at the two cooking times for white rices (WL showed the highest WU, up to 3.40 g g^{-1}, and the lowest amylose content, less than 20 %).

Figure 1 Water Uptake (g g^{-1}) at 15 and 20 min cooking of selected rices.

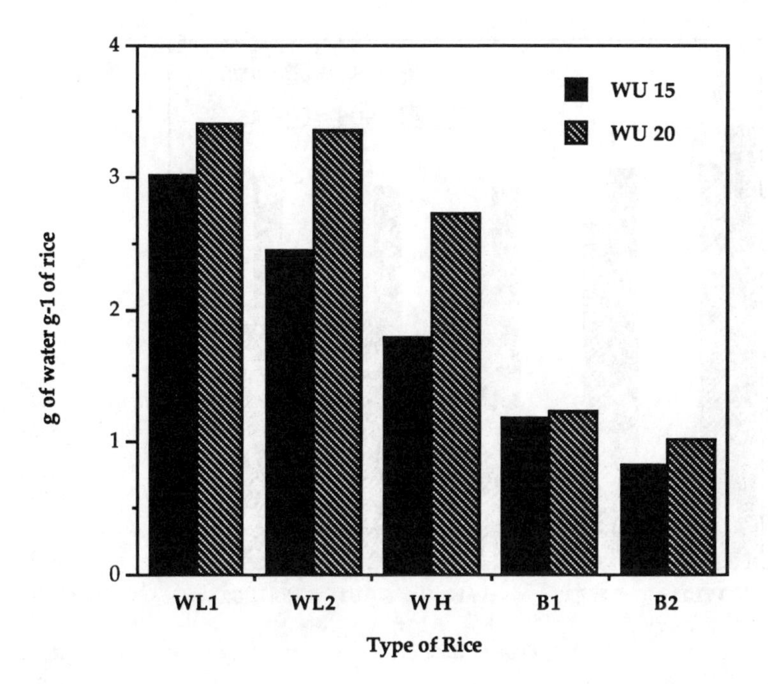

Figure 2 presents data for the TS, RDS, SDS and RS fractions of the selected rices cooked at 100 C during 20 min. Free glucose was considered as trace since it value was less than 0.02 g 100 g^{-1} and did not affected TS content. The white rices showed higher values of RDS (<77 g 100 g^{-1}) than the brown rices (<50 g 100 g^{-1}). Because brown rices inhibits swelling and the dispersion of starch[11], the opposite trend is observed in SDS and RS fractions, which showed the highest values on brown rices (between 11 and 17 g 100 g^{-1}), while in the white rices, those fractions did not go over 9 g 100 g^{-1}. Significant differences ($p<0.05$) were found between the RS fraction of WH (8.12 g 100 g^{-1}) and WL rices (<1 g 100 g^{-1}). This variability depends on retrograded amylose which is a small portion of the digestion resisting starch in the small intestine of man, but not the only one[11]. According to our results, brown rices with a high RS+SDS content and a slow SDI (<67 %) may have a better metabolic impact than whites, which showed a SDI >80 %.

Figure 2 Starch and Starch fractions (g 100 g^{-1}) of selected rices cooked at 100 C during 20 min.

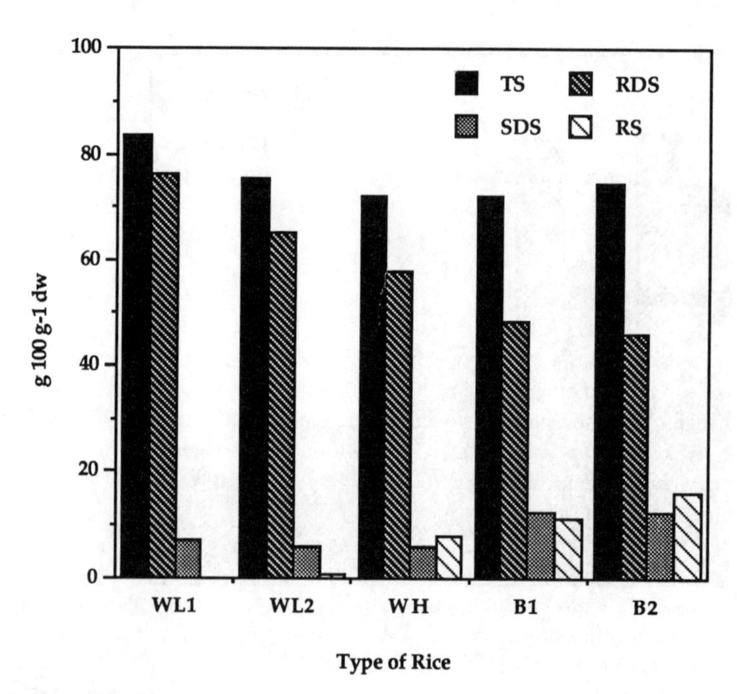

References

1. C.G. Biliaderis and B.O. Juliano, *Food Chem.*, 1993, **48**, 243-250.
2. J.H. Cummings and H.N. Englyst, *Am. J. Clin. Nutr.*, 1987, **45**, 1243-1255.
3. R.W. Kerr, 'Chemisry and Industry of Starch', Academic Press, New York, 1950, p. 165-167.
4. I. Bjorck, M. Nyman, B. Pedersen, M. Siljestrom, N-G Asp and B.O. Eggum, *J. Ceral Sci.*, 1986, **4**, 1-11.
5. R. Collinson, 'Starch and its Derivates', Radley, Chapman and Hall, London, 1968.
6. B.O. Juliano, *Cereal Sci. Today.*, 1971, **16** (10), 334-338, 340, 360.
7. H.N. Englyst, S.M. Kingman and J.H. Cummings, *Eur. J. Clin. Nutr.*, 1992, **46** (2), S33-S50.
8. K.R. Bhattacharya and C.M. Sowbhagya, *Cereal Sci. Today.*, 1971, **16** (12), 420-424.
9. V.C. Sabularse, J.A. Liuzzo, R.M. Rao and R.M. Grodner, *J. Food Sci.*, 1991, **56** (1), 96-98, 108.
10. S.L. Metcalf and D.B. Lund, *J. Food Sci.*, 1985, **50**, 1676-1679, 1684.
11. H.N. Englyst and J.H. Cummings, 'Cereals in a European context', Morton, Chichester, Ellis Horwood, 1987, p. 221-233.

COWPEA: HTC BEAN COMPOSITE FLOUR. EVALUATION OF FUNCTIONAL AND NUTRITIONAL CHARACTERISTICS

1T.N.Tunkap, [1]Carl M.F.Mbofung and [2]K.W.Waldron

[1]Department of Food Science and Nutrition, ENSAI, University of Ngaoundere Cameroon.
[2]Institute of Food Research, Norwich Laboratory, Norwich Research Park, Colney, Norwich NR4 7UA.

1.INTRODUCTION

Cowpeas (Vigna unguiculata) and common beans (Phaseolus vulgaris) are important potential sources of protein and other nutrients in developing countries especially in West and Central Africa. These legumes, particularly cowpea, are processed and eaten in a variety of forms such as Akara (a deep fried product) and Moinmoin (a steamed product). The preference for cowpea in the processing of these foods are linked to functional properties that meet the quality characteristics of the products usually made.

The storage related development of the hard-to-cook (HTC) defect in beans have, as a result of processing constrains and problems related with digestion, greatly reduce its consumption in Cameroon1. Conversion of beans into flour might increase its utilisation and consumption. There is little information on the use of Cowpea:HTC bean flours in the processing cowpea-based foods. Since cowpea-based foods are widely cherished and eaten in most of West and Central Africa, utilisation of a cowpea:HTC bean flour to produce such foods would increase the utilisation of HTC bean. The objective of this study was to study the effect of partial substitution of Cowpea flour with HTC bean flour on the functional and nutritional properties of the composite flour in comparison to that of whole cowpea flour.

2.MATERIALS AND METHODS

2.1 Legume Samples: Two varieties of Phaseolus vulgaris: large red bean (LRB) and spotted light brown bean (SBB), and one variety of Vigna unguiculata: white cowpea, were bought from a local market in Ngaoundere. The beans were cleaned of debris, diseased and broken seeds before being processed into flour as described below.

2.2 Processing into Flour: Raw flour was obtained by direct milling of cleaned seeds into a fine flour. To obtain heat processed flours, cleaned seeds were soaked at 37°C for 9 hrs, drained and dried at 50°C for 10 hrs and decorticated by impact milling. The seeds were then aspirated of their seed coats and the cotyledon milled into a fine flour (1mm) which was stored in sealed polythene bags at 4°C until required for analysis.

2.3 Extraction Rate: The % extraction rate was calculated as the ratio of the weight of beans after aspirating the loose seed coats to the weight of decorticated beans before aspirating multiplied by 100.

2.4 Composite Flour: Composite flowers were obtained by mixing (w/w) the processed cowpea and HTC-bean flour in different proportions (100:0, 90:10, 80:20, 70:30, 60:40, 50:50).

2.5 Proximate Analysis: Moisture, ash, ether soluble material, crude fibre and crude protein (N*6,25) were determined according to standard AOAC[2] methods. Total carbohydrate was determined by difference. Starch content was determined by the modified Ewers Polarimetric method[3].

2.6 Determination of minerals: Ashed samples were brought into solution by application of concentrated HC1, heat and distilled water followed by filtration before being analysed for calcium and magnesium by complexometry. Iron and phosphorus of the same solutions were determined by spectrophotometric methods[2] against a known standard.

2.7 In vitro-Carbohydrate Digestibility: This was determined according to the method described by Bjorck et al[4].

2.8 Functional and Physical Properties: Emulsion capacity and stability (EC and ES), wettability, pH and bulk density were determined according to the method described by Okezie and Bello[5]. Water and oil absorption capacities (WAC and OAC) were estimated by the method of Beuchat[6]. Cotton seed oil (Diamoar-SODECOTON, Cameroon) was used for the evaluation of OAC. Protein or nitrogen solubility (NS) was measured according to Quinn and Beuchat[7], while foaming capacity and stability (FC and FS) and gelation properties were investigated according to Coffmann and Garcia[8].

3.RESULTS AND DISCUSSION

3.1 Extraction Rate: Decorticating cowpea is an important step in the processing of cowpea-based foods and extraction rate also known as decorticating yield or recovery rate, is a simple convenient way of representing decortication efficiency[9]. The extraction rate for LRB, SBB and CP dried at 50°C were 88.05%, 88% and 90.93% respectively. These values are within acceptable range for a flour to be used commercially. The greater the extraction rate, the higher the percentage of the original nutrient content retained by the flour[10]. The removal of the seed coat also has the added advantage that it reduces the incidence of flatulence and other digestive problems very often associated with the consumption of boiled beans[11,12].

Parameter %	Raw Flour			Heat processed Flour		
	LRB	**SBB**	**CP**	**LRB**	**SBB**	**CP**
Moisture	6.07	6.04	5.74	5.92	6.95	4.87
Ash	4.61	3.57	3.21	3.64	3.36	3.12
Protein	22.98	21.10	25.31	22.9	23.86	25.40
Lipid	3.76	3.21	3.92	2.58	2.47	2.20
Crude fibre	6.45	7.73	6.42	4.78	5.31	4.32
Carbohydrate	56.13	58.35	55.34	60.18	58.05	62.31
Ca (mg/100g)	85.67	128.45	126.41	170.20	77.07	167.86
Fe (mg/100g)	3.89	6.29	5.64	1.55	1.86	1.87
Mg (mg/100g)	254.50	415.47	204.44	103.22	186.96	101.40
P (mg/100g)	727.41	906.80	432.85	1464.71	811.39	729.38
Starch (%)	40.82	41.43	47.32	44.83	43.35	48.29

The values are the means of duplicate determinations on dry weight basis

Table 1. *Proximate and mineral composition of raw and heat processed cowpea and HTC-bean flours.*

3.2 Proximate and Mineral Composition: Generally, heat processing the flour had no significant effect on the moisture and ash contents. However the level of fibre and lipid decreased, probably as a result of removal of the cell-wall and wax-rich testa. Corresponingly, the cotyledon derived components, protein, carbohydrate, starch, calcium and magnesium contents increased slightly (Table 1). This trend is similar to that reported for heat processed brown beans[13]. With particular reference to crude protein content, values obtained were similar to those obtained for raw chickpea flour[14].

3.3 Carbohydrate Digestion: The in vitro digestibility of carbohydrate in LRB and SBB was 41 and 32% respectively, lower than that in cowpea. As a result there was a progressive and significant decrease in digestibility of composite flours with successive increase in the level of HTC bean (Table 2). These differences in digestibility may be attributed to the presence of amylase inhibitors, as well as to differences in the amylose and amylopectic content of flours.

3.4 Physico-Functional Properties: Several authors have evaluated the functional properties of different legume flours13-16. No reports were available, however, comparing the functional properties of composite flours made from cowpea and HTC-bean flours. Values for pH, bulk density, wettability, EC, FC, WAC, OAC and NS for cowpea: HTC bean composite flours are presented in Table 2.

Properties such as NS, FC and EC are highly affected by pH5,6,7,8,16. In this study pH values varied between 6.40 and 6.46 and inversely with the level of substitution. Partial substitution of cowpea with HTC bean flour produced higher EC and WAC values. The composite flours also had higher foam volume and stability. Although pre-treatment conditions such as the use of heat have also been reported to influence the functional properties of bean flours13,14, the application of moderate heat at 50°C for 10 hours in this study did not have a significant effect on the functional properties. The least gelation concentration ranged between 14 and 18% with gels formed at the highest LGC showing a firmer consistency. High least gelation capacity is generally ascribed to high proportions of globular proteins.

Flour	pH	Bulk density g/ml	WAC ml/g	OAC ml/g	EC ml/g	FC %	Wetta -bility (s)	CHO digest (%)	PS (%)
CP	6.46	0.98 ·	0.99	1.10	14.1	70.0	326	100.0	79.4
LRB10%	6.44	0.99	1.14	0.75	13.9	84.6	381	89.2	77.7
LRB20%	6.43	1.03	1.19	0.76	14.0	86.3	444	80.4	77.6
LRB30%	6.42	1.03	1.23	0.76	14.1	84.1	338	64.6	77.5
LRB40%	6.41	1.03	1.33	0.86	14.4	90.0	210	58.6	77.4
LRB50%	6.40	1.03	1.33	0.88	14.0	72.5	295	57.5	76.6
SBB10%	6.44	1.02	1.25	0.84	14.3	94.6	335	97.2	82.3
SBB20%	6.43	0.99	1.28	0.93	14.5	52.6	412	76.0	78.4
SBB30%	6.43	1.01	1.34	0.99	14.3	61.1	277	69.6	77.5
SBB40%	6.42	1.03	1.61	1.01	14.3	96.0	355	68.92	77.1
SBB50%	6.40	1.01	1.64	1.13	14.0	102	463	65.01	75.6

All values are means of duplicate measurements except in the case of wettability which was measured in triplicate per method used. Digestibility Values are expressed as a percentage of that of whole cowpea flour.

Table 2. *Physico-functional properties and CHO digestibility of bean flour*

The maximum solubility of proteins is known to occur at alkaline pH[13-16]. In this study protein solubility decreased with increase in the amount of HTC-bean in the composite flours. The change in protein solubility probably reflects lower solubility of bean protein may also be related to the pH factor in the flours. The measurement of bulk density gave values much higher than those reported by other works in the literature[5,14,15]. These differences may partly be associated with the differences in amount of gelatinised starch and fibre in the samples.

4. CONCLUSIONS

The present results suggest that a composite flour obtained from partial substitution of cowpea with HTC-bean at levels of 10%, 20% and 30% have functional properties which are not too dissimilar to those of whole cowpea flour. We are currently studying the possibility of using these flours in cowpea based foods. Since cowpea-based foods are widely consumed, a successful incorporation of HTC-bean flour into these products would significantly increase HTC bean consumption and hence the nutritional and health status of consumers. Such a change in the attitude of bean consumption may equally lead to an increase production and marketing of beans.

5.0 ACKNOWLEDGEMENT

This work is part of an EEC STD3 funded programme, Contract no. TS3*-CT92-0085.

REFERENCES

1. C.M.F.Mbofung, ,T.N.Tunkap and L.L.Niba, K.W.Waldron, unpublished 1993
2. AOAC 14th ed. Association of Official Analytical Chemists, Washington, D.C. 1984.
3. S.Padmanabhan and M.Ramakrishna, *J. Food Sci. Technol.*, 1993, **30**, 313.
4. I.Bjorck, M.Nyman, B.Pedersen, M.Siljestrom, N-G., Asp and B.O.Eggum, *J. Cerea*l. 1986, 4, 1
5. B.O.Okezie and A.B.Bello, J. Fd. Sci., 1988, **53**, 453.
6. L.R.Beuchat, J. A*gric. Food Chem.*, 1977, **25**, 258.
7. M.R.Quinn and L.R.Beuchat, J. *Fd Sci.*, 1975, **40**, 475.
8. C.W.Coffmann and V.V.Garcia, *J. Food Technol.* (U.K.), 1977, **12**, 473.
9. R.D.Phillips, M.S.Chinnan, A.L.Branch, J.Miller and K.H.McWatters, *J. Fd. Sci.*, 1988, **53**, 805.
10. K.Lorenz and R.Valvano, *J. Fd. Sci.*, 1981, **46**, 1018.
11. R.D.Phillips and K.H.McWatters, *J. Fd. Sci.*, 1991, **53**, 805.
12. M.S.Chinnan, K.H.McWatters, P.O.Ngoddy and D.O.Nnanyelugo, In "Trends in food processing II", 1989, ed. A.H.Ghee, N.Lodge and O.K.Lian. p.84.
13. B.W.Abbey and G.O.Ibeh, J. Fd. Sci., 1987, **52**, 406.
14. M.Carcea Bencini, *J. Fd. Sci.*, 1986, **51**, 1518.
15. K.H.McWatters and J.P.Cherry, J. *Fd. Sci.*, 1977, **42**, 1444.
16. R.A-Halim. Ahmed and Ramanatham, *J. Fd. Sci.*m 1988, **53**, 218.

DEVELOPMENT OF HARD-TO-COOK DEFECT IN HORSEHEAD BEANS: EFFECT ON SOME PHYSICAL CHARACTERISTICS AND ON IN VITRO PROTEIN DIGESTIBILITY

[1]C.M.F. Mbofung, [1]Lorraine L. Niba, [2]M. L. Parker, [2]A. J. Downie, [2]N. Rigby, [3]C.L.A. Leakey, [2]K. W. Waldron.

[1]ENSAI, University of Ngaoundere Cameroon, [2]BBSRC Institute of Food Research, Norwich Research Park, Colney, Norwich UK, [3]Peas and Beans Ltd. Cambridge, UK.

1. INTRODUCTION

Beans (*Phaseolus vulgaris*) are an important source of proteins, energy, vitamins and minerals for millions of people particularly in the developing countries[1]. Their maximum use as a nutrient source is however generally curtailed by the development of the Hard-to-cook (HTC) defect, usually brought about by storage conditions of high temperature and humidity. This defect results in extended cooking time, increased fuel consumption, reduced palatability and nutritional value[2-4].

The nutritional quality of legumes differ with variety and is determined partly by such factors as protein content, quality and digestibility. Other factors that limit the utilisation of legumes are the presence of digestive enzyme inhibitors, haemagglutonins (lectins), phytates, flatus and tanins. It is generally accepted that the development of the HTC defects affects the nutritional quality of legumes. However the manner and the extent to which this occurs in different legume varieties is still poorly understood.

Unfortunately, storage strategies to try to improve food security tend to increase the prevalence of the HTC defect. As part of a programme which seeks to improve the utilisation of HTC beans, the purpose of this study was to investigate some changes that occur in the physical characteristics and digestibility of protein in horsehead beans that develop the HTC defect.

2. MATERIALS AND METHODS

2.1 Bean germplasm

As a basis for subsequent comparison the Horsehead cultivar was chosen for the present studies. Horsehead is a pure-line and documented bean cultivar[5] which was developed from a cross originally made at Makerere University in Uganda with subsequent selection in Cambridge, UK. Parents are the tropical Calima type of bean Diacol Nima rom Colombia and Cofinel, an anthracnose-resistant derivative of Contender, bred at INRA Versailles, France. Its seed type is similar to the large number of Calima-type dark red-mottled beans now popular over much of Eastern Africa. It is day neutral, early maturing in long days and has upright determinate growth habit with adaptation for production in temperate as well as in tropical climates. Lots of 250g cleaned freshly harvested Horsehead were wrapped in muslin cloth and stored in a desiccator containing saturated KCl giving a relative humidity of 78% at 42° C. Samples were withdrawn weekly for six weeks.

2.2 Physical Analysis

2.2.1 Bean texture. Beans were soaked overnight in distilled water at 25° C and their CT_{50} (time required to cook 50% of the beans) was monitored with a Mattson bean cooker[6].

2.2.2 Water Absorption. Water absorption pattern was determined by soaking preweighed whole beans in deionised water at 25°C for specified periods of time. At the end of each soaking period, samples were removed, blotted dry and reweighed. Differences in weight was calculated as a percentage of the presoaked weight correcting for leached solids. The determination was performed in duplicate.

2.2.3 Electrolyte and solid leakage. Electrolyte leakage during soaking was determined by measuring the specific conductivity of the soaking solution at the end of soaking using a standardized Hanna conductivity meter. In each case preweighed 20 seeds of relatively the same size were soaked overnight in 100 ml of distilled water. At the end of the soaking period the volume of the soaking solutions was brought up to 100ml before taking measurements. Following the measurement of specific conductivity, soak water was collected and freeze dried and the weight of the leached solid deposits measured.

2.2.4 Moisture and Protein (Nx6.25) Content. The moisture and protein content of beans was determined by standard AOAC[7] procedures.

2.2.5. Seed Volume and Density. Whole seeds of known weight were transferred into a 100ml measuring cylinder containing 50ml of water. The seeds were allowed to soak for one minute for equilibrium and the volume of water displaced was recorded. The mass of the beans and the recorded volume were then used to calculate seed density. This characteristic was measured for dry seeds as well as for seeds that had been soaked over night.

2.2.6. Hydration and Swelling Coefficient. These characteristics were measured by the methods described by El-Refai *et al*[8].

2.2.7. Scanning Electron Micrography. Scanning electron micrography of fresh and HTC bean sections were carried out as described by Parker *et al*[9].

2.3 Protein Solubility and Digestibility

2.3.1 In vitro Protein Digestibility. Protein digestibility was evaluated by the use of a modified *in vitro* method involving a two stage enzyme digestion technique with dialysis and simultaneous fractional collection of digests[10-11]. Digestion and collection of fractions was carried out for 6 hours and the protein content in fractions were assayed by the method of Lowry[12]. The remaining diffusate surrounding the dialysis bag was collected at the end of each digestion and also analyzed.

2.3.2 Amino Acid Analysis. Serial fractions of in vitro protein digestion were selected, prepared and analyzed for their amino acid by reversed phase HPLC methods[13-14].

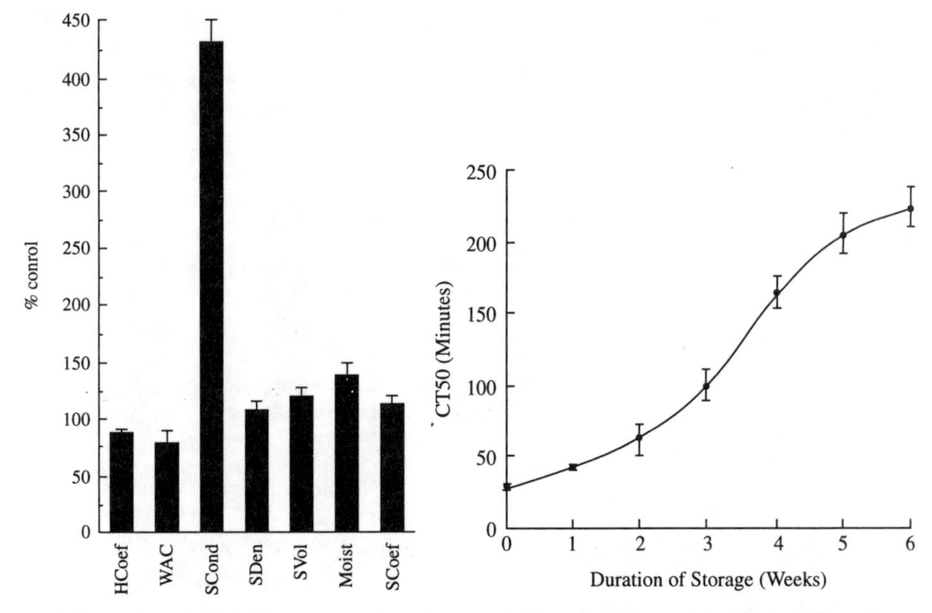

Figure 1 *Physical Characteristics of HTC Horsehead beans*

Figure 2 *Change in Cookability of Horsehead stored at 42 deg C and 78% RH*

3.0 RESULTS AND DISCUSSION

3.1 Effect of storage on some physical characteristics.

Figure 1 shows the physical characteristics of HTC horsehead expressed as a percentage of those of fresh Horsehead. It is evident from the results that the hydration coefficient and water absorption capacity decreased with the development of the hard to cook defect. On the other hand the specific conductivity of the soaking solution, bean moisture content and swelling coefficient increased. In addition cookability was significantly influenced by the duration of water imbibition.

3.1.1 Bean Texture. A plot of CT_{50} against storage time (Figure 2) depicts a sigmoidal relationship. This has been reported by several other investigators[20-21]. The slight lag observed prior to the sharp increase in the cooking time has been described as a pseudo-induction phase of the HTC defect[20]. Development of the HTC defect over six weeks resulted in more than an eight fold increase in the cooking time. As confirmed by scanning electron microscopy studies (results not shown) the extended cooking time was due to an increase in the thermal stability of cell-cell adhesion.

3.2 Development of HTC and leakage of electrolytes. Measurements of specific conductivity were expressed as ratios of specific conductivity before storage. The loss of electrolytes from seeds is known to be related to membrane degradation catalysed by lipoxygenase[18] and phospholipase[19] activity. The results show that losses in electrolyte increased with storage time and reached maximum levels three and a half weeks before the seeds were fully hard-to-cook (Figure 3).

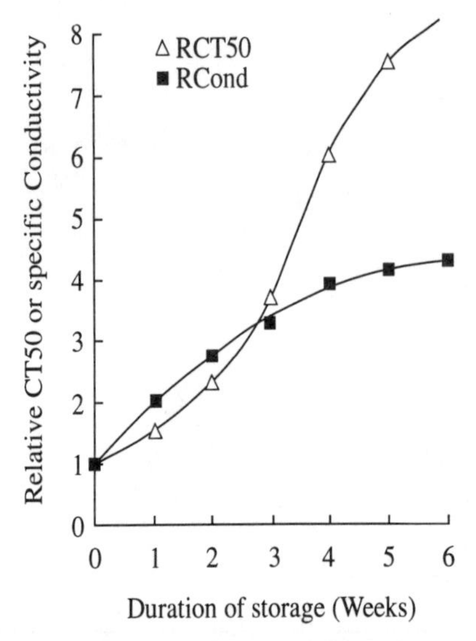

Figure 3 *Effect of duration of storage on the Relative Specific Conductivity and Cookabilty of Horsehead bean*

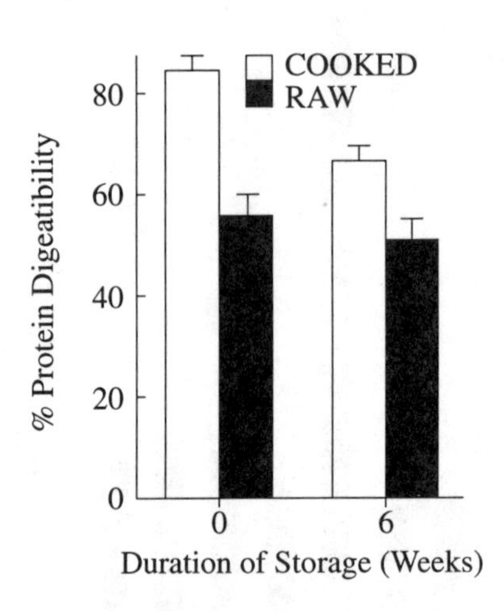

Figure 4 *Effect of Storage and Processing on the Horsehead Protein Digestibility*

3.3 Development of HTC and leaching of protein. Soaking and cooking of horsehead also resulted in the leaching of proteins into the surrounding solution. For fresh horsehead about 1.5% protein was lost in solution during an overnight soaking. Cooking for one hour led to a further loss of 9.89%. For HTC horsehead the losses during soaking and cooking were 3.77% and 14.84% respectively. Although the total protein leached during cooking was significantly higher for HTC than for fresh horsehead, the initial rate of leaching was slower for HTC than for fresh horsehead. The best equation describing the relationship between the amount of protein leached in boiling water and the degree of hard-to-cook (CT_{50}) was obtained after fifteen minutes as $Y=203.6586 - 0.32524X$; $r^2=0.99307$. Similar regression equations obtained at 5 and 10 minutes of boiling were slightly lower but significant ($P<0.01$).

3.2 Effect on Protein Digestibility

The digestibility of horsehead protein was found to significantly decrease ($P<0.01$) with the development of the HTC defect. This has also been shown to be the case for black beans[22], and pigeon pea[23]. Cooking however improved on the digestibility but not to the same degree for HTC beans as for fresh Horsehead beans (Figure 4).

Differences in protein digestibility could be due to increase in the non digestibile protein brought about by a possible formation of insoluble complexes during the development of the HTC defect and/or the increased loss of soluble proteins during soaking. Amino acid analysis (Figure 5) of representative fractions of different bean samples emphasise the effect of cooking on the susceptibility of amino acids for enzyme hydrolysis. Some research findings published in the literature suggest that the development of the of the HTC defect in some legumes actually lead to a losses in some amino acids such as lysine, methionine and tryptophan[23].

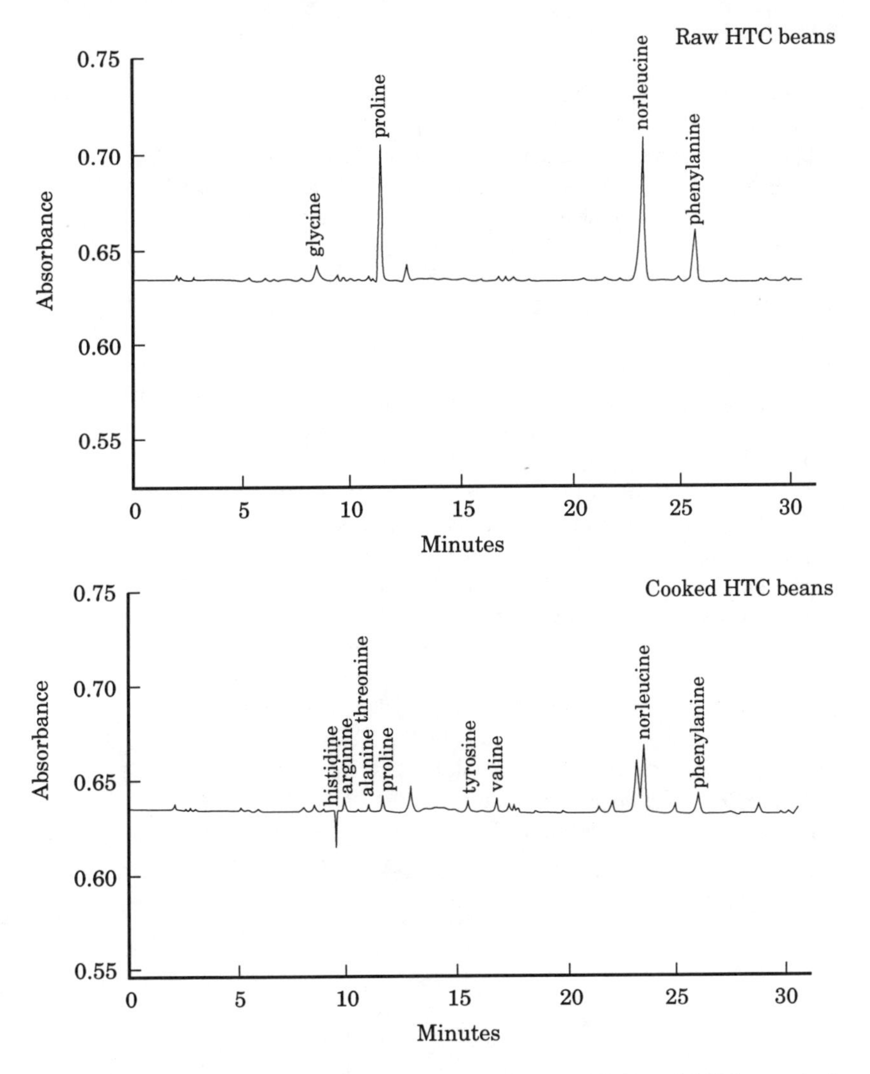

Figure 5 *Effect of cooking on amino acid release from HTC Horsehead beans during in vitro protein digestibility*

4.0 Conclusion

The development of the HTC defect in Horsehead beans leads to significant losses in minerals and soluble proteins during soaking and cooking. Protein digestibility is reduced. The use of alternative processing strategies to increase the nutritional value of HTC beans is being investigated.

Acknowledgement: This work was funded by the EC and UK office of Science and Technology via BBSRC..

4 REFERENCES

1. U. Singh and B. Singh 1992. Ecol. Bot. 46: 310-321.
2. S. Sefa-dedej and D.W. Stanley 1979. Cereal Chem 56: 379.
3. J.M. Aguilera and D.W. Stanley 1985. J. Food Process. Preserv. 9: 145-169.
4. C.M.F. Mbofung, L.L.Niba, T.N. Tunkap and K.W. Waldron 1993.(unpublished)
5. MANAS (1973)UK Ministry of Agriculture Official Registration
6. S. Mattson (1946). Acta Agric Suec. 2. 185-231.
7. AOAC Official Methods of Analysis 1984 Washington DC.
8. A. A. El-Refai, H.M. Harras, K.M. El-Nemr & M.A. Noaman 1988. Fd Chem.29:27-39
9. M.L. Parker and K.W. Waldron. J.Fd. Agric 1995, 68, 337..
10. S.F. Gauthier, C. Vachon, J.D. Jones and L. Savoie (1982) J. Nutr. 112: 1718-1725.
11. L. Savoie and S.F. Gauthier 1986. J. Food Sci. 51; 494-498.
12. O.H. Lowry 1951. J. Biol Chem. 193 265-275
13. J.A. White.,R.J. Hartand J.C Fry. 1985 Journ. Automatic Chem
14. R.L. Heirijso and Meredith S.C. (1984) Anal.Biochem. 136,
14. R.L. Heirjson and Meredith S.C. (1984) Anal Biochem. 136:65
15. Parish DJ and Leopold AC1978. Plant Physiol. 61: 365.
16. Ching TM and Schoolcraft I. 1968. Crop Sci 8: 407
17. Jackson GM and E. Varriano-Marston J. Fd Sc 46 799-803
18. Mazliak PI Post Harvest Physiology and Crop Preservation.Ed.
19. Nakayama Y., Saio K, Kito M. 1981. Cereal Chem. 58: 260.
20. Aguilera JM & Ballivian A. 1987 J.Food Sc. 52 69
21. Rivera JA, Hohlberg AI, Aguillera JM, Plhak LC & Stanley DW
22. Molina MR, De La Fuente, G and Bressani R. 1975. J.Food Sci.
23. Uma Reddy,M. and Pushpamma, P 1986 Nutr. Rep Int. 33 1021.

COMPOSITION AND *IN-VITRO* NUTRITIVE VALUE OF SOME LEAFY VEGETABLES IN ETHNOBOTANICAL FOOD USE IN NORTH EASTERN INDIA

V.N. Pandey and A.K. Srivastava

Experimental Botany Laboratory,
Department of Botany,
University of Gorakhpur,
Gorakhpur- 273 009 (U.P.)
INDIA

1. INTRODUCTION

Leafy vegetables form important constituent of human diet particularly in humid tropics. These are used as uncooked salad or in cooked form. Green leafy vegetables are cheaper and easily accessible source of food protein, ß- Carotene/Vitamin A and minerals.[1,2] Increasing the proportion of green leafy vegetables in food is useful in controlling xeropthalmia leading to night blindness in poorer sections of the society. However, leafy vegetables obtained from only few plant species are known in the food use by civilized world. Many more species are being ethnobotanically consumed in humid tropical/subtropical regions, including India. Little is known about nutritional composition of these leaves.

In the present study, 28 such uncommon leafy vegetables of ethnobotanical food use in North-Eastern Uttar Pradesh, India, were analysed for their contents of protein, lipid, fibre, total soluble sugars, amino acids, minerals and calories. The *in-vitro* nutritional quality of protein concentrates obtained from these leafy vegetables was also determined. Some of these vegetables have been found to be nutritionally promising.

2. MATERIALS AND METHODS

Fresh leaf samples were collected from plant growing wild/cultivated at the prefolwering growth stage. The samples were washed thoroughly and dried at 60 ± 1°C till constant weight. Crude protein was extracted from fresh leaves. All other analyses were done using dried samples. Nitrogen content was determined by the micro-method of Doneen[3] and protein content was estimated by multiplying protein nitrogen with factor 6.25.

Composition of leaves was estimated as Crude lipids,[4] Crude fibre,[5] total soluble sugar[6,7] and phosphorus[8]. Energy content was determined using oxygen bomb calorimeter. Crude protein was extracted form fresh leaf sample[9]. For determination of amino acid composition, 25 mg of defatted, dried crude protein was hydrolysed with

6N HCl in a sealed tube for 24h at 110°C. After hydrolysis, HCl was removed in vacuo and the residue was analysed in an amino acid analyser (LKB 4101). *In-vitro* enzymatic digestibility of crude leaf protein was assayed using pepsin followed by trypsin according to Saunders *et al*[10] and the results are expressed as per cent protein digested.

3. RESULT AND DISCUSSION

The content of protein, lipid, fibre, total soluble sugars, phosphorus and calories as well as *in-vitro* enzymatic digestibilities of 28 uncommon leafy vegetables are presented in Table 1. The protein content ranges from 8.5% (*Commelina bengalensis*) to 27.94 (*Amaranthus viridis*); lipid content 1.11% (*Brassica oleoracea*) to 9.45% (*Spinacea oleracea*); fibre content 5.35% (*Ophioglossum fibrosum*) to 26.66% (*Pisum sativum*); total soluble sugars 7.3% (*Trianthema monogyna,*) to 19.35% (*Cicer arietinum*), phosphorus content 191 mg/100 mg (*Brassica campestris*) to 967 mg/100 gm (*Cicer arietinum*), calories 0.162 Kcal/g (*Brassica oleoracea*) to 1.216 Kcal/g (*Spinacea oleoracea*) and *in-vitro* enzymatic digestibility of crude leaf protein by pepsin-trypsin enzyme system 56.20% (*Raphnus sativus*) to 68.70% (*Ipomoea aquatica*), on dry weight basis.

The leafy vegetables having more than 25% protein contents were selected for amino acid analysis viz. *Amaranthus viridis* (27.94%), *Coriantrum sativum* (25.0%), *Crotalaria sericea* (26.56%), *Lathyrus sativus* (27.38%), *Ophioglossum fibrosum* (25.30%) and *Trigonella foenum-graecum* (25.10%). The amino acid composition of leaf protein (Table 2) show that all the essential amion acids are present. The lysine content in all the samples appeals adequate from the nutritional view point. Methionine and cystein contents are, however, appreciably lower, probably due to partial or complete destruction of sulphur containing amino acids during hydrolysis prior to amino acid analysis.[11]

Alternenthra sessiles, Amaranthus viridis, A. spinosus, A. caudatus, Anethum graveolens, Basella rubra, Brassica campestris, Brassica oleoracea, Chenopodium album, C. amaranticolor, C. arietinum, Colocosia esculenta, Commelina bengalensis, Coriandrum sativum, Crotalaria sericea, Cucurbita maxima, Dolichas lablab, Ipomea aquatica, Lathyrus sativus, Medicago polymorpha, Ophioglossum fibrosum, Pisum sativum, Portulaca oleoracea, Raphnus sativus, Spinacea oleracea, Trianthema monogyna and *Trigonella foenum-graecum* appear nutritionally promising due to their higher contents of protein, lipids, fibre, total soluble sugars, phosphorus and calories as well as better *in-vitro* enzymatic digestibility of their crude protein isolate. Their amino acid composition is also nutritionally adequate. The non-toxic nature of these leaves is evident from their native use by people. However seeds of *Lathyrus sativus* are known to be toxic. Cultivation and consumption of five species viz. *Amaranthus viridis, Coriandrum sativum, Crotalaria sericea, Ophioglossum fibrosum* and *Trigonella foenum-graecum* species except *Lathyrus sativus*, which have promising nutritional potential for human use is therefore, recommended.

Table 1. Composition and *in-vitro* enzymatic digestibility of leafy vegetables

Species	Contents : % dry weight basis							*In-vitro* enzymatic
	Dry Matter	Protein	Lipid	Fibre	Total Soluble Sugars	Phos-phorus mg/100g	Calories Kcal /gm	Dige-stibi-lity*%
1. *Alternenthra sessiles* L.	14.70	17.19	4.76	11.30	8.16	333	0.299	60.80
2. *Amaranthus viridis* L.	17.20	27.94	4.65	9.80	8.30	406	0.269	58.20
3. *A. spinosus* L.	18.76	24.41	5.33	9.85	8.60	245	0.245	57.30
4. *A. caudatus* L.	16.34	16.66	6.73	10.12	9.30	306	0.324	57.40
5. *Anethum graveolens* L.	21.60	19.38	3.70	9.30	12.50	879	0.212	61.30
6. *Basella rubra*	11.10	13.30	6.30	8.90	8.82	405	0.405	57.30
7. *Brassica campestris* L.	13.60	12.89	4.41	11.20	8.90	191	0.243	56.54
8. *Brassica oleoracea* Var. capitata L.	16.70	10.85	1.11	11.00	9.58	263	0.162	58.90
9. *Chenopodium album* L.	16.30	22.69	2.45	12.20	7.36	276	0.184	67.30
10. *C. amaranticolor* Coste et Reyn.	14.62	24.62	4.78	13.30	8.25	342	0.239	61.40
11. *C. arietinum* L.	12.40	24.03	4.03	9.70	19.35	967	0.365	63.54
12. *Colocosia esculenta* (L.) Schott	21.10	18.48	7.11	16.00	8.53	592	0.265	62.45
13. *Commelina bengalensis* L.	12.68	8.50	6.34	17.00	10.34	473	0.275	67.34
14. *Coriandrum sativum* L.	13.20	25.00	4.54	9.09	11.36	469	0.364	68.49
15. *Crotalaria sericea*	15.75	26.56	5.95	15.30	8.30	698	0.444	68.30
16. *Cucurbita maxima* L.	17.30	21.30	4.05	12.13	15.60	225	0.324	63.20
17. *Dolichas lablab*	19.72	24.03	5.59	9.13	10.64	542	0.370	67.30
18. *Ipomea aquatica* Forsk	8.20	17.05	4.87	14.63	13.41	560	0.355	68.70
19. *Lactuca sativa* L.	10.80	17.59	2.77	4.63	10.19	481	0.204	62.30
20. *Lathyrus sativus* L.	16.80	27.38	5.95	12.50	14.88	595	0.405	58.30
21. *Medicago polymorpha*	17.20	21.51	8.13	7.55	9.30	529	0.279	59.60
22. *Ophioglossum fibrosum* L.	11.20	25.30	6.25	5.35	16.96	625	0.536	68.40
23. *Pisum sativum* L.	14.50	20.15	3.30	26.66	12.00	486	0.200	61.58
24. *Portulaca oleoracea* L.	10.80	18.99	2.77	7.41	13.88	370	0.185	57.34
25. *Raphnus sativus* L.	16.50	17.95	3.64	5.45	8.49	363	0.242	56.20
26. *Spinacea oleracea L.*	7.40	18.60	9.45	8.11	13.51	284	1.216	57.90
27. *Trianthema monogyna* L.	12.20	23.30	4.68	7.90	7.30	410	0.270	62.30
28. *Trigonella foenum-graecum L.*	14.30	25.10	5.59	7.69	10.48	356	0.342	59.60

* pepsin - trypsin enzyme system

Table 2. Amino acid composition of crude protein of leafy vegetables

Amino Acids	g/100g protein					
	Amaranthus viridis	*Coriandrum sativum*	*Crotalaria sericea*	*Lathyrus sativus*	*Ophioglossum fibrosum*	*Trigonella foenum-graecum*
Aspartic acid	10.8	9.8	11.0	10.4	9.8	11.6
Threonine	5.0	4.9	4.9	3.2	4.9	5.2
Serine	4.7	4.8	4.1	5.8	4.7	5.1
Glutamic acid	9.9	10.7	8.7	12.0	10.3	9.7
Proline	4.5	5.6	3.2	6.5	4.9	4.5
Glycine	5.6	5.4	9.1	8.1	5.3	6.7
Alanine	5.8	6.7	7.8	6.8	5.9	6.8
Half cysteine	1.3	1.2	1.0	1.6	1.1	0.9
Valine	6.0	6.3	6.7	6.0	5.9	5.8
Methionine	2.0	1.7	1.8	1.2	1.9	1.8
Isoleucine	5.2	5.0	5.2	4.8	4.9	4.7
Leucine	8.3	8.5	8.7	7.6	8.3	7.3
Tyrosine	3.5	2.9	2.8	2.5	3.1	2.9
Phenyl alanine	4.7	3.8	3.8	4.1	3.9	3.9
Histidine	2.0	2.3	2.2	3.1	2.1	3.0
Lysine	5.2	4.1	3.3	4.9	3.2	4.8
Ammonia	11.2	11.2	12.2	11.6	10.8	9.7
Arginine	4.2	4.3	3.6	5.3	4.1	3.6

REFERENCES

1. N.W. Pirie, Nature, 1942, 149.
2. N.W. Pirie, *Pl. Fds Hum. Nutr.*, 1985, 35, 73.
3. L.D. Doneen, *Pl. Physiol.* 1932, 7, 717
4. E.S. Charles, *Pl. Physiol.* 1928, 3, 155.
5. A.O.A.C. W. Horwitz (ed.) Washington, DC, 1975, 1018 PP.
6. F.D. Snell and C.T. Snell, Colorimetric Methods of Analysis, D. Van Nostrand Co. Inc. N.Y. 1961, 3, 219.
7. M. Somogyi, J. Biol. Chem. 1952, 195, 19.
8. E.C. Humphries, Modern Methods of Plant Analysis (eds.) K. Peach and M.V. Tracey. Springer Verlag-Berlin, 1956. p. 468.
9. V.N. Pandey, PhD Thesis, University of Gorakhpur, 1989.
10. R.M. Saunders, M.A. Connor, A.N. Booth, E.M. Bickoff and G.O. Kohler 1973, J. Nutr. 103, 530.
11. M. Byers, In Leaf Protein concentrates (eds.) L Telek and H.D. Graham, AVI Publi. Co., USA 1983 p. 135.

LEVELS OF THE ANTINUTRIENTS AND PROTEIN DIGESTIBILITY (*in vitro*) OF SOY *RABADI* - AN INDIGENOUS FERMENTED FOOD

Raj Bala Grewal

Department of Foods and Nutrition
CCS Haryana Agricultural University
Hisar - 125 004, INDIA

1 INTRODUCTION

Soybean (*Glycine Max* (L) Merril) is one of the richest and cheapest source of protein, fat and other essential nutrients. Soybeans are capable of producing the greatest amount of protein per unit land of any major plant or animal source used as food by man. It contains as high as 40-45% protein; its amino acid profile approaches the optimum pattern recommended by FAO[1]. By virtue of its chemical composition and comparatively low price it has potential for meeting nutritional requirements of population in developing countries, where animal foods are beyond the reach of majority of the population. But in addition, it also contains various antinutritional factors[2] and its typical beany flavour impedes its utilization for human nutrition. Phytic acid, saponins, polyphenols and protease inhibitors are known to decrease the digestibility of protein[3-5]. Lectins are known to inhibit growth and even cause death of experimental animals when ingested at significant concentrations[2]. Many of these factors associated with raw soybean can be eliminated or minimized by proper heat treatment and other household processes[6].

Fermentation is one of the oldest and still widely used methods of upgrading plant foods for human nutrition. *Rabadi*, an indigenous fermented food of north-western region of India, is generally prepared by fermenting pearl-millet flour with butter-milk[7]. *Rabadi* fermentation is known to improve flavour and brings about several desirable nutritional changes in the traditional product. Besides reducing the level of antinutrients[8] in pearl-millet, the fermentation improves the digestibility of proteins and carbohydrates[9]. Perspectively, use of *rabadi* type indigenous fermentation may be of interest to lower the antinutritional content and to improve protein digestibility of soybean.

2. MATERIALS AND METHODS

2.1 Materials

Soybean (*Glycine max* var. PK327) were procured from the Department of Plant Breeding, CCS Haryana Agricultural University, Hisar (India) in a single lot and cleaned of broken, cracked and wrinkled seeds, dust and other foreign materials. Skim milk powder was procured from National Dairy Research Institute, Karnal (India).

2.2 Processing of Soybeans

Soybeans (100 g) were soaked in water (1:7 w/v) at 30°C for 12 h. The unimbibed water was discarded and seeds were dehulled. The dehulled seeds were taken in conical flask, added 200 ml water, autoclaved at 1.05 kg/cm^2 for 15 min and cooled to room temperature.

2.3 Preparation of Curd

The reconstituted milk (skim milk powder and water 1:6 w/v) was heated and maintained at 40°C for 15 min, inoculated with previously prepared curd (2:100 w/v) and incubated at 37°C for five and half hours. The freshly prepared curd was used for *rabadi* fermentation.

2.4 Preparation of Soy - *Rabadi*

To autoclaved dehulled soybean, 240 g of curd and 40 ml of water was added and mixed thoroughly in an electric blender to obtain a homogeneous slurry possesing a desirable consistency. The slurry was poured in sterile conical flask and fermented at 25, 30 and 35°C for 12, 24 and 48 h.

2.5 Chemical Analysis

The samples were oven dried at 60°C for 48 h in plastic trays with polythene sheet and ground in an electric grinder (Cyclotec M/s Tecator, Hoganas, Sweden). Phytic acid content and saponins were determined colorimetrically[10, 11]. The polyphenolic compounds were extracted in methanol-HCl[12] and estimated as tannic acid equivalent[13]. Trysin inhibitor activity was assayed using casein as a substrate[14]. Lectins were determined qualitatively[15]. *In vitro* protein digestibility was assessed by employing pepsin and pancreatin[12,16]

2.6. Statistical Analysis

The data was subjected to analysis of variance and correlated coefficient in a completely randomized design according to standard statistical methods.

3. RESULTS AND DISCUSSION

3.1 Antinutrients

3.1.1 Phytic acid. Unprocessed soybean contained 1573 mg phytic acid/100 mg. Phytic acid in dehulled and autoclaved soybean was 11% less than unprocessed soybean (Figure 1). The phytic acid content of unfermented slurry was 1% and it decreased significantly ($P<0.05$) with rise in temperature and prolongation of the period of fermentation (Figure 2); higher the temperature and longer the period of fermentation, greater was the extent of phytic acid reduction. Soy *rabadi* fermented for 48 h at 25, 30 and 35°C contained 39, 41 and 50% lower phytic acid content than the control. The hydrolysis of phytic acid may be due to phytase produced by micro-organisms that ferment the slurry[17], the activity of which may be influenced by the pH changes occuring during fermentation. Decrease in phytic acid content during fermentation has been reportred in other fermented products such as tempeh[18] and pearl millet *rabadi*[8].

3.1.2 Polyphenols. The polyphenolic content of soybean was 0.3%. Removal of hull and autoclaving of soybean brought about a significant decrease in polyphenol content (Figure 1). Unfermented slurry contained 200 mg polyphenols/ 100 g. Fermentation of slurry either did not change or resulted in a gradual increase in polyphenol content of fermented product (Figure 2). The polyphenolic content of soy *rabadi* increased significantly (P<0.05) after 24 h fermentation at 30 and 35°C and 48 h fermentation at 25°C. The metabolic changes during *rabadi* fermentation may have facilitated extraction of polyphenols in methanol-HCl. Increase in poly phenol content has been reported in fermented cereal - legume mixture[19].

3.1.3. Saponins. Soybean contained considerable amount of saponins. Autoclaving after dehulling lowered the saponin content (Figure 1). Unfermented slurry contained 1.68 g saponins/100 g. The saponin content of soy *rabadi* decrerased significantly (P<0.05) with time at all the temperatures (Figure 2); the extent of reduction was greater at higher temperature and longer period of fermentation. Fermentation at 35°C for 48 h reduced the saponins to approximately half. Reduction in saponin content may be due to metabolic changes occuring during fermentation process. Tempeh has been reported to contain less than half the saponins of raw soybeans[20].

3.1.4 Trypsin inhibitor activity. Trypsin inhibitor activity was 2370 units/g in unprocessed soybeans. Autoclaving of soybeans after soaking and dehulling destroyed trypsin inhibitor activity and it could not be detected in autoclaved dehulled samples and *rabadi* mixture. Heating in presence of moisture can induce changes in disulphide bonds which appears to contribute to denaturation of inhibitors. Complete disappearance of trypsin inhibitor activity in soybeans after soaking and autoclaving has been reported[21].

3.1.5 Lectins. Lectins were present in unprocessed soybeans. Autoclaved dehulled soybeans and *rabadi* mixture did not exhibit any lectin activity. Moist heat is an effective method in eliminating the lectins[6,22]. Autoclaving prior to preparation of soy *rabadi* has eliminated the heat labile factor, lectin, completely.

3.2 Protein Digestibility (*In vitro*)

Protein digestibiltiy of soybean (54%) increased significantly (P<0.01) by 25% when soaked soybeans were dehulled and autoclaved (Figure 1). Heat processing increase protein digestibility of soybean most likely by destroying protease inhibitors[3], opening protein structure through denaturation and by denaturing globulins, highly resistant to proteases in the native state[5].

Unfermented slurry possessed 71% protein digestibility. Supplementation of soybeans with milk protein in form of curd may be responsible for higher protein digestibility. Indigenous fermentation further improved the protein digestibility. It enhanced gradually and significantly (P<0.05) with increase in the temperature and period of fermentation (Figure 2). Both temperature and time of fermentation have a cumulative effect on improvement of protein digestibility. Proteolytic enzymes produced during fermentation may be responsible for increased protein digestibility. An increase in the amino nitrogen by fermentation signifies partial breakdown of protein to peptide and aminoacids, thereby improving protein digestibility[23]. Antinutrients known to inhibit the proteolytic enzymes[3-5] are considerably reduced during *rabadi* fermentation (Figure 2), which may partly explain the enhancement to protein digestibility. A significant (P<0.01) and negative correlation has been found between phytic acid (0.9594) as well as saponins (0.9358) with protein digestibility of soybean.

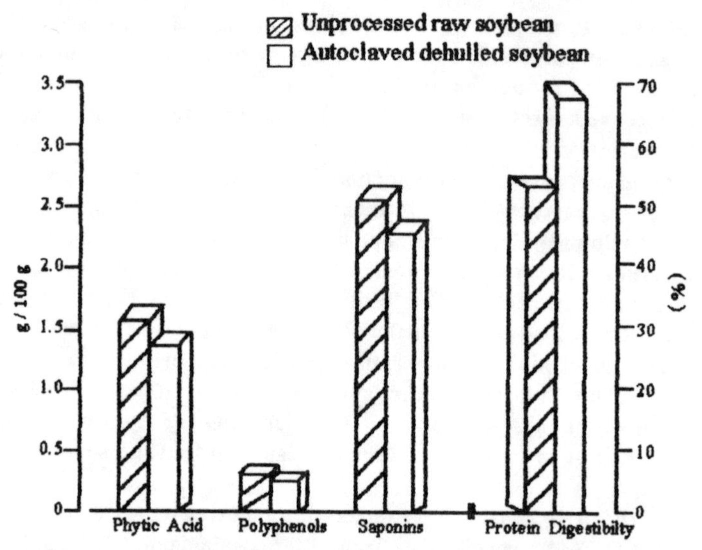

Figure 1 *Change in the level of antinutrients and in vitro Protein digestibilty (%) by autoclaving and dehulling soybeans*

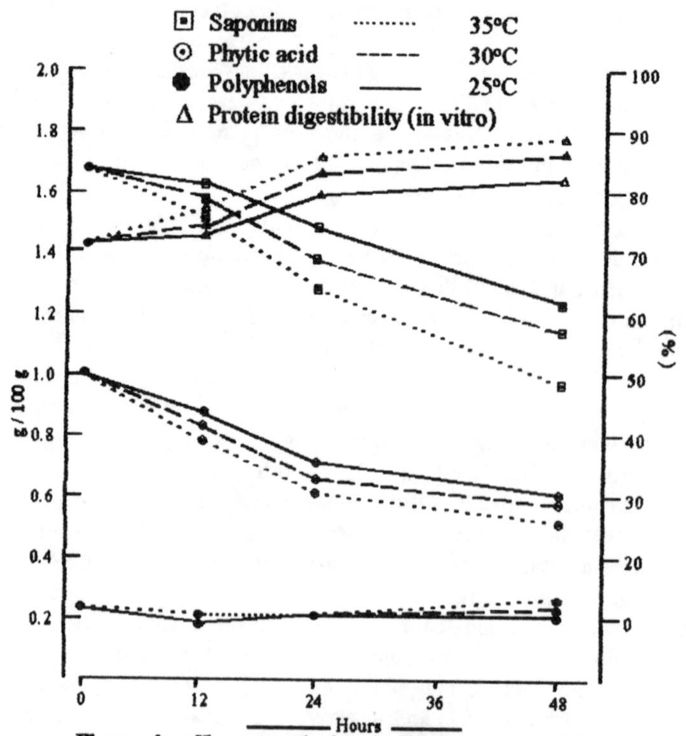

Figure 2 *Change in the level of antinutrients and in vitro protein digestibility (%) in soy rabadi at different temperatures and period of fermentation*

Therefore, *rabadi* fermentation is a unique and useful indigenous method of reducing the levels of antinutrients and improving the nutritive value of soybean.

REFERENCES

1. L.K. Ferrier, Int. Soy Series, 1976, **10**.
2. G. Grant, Progress Food Nutr. Sci., 1989, **13**, 317.
3. I.E. Liener, J. Food Sci., 1976, **41**, 1076.
4. B.E. Knuckles, D.D. Kuzmicky, A.A. Betschart, 1985, **50**, 1080.
5. Y.P. Gupta, Plant Foods Human Nutr., 1987, **37**, 201.
6. R. Ayyagari, B.S.N., Rao, and D.N. Roy, Food Chem, 1989, **34**, 229.
7. N. Dhankher and B.M. Chauhan, Int. J. Food Sci. Technol., 1987, **22**, 173.
8. N. Dhankher and B.M. Chauhan, J. Food Sci., 1987, **52**, 828.
9. N. Dhankher and B.M. Chauhan, J. Food Sci., 1987, **52**, 489.
10. N.T. Davies and H. Reid, Brit. J. Nutr., 1979, **41**, 579.
11. B. Gestetner, Y. Birk, A. Bandi and Y. Tencer, Phytochem., 1966, **5**, 803.
12. U. Singh and R. Jambunathan, J. Food Sci., 1981, **46**, 1364.
13. J. Swain and W.E. Hills, J. Sci. Food Agric., 1959, **10**, 63.
14. D.N. Roy and P.S. Rao, J. Agric. Food Chem., 1971, **19**, 257.
15. I.E. Liener, Arch. Biochem. Biophys., 1955, **54**, 223.
16. W.E. Akeson and M.A. Stahmann, J. Nutr., 1964, **83**, 257.
17. Y. Lopez, D.T. Gordon and M.L. Field, J. Food Sci., 1983, **48**, 9553.
18. Sutardi and K.A. Buckle, J. Food Sci., 1985, **50**, 260.
19. R. Goyal, M.Sc. Thesis, CCS Haryana Agricultural University, Hisar. 1991.
20. D.E. Fenwick and D. Oakenfull, J. Sci. Food Agric., 1983, **34**, 186.
21. M. Friedman, D.L. Brandon, A.H. Bates and T. Hymowits, J. Agric. Food Chem., 1991, **39**, 327.
22. V.A. Aletor and D.I. Ojo, Die Nahrung, 1989, **33**, 1009.
23. C. Kao and R.J. Robinson, Cereal Chem., 1978, **55**, 512.

DEHUSKING IMPROVES NUTRITIONAL VALUE OF PEARL MILLET GRAINS

Darshan Punia and S. Chowdhury

Department of Foods and Nutrition
Haryana Agricultural University
Hisar - 125 004 India

1 INTRODUCTION

Pearl millet (*Pennisetum glaucum*) is an important crop and a nourishing food in arid and semi-arid regions of Asia, Africa and Middle East. Pearl millet has a better nutrient profile but less acceptable, among rice and wheat eaters, because of its coarseness and colour. Dehusking of pearl millet renders its food products palatable and acceptable. Dehusking may not only affect the colour but also the nutrient and anti-nutrient level of the grain. Polyphenolic pigments present in bran impart grey colour and also limit minerals and protein utilization[1]. Phytate which may be present in pearl millet husk in a considerable concentration reduces protein[2] and starch digestibility[3] and mineral bioavailability[4]. This study deals with the effect of dehusking on nutrient and antinutrient composition, protein and starch digestibility and mineral availability of pearl millet grains.

2 MATERIALS AND METHODS

Pearl millet grains of two cultivars (HC-4 and HHB-67) were procured from the Department of Plant Breeding, CCS Haryana Agricultural University, Hisar, India. The seeds of millet grains were cleaned, made free of dust, dirt and foreign materials and stored in air tight plastic containers for further use.

2.1 Dehusking

Distilled water (30 ml) was sprinkled on grains (100 g) and kept at room temperature for half an hour. The moistened grains were hand pounded in a steel pestle and mortar till most of the grains were pearled. The husk was separated by winnowing with a closely woven basket.

The whole grains, dehusked grains and husk were oven dried to a constant weight at 60°C, ground to a fine powder in an electrical grinder and used for various determinations.

2.2 Chemical Analysis

Crude protein, Ash and Fat were determined by Standard methods; true protein by method of Osborne and Voogt[5];phytic acid, colorimetrically[6] ; polyphenolic compounds as tannic acid equivalent[7]; in vitro digestibility of protein by modified method of Singh and Jambunathan[8] ; starch digestibility by employing pancreatic amylase[9]; total minerals by Atomic Absorption Spectrophotometer[10]; available minerals, extracted using pepsin-pancreatin, by Atomic Absorption Spectrophotometer[10].

2.3 Statistical Analysis

The data were subjected to analysis of variance by standard method.

3 RESULTS AND DISCUSSION

3.1 Gross Chemical Composition

Gross chemical composition of two varieties varied significantly. HC-4 appeared to be nutritionally superior to HHB-67 as it contained more of protein,fat and ash (Table 1). The grain husk was specially rich in fat and ash. Dehusking in both the varieties resulted in a significant decrease in fat, ash and true protein content whereas crude protein content remained unaffected. Dehusking of whole grains was reported to lower the level of protein, fat and ash[11,12].

3.2 Antinutrient content

Phytic acid content of two varieties varied significantly whereas, the polyphenol content was almost similar (Table 2) and the values were within the range reported by various workers[13,14] . Husk contained significantly higher amount of these antinutrients in both the varieties and dehusked grains, therefore, contained relatively less antinutrients as compared to whole grains.

Table 1 *Gross Chemical Composition of Whole Pearl Millet Grain, Dehusked Grain and Husk (g/100 g)*

Proximate principles	*Variety*	*Whole grain*	*Dehusked grain*	*Husk*	*CD (P<0.05)*
Crude protein	HC-4	12.66	12.67	12.59	0.29
	HHB-67	10.30	10.29	10.76	
True protein	HC-4	12.21	11.38	12.00	0.39
	HHB-67	9.96	9.68	9.48	
Fat	HC-4	5.90	4.57	10.56	0.07
	HHB-67	5.60	4.50	10.32	
Ash	HC-4	2.19	1.58	2.80	0.03
	HHB-67	1.93	1.47	2.79	

Table 2 *Level of Phytic Acid and Polyphenols (mg/100 g) in Whole Pearl Millet Grain, Dehusked Grain and Husk*

Antinutrient	*Variety*	*Whole grain*	*Dehusked grain*	*Husk*	*CD (P<0.05)*
Phytic acid	HC-4	713	708	773	7
	HHB-67	726	722	787	
Polyphenols	HC-4	714	713	807	5
	HHB-67	713	709	822	

3.3 *In vitro* Digestibility

The grains of two varieties varied significantly in starch digestibility and non-significantly in protein digestibility (Table 3). The coarse grains are known to possess low protein digestibility[15,16], which may partly be attributed to the high content of phytic acid and polyphenols in these grains.

Dehusking of grains of both the varieties improved protein as well as starch digestibility significantly. This is perhaps due to the removal of less digestible protein and starch and antinutrients in the husk.

3.4 Total and Available Minerals

The variety HHB-67 had significantly higher level of Zn whereas HC-4 was having higher amount of total Ca and Fe (Table 4). In comparison to whole pearl millet grains its husk contained much higher concentration of Ca, Fe and Zn. Since husk formed a small fraction of grain (7%), dehusking did not cause a significant reduction in the level of these minerals in both the varieties.

About one third of Ca and one fourth of Fe and Zn of whole grains seemed to be available. Cereals and other plant foods are known to have low divalent minerals availability mainly because these foods have a considerable concentration of antinutrients including phytic acid and polyphenols. Processing treatments which cause reduction in the level of these antinutrients may enhance the availability of these minerals.

Though husk contained much higher concentration of Ca than whole grain but its availability was less than that of whole grain (Table 4). Dehusking improved Ca availability. Husk contained a major portion of phytic acid and polyphenols of the grain. Removal of these outer structures of the grain can, therefore, contribute towards improved Ca, Fe and Zn availability. An increase in the availability of these minerals as a result of dehusking was also reported earlier[17].

Table 3 *In vitro Digestibility of Protein(%) and Starch(mg maltose released/g) of Whole Pearl Millet Grains, Dehusked Grain and husk*

Digestibility	*Variety*	*Whole grain*	*Dehusked grain*	*Husk*	*CD(P<0.05)*
Protein digestibility	HC-4	58.9	68.6	27.4	2.6
	HHB-67	58.3	70.5	29.8	
Starch digestibility	HC-4	18.1	29.8	4.0	0.6
	HHB-67	15.7	29.1	2.5	

Table 4 *Content(mg/100 g) and in vitro Availability(%) of Calcium, iron and Zinc in Whole Pearl Millet Grain, Dehusked Grain and Husk*

Mineral content	*Variety*	*Whole grain*	*Dehusked grain*	*Husk*	*CD(P<0.05)*
Calcium	HC-4	58.17	56.53	80.00	1.99
	HHB-67	52.42	50.58	76.92	
Iron	HC-4	8.96	9.18	12.00	0.57
	HHB-67	8.58	8.39	11.14	
Zinc	HC-4	3.63	3.58	4.29	0.12
	HHB-67	4.40	4.10	5.89	
In Vitro availability					
Calcium	HC-4	38.4	41.4	22.9	3.79
	HHB-67	39.9	44.8	22.9	
Iron	HC-4	23.9	28.7	15.5	2.03
	HHB-67	23.2	29.7	15.7	
Zinc	HC-4	28.0	31.4	18.8	1.96
	HHB-67	23.6	25.9	17.7	

References

1. T.L. Aw and B.G. Swanson, *J. Food Sci.*, 1985, **50**, 67.
2. M.R. Serraino, L.U. Thompson, L. Savoie and G. Pasent, *J. Food Sci.*, 1985, **50**, 1689.
3. J.H. Yoon, L.U. Thompson and D.J.A. Jenkins, *Am. J. Clin. Nutr.*, 1983, **38**, 835.
4. A. Wise, *Nutr. Abst. Rev.*, 1983, **53**, 791.
5. D.R. Osborne and P. Voogt, 'The Analysis of Nutrients in Foods', A.I. Acad. Press, New York, 1978.
6. W. Haug and H.J. Lantzsch, *J. Sci. Food Agric.*, 1983, **34**, 1423.
7. J. Swains and W.E. Hills, *J. Sci. Food Agric.*, 1959, **10**, 63.
8. U. Singh and R. Jambunathan, *J. Food Sci.*, 1981, **46**, 1364.
9. U. Singh, M.S. Kherdekar and R. Jambunathan, *J. Food Sci.*, 1982, **47**, 510.
10. W.L. Lindsey and M.A. Norwell, *Agron. Abst.*, 1969, **61**, 84.
11. R.D. Reichert and C.G. Youngs, *Cereal Chem.*, 1977, **54**, 174.
12. A.B. Bello, L.W. Rooney and R.D. Waniska, *Cereal Chem.*, 1990, **67**, 20.
13. N. Khetarpaul and B.M. Chauhan, *J. Sci. Food Agric.*, 1991, **55**, 189.
14. A.J. Aggarwal, Ph.D. Thesis, Haryana Agricultural University, Hisar, India, 1992.
15. N.Khetarpaul and B.M. Chauhan, *J. Food Sci.*, 1990, **55**, 883.
16. A. Kumar and B.M. Chauhan, *Cereal Chem.*, 1995 (in press).
17. B.L. Carlson and D.D. Miller, *J. Food Sci.*, 1983, **48**, 1211.

DIVERSIFICATION IN NUTRITIONAL VALUE OF VARIOUS PEA VARIETIES SEEDS

Z. Zdunczyk, I. Godycka and R. Amarowicz

Division of Food Science, Centre for Agrotechnology & Veterinary Science
Polish Academy of Sciences
10-718 Olsztyn, 10 Tuwima Street, Poland.

1 INTRODUCTION

In many European countries pea seeds are an alternative to imported soybeans source of protein in humans and monogastrics nutrition. With regard to crude protein peas are between soybeans and cereals, within the broad range of 20-30%[6, 7]. Along with protein content accepted as cultivar selection criterion in many countries[3] the content of other components, including amino acids, may vary. In seeds of higher crude protein content total amino acids content is usually higher[10, 11] at diversified share of particular amino acids in protein[14]. The objective of the study was to determine the differences in chemical composition of 9 Polish pea cultivars (Pisum sativum L.) positively distinguished in protein crop during field experiments.

2 MATERIAL AND METHODS

Seeds of 9 best cropping pea (Pisum sativum) cultivars from breeding and acclimatization stations were studied. Chemical composition of seeds was determined by standard methods[1]. Amino acid composition of protein was determined on automatic analyser Beckman 6300 following samples preparation according to Hirs[9]. Hydrolysis of sulphur amino acids was prepared according to Schram[15]. Tryptophan was measuered spectrophotometrically according to Opienska-Blauth[13], while saccharose and α-galactosides (raffinose, stachyose and verbascose) by HPLC following sample preparation according to Muzquiz[12].

3 RESULTS

The seeds of the cultivars compared had similar energy value (17 623 to 18 608 kJ gross energy) (Table 1). Somewhat greater differences were found in crude protein content (22.09-25.56%) which was typical for Polish pea cultivars reported by many other authors[5, 6]. In other Polish cultivars analysed by Gdala et al.[8] total protein content was higher (22.1-27.7%) and was close to the average protein share in European pea seeds (about 25%)[8]. The content of dietary fibre was 16.15 to 20.99% of dry matter of seeds at considerable differences in the content of soluble (2.55±1.75) and insoluble (15.68±2.03%) fraction. A negative, almost statistically significant relation was found between protein content in the seeds and dietary fibre content (r = -0.634 at limiting value 0.666). Relatively low correlation coefficient results from the small number of samples analysed[9]. Reichert & MacKenzie[14], who studied peas of various total protein content, reported a highly significant negative correlation (r = -0.96) between protein content and neutral detergent fibre content. The content of saccharose was 1.17-2.06% and total of raffinose, stachyose and verbascose 3.22-4.38%. Unlike in reports by Reichert & MacKenzie[14] in the varieties of higher protein content lower α-galactosides share was found. A negative, almost statistically significant relation was found between protein content in the seed and α-galactosides content (r = -0.530). This advantageous tendency shows that selection of cultivars of higher protein content does not cause increase in indigestible components of dietary fibre and α-galactosides.

Table 1. Chemical comopsition (% DM) and gross energy content (kJ/kg) in various pea cultivar seeds

Cultivar	Dry matter	Gross energy	Crude protein	Ether extract	Dietary fibre	Saccharose	α− galactosides
Ametyst	89.62	17 967	23.25	0.97	19.39	1.81	3.33
Diament	87.98	18 140	24.48	0.97	18.81	1.27	3.38
Ergo	89.40	18 412	25.05	0.99	18.75	1.88	4.08
Hermes	87.91	18 196	22.31	1.01	17.52	1.38	3.97
Karat	89.38	18 306	24.39	0.98	16.92	1.17	3.74
Koral	88.45	18 322	23.36	0.75	19.30	2.06	3.63
Rodan	91.18	17 917	22.09	0.98	20.99	1.52	4.38
Rubin	87.71	17 623	25.56	0.84	16.15	1.78	3.39
Szafir	87.73	18 608	24.76	1.21	16.36	1.81	3.22
x	88.82	18 166	23.92	0.97	18.24	1.63	3.68
s	6.76	295	1.22	0.13	1.61	0.31	0.39

Amino acid composition of protein of all the cultivars was similar which is indicated by similar essential amino acids index (EAAI) of about 69 and PER of about 1.95 (Table 2). In all the cultivars the limiting amino acid of protein was methionine with cysteine (on average 41.4% of standard protein). Lysine content was slightly lower compared with the values observed for European pea cultivars[7]. This might have resulted from somewhat different conditions of samples hydrolysis and use of other amino acids analyser.

Like in the experiments by other authors[10, 11] the content of most amino acids grew along with the protein content in dry matter of seeds. Also the share of particular amino acids in protein changed. In the pea seeds compared (Table 3) lysine content positively correlated with total protein content in a degree close to statistical significance. A similar relation was found for cysteine for which, however, threonine content decreased along with growth of total protein content (highly significant drop in content) as well as for tryptophan (drop in content close to significance). On analysing seeds of diversified protein content Reichert a& MacKenzie[14] found a negative correlation between protein and the share of threonine, cysteine, methionine and lysine in it, and a positive correlation between protein content and glutamic acid and arginine.

Table 2. Amino acid composition of protein, g/16 g N

Cultivar	Lys	Met	Cys	Thr	Trp	Ile	Leu	Phe	Tyr	Val	EAAI	CS[1]	PER[2]
Ametyst	6.81	1.08	1.46	3.30	0.98	4.24	7.20	4.32	3.13	4.59	69.3	39.7	1.93
Diament	6.67	1.09	1.68	3.16	1.00	4.24	7.20	4.25	3.09	4.76	69.8	43.3	2.02
Ergo	6.87	1.13	1.70	3.01	1.03	4.20	6.91	4.08	3.07	4.83	69.5	44.2	1.83
Hermes	6.51	1.07	1.43	3.24	1.08	4.17	7.16	4.28	3.08	4.74	69.4	39.1	2.00
Karat	6.63	1.10	1.52	3.16	0.97	4.24	7.27	4.64	3.10	4.71	69.5	40.9	2.02
Koral	6.19	1.10	1.55	3.36	0.99	4.13	7.09	4.27	3.09	4.63	68.7	41.4	1.92
Rodan	6.55	1.19	1.49	3.31	1.04	4.24	7.11	4.21	3.11	4.76	69.9	41.9	1.98
Rubin	6.96	1.06	1.62	3.06	0.98	4.14	7.10	4.18	3.09	4.50	68.6	41.9	1.88
Szafir	6.72	1.09	1.46	3.06	0.95	4.26	7.30	4.31	3.19	4.73	68.8	39.8	1.98
x	6.66	1.10	1.55	3.18	1.00	4.21	7.15	4.28	3.11	4.69	69.3	41.4	1.95
s	0.23	0.04	0.10	0.13	0.04	0.05	0.12	0.15	0.04	0.10	0.47	1.7	0.07

[1]Methionine with cysteine, [2]Calculated with regression coefficient acc. to Alsmayer et al.[2]

Table 3. Regression equations characterizing relations between crude protein and selected amino acids in protein

Amino acid	Amino acid content in protein (Y) as function of nitrogen content in seeds (x)	Correlation coefficient
Lys	$Y = 3.920 + 0.7155\,x$	0.615
Thr	$Y = 5.227 - 0.5346\,x$	- 0.817**
Cys	$Y = 0.279 + 0.3318\,x$	0.654
Met	$Y = 1.401 - 0.0786\,x$	- 0.396
Trp	$Y = 1.473 - 0.1236\,x$	- 0.598

*, **: significant level r>0.666 (p=0.05) and >0.798 (p=0.01)

4 SUMMARY

The seeds of the cultivars grown in Poland contained 17.6 to 18.6 MJ gross energy and 22.1 to 25.5% crude protein in dry matter. The seeds of higher crude protein content contained also higher level of threonine and tryptophan. Essential amino acids index (EAAI) was alike for all the cultivars (69.3 +-0.47). The content of oligosaccharides was 4.64 to 5.96%, in this 3.22-4.38% of total raffinose, stachyose and verbascose. The content of dietary fibre was 16.15 to 20.99% of dry matter. A negative, almost statistically significant relation was found between protein content in the seed and dietary fibre content (r = -0.634 at limiting value 0.666), and between protein and α-galactoside content (r = -0.530). This advantageous tendency shows that selection of cultivars of higher protein content does not cause increase in indigestible ingredients (dietary fibre and α-galactosides).

5 REFERENCES

1. AOAC (Official Methods of Analysis of the Association of Official Analytical Chemists) 15th Edition, 1990.
2. R.H. Alsmayer, A.E. Cunningham and M.L. Happich, Food. Technol. **28** (7), 34.
3. F. Angevin, Grain Legumes, 1994, **1**, 16.
4. N.G. Asp, C. G. Johansson, H. Hallmer and M. Siljeström, J. Agric. Food Chem., 1983, **31**, 476-482.
5. A. Brenes, B.A. Rotter, R.R. Marqardt and W. Guenter, Can. J. Anim. Sci., 1993, **73**, 605.
6. M.Z. Fan, W.C. Sauer and S. Jaikaran, J. Sci. Food. Agric., 1994, **64**, 249.
7. F. Gatel and F. Grsjean, Livest. Prod. Sci., 1990, **26**, 155.
8. J. Gdala, L. Buraczewska and W. Grala, J. Anim. Feed Sci., 1992, **1**, 71.
9. C.H. Hirs, J.A. Fernandez and H. Jorgensen, J. Biol. Chem., 1954, **211**: 9411-9500.
10. N.W. Holt and F. Sosulski, Can. J. Plant. Sci., 1979, **59**, 653.
11. J.C. Huet, J. Baduet and J. Mosse, Phytochemistry, 1987, **26** (1), 47.
12. M. Múzquiz, C. Rey and C. Cuadrado, J. Chromatogr.,1992, **607**, 349.
13. J.Opieńska-Blaut, M. Charezińska and H. Berbeć, Anal Bioch., 1963, **6**, 60.
14. R.D. Reichert and S. L. MacKenzie, J. Agric. Food Chem., 1982, **30**, 312.
15. E.E. Schram, E. Moore and J. Bigwood, J. Biochem.,1954, **54**, 33.

Section 9: The 1994 Royal Society of Chemistry Food Chemistry Group Senior Medalist

MY LIFE IN PLANT CARBOHYDRATE RESEARCH

Robert R Selvendran

Institute of Food Research
Norwich Research Park
Colney
Norwich NR4 7UA

My life in carbohydrates and carbohydrate-conjugates of higher plants spanned about thirty years (1964 - 1994). During this period I worked on four main areas of research, which I shall outline in sequence. Over the years, I found the need to develop new and improved methods of analysis to meet the requirements of a particular problem challenging and engaging, and provided the necessary stimulus for research. In what follows, attention will be drawn only to some of the more notable developments. For brevity, most of the publications cited refer to my and my colleagues work. However, the literature coverage in the selected reviews is fairly extensive and so it should be easy to find related work.

The first area of research was concerned with the isolation, characterization, and metabolism of phosphate esters and nucleotide-sugars in strawberry leaves and tea leaves. In mature strawberry leaves the uridine diphosphate-sugars were present in appreciable amounts (~15µmoles/100g fresh weight)[1,2]. Mild acid hydrolysis of the UDP-sugars fraction released xylose, arabinose, fructose, glucose and galactose in the ratio of 1:1.3:2.9:10.8:2.5 respectively. The absolute configurations of glucose and fructose present in the mixture was established enzymically, because the necessary enzymes were available commercially. However, no attempt was made to characterize the other sugars rigorously. This was because, in the mid-sixties, the identity of the sugar moieties associated with the UDP-sugar fraction had been reasonably well established for several higher plants.

The situation was quite different for the sugars moieties associated with the GDP-sugar fraction. The GDP-sugar content of strawberry leaves was very low (~0.5µmole/100g fresh leaves). To rigorously characterize the sugars released on mile acid hydrolysis, an isotope-dilution procedure was developed[1]. Each isolated sugar fraction, after paper chromatography, was condensed with excess of $Na^{14}CN$, the epimeric [1-^{14}C] nitriles were hydrolysed to aldonic acids with one more C atom than the original sugar and then, after conversion into a suitable derivative (mostly the lactone), one of the radioactive epimers was isolated in crystalline form by co-crystallization with non radioactive carrier. This was recrystallized until its specific activity was constant. Usually two recrystallizations were sufficient if the fraction contained only the expected sugar. Using this method, the presence of D-mannose and D-xylose was well established[1]. With hindsight, the identity of the other sugars (D-glucose and D-galactose) which were present in very small amounts, seems less certain.

A related study concerned the concentrations of the intermediates which were likely to be involved in sucrose biosynthesis. Whilst enzymic methods were available to estimate most of the sugar phosphates, none was available for sucrose-6-phosphate (S-6-P) (α-D-glucopyranosyl-(1→2)-ß-D-fructofuranosyl-6-phosphate). The method developed to estimate S-6-P involved the following three steps. (1) The purified extract was heated with dilute NaOH for 20 min at 100°C. (2) The resulting solution was passed through a cation exchange column (H+ form), and the effluent was hydrolysed with dilute acid for 10 min and neutralized. (3) The fructose-6-phosphate content of the resulting solution was determined enzymically, and taken as a measure of S-6-P. The first step destroyed reducing sugar phosphates, but not S-6-P. The second step released fructose-6-phosphate from S-6-P, which was determined enzymically[3].

The above method helped us to monitor the changes in S-6-P (and other intermediates) in strawberry leaves during a dark-light-dark transition[4]. The results suggested that the S-6-P formed from UDP-glucose and fructose-6-phosphate was rapidly hydrolysed to sucrose, by S-6-P phosphatase, and was the predominant pathway for sucrose biosynthesis.

The work at the Tea Research Institute of Ceylon (Sri Lanka) was concerned with the biochemical changes in tea plants during post-prune growth and during fertilizer N (as ammonium sulphate) uptake, and the effects of residual N in the soil on recovery from pruning. At the time I started this work, it was fairly well established that the root starch reserves were rapidly depleted during recovery from pruning, but little was known of the biochemical changes involved. I reasoned that if one could monitor the changes in the concentrations of the metabolites translocated from the roots to the aerial parts of the plant during post-prune growth, then a better appreciation of the processes involved could be obtained.

To do this effectively, a method was developed to collect xylem sap from potted tea plants, by applying suction to the cut end(s) of pruned plants, using a filter pump[5]. Preliminary experiments showed that all leaf-bearing branches had to be removed to make the suction from the filter pump effective. The rate at which the sap was collected decreased during the first 3 - 4 weeks after pruning, and increased noticeably after bud-break, which became quite marked once the developing shoots had been well established. The amino acid composition of the sap decreased significantly during the first 3 - 4 weeks after pruning, and there was a marked increase in the amino acids, particularly of glutamine and to a lesser extent theanine, with bud-break[5]. Studies with acidic dyes (fuschin and eosin) showed that the movement of the dyes in the xylem vessels of the roots and stems increased significantly with bud-break. From these and related studies[6], we saw that the root starch was metabolised and translocated mainly as amino acids, and not as sugars, to support the growth of developing shoots.

To monitor the changes in the xylem sap of field plants during post-prune growth, a slightly modified method was developed to collect the xylem sap. In this procedure, an evaculated flask replaced the filter pump[7], and the method was found to work very well and the overall conclusions were similar to those with potted plants.

A parallel study was made on the uptake of fertilizer N by young potted plants[8]. This work showed that the uptake of N and translocation of the N, mainly as glutamine and theanine in the sap, was closely associated with the rapid decline in the starch reserves of the roots. The fertilized plants were 'clean-pruned' at weekly intervals for up to three months, and all the plants which were pruned during the first 6 - 7 weeks failed to recover from pruning because of the residual N in the soil. Failure to recover from pruning was not a problem with the unfertilized potted plants. The knowledge gained from these studies proved very useful when recommending the timing of fertilizer application, both before and after pruning, particularly in high yielding estates, some of which suffered severe loss because of the failure of the plants to recover from pruning.

My third area of research was concerned with cell walls of edible plant organs and dietary fibre (DF). At the time of my joining the Institute of Food Research (IFR) in Norwich, in September 1972, there was some confusion in our state of knowledge on the structure, chemistry, and properties of cell walls of edible plant organs and DF. Through the implementation of an integrated research programme, we developed greatly improved methods for the isolation and detailed chemical analysis of cell walls from starch-, protein- and oil-rich, vegetable, fruit and cereal products. The developmental aspects of our work and their applications will be outlined briefly, for a better appreciation of the problems encountered and how they were overcome.

In most of the early work on cell wall analysis, the alcohol-insoluble residue (AIR) was used as the starting material. Our work on cell walls of tea leaves and various tissues of the tea plant, using AIR[6,9], clearly showed that such studies are fraught with problems because of co-precipitation effects. Hot aqueous alcohol precipitates the bulk of the intracellular proteins, nucleic acids, starch and some polyphenols. Also, dehydration with alcohol is known to affect the solubility characteristics of polysaccharides (particularly of pectins and wall glycoproteins), so generating artefacts. The problem is compounded for edible plant organs, such as potatoes and legume seeds, because the starch and intracellular protein content are high compared with the amount of cell wall material (CWM) present. In the case of potatoes, the ratio of starch to CWM is about 15:1[10]. Further, the AIR is not suitable for studies on cell wall proteins. The intracellular protein contamination of the AIR could vary from as low as 5-6% (w/w) for apple parenchyma, to 10-15% for runner bean pods, to well over 50% for mung bean cotyledons.

The method which we developed avoided co-precipitation effects by using aqueous inorganic solvents instead of alcohol. To minimize the formation of oxidation products of polyphenols, 5mM sodium metabisulphite was incorporated in the extraction medium. To achieve quantitative removal of proteins and starch, it was necessary to ensure 'complete' disruption of tissue structure and then to use solvents that have a high affinity for these compounds. The first objective was achieved by wet ball-milling the tissue triturated in an aqueous detergent. The proteins were removed by trituration with 1 - 1.5% aqueous sodium dodecyl sulphate (SDS) and subsequent extraction with phenol : acetic acid : water (PAW) (2 : 1 : 1, w/v/v), and the starch was removed with dimethylsulphoxide containing 10% water (90% aqueous DMSO). This method enabled us to prepare gram quantities of CWM from a range of plant foods, such as runner bean pods[11], potatoes[10], onions[12], olive pulp[13], wheat bran[14,15] and mung bean cotyledons[16].

In our early work, for extracting pectic polysaccharides from the CWM, we used hot aqueous solutions of chelating agents, such as ammonium oxalate[17-20], but this was shown to cause significant degradation of the pectic polysaccharides, but not of the xyloglucans and wall glycoproteins. This resulted in the release of much 'neutral' galactans from the CWM of potatoes[17], arabinans from cabbage[19,20], and arabino-galactans from carrots[21]. The release of much arabinans from legume seed cell walls under comparable conditions is well documented[22,23].

In our improved method to extract pectic polysaccharides with minimum degradation, we used 50 mM trans-1,2-cyclohexane diamine tetra-acetate (CDTA, Na salt) at room temperature (20°C), and this was followed by sequential extraction of the residue with 50 mM Na_2CO_3 at 1°C for 18h, and then at 20°C, and finally with 0.5 M KOH at 20°C[12]. CDTA at 20°C completely abstracts Ca^{++} from the CWM and most of the pectic polysaccharides held in the walls by ionic cross-links are solubilised. The use of CDTA at 20°C for the extraction of pectic polysaccharides was first reported by Jarvis[24]. CDTA cannot be completely removed by dialysis[13]; but this is not usually a problem if the polysaccharides are further fractionated by precipitation with graded ethanol and then by anion-exchange chromatography[13]. Extraction with Na_2CO_3 in the cold caused preferential hydrolysis of pectin methyl esters and ester cross-links, and this minimized ß-eliminative degradation of the pectic polysaccharides on subsequent extraction at 20°C. The mild alkali treatment of course caused de-esterification of the pectic polysaccharides and increased their solubility, but a significant amount remained in the insoluble residue.

The hemicellulosic polysaccharides present in the 0.5 M KOH insoluble residue can be sequentially extracted with 1 M and 4 M KOH containing $NaBH_4$, and from the resulting residue a significant amount of hydroxyproline (hyp)-rich glycoproteins can be solubilised with 4 M KOH + boric acid, to leave a cellulose-rich residue[12,13,25,26]. Interestingly, in the case of potatoes, a significant amount of the galactose-rich pectic polysaccharides, which would normally have been degraded and solubilised from the CWM, had it first been extracted with hot aqueous chelating agent, was found to be associated with the cellulose-rich residue[25]. Neutralization of a suspension of the residue was found to release some of the associated pectic polysaccharides[13].

With CWM from parenchymatous tissues of mature runner bean pods, the cellulose-rich residue was found to contain much hyp-rich wall glycoproteins, from which the glycoproteins could be released by a short treatment with chlorite/acetic acid[26]. A novel procedure for solubilizing hyp-rich wall glycoproteins, from the depectinated CWM of runner bean, was first described by us in 1975[27] and further refinements were reported subsequently[26,28].

Using the improved procedures, we have isolated and characterized cell wall polymers from a range of plant foods - runner bean pods[26], potatoes[25], onions[12], olives[13] and mung bean cotyledons[16]. It is of interest to note that with the refined methods, little or no 'neutral' galactans, arabinans or arabinoglactans were detected in any of the materials examined. These studies enabled us to show the following:

(1) The pectic polysaccharides of onions[12], runner beans[26], potatoes[25], olives[13] and mung beans[16] exhibit heterogeneity, and there are two main groups of pectic polysaccharides, which are slightly- and highly-branched.

(2) The xyloglucans of apples[29], runner bean[26] and potatoes[25] are also heterogeneous, and there are two main groups of xyloglucans, which are highly- and very highly-branched.

(3) The hyp-rich wall glycoproteins of runner bean parenchyma exhibit heterogeneity[26,28], but this is of doubtful significance, because the conditions of extraction used may have caused some degradation of the proteins.

(4) The isolation of polysaccharide-complexes from CWM of both parenchymatous[30-34] and lignified tissues[35-37]. Although some of the complexes may be artefacts, the property of some of the polymers to associate into complexes which cannot readily be separated by anion-exchange chromatography, may give useful insights into their role(s) *in vivo*. For example, during maturation of legume seeds, and consequent dehydration of the seeds and pods during the final stages.

In related studies on lignified tissues, we have found the glucurono-xylans of parchment layers of mature runner bean pods[35] and olive seed hulls[36], and the glucurono-arabinoxylans of beeswing wheat bran[37], to exhibit heterogeneity. The heterogeneity may be partly due to the fact that the polysaccharides are laid down at different times during the development of the wall.

The refined methods have been used to monitor cell wall changes during processing of the following plant materials - asparagus shoots during maturation and storage[34,38]; ripening of tomatoes[39] and pears[40,41]; production of green table olives[42]; and the effects of extrusion cooking on mung beans[43]. Our work on the cell wall polysaccharides present in the various tissue types (e.g. parenchymatous and lignified tissues) of the above plant organs, made the results of the investigations more meaningful[34,38,41,43].

To elucidate the fine structure of the cell wall and related polymers, published methods were largely used. However, some of the methods required critical assessment and others were refined to meet the requirements of our work. A few of our contributions to this area of research are outlined below.

(1) A critical assessment of the methods used for methylation analysis of oligo- and polysaccharides[44-46], and the development of a method to methylate, completely, resistant starch[46].

(2) The use of partial acid hydrolysis[17,35,47-49] and acetolysis[18] to determine the fine structure of polysaccharides.

(3) The use of purified polysaccharide degrading enzymes to release defined fragments from polysaccharides[17,50]. Well defined oligogalacturonide fragments from pectins were radio labelled and used to study their translocation and metabolism in seedling plants[51,52], in relation to their elicitor properties[53].

(4) The extensive use of GC-MS[17,18,50] and HPLC-MS[48-50] to characterize oligosaccharides derived from polysaccharides.

(5) The use of ^{13}C NMR to determine specific structural features of sugar residues in polysaccharides[13,25,26,39,54].

(6) In addition to cell wall polysaccharides, we have also elucidated the fine structure of lectins[55], kiwi fruit tree mucilage[48,49], and some extracellular bacterial polysaccharides[56,57] which may have application in the food industry.

Our detailed studies on cell walls of parenchymatous and lignified tissues of a whole range of edible plant organs, coupled with light and electron microscopy studies of the products, helped us to take a holistic view of the cell wall in relation to dietary fibre (and texture), and its effects in the human alimentary tract[58-60]. Because DF is ingested in the diet mostly as clusters of cells rather than as extracted purified polymer, the analysis of extracted polymers from purified CWM can go only part of the way to explaining the properties of DF. Several levels of organisation can be described for cell walls and their constituent polymers within plant organs. The higher levels of organisation have important implications which are not generally well appreciated[59,60]. For example, the maintenance of tissue structure within a food particle is important because of the role this plays in water retention. It is the retention of this structure that is a major contributing factor to the faecal bulking effect of wheat bran[58,61]. Our studies have clearly shown that the effects of DF result from the composition, structure and physicochemical properties of the cell wall matrix within a plant tissue. This aspect has been discussed at length in some of our reviews[59,60].

The experience gained from the above studies helped us to assess critically and realistically the published work on a range of subjects and we were invited to write detailed reviews by the editors of plant biochemical, food and nutritional journals. Writing reviews was time-consuming and required much extra work. However, I found this part of our research work very rewarding, because it gave us an opportunity to share our research experience with others. Some of the important topics which we wrote reviews are given below:

(1) Plant glycoproteins[62]
(2) Methods for the isolation and analysis of plant cell walls[63,64]
(3) Methods for the analysis of dietary fibre[65,66]
(4) Developments in the chemistry of cell wall polymers[23,67]
(5) Cell walls in relation to DF and human nutrition[58,68,69]

Over the years, I have been invited to give keynote and plenary lectures in many different countries, particularly in Europe. I have also been awarded scholarships by the Agricultural and Food Research Council (AFRC, now BBSRC), the British Council, the Royal Society of Chemistry, and the Royal Society to visit and give lectures on specialist topics in Universities and Research Institutes in several East European Countries, Israel, India and Sri Lanka. I do hope that our work on plant carbohydrates and carbohydrate-conjugates would serve as eye-openers and as a source of stimulus for researchers all over the world. It is important to bear in mind that a research worker remains a student all his life, because one has to keep abreast with the growth of knowledge. My prayer is that in addition to their academic interest, some useful practical benefits would accrue from our research efforts.

References

1 R R Selvendran and F A Isherwood, Biochem. J., 1967, **105**, 723-728.
2 F A Isherwood and R R Selvendran, Phytochem., 1970, **9**, 2265-2269.
3 R R Selvendran and F A Isherwood, Phytochem., 1970, **9**, 533-536.
4 F A Isherwood and R R Selvendran, Phytochem., 1971, **10**, 579-584.
5 R R Selvendran, Ann. Bot., 1970, **34**, 825-833.
6 R R Selvendran and S Selvendran, Phytochem., 1972, **11**, 3167-3171.
7 R R Selvendran and S Sabaratnam, Ann. Bot., 1971, **35**, 679-682.
8 R R Selvendran and S Selvendran, Ann. Bot., 1973, **37**, 453-461.
9 R R Selvendran, B P M Perera and S Selvendran, J. Sci. Fd. Agric., 1972, **23**, 1119-1123.
10 S G Ring and R R Selvendran, Phytochem., 1978, **17**, 745-752.
11 R R Selvendran, Phytochem., 1975, **14**, 1011-1017.
12 R J Redgwell and R R Selvendran, Carbohydr. Res., 1986, **157**, 183-199.
13 M A Coimbra, K W Waldron and R R Selvendran, Carbohydr. Res., 1994, **252**, 245-262.
14 R R Selvendran, S G Ring, M A O'Neill and M S DuPont, Chem. & Ind., 1980, 885-888.
15 R R Selvendran and M S DuPont, Cereal Chem., 1980, **57**, 278-283.
16 J Gooneratne, P W Needs, P Ryden and R R Selvendran, Carbohydr. Res., 1994, **265**, 61-77.
17 S G Ring and R R Selvendran, Phytochem., 1981, **20**, 2511-2519.
18 M A O'Neill and R R Selvendran, Carbohydr. Res., 1983, **111**, 239-255.
19 B J H Stevens and R R Selvendran, Phytochem., 1984, **23**, 107-115.
20 B J H Stevens and R R Selvendran, Phytochem., 1980, **19**, 559-561.
21 B J H Stevens and R R Selvendran, Carbohydr. Res., 1984, **128**, 321-333.
22 J-M Brillouet and B Carré, Phytochem., 1983, **22**, 841-847.
23 R R Selvendran, 'Developments in the chemistry and biochemistry of pectic and hemicellulosic polymers', in "Proceedings of the Sixth John Innes Symposium", J. Cell Sci. Suppl. **2**, 1985, pp 51-88.
24 M C Jarvis, Planta, 1982, **154**, 344-346.
25 P Ryden and R R Selvendran, Carbohydr. Res., 1990, **195**, 257-272.
26 P Ryden and R R Selvendran, Biochem. J., 1990, **269**, 393-402.
27 R R Selvendran, Phytochem., 1975, **14**, 2175-2180.
28 M A O'Neill and R R Selvendran, Biochem. J., 1980, **187**, 53-63.
29 P Ruperez, R R Selvendran and B J H Stevens, Carbohydr. Res., 1985, **142**, 107-113.
30 B J H Stevens and R R Selvendran, Carbohydr. Res., 1984, **135**, 155-166.
31 M A O'Neill and R R Selvendran, Biochem. J., 1985, **227**, 475-481.
32 B J H Stevens and R R Selvendran, Phytochem., 1984, **23**, 339-347.
33 M A Coimbra, N M Rigby, R R Selvendran and K W Waldron, Carbohydr. Polymers, 1995 (in press).
34 K W Waldron and R R Selvendran, Phytochem., 1992, **31**, 1931-1940.
35 R R Selvendran and S E King, Carbohydr. Res., 1989, **195**, 87-99.
36 M A Coimbra, K W Waldron and R R Selvendran, Carbohydr. Polymers, 1995 (in press).
37 M S DuPont and R R Selvendran, Carbohydr. Res., 1987, **163**, 99-113.
38 K W Waldron and R R Selvendran, Physiologia Plantarum, 1990, **80**, 568-575.
39 G B Seymour, I J Colquhoun, M S DuPont, K R Parsley and R R Selvendran, Phytochem., 1990, **29**, 725-731.
40 M A Martin-Cabrejas, K W Waldron and R R Selvendran, J. Plant Physiol., 1994,

144, 541-548.

41 M A Martin-Cabrejas, K W Waldron, R R Selvendran, M L Parker and G K Moates, Physiologica Plantarum, 1994, **91**, 671-679.

42 M A Coimbra, K W Waldron, I Delgadillo and R R Selvendran, Carbohydr. Polymers (in press).

43 J Gooneratne, G Majsak-Newman, J A Robertson and R R Selvendran, J Agric. Food Chem., 1994, **42**, 605-611.

44 P W Needs and R R Selvendran, Carbohydr. Res., 1993, **245**, 1-10.

45 P W Needs and R R Selvendran, Phytochem. Analysis, 1993, **4**, 210-216.

46 P W Needs and R R Selvendran, Carbohydr. Res., 1994, **254**, 229-244.

47 R R Selvendran and B J H Stevens, 'Applications of mass spectrometry for the examination of pectic polysaccharides', in 'Modern Methods of Plant Analysis', New Series Vol **3**, 1986, pp. 23-46.

48 R J Redgwell, M A O'Neill, R R Selvendran and K R Parsley, Carbohydr. Res., 1986, **153**, 97-106.

49 R J Redgwell, M A O'Neill, R R Selvendran and K P Parsley, Carblhydr. Res., 1986, **153**, 107-118.

50 M A O'Neill and R R Selvendran, Carbohydr. Res., 1985, **145**, 45-58.

51 A J MacDougall, N M Rigby, P W Needs and R R Selvendran, Planta, 1992, **188**, 566-574.

52 N M Rigby, A J MacDougall, P W Needs and R R Selvendran, Planta, 1994, **193**, 536-541.

53 M H Doherty, R R Selvendran and D J Bowles, Physiological and Molecular Plant Pathology, 1988, **33**, 377-384.

54 P Ryden, I J Colquhoun and R R Selvendran, Carbohydr. Res., 1989, **185**, 233-237.

55 D Ashford, N N Desai, A K Allen, A Neuberger, M A O'Neill and R R Selvendran, Biochem. J., 1982, **201**, 199-208.

56 M A O'Neill, R R Selvendran and V J Morris, Carbohydr. Res., 1983, **124**, 123-133.

57 M A O'Neill, R R Selvendran, V J Morris and J Eagles, Carbohydr. Res., 1986, **147**, 295-313.

58 R R Selvendran, B J H Stevens and M S DuPont, 'Dietary Fibre: Chemistry, Analysis and Properties', in 'Advances in Food Research', Vol **31**, 1987, pp. 117-209.

59 R R Selvendran and J A Robertson, 'The chemistry of dietary fibre: An holistic view of the cell wall matrix', in 'Dietary Fibre: Chemical and Biological Aspects', Royal Society of Chemistry, London, 1990, pp. 27-43.

60 R R Selvendran and A J MacDougall, European Journal of Clinical Nutrition Suppl., 1995 (in press).

61 B J H Stevens and R R Selvendran, Carbohydr. Res., 1988, **183**, 311-319.

62 R R Selvendran and M A O'Neill, 'Plant Glycoproteins', in 'Encyclopaedia of Plant Physiology', Vol **13A**, Springer-Verlag, Heidelberg, 1982, pp. 515-583.

63 R R Selvendran and M A O'Neill, 'Isolation and analysis of cell walls from plant material', in 'Methods of Biochemical Analysis', Vol **32**, John Wiley and Sons Inc., 1987, pp. 25-153.

64 R R Selvendran and P Ryden, 'Isolation and analysis of plant cell walls', in 'Methods of Plant Biochemistry, Vol **2**, Carbohydrates', Academic Press, 1990, pp. 549-579.

65 R R Selvendran and M S DuPont, 'Problems associated with the analysis of dietary fibre and some recent developments', in 'Developments in Food Analysis Techniques - 3', Applied Science Publishers, 1984, pp. 1-68.

66 R R Selvendran, A V F V Verne and R M Faulks, 'Methods for analysis of dietary fibre', in 'Modern Methods of Plant Analysis, New Series Vol **10**', Springer-Verlag, Berlin, 1988, pp. 234-259.

67 R R Selvendran, 'The chemistry of plant cell walls', in 'Dietary Fibre', Applied Science Publishers, London, 1983, pp. 95-147.

68 R R Selvendran, 'The plant cell wall as a source of dietary fibre: chemistry and strucutre', Amer. J. Clin. Nutr., 1984, **39**, 320-337.
69 R R Selvendran, 'Dietary fibre, chemistry and properties', in 'Encyclopaedia of Human Biology, Vol **3**', Academic Press Inc., 1991, pp. 35-45.

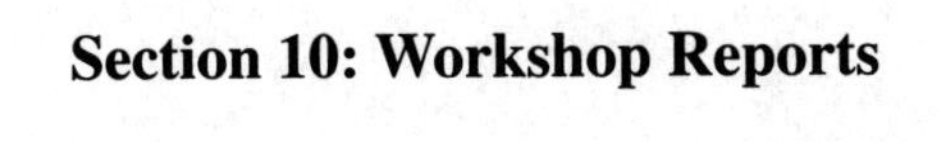

Section 10: Workshop Reports

Consumer attitudes to quality. What are the relevant issues in consumer research?

Workshop A Report

Dr L.J. Frewer and Dr H.J.H. MacFie, Co-Chairs

Department of Consumer Sciences;
Institute of Food Research,
Earley Gate,
Whiteknights Road, Reading, RG6 2EF

The focus of discussion in the workshop was on both the extrinsic and intrinsic product attributes which determine consumer acceptance of products. The workshop addressed issues of relevance both to those products with an established market, and those products which are currently under development.

1 EXTRINSIC FACTORS

Various extrinsic factors were discussed as being important in the assessment of product quality and subsequent consumer acceptance of products. Extrinsic factors were defined as those which did not relate to the primary sensory qualities of the product itself, but rather to the social environment in which the consumer made decisions about purchase, and to safety and ethical considerations in acceptance.

Perceptions of Risk

Consumer risk perceptions, particularly with regard to the products of emerging technologies such as genetic engineering, were discussed. Public perceptions regarding new hazards, as opposed to those which had long been established, were regarded as more of a barrier to consumer acceptance because of uncertainty regarding both occurrence of the hazard and the severity of consequences should the hazard occur. The aggregation of risk, where many people are affected, was also discussed as being of great relevance to public acceptance of products. It was pointed out that the occurrence of one incidence with negative consequences for products where there was public uncertainty regarding safety was likely to have a great impact on consumer acceptance, even though statistically safety was greater than for some other food-related risks. Public acceptance was likely to be increased by the establishment of an effective dialogue between producers and manufacturers on one hand, and the public and their representatives (for example, consumer groups) on the other. It was suggested that ultimately, a solution might be offered by making the consumer responsible for their food choices, although clearly this requires proactive information provision by those responsible for the food supply to the consumer. Further discussion focused on the role of labelling in consumer choice. As well as increasing public trust in the proactive manufacturer or producer, this allows the consumer to feel in control of food choice, and is likely to enhance acceptance of novel products and processes.

Perceptions of Naturalness

The question of “what does the consumer see as natural” was raised, a particularly pertinent issue given the current trend for enhanced consumer acceptance of “natural” products. Conflict between rationality and affect decision making was seen as a problem in product acceptance, and again the issue of how to effectively communicate with the consumer was discussed.

It was suggested that some new processes are likely to increase perceptions of naturalness, or at least will provide benefits to the environment in the form of reduced use of "unnatural" chemicals. The case of virus resistant potatoes was cited, and linked with increased consumer acceptance due to reduced chemical use. However, debate also focused on the concept of perceived "naturalness", and whether the tendency of modern society to equate it with "goodness" was open to fundamental questioning. It has been suggested that the agricultural and food industries should be taking a more pro-active role in pointing out that "naturalness" may be related to underdevelopment of resources [1]. Problematically, research has shown that one of the major barriers to consumer acceptance of the products of genetic engineering is likely to be consumer perceptions that the products are "unnatural" [2 3], and it is suggested that the most effective route to consumer acceptance is likely to be emphasis on this aspect of novel products. Trust in the safety measures taken by producers were thought to be an important determinant of acceptance, and indeed this issue has been extensively discussed elsewhere [4 5 6].

Taking New Technologies into the Market Place

It was also suggested that market-testing the products of new technologies on a case-by-case basis was the most appropriate way to assess consumer acceptance, rather than flooding the market with many new products simultaneously. This would enable the consumer to fully recognize the advantages of both the new products and the technology. It was important to stress the benefits of new products to the consumer and to the environment, rather than focus on the minimal risks associated with the technology.

An alternative view was expressed that exposing the consumer to several products simultaneously would facilitate consumer acceptance of the technology, as multiple advantages could be recognised. It was pointed out that if this were to occur, the products would all have to have direct and tangible benefits to the consumer, which should not be related to pricing factors. Research has indicated that reduced price alone is unlikely to facilitate consumer acceptance of novel products [7], although it is not yet known whether price in combination with other benefits may interactively enhance consumer acceptance. Against this, the importance of market testing was also felt to be great because the consumer does not always do what he or she says they will do. Under these circumstances, price might prove to be of increased importance.

2 INTRINSIC FACTORS

The role of price in consumer choice and as a determinant of quality was discussed in depth. It was agreed that the most helpful way to view price and cost was to view it as a constraint on consumer choice, but not to use it as a quality indicator. Different views of the different sectors were put forward:

The Retail Sector

Concerns in the retail sector were focused on the development of a product with carefully specified sensory properties, in conjunction with minimizing costs by using a range of materials from cheap sources. Thus the product should be acceptable in terms of its intrinsic qualities, avoiding the need for sophisticated marketing techniques. An example was given of the utilization of a brand name (e.g. "Superplonk"), a product with desirable intrinsic properties and a low pricing policy. Guaranteed product trial was ensured by the retailer by the provision of adequate shelf space in the retail environment. This was presumed sufficient to increase sales to appropriate levels.

The Branded Manufacturer

The position of this sector was becoming more difficult as retailers were able to reproduce the exact sensory profiles of products quickly and accurately. The solution was to develop and patent some technological advances which would hold the retailers at bay, or keep the brand and its extensions moving forward at a pace which ensured the retailer was always "one step ahead".

The Fresh Produce Manufacturer
The need to innovate in this area was widely recognised. The success depended not so much on technical knowledge, because this was widely distributed throughout the industry, but on careful analysis of how the proposed product would be viewed by the target consumer sector. This type of consumer research was cheaper than the cost of new product failure which was characteristic of new product development today.

Breeders
The difficulty of predicting consumer preferences over the 10 - 15 year time scale that was required to produce new fruit or vegetable varieties was emphasised. The apparent solution was to develop a large number of varieties simultaneously and hope that consumers would develop brand loyalty to one or two. This view was not universally accepted by the breeders present although the need for constant perceived improvement was accepted.

3 EXTRINSIC *VERSUS* INTRINSIC PROPERTIES?

With the current highly competitive retail sector, the intense interest in methods of production (for example, production methods with low economic inputs), the high marketing budgets of some products (e.g. coffee) and the increasing interest of the consumer in the long term health outcomes of consuming particular foods, the hypothesis was put forward that in many products the sensory properties did not matter. This view was soundly rejected by the manufacturers and retailers present. It was still the critical factor in determining repurchase after the first product trial. However, it was also important to consider the extrinsic factors relevant to product development. Particular emphasis was placed on using consumers in the determination of the importance of relevant extrinsic and intrinsic factors, and their role in acceptance of products [8].

Finally, it was concluded that there was a need to develop the methodologies to assess what is likely to determine the qualities which will lead to product acceptance. New methodologies which conjoin sensory factors with extrinsic factors are needed in product development. Detailed examination of why products have failed to be accepted is likely to lead to the development of methods to predict why new products will be accepted.

4 REFERENCES

1. C. L. A. Leakey, Personal Communication.
2. L. J. Frewer, R. Shepherd and C. Howard, Genetically engineered food: The effects of product exposure on consumer acceptability (This volume).
3. L.J. Frewer, C. Howard and R. Shepherd (submitted). Public concerns about general and specific applications of genetic engineering: Risk, benefit and ethics.
4. L. Rothenberg, *Tibtech*, 1994, **12**, November, 435.
5. K. L. Dittus and V. N. Hilliers, *Food Technology*, 1993, **July**, 87-89.
6. L.J. Frewer, C. Howard, D. Hedderley and R. Shepherd. What determines trust in information about food related risks? Underlying psychological Constructs (submitted).
7. L.J. Frewer, C. Howard and R. Shepherd, The Influence of realistic product exposure on attitudes towards genetic engineering of food, *Food Quality and Preference*, in press.
8. H. J. H. MacFie and D. M. H. Thomson, Measurement of Food Prefences, 1994, Blackie, Glasgow, UK.

THE QUALITY OF FOOD PRODUCED IN DEVELOPING COUNTRIES:

Implications for importers and consumers.

S Khokhar
Department of Foods and Nutrition, CCS Haryana Agricultural University, Hisar, India

B Axtell
Midway Technology Ltd, Rugby, UK

The area of discussion of this workshop proved, as expected, to be both wide-ranging and stimulating. The discussion covered technical, political, economic, social and trade issues.In the short time available, it was not possible for the participants to address these issues in the depth they deserved, but nevertheless there was a good, and on occasion, very direct expression of views. The Workshop benefited from having participants from nine countries, including Cameroon, Kenya, India, Nigeria and South Africa. Unfortunately, the workshop suffered from an absence of input from the [multi-national] industrial sector.

After participants had briefly introduced themselves, the Co-chairs outlined a broad suggested base for discussion of issues of food quality and legislation of various groups involved in the use, sale and production of food products in developing countries.This is shown below and was accepted as a basis for discussion:

EXPORT MARKETS	REGIONAL EXPORT
	EXTRA-REGIONAL EXPORTS
NATIONAL MARKETS	MIDDLE-CLASS CONSUMERS
	THE VERY POOR

The perception and concerns of a typical UK retailer were then presented. It was explained that such companies had to comply with the regulations emanating from Brussels and that typical key areas of concern were :-

COST	product had to compete
MARKET DEMAND	there had to be good demand
REGULATIONS	product had to meet all regulations
DUE DILIGENCE	the retailer is wholly responsible for own label product

In order to assume that the regulations are being met larger retailers have QA inspectors who regularly inspect supplies. Such systems have regularly led to the development of the various HACCP, BS, ISO, and EU systems. The trend is for suppliers to become increasingly self-auditing. the high cost of external auditing of overseas suppliers was pointed out. the central element wsa the need for trust between suppliers and retailers.

The industrial speaker moved on to recent consumer-driven standards, ethical retailing, fair trade, animal welfare and environmental concerns. It was emphasised that such concerns had to be taken seriously if market shares were to be maintained.

Speakers from Cameroon, Kenya and Nigeria then responded. The commonly held view was that importers wanted it all their own way by defining both the quality required and the price to be paid. The speakers felt that this took little account of local costs and costs involved in meeting the quality standards. It was stated that in many cases, for example, small coffee and cocoa farmers, the return received was so low as to provide little incentive to raise quality. The general view was that the only aim of large companies is profit rather than the provision of markets to generate two-way trade.

Reference to national markets was made and the view expressed that local standards should take human and social aspects into account.

One speaker posed the question "Do you want to continue forever seeing us as a cheap supplier and continue to provide aid for health, food etc.?"

At various times during the discussion the subject of "new or non-traditional" foods was raised. Such commodities might provide a means of adding value.Conflicting views were expressed, and no conclusion was reached. There was, nevertheless, a broad consensus that one way forward would be to invest more in effective marketing strategies.

The situation in South Africa, a country having both developed and developing country characteristics, was described. There was no difficulty in the ability of its large modern food manufacturing sector to meet European standards. At the same time serious problems exist with supplying affordable, safe foods to poorer sections of the community.The possible strategies for implimentation of small de-centralised production systems were described.

The participants felt the need of exporting finished products which add value to produce for example, export of bread instead of wheat. But the complications arise how the EC countries could help to provide the developing nations with the technology; for instance, processing of wheat to make bread. However, it was emphasised that this has been successfully followed in area of non-food commodities such as leather, textiles etc. ISAAA is working in this area by approaching large companies to give the technology to developing countries.

Finally the potential for genetically-improved varieties was considered. Trials of such varieties are growing in both developed and developing countries. However, the problem with new technologies are arising because the food habits of people are not considered while planning for genetically modified foods. An example of unsuccessful introduction of high-lysine corn in Cameroon was quite obvious. The high yielding varities do attract the farmers to get more cash. In India, the farmers continue to grow the old varieties of wheat because of preferential taste and thus there is no absolute benefit of genetically modified better-quality foods on health status of populations. It was suggested that this development would, in future, hold out possibilities for all countries. Serious problems concerning international harmonisation, sharing of experiences and dissemination of data to developing countries remain to be effectively and equitably addressed.

The Workshop concluded with a unanimous recommendation that consideration be given to the planning of a larger meeting at which such issues could be discussed across a wider range of participants than was able to be present within the present workshop.

The participants welcomed the involvement of Professor T. Blundell, Chief Executive, BBSRC, and his subsequent comments on the significance of agri-food quality in a global context.

REPORT ON THE AGRI-FOOD CONFERENCE WORKSHOP ON "THE SIGNIFICANCE OF ANTIOXIDANTS IN PLANTS AND IN MAN"

G. Williamson and *P. Mullineaux

Institute of Food Research, and *John Innes Centre, Norwich Research Park, Colney, Norwich, UK.

1 INTRODUCTION

The workshop considered two themes: the effect of antioxidants in the diet and the effect of antioxidants as protective agents in the plant. About 35 people attended the workshop, and most of these were either research scientists interested in phenolics, secondary plant metabolites and nutrition aspects, or members of funding bodies such as The Ministry of Agriculture, Fisheries and Food (MAFF), UK. The following were discussed, and we have tried to list the main points below. Obviously, for a subject as large as antioxidants, it was not possible to comprehensively cover every aspect. However, the issues raised were of importance to the attenders at the Agri-Food conference, and illustrate some of the current "hot topics" in this area.

2 DISCUSSION POINTS

2.1. Potential significance of antioxidants.

There are a considerable number of epidemiological (1), biological (2) and chemical (3) studies on antioxidants. Funding bodies have placed considerable emphasis on this subject, and MAFF have a special emphasis programme on antioxidants. Although the health effects are not proven, there is substantial evidence to suggest that there are considerable benefits to be obtained by optimising consumption of antioxidants; thus providing evidence to support the "antioxidant hypothesis".

2.2. Assessment of an antioxidant.

There is no general in vitro test to assess the properties of an antioxidant, but a review by Halliwell (4) has listed many of the methods currently available. Assays usually contain a target molecule (lipid, protein, DNA or carbohydrate), a free radical generator (many exist and include iron containing materials with an oxidant or reductant, azido generators, hydrogen peroxide, superoxide generators, etc. etc.), in a suitable medium (organic, aqueous, detergent). The products of peroxidation are measured by one of a large number of methods (TBARS (thiobarbituric acid reactive substances) methods are the most common). Consequently, an antioxidant may be shown to protect a protein but not a lipid from free radical mediated damage, for example. The relevance of these assays to the *in vivo* situation is not yet proven, and an important issue in this context is bioavailability (see below). It is important to study also the basic chemical properties of antioxidants, such as redox potential, since many of the standard antioxidant assays assess multifactorial properties of the test compound. For example, inhibition of lipid peroxidation can be due to one or a combination of several processes: prevention of initial contact between radical generating agents; inhibition of production or scavenging of free radicals produced; inhibition of the propagation stage of peroxidation; increase in the termination rate of peroxidation; repair or inactivation of products of peroxidation.

2.3. Absorption of antioxidants.

The absorption of dietary antioxidants is crucial, but poorly studied in man. The most controversial aspect is whether flavonoids are absorbed. There is known to be substantial variation between individuals, even in the absorption of vitamins such as α-tocopherol, and so variations in absorption would be predicted for other dietary components. Absorption of any compound is also known to be affected by the presence of others. It is possible that the proteins in beans may exert an "antinutritional effect" by binding to phenolics and so reduce bioavailability, and that milk proteins might bind to phenolics in tea and so affect antioxidant activity *in vivo*.

2.4. Inter-relationships between antioxidant levels.

One characteristic of endogenous antioxidants is that alteration in the levels of one will affect the levels of others. For example, permanent transfection of mammalian cells with human copper/zinc superoxide dismutase alters cellular levels of glutathione peroxidase, manganese-superoxide dismutase, glutathione reductase and NADPH reductase (5). A comparable phenomenon has also been observed in plants. For example, alteration of phenolic antioxidant levels depletes the glutathione pool in the leaves of such plants (6). Increased levels of glutathione also elevates the levels of ascorbate in some situations and allows for improved maintenance of the redox state of the ascorbate pool under stress. However, central to many of the processes that occur *in vivo* is the tripeptide glutathione, which is both redox active and also a cofactor for many enzymes which are involved in modulation of oxidative stress.

2.5. Hydroxycinnamic acids, such as ferulic and caffeic acids.

The study of flavonoids is currently the subject of much interest. However, the level of hydroxycinnamic acids is present at a much higher in many foods. Coffee and apple juice, for example, contain high levels of chlorogenic acid (7), which is an effective antioxidant (8). Cereals also contain high contents of these acids (9).

2.6. Effect of antioxidants on human health.

Anticarcinogenic activity of antioxidants has been proven in animal models, and many compounds have been shown to be protective against a wide range of insults in such systems (10). However, the challenge is now to determine if these mechanisms are operative in humans by a combination of approaches, using *in vivo* studies on humans and on human cells.

2.7. Effect of antioxidants in plants.

It is possible to alter the levels of components of antioxidant metabolism in plants, but virtually nothing is known about what effects this will have on post-harvest performance and the nutritional value of any food derived from such a crop.

References.

1. G. Block, B. Patterson and A. Subar, *Nutrition and cancer*, 1992, **18**, 1-29.
2. N. I. Krinsky, *Proc. Soc. Exp. Biol. Med.*, 1992, **200**, 248-254
3. B. Halliwell, *Am. J. Med.*, 1991, **91**, S14-S22
4. B. Halliwell, *Free Rad. Res. Comm.*, 1990, **9**, 1-32
5 M. J. Kelner, R.Bagnell, M. Montoya, L. Estes, S. F. Uglik and P. Cerutti, *Free Radical.Biol.Med.*, 1995, **18**, 497-506.
6. P. Mullineaux, unpublished observations
7. K. Herrmann, *Crit Rev Food Sci Nutr.*, 1989, **28**, 315-347
8. J. Zhou, F. Ashoori, S. Suzuki, I. Nishigaki and K. Yagi, *J. Clin. Biochem. Nutr.*, **15**, 119-125.
9. B. Bartolome, C. B. Faulds, M. Tuohy, H. Gilbert, G. Hazlewood and G. Williamson, *Biotechnol. Appl. Biochem.*, 1995, **22**, 65-73.
10. L.W. Wattenberg, *Cancer. Res.*, 1985, **45**, 1-8.

Subject Index